Geological Applications of Wireline Logs

Geological Society Special Publications

Series Editor K. COE

GEOLOGICAL SOCIETY SPECIAL PUBLICATION NO 48

Geological Applications of Wireline Logs

EDITED BY

A. HURST
Statoil
Stavanger
Norway

M. A. LOVELL
Geology Department
University of Nottingham
UK

A. C. MORTON
British Geological Survey
Nottingham
UK

1990

Published by

The Geological Society

London

THE GEOLOGICAL SOCIETY

The Geological Society of London was founded in 1807 for the purposes of 'investigating the mineral structures of the earth'. It received its Royal Charter in 1825. The Society promotes all aspects of geological science by means of meetings, special lectures and courses, discussions, specialist groups, publications and library services.

It is expected that candidates for Fellowship will be graduates in geology or another earth science, or have equivalent qualifications or experience. All Fellows are entitled to receive for their subscription one of the Society's three journals: *The Quarterly Journal of Engineering Geology*, the *Journal of the Geological Society* or *Marine and Petroleum Geology*. On payment of an additional sum on the annual subscription, members may obtain copies of another journal.

Membership of the specialist groups is open to all Fellows without additional charge. Enquiries concerning Fellowship of the Society and membership of the specialist groups should be directed to the Executive Secretary, The Geological Society, Burlington House, Piccadilly, London W1V 0JU.

Published by the Geological Society from:
The Geological Society Publishing House
Unit 7
Brassmill Enterprise Centre
Brassmill Lane
Bath
Avon BA1 3JN
UK
(*Orders:* Tel. 0225 445046)

First published 1990

British Library Cataloguing in Publication Data
Hurst, A. *1953–*
Geological applications of wireline logs.
1. Engineering geology
I. Title II. Lovell, M. A. 1955– III. Morton, A. C. 1954– IV. Geological Society of London V. Series
624.1'51

ISBN 0–903317–45–1

Printed in Great Britain at the Alden Press, Oxford

Contents

Identification of faults and fractures

Physical properties

Introduction

A. C. MORTON, A. HURST & M. A. LOVELL

For two days in June 1988, geoscientists from around the world converged on Burlington House, the home of the Geological Society of London. The meeting, entitled 'Geological Applications of Wireline Logs' and held under the auspices of the Petroleum Group of the Geological Society, is believed to be the first of its kind. The meeting was attended by 178 delegates, and 38 papers were presented, four as posters. 142 of the delegates were from the UK, 21 from the remainder of Europe (Norway, France, The Netherlands, West Germany and Switzerland) and 12 from North America (USA and Canada). There were also delegates from Japan, Dubai and New Zealand. Attendance covered a wide spectrum of professional environments, including industry, academia, geological surveys and service companies.

The use of wireline logs is diverse, and has a history dating back to the experiments in the Pechelbronn oilfield in Alsace where, in the late 1920s, the first measurements were made. However, application of wireline logs to geological studies is probably still in its infancy. This not only reflects a reluctance on behalf of geologists to interpret data that were not gathered by traditional observational methods, but also the high level of computer-based technology associated with collection and interpretation of log data foreign to many geoscientists. Despite this, logs make measurements on formations that are never likely to be accessible for traditional forms of geological study and thus they provide a unique source of data that should not be ignored.

So that the meeting and this subsequent publication would accurately reflect the areas of geology in which wireline logs are applied, we placed few restrictions on the subjects suggested by contributors. Perhaps inevitably, almost two-thirds of the papers focus on problems directly associated with the exploration for and production of hydrocarbons. Of the 28 papers, 8 come from the oil industry, 8 from service companies, 7 from academia and 5 from geological survey organisations. This is similar to the balance of delegates at the meeting: industry provided 77 delegates, again mostly from companies concerned with hydrocarbon exploration and production, although coal and other mining interests were also represented, with 49 from service companies, 36 from academic institutions and 15 from geological surveys. As the volume deals with geological applications of wireline logs, the papers are grouped together according to their geological topics, rather than in terms of the different logging tools. The topics identified as of major interest are *sedimentology*, *mineralogy and geochemistry*, *stratigraphic correlation*, *fault and fracture identification* and *physical properties*.

Sedimentology. Nine papers deal with the applications of logging to sedimentological study. All are principally concerned with sedimentology of potential or actual hydrocarbon-bearing sequences, and many are concerned with the development of accurate reservoir models. Most deal with clastic sediments, but two (**Nurmi *et al.*** and **Dorfman *et al.***) describe the application of logs in the evaluation of carbonate sequences, a topic probably unfamiliar to most petroleum geologists working in the NW European area.

The remaining seven cover a wide variety of sedimentological topics from a number of different viewpoints. Two (**Luthi** and **Harker *et al.***) deal with the application of high-density microresistivity measurements to identify sedimentary structures in the borehole environment. Four of the papers stress the importance of utilizing geological data in the interpretation of wireline logs. **Rider** describes a number of instances where classical interpretation of gamma-ray patterns is incorrect, and shows the need for calibration with core and other data; **Herweijer *et al.*** show how dipmeter interpretation of sedimentary sequences is enhanced by studying analogous rock outcrop sequences; and **King** shows how integration of geological data with log measurements can add extra sophistication to lithological analysis. Similarly, **Moss** describes a stochastic method of reservoir description that incorporates log interpretation with sedimentological data to build a complete reservoir model. On a more unusual note, **Griffiths** applies the technique of syntactic analysis of wireline log measurements to sedimentological problems.

Stratigraphic correlation. Correlation of sedimentary sequence by means of wireline logs is perhaps the most well known application of wireline log data, and for this reason papers on this topic were deliberately not solicited. However, a number of papers that dealt either with new approaches to correlation using wireline log data or with unusual examples of correlation of rock sequences were presented at the meeting. This section comprises three of those papers. **Worthington** describes how cyclicity can be recognised on a regional basis despite the diversity of geological environments, and **Griffiths & Bakke** combine numerical lithologies derived from log data with gene-typing techniques to correlate sequences on a well-to-well basis. Finally in this section, **Buckley & Oliver** describe how gamma-ray logging of water exploration boreholes in the basaltic sequences of the Deccan Traps identified high-gamma basaltic flows that could be correlated over long distances.

Mineralogy and geochemistry. One of the major recent advances in wireline logging is the advent of logs that allow downhole measurement of geochemical parameters, and as a consequence many of the papers read at the conference and included in this volume deal with this topic. Two of the papers (**Herron & Herron** and **Anderson *et al.***) describe applications of these new geochemical logs to the full range of lithological types (sediments, basalts, granites and metamorphic rocks), demonstrating the immense potential of these tools in subsurface geological studies. However, **Brewer *et al.*** describe some problems associated with these logs run in basaltic basement in the deep ocean floor, and **Primmer *et al.*** note that it is preferable to avoid the 'global' strategy of interpretation of such logs, but to develop mineralogical models that apply to individual basins. Similarly, **Hurst** and **Humphreys & Lott** both note the importance of acquiring accurate geochemical data on minerals present in the sequences under investigation before quantification of bulk sediment compositions is made. Finally in this section, **Stocks & Lawrence** discuss an organic geochemical problem, namely the evaluation of petroleum source rocks on the basis of wireline log data.

Identification of faults and fractures. Four of the five papers in this section describe the use of a number of logs, particularly the circumferential acoustic (sonic) log and the dipmeter log, in identifying faults and fractures, with examples from a number of different lithologies from North America (**Fertl**), and in chalk (**Lehne**) and clastic rocks (**Rønningsland** and **Devilliers & Werner**) from NW Europe. The fifth (**Goldberg *et al.***) describes the use of hydraulic pulse testing and sonic logs in the assessment of fracture permeability.

Physical properties. A diverse series of topics are presented in this final section. Two of the four (**Bell** and **Evans & Brereton**) deal with identification of borehole breakouts from well logs and their value in interpreting the state of stress in the crust. In contrast, **Lovell & Pezard** show how downhole measurements can be used to assess pore morphologies in basaltic oceanic basement, and **Entwisle & McCann** assess the validity of the application of Christensen's equation in determination of engineering parameters such as the formation dynamic elastic moduli, concluding that it should not be universally applied to all rock types.

Summary. Wireline log interpretation has great potential for solving a diversity of geological problems in a range of different subsurface geological environments. With the immense cost and practical problems involved in recovering core from deeply buried sequences, the major advances discussed in this volume to geological research, both within academia and industry, are clearly of great importance, and the field remains one in which research and development of new tools and interpretative techniques will be keenly followed. Many of the papers in the volume, however, stress the importance of correlating and calibrating log-derived data with traditional geological observations, and of incorporating such geological data to produce a meaningful geological interpretation of the log measurements. We hope that this volume will act as a small milestone that will mark the status of geological applications of wireline logs in the late 1980s, and that it will be the first in a series resulting from further meetings that will follow the progress in this important frontier field of the geological sciences.

Finally, we would like to thank the following for reviewing the manuscripts submitted for this volume and helping us to maintain a high quality throughout: J Archer, J Adams, K Becker, S Begg, S Bell, A Buller, N Brereton, R Brooks, R Cull, D Cant, R Dove, M Dusseallt, S Eggen, S Ehrenberg, M Eller, W Fertl, L Fält, R Gabrielsen, D Goldberg, C Griffiths, R Hardman, M Herron, S Herron, J Herweijer, B Humphreys, J Jensen, H Jøranson, C Jourdan, J Ineson, H Irwin, M King, T Knai, A Koestller, S Larter, K Lehne, M Longman, S Luthi, D McCann, A McDonald, B Mair, B Moss, K Myers, P Nadeau, E Nielsen, T Olsen, H Omre, A Parker, J Pearce, I Penn, R Peveraro, M Rider, A Roberts, H Rueslatten, O Serra, B Sellwood, A Spencer, H Steward, N Trewin, D van der Wel, P Worthington and M Worthington.

Sedimentology

Sedimentary structures of clastic rocks identified from electrical borehole images

STEFAN M. LUTHI

Schlumberger-Doll Research, Ridgefield, CT 06877, USA

Abstract: High-resolution electrical borehole images are obtained from the Formation Micro Scanner* (FMS) by scanning the borehole wall with arrays of small electrodes pressed against the borehole surface. The images provide detailed sedimentary structural information from clastic rocks which is of great value for geological and reservoir modelling as illustrated by two case studies. Analysis of electrical borehole images through a Pennsylvanian deltaic cycle from Texas shows different types of bedding surfaces, including channel base scours, cross-bedding, slump folds and growth-faulting. The ensemble of the data constrains the subsurface geometry of the reservoir and helps to establish a depositional model. Thresholding may produce a segmentation of sand/shale images. In the aeolian Rotliegendes of the North Sea, electrical borehole images provide a vertical succession of facies, dip and azimuth measurements of interdune layers, bounding surfaces and foresets, as well as thickness measurements of the cross-bedded sets. Statistical modelling using the data indicates scale-invariant crescentic bedforms, and describes sizes, shapes and layering of the cross-bedded sandstones. Computer-aided design tools are used for forward-modelling of structures observed on borehole images in order to test possible interpretation hypotheses. The use of graphic workstations greatly facilitates the interpretation process.

Figure 1 shows images recorded on the wall of a borehole at a depth of approximately 5000 ft (1500 m) with the FMS tool which is equipped with a large number of densely spaced, small measurement electrodes. The operating principle of the FMS tool is described by Ekstrom *et al.* (1987). Darker grey levels in Fig. 1 represent more conductive parts of the rock, or, more precisely, zones where the tool's electrodes record a higher formation conductance. Unlike most resistivity tools, the FMS tool does not have voltage-monitoring electrodes since its main purpose is to provide good variations of the individual electrode measurements by suitably adjusting voltage and total current. Whereas in dipmeters the same curves are used for cross-correlations to obtain layer dips, in the FMS tool they are used to assemble a grey-scale image. This is made possible by the much larger number of data points obtained by the arrays of electrodes implemented on the FMS. The number of such arrays has increased from one on the experimental prototype, to two and, more recently, to four on the commercial versions of the tool, resulting in an equal number of image stripes. The arrays are arranged at right angles to each other, and they are mounted onto conductive metal pads which are pressed against the borehole wall while the tool is being pulled up the hole and recording takes place.

The electrical images in Fig. 1 show at the base a conductive (dark) zone overlain by a more resistive sequence. This upper sequence also shows intriguing details of layering, most notably a zone that seems to be highly distorted, discussed in more detail below. These two features are enhanced and enlarged in Fig. 1 for clarity. From the core photographs through the interval (Fig. 2) it becomes obvious that the sudden change in resistivity corresponds to a change from dark shales to a conglomerate grading into sandstones. The distorted layers are found higher up in the sandstones as faint laminations. Two immediate — though preliminary — sedimentological interpretations can be made at this stage: firstly, that the conglomerate/sandstone sequence represents a channel cut into the underlying shales; secondly, that the distorted layers were produced by soft sediment deformation as caused, for instance, by sediment sliding or slumping. Once the correspondence of the resistive zone with the conglomerate/sandstone sequence is recognized, the interpretation is likely to remain the same regardless of whether it is based on core photographs or electrical images.

FMS borehole images are conveniently interpreted on displays where they are oriented in an azimuth/depth coordinate system (Fig. 1). Accurate depth control is provided by the tool's triaxial accelerometer unit, while azimuthal orientation of the images is measured by a triaxial fluxgate magnetometer (Ekstrom *et al.*

* Mark of Schlumberger.

From Hurst, A., Lovell, M. A. & Morton, A. C. (eds), 1990, *Geological Applications of Wireline Logs* Geological Society Special Publication No. 48, pp. 3–10

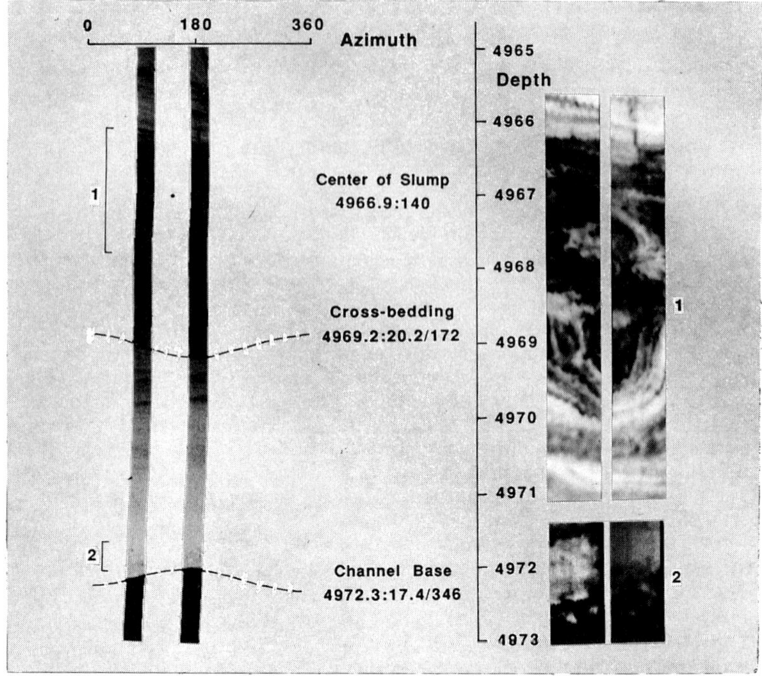

Fig. 1. Electrical borehole images obtained with the FMS tool over part of a Pennsylvanian deltaic cycle. On the left are electrical images in a depth/azimuth co-ordinate system (depth in feet). 1 and 2 denote intervals which are enhanced and enlarged on the right. Geometric measurements include depth and azimuth for the centre of the slump fold, and depth/dip/azimuth for cross-bedding and channel base.

1987). Representing borehole images in an azimuth/depth coordinate system corresponds to projecting, or unwrapping, a cylinder onto a plane. A dipping plane in such a projection takes the form of a sinusoid, an example of which is provided by the segments of the channel base in Fig. 1. Measuring the sinusoid by template matching, least-square fitting or some other method will give the dip and azimuth of the feature (for details see Plumb & Luthi 1986). Together with the depth, these data completely define a plane representing the local slope of a feature intersecting the borehole. Necessary trigonometric adjustments have to be made for cases where the borehole is deviated, or where an underlying structural tilt has to be compensated for.

As illustrated by the foregoing discussion of Fig. 1, two principal areas of interest in the interpretation of electrical borehole images emerge: the correspondence of rock types to conductivities or conductivity changes, and geometric measurements on the images. The first subject will be loosely referred to as ‘image texture’, which is not the same as rock texture, a term applied to the microgeometry of rocks. These two topics are examined in more detail below using examples from a Pennsylvanian deltaic cycle in North Central Texas and from the aeolian Rotliegendes in the North Sea.

Image texture and rock types

Visual image texture is defined as a repetitive arrangement of a basic pattern with some degree of randomness (Hall 1979). According to Rosenfeld & Kak (1982), image texture includes properties such as first-order statistics of absolute grey levels (or conductivities) at various distances, second-order statistics of pairs of grey levels at various distances, and frequency-related properties. In aerial photography, image textures are used to identify crop patterns and other land uses. In electrical images from the FMS tool, grey level changes occur where porosity or clay content give rise to changes in rock conductivity. Since image pixels are spaced at 2.5 mm both laterally and vertically, they generally represent rock volumes considerably larger than a grain, except for conglomerates.

Image texture therefore, is a macroscopic expression of spatial changes in rock texture and is directly related to sedimentary structures. We use the term sedimentary structures here in a broad sense that includes any macroscopic geometric aspect of clastic rocks that is either caused by deposition, erosion or post-depositional soft-sediment deformation.

Since sedimentary structures span a wide range of scales, it is important to identify the lower limits of resolution of the FMS tool. In order to do so, the spacing of mud drapes on a thick cross-bed from the Cretaceous Travis Peak Formation was analysed (Fig. 3). Whenever the mud drapes were closer than one centimetre, they merged into one conductive streak, but above that spacing there was a clear separation. The value for the spatial resolution of the tool is likely to depend on other factors such as the ratio of formation to mud resistivity, or the amount of mudcake built up on the borehole wall. Although it is realized that a whole class of sedimentary structures with length scales below one centimetre will be blurred on the images, no other wireline geophysical measurement is at present capable of depicting sedimentary structures at this resolution.

In a study of the Permian Rotliegendes in the North Sea, Luthi & Banavar (1988) have found an excellent visual correlation of borehole image texture and core lithofacies. Examples of these comparisons are shown in Fig. 4. Some of the image textures can be readily related to sedimentary structures; examples include the high-angle and low-angle cross-bedded facies. Others, such as the speckled image texture, have to be carefully correlated with core. It was found that the heterogeneous electrical appearance of this image texture corresponds visually to a massive, rather homogeneous bedding type observed in the core.

Once such relationships are established it is possible to subdivide the well vertically into facies zones, and to measure thicknesses of the different rock types. Figure 5 shows an interval from the study of Luthi & Banavar (1988) of the Rotliegendes in the North Sea. The zonation is obtained through initial correlation of the electrical images with cores, and subsequent interactive scrolling of the images on a graphic workstation and recording of the vertical facies distribution. The log-derived porosity curve in Fig. 5, whose wiggles would be difficult to understand without further information, shows an interesting correlation with the various facies types. By far the lowest porosities are found in the ‘massive’ facies, a cemented interdune sandstone with adhesion ripples. The highest porosities occur at the base of the foresets, presumably due to an abundance of originally loosely packed grainflows (Luthi & Banavar 1988). Such an integration of geological with petrophysical properties provides a powerful link between sedimentological–depositional and reservoir models.

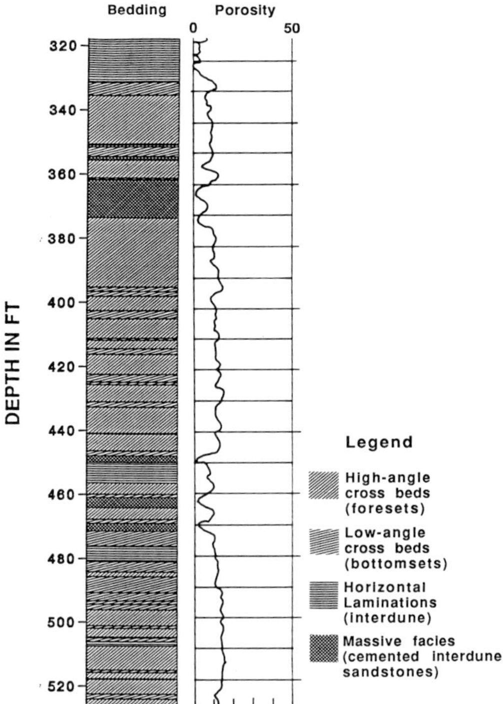

Fig. 5. Vertical sequence of bedding types illustrated in Fig. 3 as obtained from electrical images in the Rotliegendes. Note composite cross-bedding with set thicknesses up to 20 ft alternating with interdune deposits. Correlation of bedding types with porosity is discussed in the text (after Luthi & Banavar 1988).

As the visual zonations are rather subjective, it is desirable to find algorithms that segment borehole images into textural regions in a more quantitative fashion. A very simple approach is based on thresholding the conductivity or grey levels. A prerequisite for this is that no prior adaptive grey level enhancement has been performed, i.e. that a particular grey value has the same meaning at all depth levels. Zones where a change in pore fluid occurs should be avoided. If the thresholds are chosen judiciously, the result can be a fast segmentation into sands and shales including possible intermediate lithologies if multiple thresholds are used. This is of particular interest in cases where thinly bedded sandstones and shales with good resistivity contrasts occur, and where the typical bed thicknesses cannot be resolved by other, more mineralogy-sensitive logs. In Fig. 6 the images on the left are segmented into binary images on the right, containing only 'sands' (white) and 'shales' (black), using a single threshold selected from histogram analysis. At our laboratory, experiments are currently being performed with more elaborate textural segmentation approaches using template matching and other methods applied to such problems (Duda & Hard 1973). In view of the large number of possible image textures of rocks, interactive user input is anticipated to be vital in constraining the procedure. Expert systems may provide further help, in view of their reported usefulness for augmenting the quality of segmentation of seismic sections (Simaan *et al.* 1987).

Geometric measurements

The importance of geometric measurements for the reconstruction of hydrocarbon reservoir geometries has been amply documented in the classical dipmeter literature (e.g. Gilreath & Maricelli 1964). But while conventional dipmeter interpretation is based on the analysis of dip patterns and dipmeter curve characteristics, the advent of high-resolution borehole images lets the interpreter identify the structures first before making any geometric measurements. This is illustrated by the channel base and cross-bedding measured with a least-square sine-fitting program in Fig. 1. Since the nature of both structures can be determined directly from the images, only a relatively small number of well-directed geometric measurements have to be made. In Fig. 7 an attempt is made to outline schematically the different working approaches of dipmeter and borehole image interpretation. The two following examples serve to illustrate the concept of bedding identification and measurements from borehole images in two different clastic depositional environments.

In a study of borehole images from the aeolian Rotliegendes in the North Sea, Luthi & Banavar (1988) have shown that several types of bedding surfaces can be identified, measured and used to enhance depositional models. Figure 8 illustrates first- and second-order bounding surfaces (Brookfield 1977), foreset laminae, grain-flow wedges and a possible reactivation surface, recognized on both cores and electrical borehole images. Since the reservoir covered almost 700 ft, a large number of such measurements could be made and subsequently analysed statistically. The results are discussed in more detail in the next section.

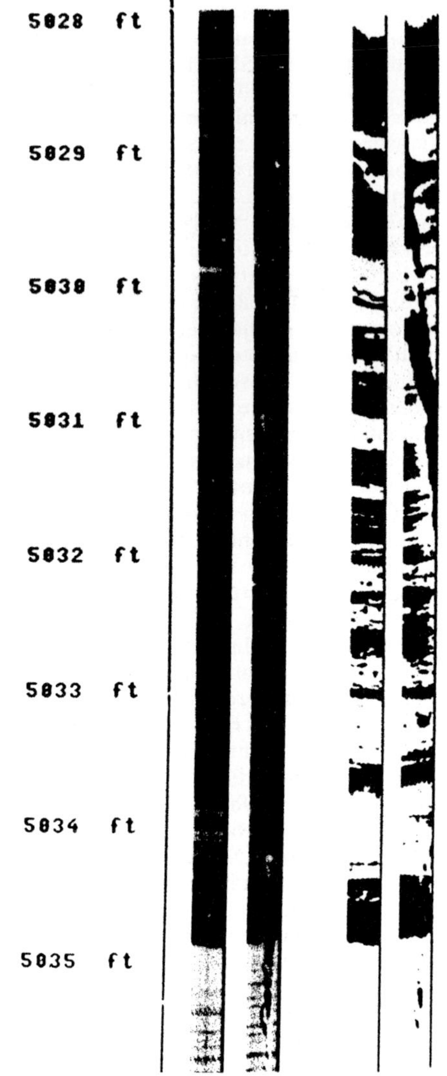

Fig. 6. Binarization of electrical images using a single threshold. Original FMS images are on the left, binarized images on right. Black on binarized images represents resistivities below threshold and corresponds to argillaceous sand- and siltstones. White corresponds to fine-grained sandstones. A simplified sand count is obtained by dividing the number of white pixels by the total number of pixels. A prominent fracture interferes on the right-hand image.

In the Pennsylvanian deltaic cycle (Fig. 1), careful image processing and analysis facilitates the recognition and measurement of such

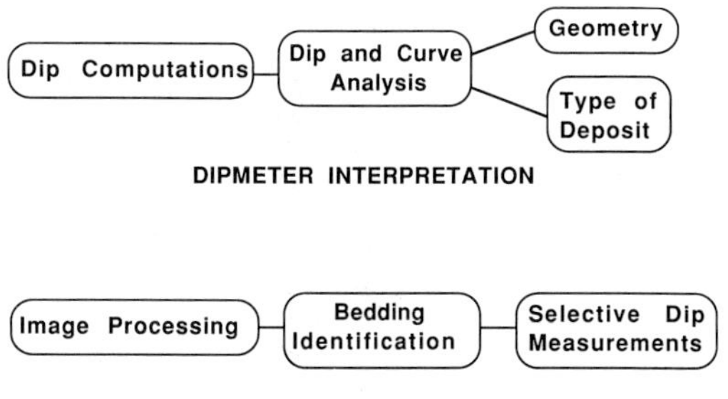

Fig. 7. The different approaches for dipmeter and borehole image interpretation.

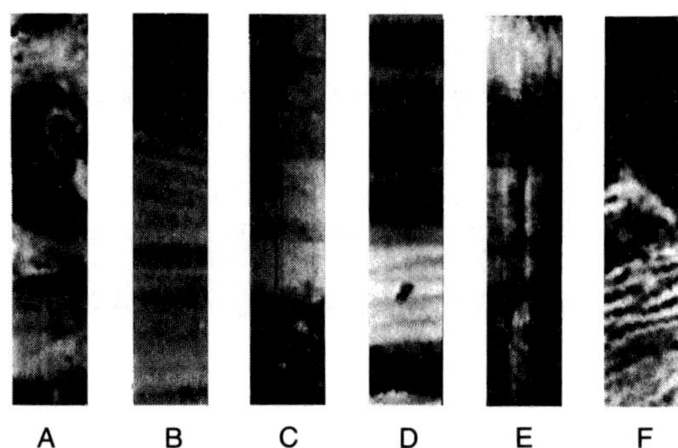

Fig. 9. Various structural features recognized through an approximately 100 ft interval of a Pennsylvanian deltaic cycle. Dip measurements of the features are shown in Fig. 11 together with a sketch of their position in the sequence. A: circular feature interpreted as a slump fold (cf. Fig. 10), occurs just above the section shown in Fig. 1.B: low-angle cross-bedding in the channel. C: channel cut with base lag conglomerate (cf. Fig. 2). D: siltstone/shale alternations with low apparent dip. E: small-scale trough cross-bedding in delta front sandstones. F: growth-fault cutting across bedding, juxtaposing fine-grained sandstones with shales; in the core, the fault has slickensides and affects the bedding attitude both above and below. Each image represents 1.2 ft vertical section; no vertical exaggeration.

diverse features as the channel base, cross-strata, shale layers, slump fold axes, fractures and a growth fault over a sequence of only approximately 100 ft. Fig. 9 shows intervals of electrical images containing these features. The cross-strata, shale layers and the growth fault are measured in the same way as the channel base, by assuming that they can be locally approximated by a plane. The fractures were found to be sub-vertical and often discontinuous, and their measurement, therefore, includes only a strike direction. Measurement of the slump fold-axis is more involved. First, the distorted layers in the uppermost part of Fig. 1, or in Fig. 9A, have to be recognized as being part of a fold. For this, a computer-aided design tool based on the forward-deformation of solid primitives (Barr 1984) was used. A simple geometric object with some texture is deformed in a hierarchical structure until the desired geometric shape is obtained. The computational procedure consists of applying transformation matrices for each deformation step to the surface-normal vector of the undeformed object. A horizontally layered block was chosen as the primitive and folded concentrically by about 140 degrees over a limited width. The object was then rotated into a position where its fold plane was horizontal and superimposed onto another fold to simulate the occurrence of two such feature on the images (only one of which is shown in Fig. 1). Finally, the deformed structure was intersected with a cylinder representing the borehole (Fig. 10). It is obvious that despite the simplicity of the model there is much similarity between the image stripes in Fig. 1 and the layering shown in some views of the synthetic core. An overturned fold, even though its geometric construction is simple, gives rise to a rather complex layering pattern on a cylindrical surface. A view parallel to the fold axis plane into the interior of the fold (i.e. onto the concave parts of the fold) exhibits somewhat hyperbolic layer shapes, while seen from the 'outside' concentric ellipses appear. The centre of the ellipses is perpendicular to the fold axis. Assuming the fold axis to be approximately horizontal, point measurements were made on two superimposed folds (Fig. 1, top), implying that the fold axes strike NE–SW.

Point measurements consist of an azimuth and a depth value. They may be made for any locally linear feature intersecting the borehole, such as cross-bed trough axes, fold axes etc. Other types of measurements include line segments and areal measurements and are discussed in Plumb & Luthi (1986). They are of particular use in characterizing discontinuous fractures and breakout zones on the borehole wall, and they can gain significance when larger numbers of them are analysed statistically.

Structural modelling

The classic Pennsylvanian deltaic cycles (Fig. 11, after Brown, 1979) consist of superimposed facies deposited during progradation, abandonment, marine reworking and marine transgression of cratonic delta lobes. Important hydrocarbon reservoirs occur in delta-front

Fig. 10. Synthetic cores constructed by a computer-aided design system through sequential deformation of textured solid primitives. The structures represent two overturned folds with horizontal fold planes but slightly different fold axes lying on top of each other. The second and fourth views correspond approximately to the two views obtained by the FMS in Fig. 1. Viewing angles increment by about 70 degrees.

sandstones and to a lesser degree in the channel fills. The latter include distributary channel-fills, meanderbelt point-bar sands and valley-fills (Brown *et al.* 1973; Brown 1979), but the differentiation between them is beyond the scope of this paper. Most of the characteristic structures found in the type sequence (Fig. 11) were identified in the analysed example. Their appearance on the electrical images is illustrated on Figs 9 A–F and their geometric measurements indicated on the stereoplot in Fig. 11. In the channel deposit three types of measurements were made, the channel base, cross-bed directions and sediment fold axes (which were translated into a slumping direction assuming the convex side of the recumbent folds point downslope). The cross-bed direction is assumed to be the most reliable measurement for constraining the channel geometry, while the channel base and the slump fold measurements may be affected by local disturbances. Cross-beds dip on average towards SSW, supporting a palaeoflow in the same direction and a local channel axis trend from NNE towards SSW. In light of this, the NNW-dipping channel base and the assumed slumping directions from the ESE suggest that the channel was penetrated toward its southeastern margin. Thus, while individual interpretation of each of the measurements has clearly a significant error margin, the coherent nature of the three structures together results in a considerably higher confidence of interpretation. The conceptual sketch (Fig. 12) aims to capture the resulting channel geometry, and suggests that offset wells should be placed towards the NNE and WSW where the channel sands are likely to be thickest.

Another approach to structural modelling can be taken in the presence of a large number of beds thought to have geometric similarity. In the aeolian Rotliegendes (Luthi & Banavar 1988), a large number of cross-bedded sets was measured for foreset dip and azimuth as well as set thickness. If a basic shape for the sets is assumed, the size distribution of the three-dimensional sedimentary bodies can be reconstructed from the thickness distribution observed in the well by making use of stereological theorems. The internal layering, expressed by the foreset attitudes, can equally be reconstructed. In this case, a scale-invariant crescentic crestline model gave an excellent fit to the observed data. From this and the relationship of second-order bounding surfaces to foresets, a depositional system was proposed which consisted of main dunes with superimposed barchanoid ridges migrating at an oblique angle to their main elongation (Fig. 13). Between the main dunes are interdune areas with predominantly damp surfaces. Besides this depositional model, a reservoir model emerges with predictive power in the vicinity of the wellbore. It quantifies extents and porosities of the dune and interdune deposits and, applying Weber's (1986) methodology, can be used for radial inflow performance estimates, computations of average permeabilities and may serve as an input to reservoir simulators.

The common basis for these approaches to structural modelling is the proper recognition of different bedding surfaces, allowing selective feature measurements for subsequent use in sedimentological interpretations and structural reconstructions.

Graphic workstations

Graphic workstations were used extensively because they allow fast and interactive image

Fig. 2. Core photographs corresponding to the section shown in Fig. 1, covering a Pennsylvanian channel deposit from north-central Texas.

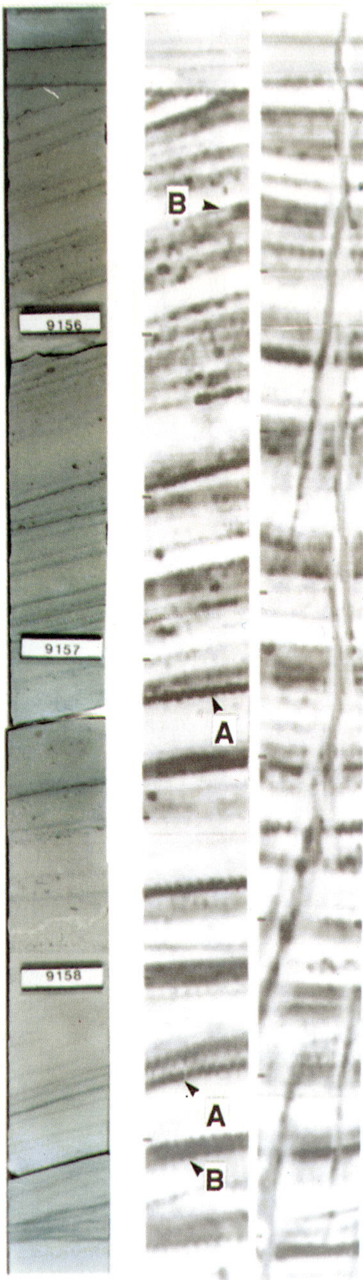

Fig. 3. Cross-bedded sets from the Cretaceous Travis Peak Formation in northeastern Texas. Electrical images (middle and right) and core photograph (left). When separation of mud drapes is above one centimetre two distinct layers are seen on the electrical image (A), but below one centimetre spacing they merge into one conductive streak (B) giving a measure of the tool's resolution. Right image is strike view and shows abundant drilling-induced fracturing, which may be used to determine image resolution. The Travis Peak is generally interpreted as fluvial (Finley *et al*. 1985), but tidal influence may locally be important. Scale in feet; no vertical exaggeration.

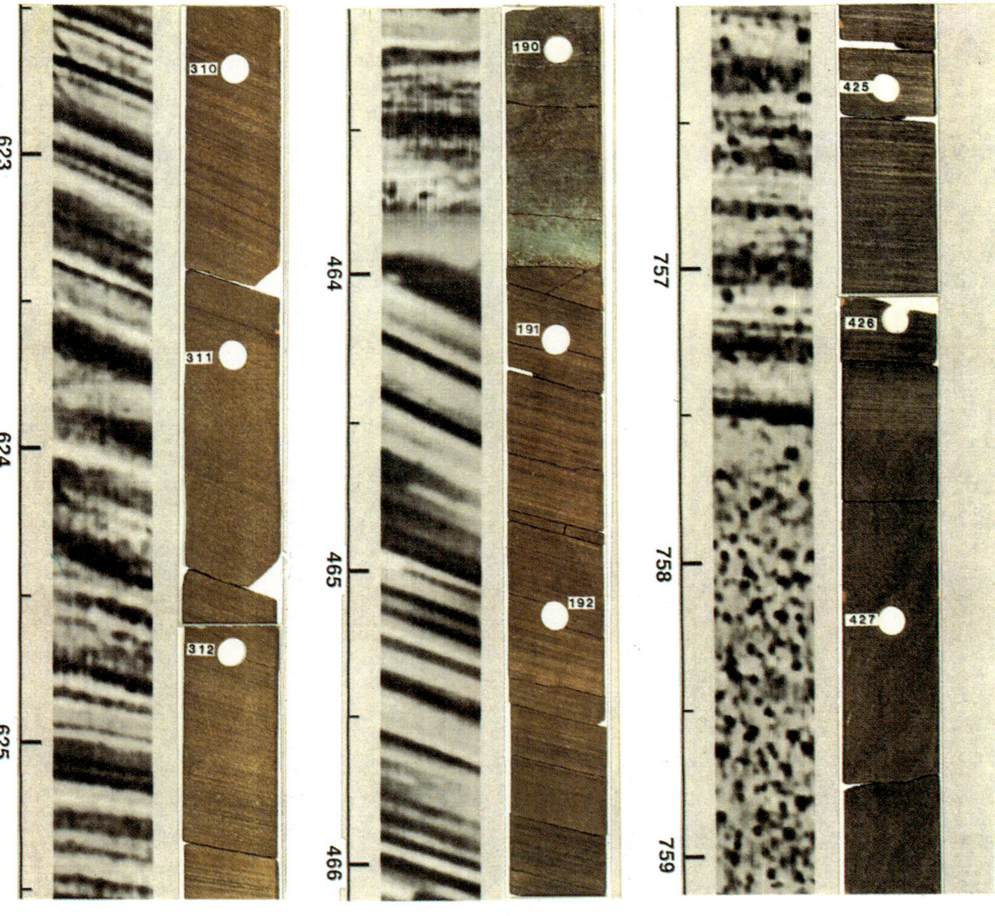

Fig. 4. Comparison of electrical images with core in the aeolian Rotliegendes. Note identical bedding attitudes (due to coincidence of the two viewing angles) and relationship of bedding types to image textures. Left: cross-bedded sandstone with high-angle cross-bedding (foreset) in upper two thirds underlain by low-angle cross-bedding (bottomset). Middle: poorly bedded ('massive'), cemented interdune sandstone overlying cross-bedded dune sandstone. Right: horizontally laminated interdune sandstone overlying dark, massively bedded argillaceous sandstone with small mudclasts ('speckled' facies) interpreted to be fluvially reworked (from Luthi & Banavar 1988). Depth in feet, no vertical exaggeration.

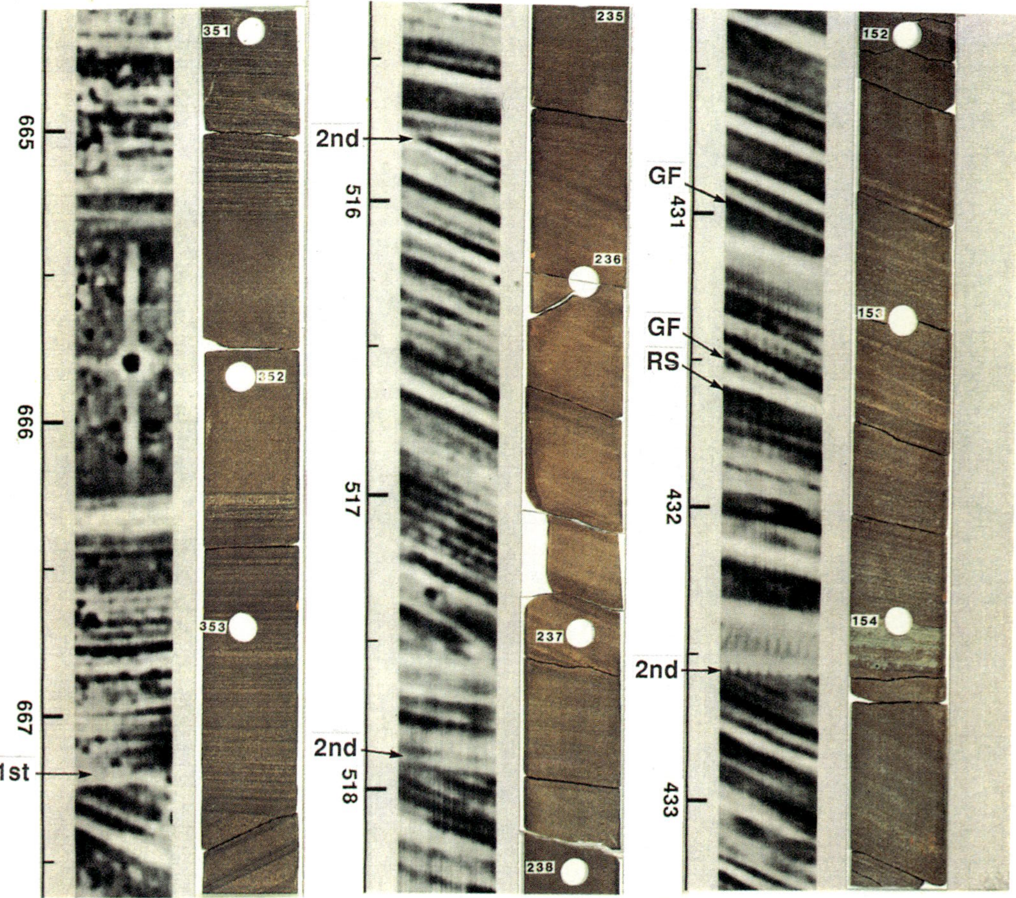

Fig. 8. Aeolian bounding surfaces as defined by Brookfield (1977) and recognized on electrical images. 1st = first order bounding surfaces (interdune over dune deposit); 2nd = second order bounding surface (dune deposit over dune deposit); GF = grainflow wedges, RS = possible reactivation surface within dune deposit (third order bounding surface).

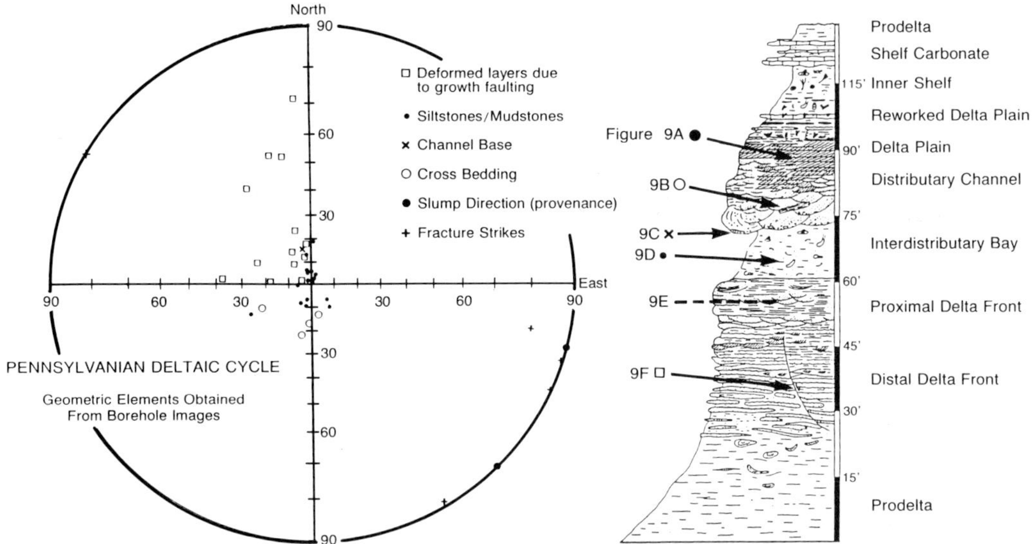

Fig. 11. Polar stereoplot (upper hemisphere) of measurements using electrical borehole images over an entire Pennsylvanian deltaic cycle. Sketch of idealized sequence at right (after Brown 1979) indicates at what levels the various structural features were measured and shows positions of features illustrated in Fig. 9.

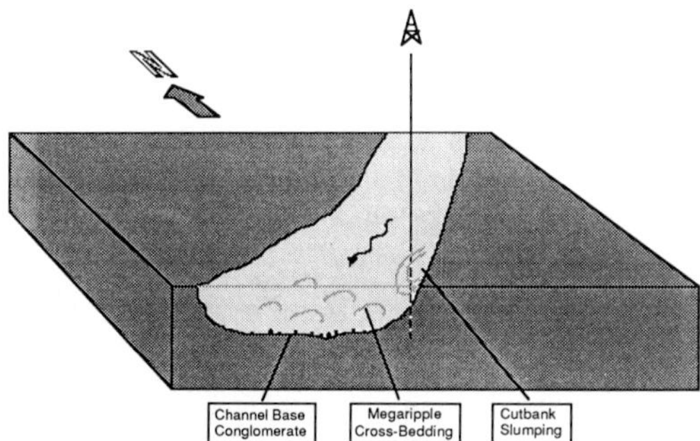

Fig. 12. Diagrammatic sketch of the geometric relationships of the channel deposit in the uppermost part of the Pennsylvanian deltaic cycle.

manipulation, processing and interpretation of borehole images and their high-resolution colour monitors allow recognition of many details which on paper would often pass unnoticed. The images presented in this paper often do not reveal the same amount of detail as do high-resolution monitors. Many standard image processing algorithms can be applied readily to borehole images. The only peculiarity is that borehole images are stored as very long but narrow two-dimensional arrays, causing their perimeter-to-area ratio to be very high. Consequently, when operations with finite size masks or filters are performed there is a considerable number of pixels at the edges for which the

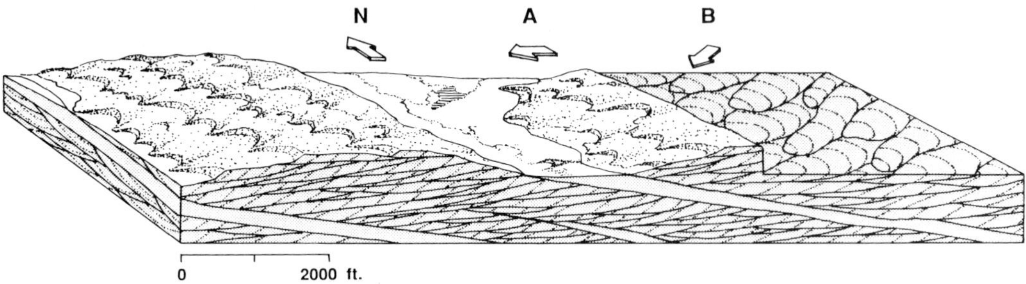

Fig. 13. Diagrammatic sketch of depositional situation for the aeolian Rotliegendes after analysis of electrical borehole images (from Luthi & Banavar 1988). Model is heavily influenced by Glennie (1983) and Clemmensen & Abrahamsen (1983). N = Geographic north; A = General foreset direction (towards W); B = Computed direction of migration (towards SW).

computations cannot be done properly. During the course of the image interpretations the most frequently used routines were found to be local image enhancements, image segmentation by thresholding, and measurement of bedding features using a sine-fitting program.

The machine used in this case was networked to a mainframe computer on which all the preprocessing and image restoration was done, but a numerically fast workstation may not require such a link. Data files such as the ones containing geometric measurements can be transferred to the host computer for data analysis. Hard-copy output is obtained by screen-dumping to a high-resolution colour plotter or with a camera.

Conclusions

Electrical images allow a different approach to the detailed structural interpretation of sedimentary rocks in boreholes. The two-dimensional nature of the images often enables the interpreter to identify directly sedimentary structures and bedding surfaces. Different bedding types are reflected in the image texture. Measurements of layer thicknesses, even of very thin layers, are possible, and dips and azimuths of bedding surfaces can be measured selectively and accurately. The data enhance sedimentological interpretations as well as reservoir characterizations.

Jay Banavar, now at Penn State University, performed most of the statistical dune modelling. Andy Witkin and Kurt Fleischer at Schlumberger Palo Alto Research kindly let us have their geometric modelling system. Rodney Adams, Steve Wong and Phil Frulla developed an excellent image analysis system. Mike Grace, Gordon Pirie, Min-Yi Chen, Rusty Frisinger and Richard Wells were instrumental in acquiring and processing the data. Special thanks are due to the following oil companies who gave us permission to use their data: Exxon and Shell U.K., Mitchell Energy and the Gas Research Institute (GRI). Thanks are also due to R. Finley and S. Dutton of the BEG Texas. Discussions with many colleagues helped clarifying the interpretations presented herein, but the responsibility for all conclusions lies entirely with the author.

References

Barr, A. H. 1984. Global and local deformations of solid primitives. *Computer Graphics*, **18**, 21–30.

Brookfield, M. E. 1977. The origin of bounding surfaces in ancient aeolian sandstones. *Sedimentology*, **24**, 303–332.

Brown, L. F. Jr. 1979. Deltaic sandstone facies of the mid-continent. *In: Pennsylvanian Sandstones of the Mid-Continent*. Tulsa Geological Society Special Publication **1**, 35–63.

—— Cleaves, A. W. II & Erxleben, A. W. 1973. *Pennsylvanian depositional systems in north-central Texas – a guide for interpreting terrigenous clastic facies in a cratonic basin*. Texas University Bureau of Economic Geology Guidebook 14.

Clemmensen, L. B. & Abrahamsen, K. 1983. Aeolian stratification and facies association in desert sediments, Arran basin (Permian), Scotland, *Sedimentology*, **30**, 311–339.

Duda, R. O. & Hart, P. E. 1973. *Pattern Classification and Scene Analysis*. John Wiley, New York.

Ekstrom, M. P., Dahan, C., Chen, M.-Y., Lloyd, P. & Rossi, D. J. 1987. Formation imaging with microelectrical scanning arrays. *The Log Analyst*, **28**, 294–306.

Finley, R. J., Dutton, S. P., Ling, Z.-S., & Saucier, A. E. 1985. *The Travis Peak (Hosston) Formation: Geologic Framework, Core Studies, and Engineering Field Analysis*. Bureau of Economic Geology, Topical Report Prepared for the Gas Research Institute.

Gilreath, J. A. & Maricelli, J. J. 1964. Detailed stratigraphic control through dip computations. *Bulletin of the American Association of Petroleum Geologists*, **48**, 1902–1910.

Glennie, K. W. 1983. Early Permian (Rotliegendes) palaeowinds of the North Sea. *Sedimentary Geology*, **34**, 245–265.

Hall, E. L. 1979. *Computer Image Processing and Recognition*. Academic, New York.

Luthi, S. M. & Banavar, J. R. 1988. Application of borehole images to the 3-D modelling of eolian sandstone reservoirs, Permian Rotliegende, North Sea. *Bulletin of the American Association of Petroleum Geologists*, **72**, 1074–1089.

Plumb, R. A. & Luthi, S. M. 1986. *Application of borehole images to geologic modeling of an eolian reservoir*. Society of Petroleum Engineers, Paper 15487.

Rosenfeld, A. & Kak, A. C. 1982. *Digital Picture Processing*. 2nd edn, Vol. 2, Academic, New York.

Simaan, M., Zhang, Z. & Love, P. L. 1987. Artificial intelligence and expert systems for seismic data. *In*: Aminzadeh. F. (ed): *Pattern Recognition and Image Processing, Handbook of Geophysical Exploration* **20**, 389–425.

Methodology of Formation MicroScanner image interpretation in Claymore and Scapa Fields (North Sea)

S. D. HARKER[1], G. J. McGANN[1], L. T. BOURKE[2] & J. T. ADAMS[2]

[1]*Occidental Petroleum (Caledonia) Limited, 1 Claymore Drive, Bridge of Don, Aberdeen, AB2 8GB, UK*

[2]*Schlumberger Log Services B. V., Woodlands Drive, Dyce, Aberdeen, AB2 OES, UK*

Abstract: The Formation MicroScanner (FMS*) tool is an advanced dipmeter tool capable of producing high-resolution resistivity images of a borehole wall. When integrated with core and open hole logs, FMS images are of great utility in evaluating the thin bedded reservoir sequences of the Claymore and Scapa North Sea oil fields. Sedimentary features as small as 0.5 inches (1.2 cm) can be resolved on FMS images. Such detail is employed in the distinction of massive, parallel laminated, ripple cross-laminated, bioturbated and conglomeratic bedding types. The application of the FMS tool to reservoir description is not limited to bedding type identification. Bed boundaries, as well as small and large scale structural features, can be distinguished, including erosional contacts, slumps, fractures and faults. In the hydrocarbon-bearing zone permeability distribution outwith cored intervals can inferred. This can be confirmed by relating resistivity changes to core measured permeability values. The proposed approach to FMS interpretation involves the integration of all available open hole logs with detailed local geological knowledge and core data. Aspect and image quality of the FMS log are important considerations. By imaging the borehole wall, the aspect of the FMS tool provides a direct method of relating features detected using open hole logs to those observed in cores. Image quality is largely dependent on resistivity contrast in the formation and correlation with cores is imperative to establish confidence limits for extrapolation into uncored intervals.

Scope of work

The Formation MicroScanner (FMS) tool has added a new dimension to reservoir description and has been used principally for fracture identification (Nurmi *et al.* 1987). It is demonstrated herein that high resolution resistivity images improve perception of reservoir characteristics. The ability to resolve small sedimentary features is of great value when applied to studies of clastic depositional environments such as fluvial, deltaic, shelfal marine and turbidite sequences. This approach can be adopted as a complement to an existing coring program.

The FMS logs run on Claymore and Scapa have been used with cores, open hole logs and conventional dipmeter in an integrated approach to reservoir description, (they were run in development wells during 1986 and 1987). This paper presents an account of this approach, together with examples of sedimentary and structural feature recognition. A confidence limit scheme is proposed for evaluating the quality of image characterization. Once the confidence limits for the FMS images are established using core, then the FMS interpretation can be extended into uncored intervals to evaluate the total reservoir sequence.

The direct interpretation of borehole images are an advantage to the interpreter over conventional dipmeter arrow plots. In addition, specialist processing knowledge is not essential.

Geological setting

Claymore and Scapa are located about 100 miles northeast of Aberdeen in UK Block 14/19 (Fig. 1). The majority of the oil production from Claymore is from Upper Jurassic sandstones of the Main Area (Maher & Harker 1987). Scapa produces oil from Lower Cretaceous sandstones located in the syncline adjacent and to the south of the Claymore structure. The main area Claymore Reservoir (MACR) is represented at the base by the Sgiath Formation, an Oxfordian paralic sequence of thin sandstones, mudstones and coals. The succeeding Oxfordian to Kimmeridgian Piper Formation consists of marine shelf, silty

* Mark of Schlumberger.

From Hurst, A., Lovell, M. A. & Morton, A. C. (eds), 1990, *Geological Applications of Wireline Logs* Geological Society Special Publication No. 48, pp. 11–25

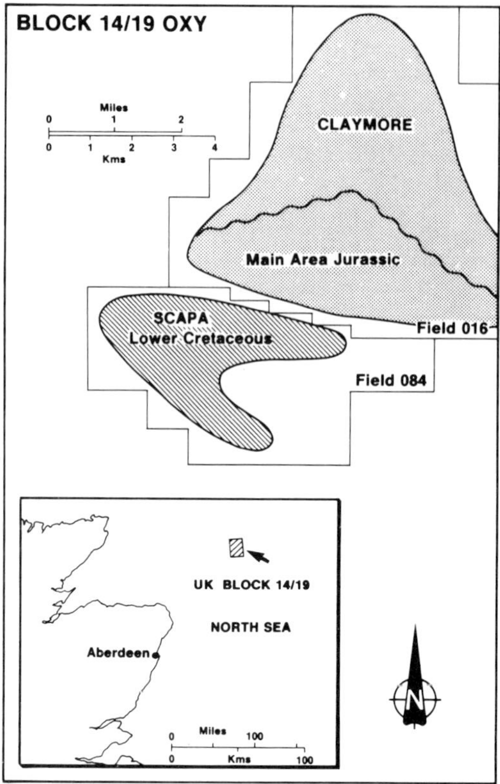

Fig. 1. Location map for Claymore and Scapa fields.

mudstones with bioturbated, argillaceous sandstones restricted to the eastern flank. The majority of the MACR comprises Late Kimmeridgian to Mid Volgian, thin bedded, organic rich, turbidite sandstones of the Claymore Sandstone Member of the Kimmeridge Clay Formation. The Scapa Sandstone Member is a thin bedded turbidite sand and conglomeratic unit, with common carbonate concretions, within the limestone and marl sequence of the Neocomian Valhall Formation. More detailed descriptions of the reservoir units are found in Harker & Maher (1988) and Harker *et al.* (1987).

Tool characteristics and processing

The FMS tool is a hybrid development of the stratigraphic high resolution dipmeter tool (SHDT*) and records simultaneously ten microresistivity curves, (two per pad plus speed buttons), but in addition high-resolution data is acquired from two orthogonally oriented pads (Ekstrom *et al.* 1987). Pads three and four have an additional array of 27 'buttons', each of 0.5 cm diameter (Fig. 2). Data are sampled every 0.1 inch (0.25 cm) both vertically and laterally. The tool configuration provides additional resistivity measurements from two 2.7 inch (6.9 cm) wide tracks of the borehole wall. In an 8.5 inch (21.6 cm) borehole some 20% of the borehole wall is covered in a single pass.

The tool incorporates inclinometry and accelerometry packages for tool orientation and speed correction. During processing (Table 1) compensation is made for variation of the focussing voltage, known as the EMEX* voltage, and the 27 curves are equalised for gain and offset to compensate for the different electrical characteristics and therefore dynamic ranges of each button. The resulting resistivity data are displayed either as raw curves ('wiggle traces') or as grey scale or colour images. Seventeen levels of grey tone are assigned to the measuring range to produce the grey scale images from the

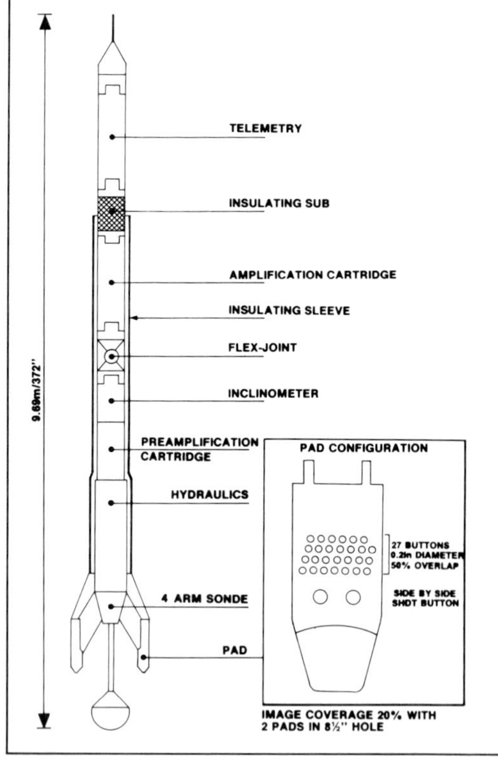

Fig. 2. Schematic diagram indicating the major components of the FMS tool.

* Mark of Schlumberger.

resistivity traces. Light tones are resistive and dark tones indicate increasing conductivity.

The images can be processed to emphasise gross lithological changes or to enhance specific features. In the case of the Scapa and Claymore data set, processing involved 'static' normalization over a large interval (approximately 400 feet) and 'dynamic' normalisation over a 2 feet sliding window. Multiple passes over a zone of interest can be presented as merged and azimuth oriented displays (Fig. 3).

Interpretation approach

The FMS tool provides the fine bed resolution and borehole coverage necessary to distinguish subtle bedding and textural features. It is crucial, however, to integrate open hole log and core data in FMS interpretation so as to establish lithological changes and log trends. The general interpretation approach adopted was as follows.

(a) Major depositional units were zoned using conventional open hole logs.

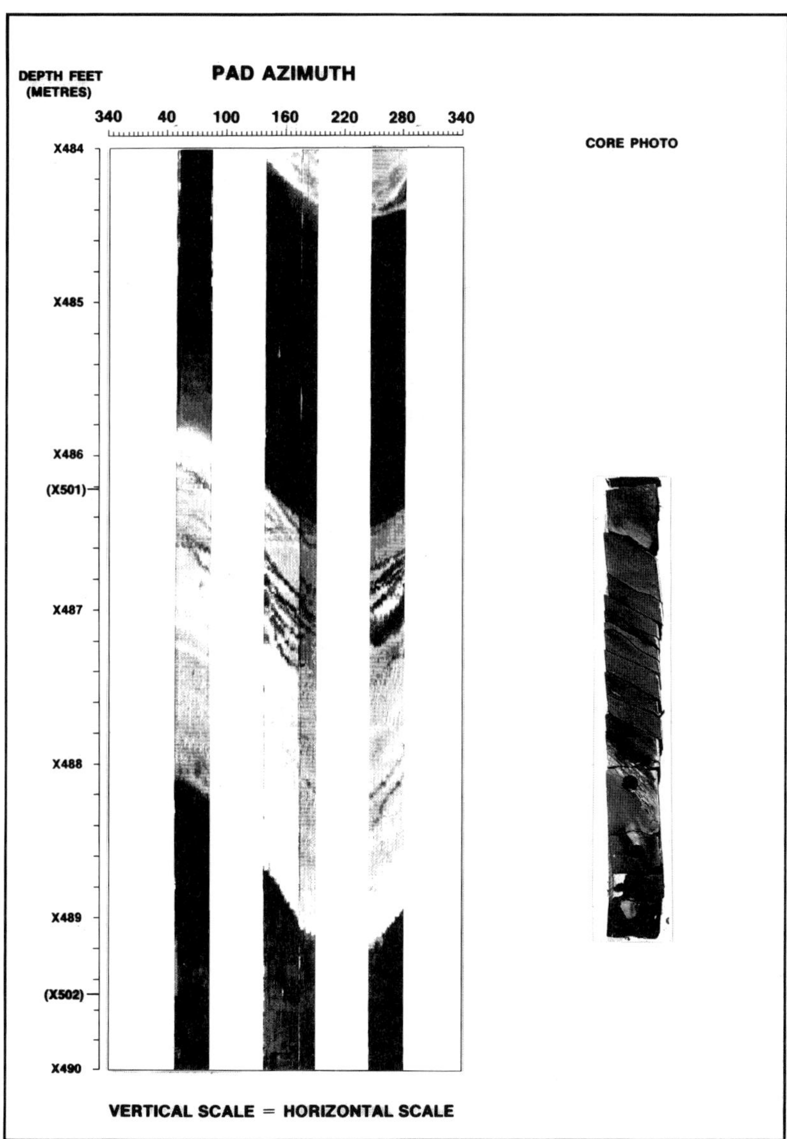

Fig. 3. Merged azimuth-oriented FMS images of Claymore Sandstone Member with thin mudstone interbed.

Table 1. *FMS Processing*

Objective: to represent formation conductivity changes as changes in grey-scale or colour shades.

1. Data Restoration: Speed correction to correct for variations in tool speed during logging.
   — Correction for variation of 'EMEX' focussing current.
   — Equalisation of curves to the same gain and offset.
2. Image Generation: Each datapoint assigned to one of 17 grey-levels or one of 256 colour shades. Light = resistive; Dark = conductive.
3. Image Enhancement
   — Static normalization: normalization of grey-scales over a large interval (400–500 ft). Shows gross resistivity variations.
   — Dynamic normalization: normalization of grey-scale within a short (1–3 ft) sliding window. Enhances small resistivity variations.
   — Dynamic histogram normalization: normalization of the full histogram of the data in a sliding window. Provides more detailed enhancement of resistivity variations.

(b) Dipmeter arrow plots were checked to identify tectonic features and major sedimentary discontinuities (e.g. unconformities). Local knowledge of the structural geology coupled with seismic evidence constrains the range of possible interpretations. Large faults, slumps and discontinuities are often easy to interpret using conventional dipmeter but can be difficult to identify in cores.

(c) Hydrocarbon zones and tightly cemented horizons were identified. The tool measures resistivity and is not a direct indicator of lithology.

(d) Core was described in detail. Careful examination of sedimentary features in a 4 inch (10.2 cm) core and on FMS borehole wall images from an 8.5 inch (21.6 cm) hole provides a means of assessing lateral variation. The combination of data can greatly improve interpretation from both core and images. Minor faults and slumps are often not seen or are ambiguous on conventional dipmeter and can be too large to interpret from core alone. Thus, the scale and borehole aspect of FMS data complement and enhance observations normally only obtainable from core.

(e) Constraints were placed on possible interpretations by local knowledge. If individual facies or depositional units can be recognised from conventional logs, then the interpreter will probably also have knowledge of types of sedimentary features which can be expected in particular facies. Local knowledge thereby constrains the possible interpretations and makes the task of identifying specific features from images much easier. Such knowledge also allows the interpreter to assess the credibility of his interpretation. An excellent parallel can be found in medical science: in ultrasonic scanning techniques, the technician knows he is working on the human anatomy, and has also localised the investigation to a particular part of the body. Thus, the ultrasonic image of a healthy kidney and the types of images expected in unhealthy or necrotic tissue are distinguishable. Similarly, in all but the most remote wildcat locations, it is usually possible to apply some local constraints to the depositional model. In the case of field development, the constraints can be quite exacting; indeed in such studies the interest will lie in establishing detailed reservoir characteristics.

(f) Log quality control was monitored to maintain interpretation confidence levels. Image quality is of paramount importance when interpreting small scale features, such as sedimentary structures. Poor image quality usually results from poor resistivity contrast such as masking or primary features by conductive or resistive cement.

Confidence limits

FMS images are correlated to core so that confidence limits can be assigned to specific features. FMS image interpretation may then be applied to the uncored part of the reservoir. The following FMS image quality grading is proposed for application to reservoir description.

Grade 1. Features which can be categorically indentified and interpreted from FMS images.

Grade 2. Features which can be clearly identified from FMS images, but do not have a unique interpretation.

Grade 3. Features with an ambiguous interpretation.

Implications of the confidence limit scheme are summarised in Table 2 and examples given in the following applications section.

Applications

The interpretation approach outlined above was used to obtain maximum sedimentological information from core and FMS logs in recent Claymore and Scapa field wells. The combined study of core and FMS images has established a number of applications to reservoir description which are an improvement on conventional methods.

Major features/boundary delineation

FMS images can delineate formation boundaries precisely, such as the contact between the bioturbated shelf sands of the Piper Formation and the overlying parallel laminated, turbidite sands and silty mudstones of the Claymore Sandstone Member of the Kimmeridge Clay Formation (Fig. 4). The FMS images of Grade 2 confidence limit show clearly the difference in bedding types between the two units. Bed boundary definition is possible with other fine resolution tools, such as microresistivity logs; however, the FMS images, details of the bed contact can be determined. The mudstone bed shown in Fig. 3 has a sharp top and base, with sandy laminae at the top. In the core, sand laminae are not present at the top of the bed, but are present at the base.

By comparing the core with the images, it is clear that the sandy laminae are discontinuous, isolated features. For a comparison of the bed resolution of common open hole logs SHDT

Table 2. *Summary of the interpretation implications of the confidence limits grading scheme proposed in this paper*

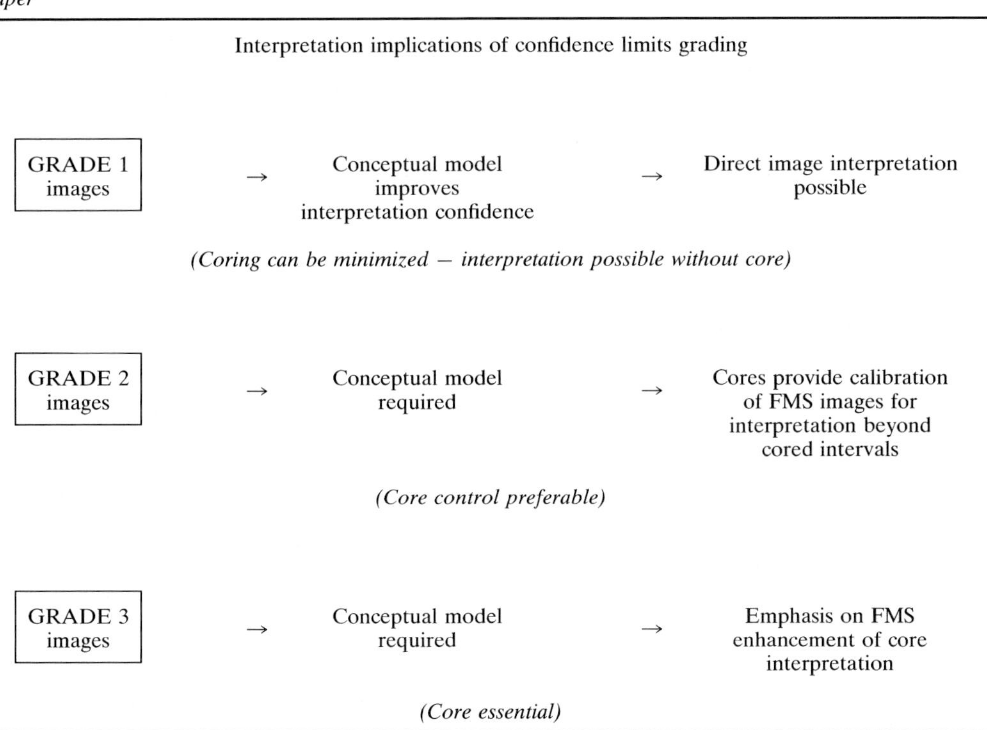

Notice that comment has been made on core requirement in relation to FMS image interpretation.

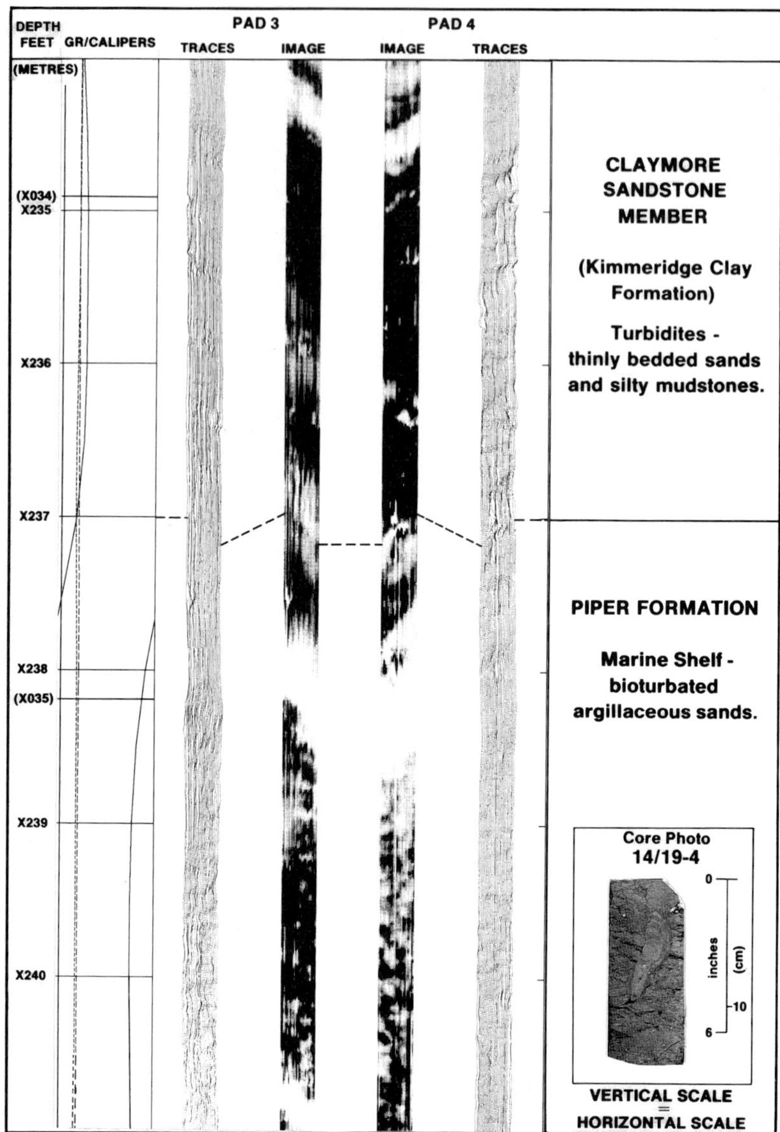

Fig. 4. Dynamically normalised FMS images over the Claymore/Piper contact.

and FMS see Höcker *et al.* (1989) and McGann *et al.* (1988).

An erosional contact is clearly shown in the Claymore Sandstone Member (Fig. 5). A massive sand bed has cut down into the underlying, parallel laminated sand bed. The correlation with core is very good and the FMS image is a Grade 2 confidence limit. The boundary represents the downcutting of a higher energy (high permeability) Bouma type A massive sand facies into a lower energy (lower permeability) Bouma type B facies.

Small to large scale slumps

The MSD*† dipmeter arrow plot (Fig. 6) shows a discontinuity within the Kimmeridge Clay Formation which could be interpreted as an unconformity or as a fault, without drag on the hanging wall. The CSB*‡ processing shows more detail of bedding irregularities than the MSD plot but interpretation remains ambiguous. Two discontinuities are evident at ×646 ft (×244.9 m) and at ×654 ft (×247.4 m), but further information is required to enhance the

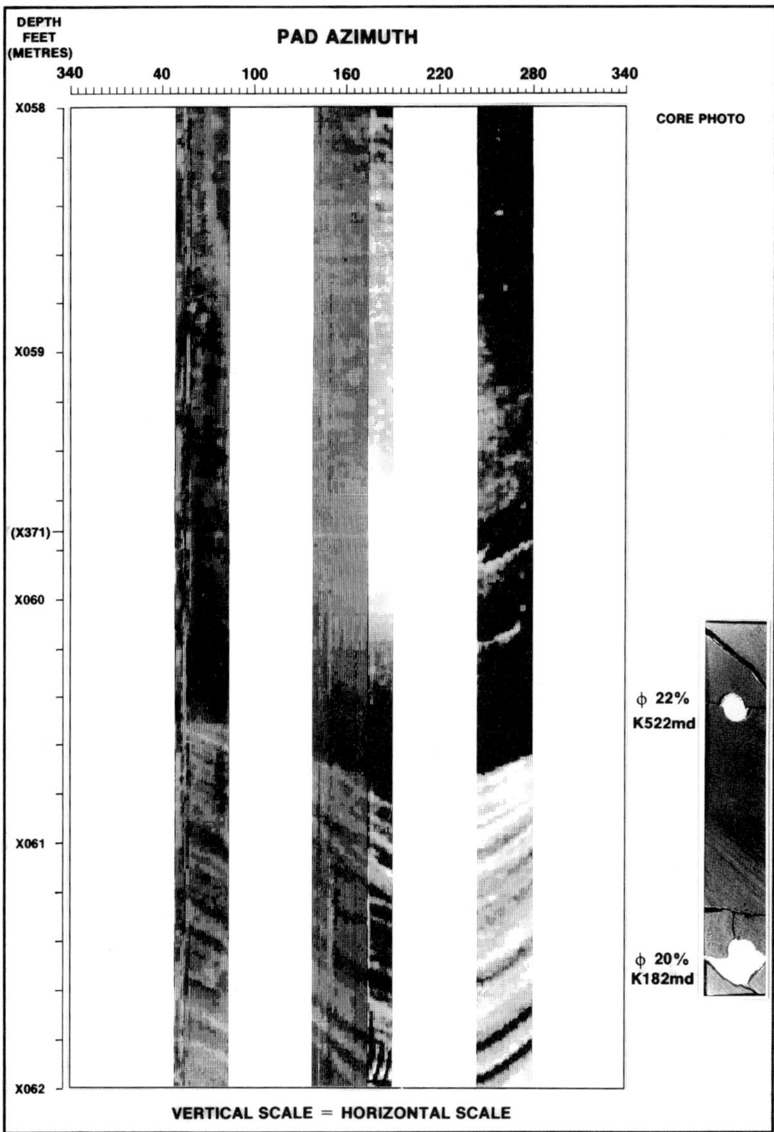

Fig. 5. Merged azimuth-oriented FMS images of erosional contact within the Claymore Sandstone Member.

(* Mark of Schlumberger).

† MSD (Mean Square Dip): large-scale interval correlation method, using all eight SHDT curves from the FMS tool. Typically uses a 4 ft correlation interval, 2 ft step distance.

‡ CSB (Continuous Side-by-Side): small-scale interval correlation method, correlating the two SHDT curves from each pad, and computing a true dip from the apparent dips from orthogonal pads. It typically uses a 4″ correlation interval, 3″ step distance. (Mair & Adams 1986.)

interpretation. The FMS images clarify the internal bedding detail of a slump deformation. From cores it has been established that slumping is common within the Kimmeridge Clay Formation in the Claymore Field as downslope gravity readjustment and accompanying soft-sediment deformation resulting from southerly rotation of the Claymore block in late Jurassic times (Harker & Maher 1988). A Grade 1 confidence limit is assigned to this example.

A small-scale slump is identified within the Scapa Member sandstones of the Valhall

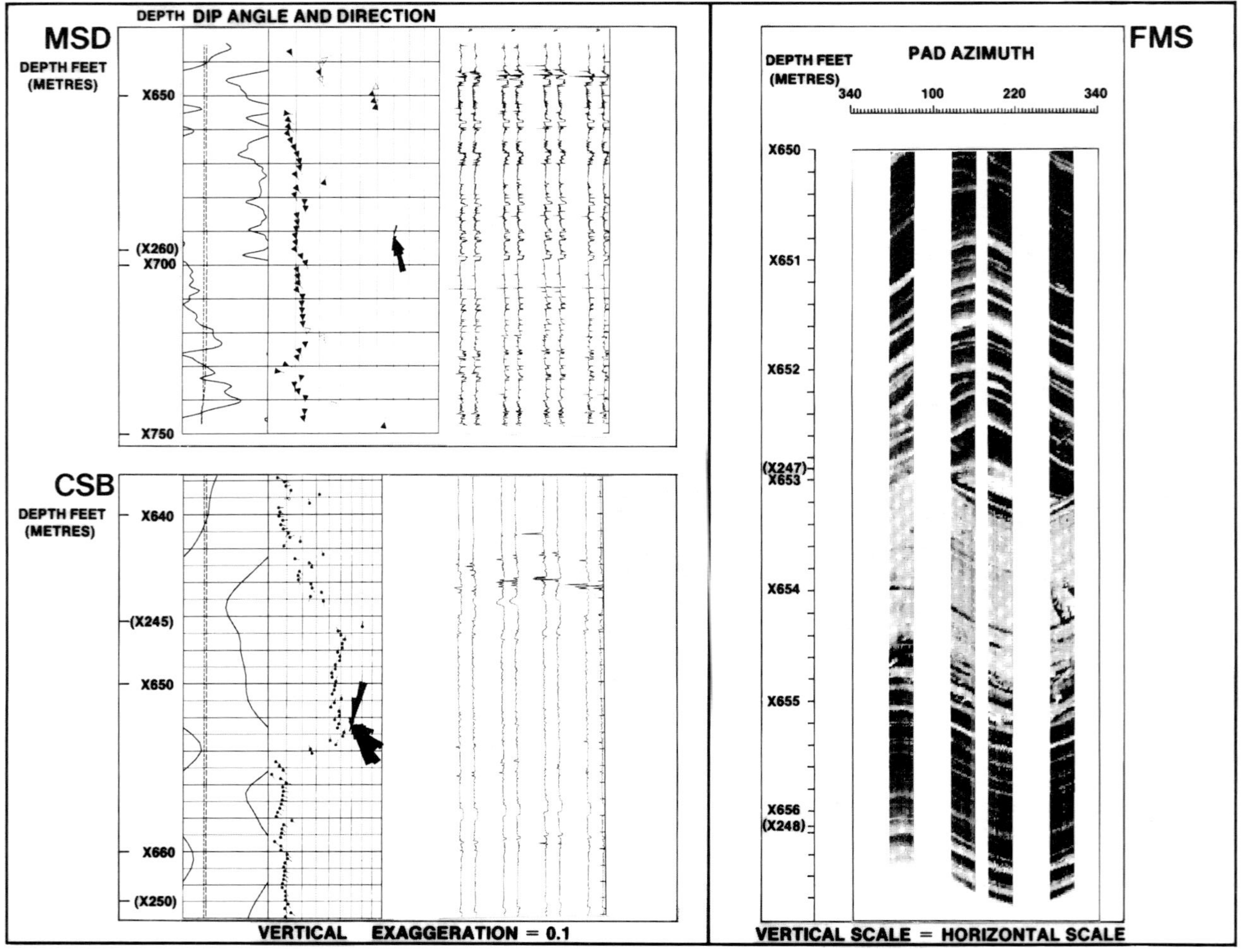

Fig. 6. Composite dipmeter plot of slump feature in Kimmeridge Clay Formation.

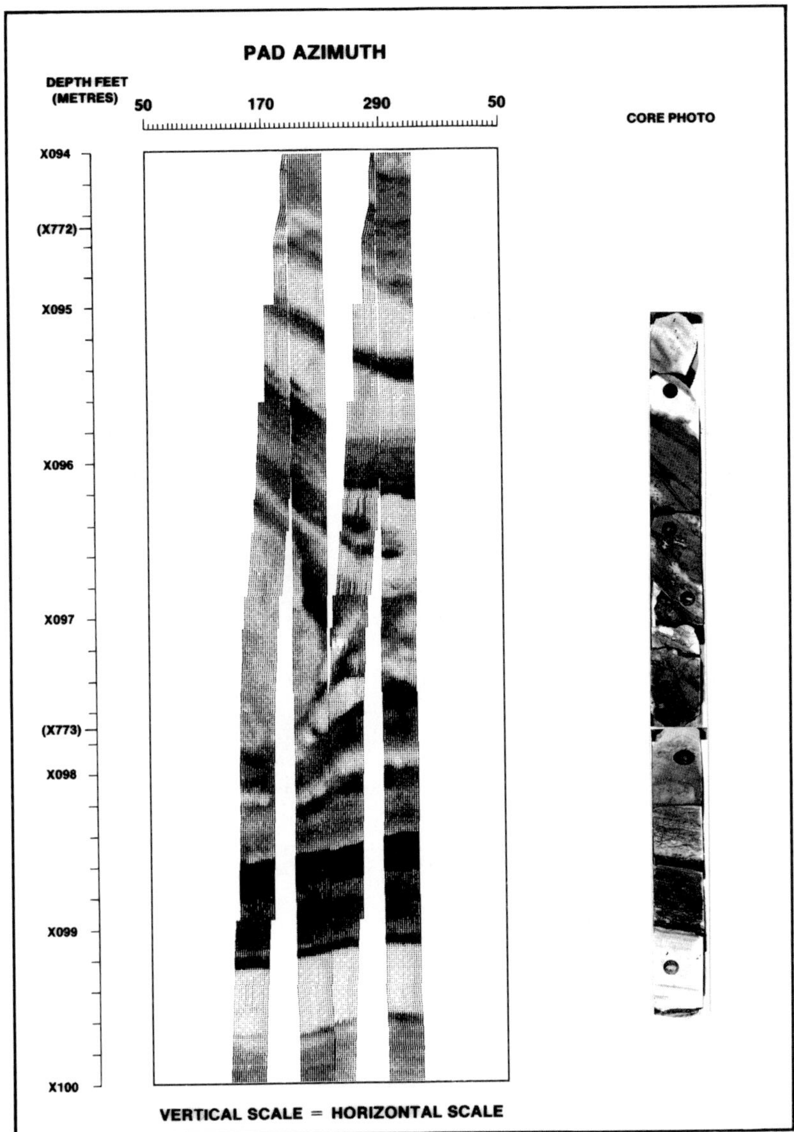

Fig. 7. Merged azimuth-oriented FMS images of a small slump in the Scapa Sandstone Member.

Formation in Scapa Field (Fig. 7). From core examination it was difficult to diagnose this structure as a slump. It would, at best, be identified as sands containing water escape structures. No characteristics typical of slumping were recognisable on the MSD dipmeter arrow plot over this interval.

Faults

Faults with little or no associated distortion of sedimentary layers are frequently difficult to identify on conventional dipmeter plots. The greater borehole perspective provided by FMS images is particularly amenable to fault identification.

A sharp break of 976 psi in the RFT pressure profile was observed in a Claymore well (Fig. 8), and the MSD dipmeter plot indicated a change of dip in proximity to the break. A cemented high-resistivity zone, 9 inches (23 cm) wide appears to constitute a permeability barrier at ×269 ft (×434 m). Above this zone there is

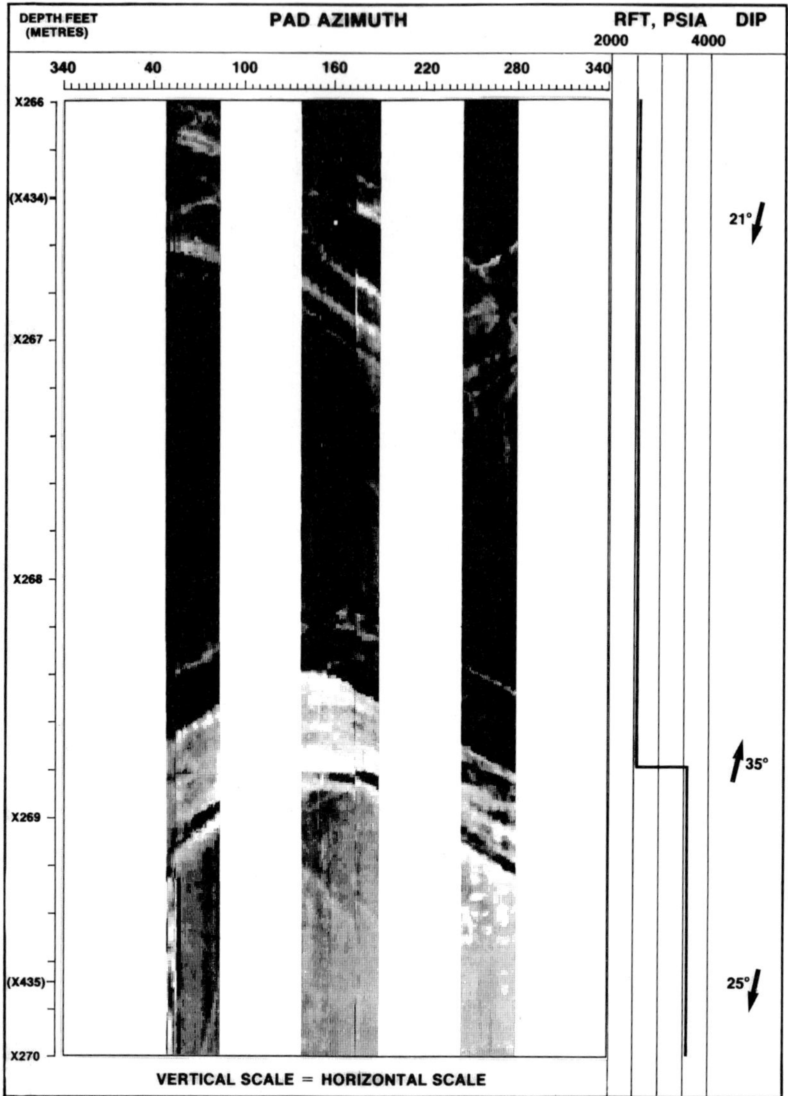

Fig. 8. Merged azimuth-oriented FMS images of a high resistivity cemented fault zone in the Claymore Sandstone Member.

little change in image quality, indicating uniformity of reservoir quality. Similarly below this zone reservoir quality appears homogeneous.

The highly resistive zone dips N at 35°, whereas the formation dip increases from 21–25° S across this zone. Data from other wells proves the presence of a listric fault zone with a throw of approximately 30 ft (9.1 m) and at that depth is virtually undetectable on seismic. This is believed to be a mineralized fault zone that constitutes a major transmissibility barrier.

Diagenetic features

It is difficult to discern FMS images of diagenetic features without the benefit of core control or local knowledge, except perhaps where strongly cemented calcareous zones produce marked resistivity contrast with the host sands (Fig. 9). The benefit of borehole imaging is that a large volumetric sample can be examined to assess the lateral variability of cement distribution. Interpretation becomes ambiguous

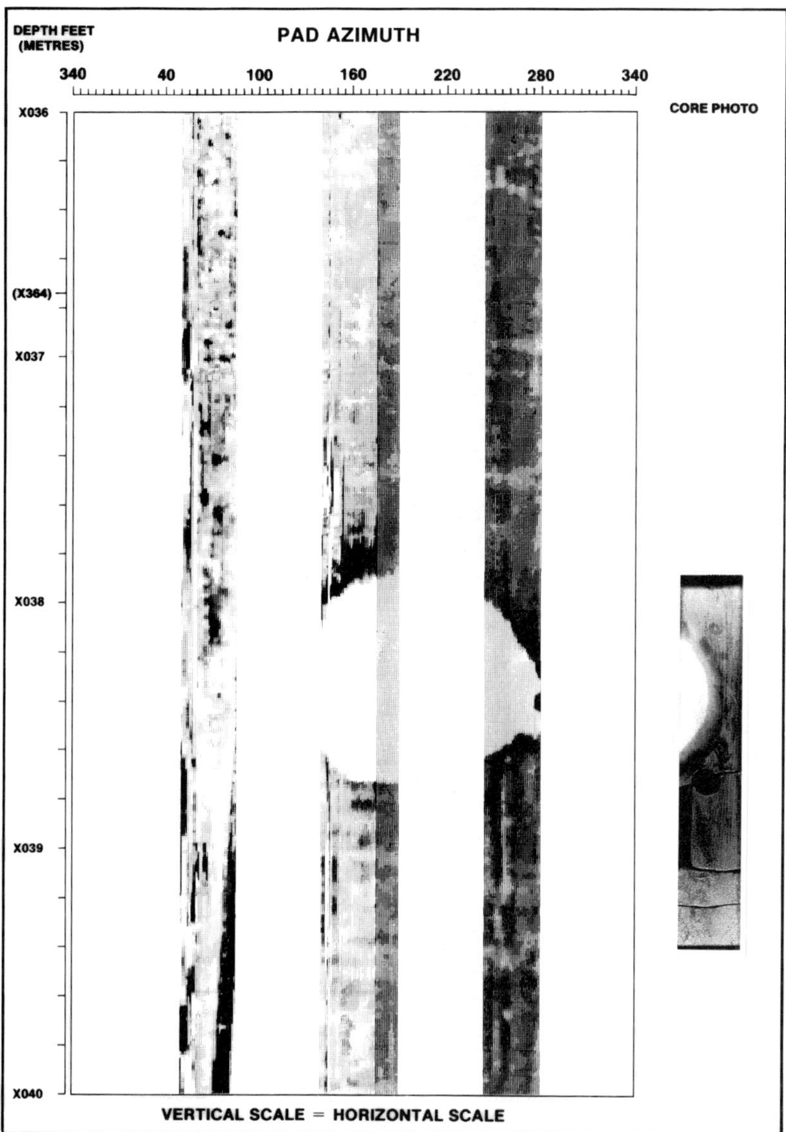

Fig. 9. Merged azimuth-oriented FMS images of a calcareous nodule in the Claymore Sandstone Member.

where cement distribution is patchy (Fig. 10). This results in images becoming mottled (Grade 3 confidence limit) and the interpreter, without core control, may be misled into diagnosing features such as bioturbation or conglomeratic bedforms.

Textural recognition – conglomerate

Conglomerate identification using FMS images is clear where the resistivity contrast between clasts and matrix is strong. In the Scapa Field, conglomerates meet this criterion and clasts from 0.5 inches (1.3 cm) to approximately 3 feet (1 m) in diameter can be identified (Grade 2 confidence limit). A debris flow sequence in Fig. 11 from ×118 ft (×781.6 m) has a range in clast size from 0.5 inches (1.3 cm) to 12 inches (30.5 cm). The clasts are mostly matrix supported and it is apparent from the images that the sequence is reverse graded, though this is not discerned from core.

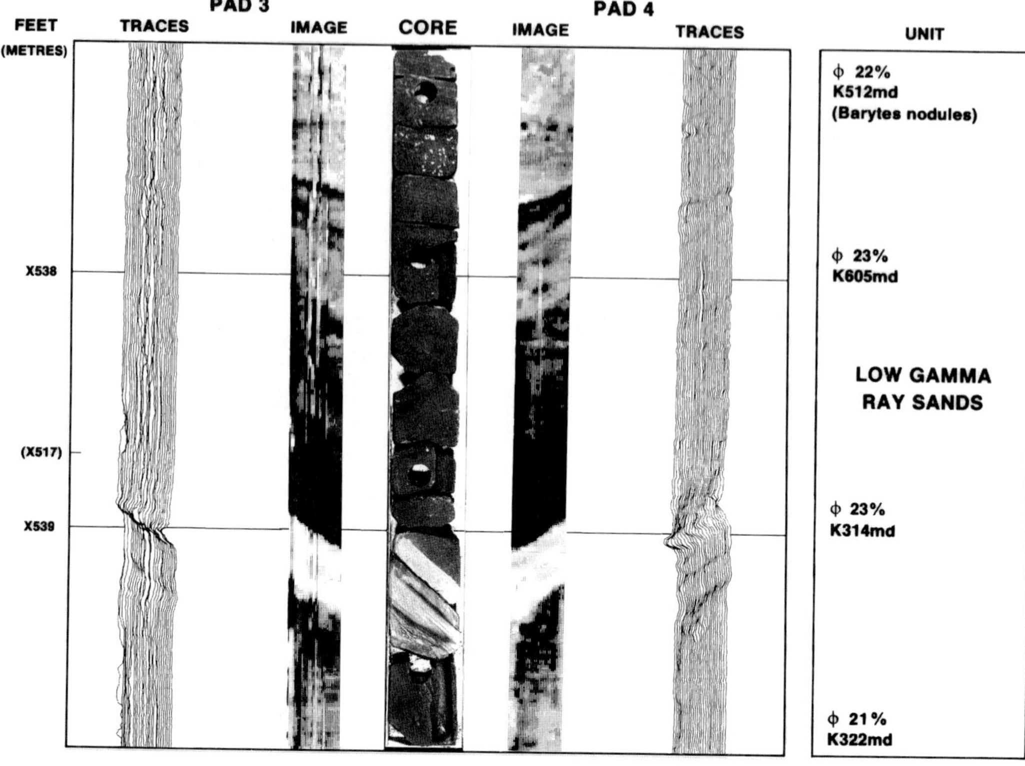

Fig. 10. Dynamically normalised FMS images of patchy cementation in the Low Gamma Ray Sands in the Claymore Sandstone Member.

The sequence from ×117 ft to ×118ft comprises 0.5 inch (1.3 cm) to 1 inch (2.8 cm) clay pebbles that are clast supported. The clay pebbles are densely interlocked and resistivity contrast is poor, resulting in poor clast definition on the images.

The conglomerates are easily distinguished from the calcareous nodules between ×126ft and ×128 ft, where the shape, orientation, bed boundary sharpness and resistivity character are diagnostic.

Sedimentary structures

Sedimentary structures such as parallel and ripple cross lamination can be inferred from FMS images. In Fig. 12, the thinner bedded High Gamma Ray Sands reservoir division of the Claymore Sandstone Member is easily distinguished from the more massively bedded Low Gamma Ray Sands (Harker & Maher 1988). The laminae are picked out by the resistivity contrast (Grade 2 confidence limit) resulting from textural differences and fluid content. Ripple foreset planes are not discernable, though the bounding surfaces are seen as flattened lenses.

Bioturbation is recognizable due to the intermixing of argillaceous sediment with the host sand, which results in a vague resistivity contrast. Bioturbation appears as vague mottling on FMS images, and often displays wavy bed boundaries (Fig. 4). However, individual burrows are occasionally distinguishable.

Permeability

Resistivity measurements made by the FMS tool are sensitive to subtle changes of lithology and fluid saturation. In core studies, high energy clean sandstones tend to provide the better reservoir units in contrast with their shaly counterparts. Variations in mineralogy, grain

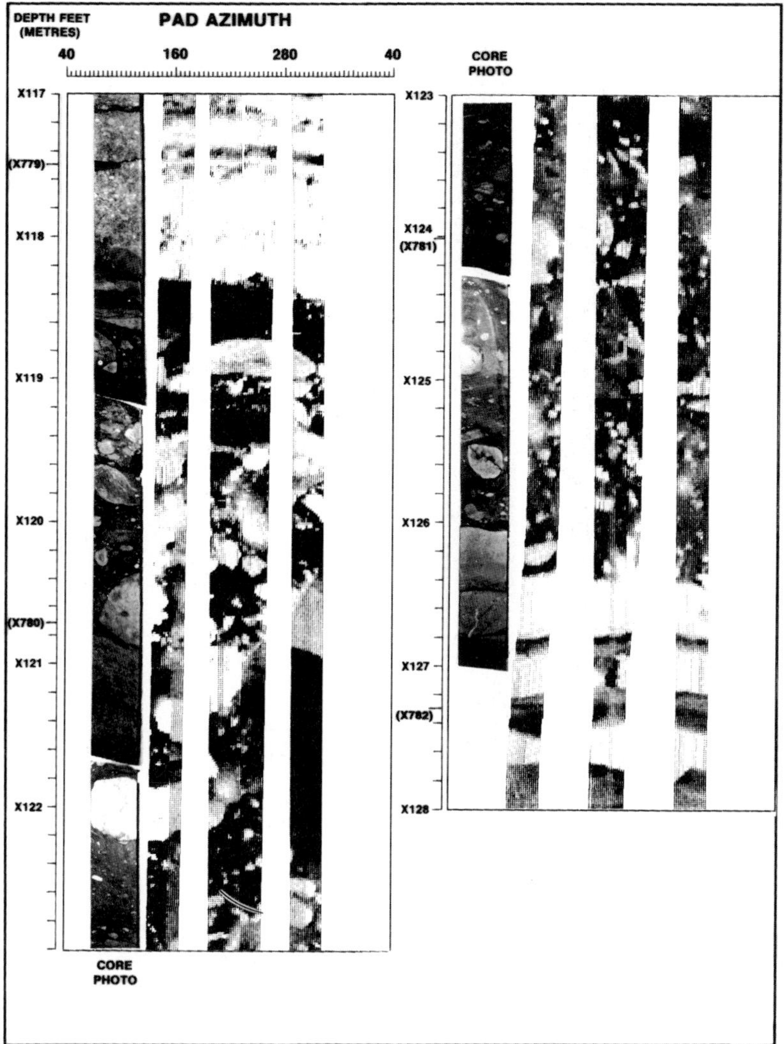

Fig. 11. Merged azimuth-oriented FMS images of a matrix supported conglomerate from the Scapa Field.

size, grain rugosity etc., affect the conductivity of the rock and the permeability.

Care must be taken to identify the native formation fluid and the composition of the drilling mud to have confidence in inferring permeability distribution. A permeability contrast between the darker images of the good quality Low Gamma Ray Sands and the lighter images of the poorer quality High Gamma Ray Sands is shown in an oil zone (Fig. 12). The more permeable Low Gamma Ray Sands are more conductive and consequently darker because saline mud filtrates have displaced the native formation oil within the flushed zone. The less permeable High Gamma Ray Sands are more resistive and produce a lighter FMS image. A similar difference is seen where a more permeable massive sand overlies a less permeable parallel laminated sand (Fig. 5). Uniform permeability distribution is indicated by an even tone (Fig. 3). Units of good reservoir characteristics can be distinguished using statically or dynamically normalised plots (within the defined window length) in conjunction with core data.

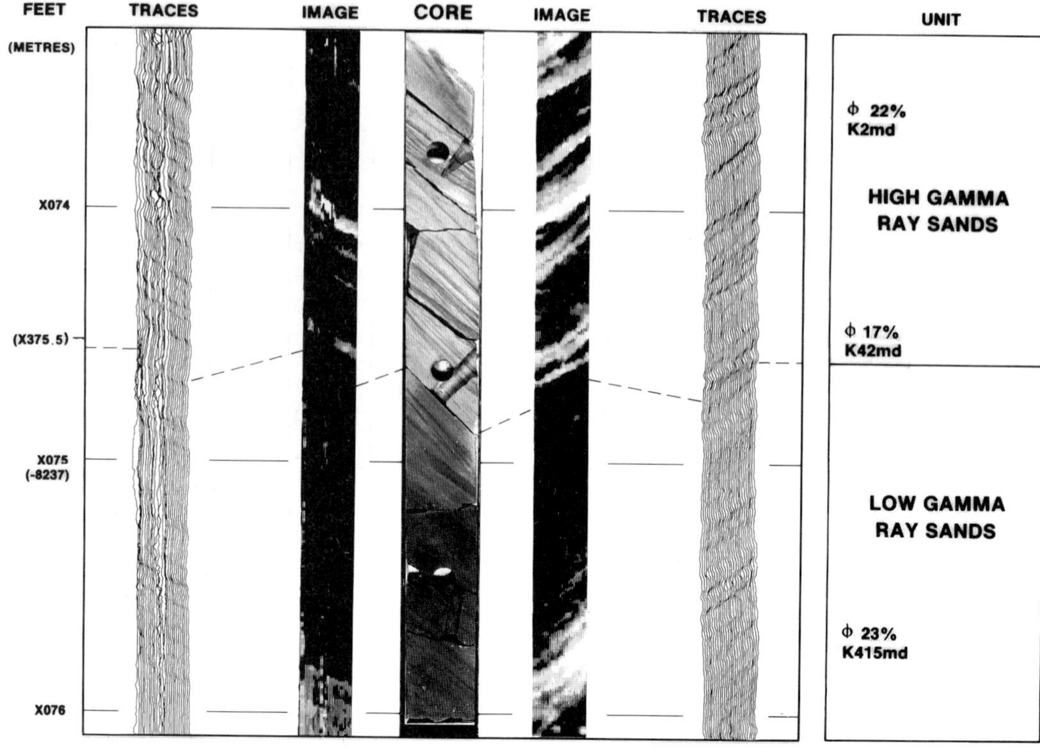

Fig. 12. Dynamically normalised FMS image of the High/Low Gamma Ray Sands contact in the Claymore Sandstone member.

Conclusions

(1) For detailed reservoir description, the integration of local geological knowledge, cores, open hole logs and conventional dipmeter as part of FMS image interpretation is critical. Extrapolation of interpretation into uncored intervals is dependent on this integration process and a confidence limit grading scheme is proposed.

(2) The imaging capability of the Formation Microscanner tool can be used for bed and formation boundary delineation, the identification of slumps, faults, diagenetic features, conglomerates and sedimentary features, and to aid permeability inference. All of these enhance reservoir description in Claymore and Scapa fields.

(3) In contrast to conventional dipmeter, a detailed knowledge of processing techniques is not necessary for the user to make useful interpretations from FMS images.

The writers wish to thank Occidental Management, the Fourth Round Consortium partners and Schlumberger for permission to publish this paper. The interpretations are solely the responsibility of the authors. In addition, we would like to thank all those who helped in the preparation of the paper.

References

EKSTROM, M. P., DAHAN, C. A., CHEN, M. Y., LLOYD, P. M. & ROSSI, D. J. 1987. Formation imaging with microelectrical scanning arrays. *Log Analyst*, **28**, 294–306.

HARKER, S. D., GUSTAV, S. H. & RILEY, L. A. 1987. Triassic to Cenomanian stratigraphy of the Witch Ground Graben. *In*: BROOKS, J. & GLENNIE, K. (eds), *Petroleum Geology of Northwest Europe*. Graham & Trotman, London, 809–818.

—— & MAHER, C. E. 1988. Late Jurassic sedimentation and tectonics, Main Area Claymore Reservoir, North Sea. *Giant Oil and Gas Fields, S.E.P.M. Core Workshop No. 12*, 395–458.

Höcker, C. F. W., Eastwood, K. M., Herweijer, J. C. & Adams, J. T. 1989. The use of SHDT data in sedimentological studies, *Bulletin of the American Association of Petroleum Geologists*.

McGann, G. J., Riches, H. A. & Renoult, D. C. 1988. Formation evaluation in a thinly bedded reservoir. A case history: Scapa Field, North Sea, *Transactions of SPWLA 29th Annual Logging Symposium, San Antonio, June 5–8*.

Maher, C. E. & Harker, S. D. 1987. Claymore Oil Field. *In*: Brooks, J. & Glennie, K. (eds), *Petroleum Geology of Northwest Europe*, Graham & Trotman, London, 835–845.

Mair, B. F. & Adams, J. T. 1986. An approach to the sedimentological interpretation of dipmeter results. *Transactions of SPWLA 10th European Formation Evaluation Symposium, Aberdeen, Paper N.*

Nurmi, R., Taha, M., Waterhouse, M. & Chahara, M. 1987. Fracture detection and geometrical analysis. *Seventh Biannual Petroleum Conference of Turkey, Ankara*.

Gamma-ray log shape used as a facies indicator: critical analysis of an oversimplified methodology

M. H. RIDER

Rider-French consulting, PO Box 1, Rogart, Sutherland, IV28 3XL, UK

Abstract: Gamma-ray log shapes are often used by geologists to determine sandstone grain size trends and hence depositional facies. However, for the simple methodology of relating shapes to facies to be valid, there must be consistent relationships between gamma-ray log values and clay content and between clay content and grain size. Log to core comparisons and sandstone textural analyses show that these relationships are naturally variable. Consequently, correlation of log shape with grain size trend is tenable only under very limited (and defineable) conditions. A universal application of gamma-ray log shape to grain size trend and depositional facies is wrong.

Three principal gamma-ray log shapes are frequently used as a basis for identifying depositional facies, especially of sandstones. The three are the bell shape (persistently upwards increasing gamma-ray value), the funnel shape (persistently upwards decreasing values) and the cylindrical or blocky shape (stable low values between high shoulders) (Fig. 1). The principal shapes are related to basic depositional sequences which have important sedimentological significance (Fons 1969). The bell shape, for instance, indicates the fining upwards grain size trend typical of a channel sandstone or transgressive marine sand, while the funnel shape indicates the coarsening upwards grain size trend of a barrier bar or delta front deposit (Fig. 1).

The classic syllogism of philosophical logic suggests that two premises can be joined to make a conclusion. It takes the form: 'all S are M, all M are P, therefore all S are P'. The mistakes that can be made in such a formula are part of philosophical folklore.

The use of gamma-ray log shapes to suggest sandstone depositional facies uses just such a syllogism and seems to fall into just such a trap as pleases the philosophers. The formula is: all (S) shaped gamma-ray curves indicate (M) lithological sequences, all (M) lithological sequences indicate (P) depositional facies, therefore all (S) shaped gamma-ray curves indicate (P) depositional facies. For example, all bell-shaped gamma-ray log shapes are indicative of upwards decreasing grain size sequences: all upwards decreasing grain size sequences indicate a channel facies or a transgressive marine facies: therefore, all bell-shaped gamma-ray curves indicate channel facies or transgressive marine facies. Such a stark statement as this clearly creates doubt about a methodology which uses such 'logic'. The philosophical point should not be laboured, but it does show how the methodology can be cut into constituent parts and examined.

The obvious relationship being suggested when gamma-ray log shapes are used in facies analysis, is that gamma-ray changes, besides indicating clay content, mirror grain size-changes. Low log values indicate a coarse grain size while high values indicate fine grain size (Fig. 2). Although this relationship was originally illustrated using the SP (Fons 1969) it was later extended to include the gamma-ray (Serra & Sulpice 1975, Selley 1976). However, the gamma-ray tool does not measure grain size: it measures natural radioactivity. Using the log shapes therefore requires implied relationships. There are three principal ones: (i) the relationship of gamma-ray log response to clay content, (ii) the relationship of clay content to grain size and (iii), the relationship of grain size sequence to depositional facies. The third relationship will not be examined in detail as it is entirely in the realm of sedimentology and will only be considered in the final discussion. The first and second relationships, however, are fundamental to the suggested use of the gamma-ray log shapes and must be examined for proper comment on the methodology (Fig. 2). To do this the gamma-ray log response to clay content and the relationship of clay content to grain size will be considered. To complete the loop, the direct, empirical relationship of the gamma-ray log response to grain size change will be illustrated.

Gamma-ray response and clay content of sandstones (sandstone and clay radioactivity)

The gamma-ray log is a record of a formation's natural radioactivity which comes spon-

From HURST, A., LOVELL, M. A. & MORTON, A. C. (eds), 1990, *Geological Applications of Wireline Logs* Geological Society Special Publication No. 48, pp. 27–37

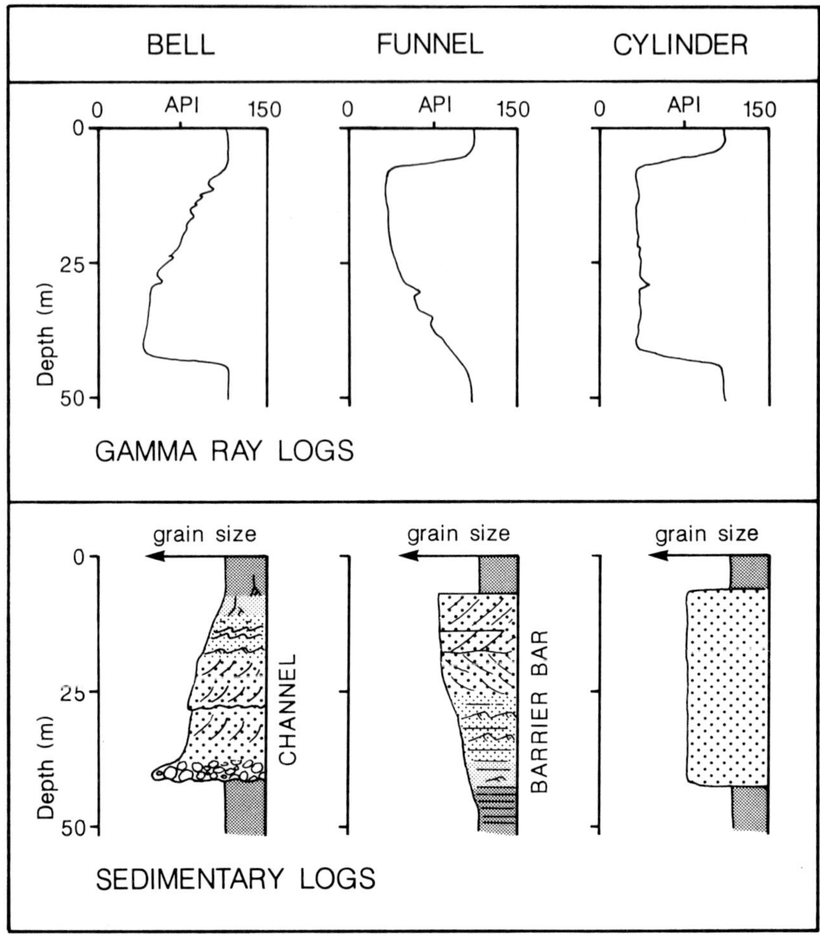

Fig. 1. The three principal gamma-ray log shapes and their corresponding sedimentary interpretation (after Serra & Sulpice 1975).

taneously from naturally occurring uranium (U), thorium (Th) and potassium (K) (Russell 1941). The naturally radioactive elements generally have a far greater concentration in shales than in most other sedimentary rocks. On the contrary, quartz, the principal component of sandstones, does not contain either U, Th or K. Therefore, in sedimentary rocks formed of a mixture of quartz sand and clay, high gamma-ray readings are usually taken to indicate high clay content. Regrettably this is often not the case. All that is radioactive is not necessarily clay and clay is not always radioactive.

Before describing the distribution of radioactivity in sand and clay, it is necessary to define how the word 'clay' is used, there being a certain amount of ambiguity in the term when sediments are described. Here, 'clay' is used in the textural sense, meaning detrital grains with a size of less than 4 μm. However, it is implied that this fraction is essentially made up of clay minerals, as the mineralogical composition controls the natural radioactivity. The use of the word 'clay' will therefore indicate clay sized particles of clay minerals. Any ambiguity created by this dual sense will be clarified in the text where necessary. Shale will be used to indicate lithologies, composed of clay-sized material.

When interpreting gamma-ray log shapes of sandstones, difficulties arise because naturally radioactive silt and sand sized detrital grains are mixed with the non-radioactive quartz grains. In a typical sandstone, quartz will be mixed with feldspars and rock fragments, as well as with clay. Indeed, the usual way of classifying

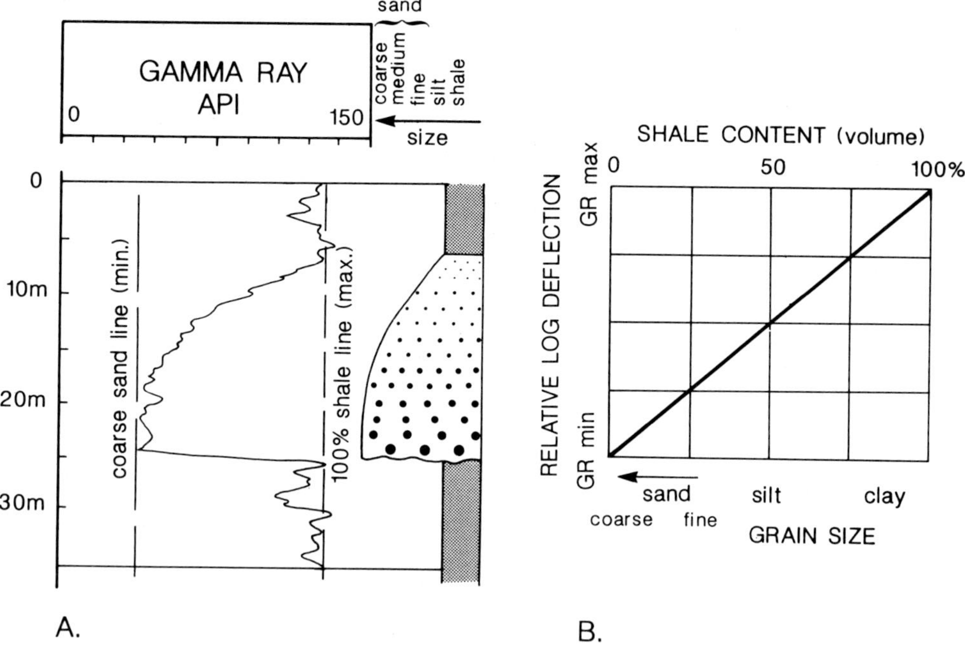

Fig. 2. Schematic representation of the use of the gamma ray log in facies analysis. A. Grain size changes reflected in the gamma ray log. B. Graphical presentation of the semi-quantitative relationships implied (modified from Rider 1986).

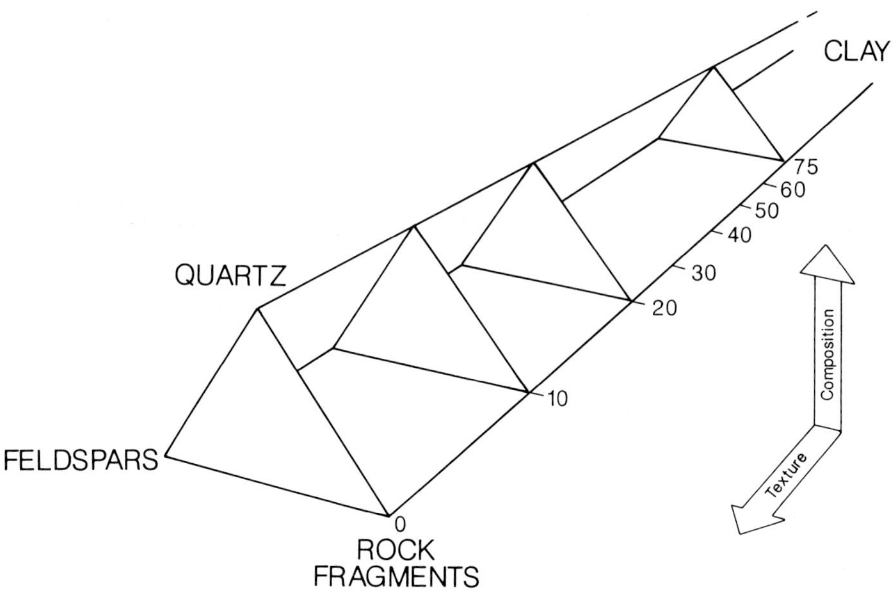

Fig. 3. Diagramatic representation of the elements of sandstone classification, based on both composition and texture (modified from Folk 1954).

sandstones is to express the composition in terms of quartz, feldspar and rock fragments on a triangular graph against detrital clay as a third dimension (Fig. 3). Importantly, this recognises the semi-independent interplay between texture and composition (Folk 1954). The common, naturally radioactive detrital components include potassium feldspars, because of their high potassium content (10–15% in potassic varieties), rock fragments when composed of micas (high potassium content, see Hurst 1990), volcanic clasts (thorium and potassium present but variable) and large clay clasts. Sand sized detrital components cannot therefore be considered as exclusively non-radioactive.

Moreover, clays cannot be considered as exclusively radioactive because clay minerals have a wide range of compositions. In terms of potassium (K), for example, although the average shale contains 2.7% K, the range is from over 8% K in illites to 0% in chlorite. With thorium (Th) and uranium (U) the range is as great (Hurst 1990). The thorium ranges are especially significant since the radioactivity from this source probably contributes about half of the total count (Dypvik & Eriksen 1983). Empirically in terms of total radioactivity, illite will have a much greater contribution than chlorite or kaolinite (Reverdy *et al.* 1983). However, the potassium concentration of illite is normally insignificant in terms of the bulk potassium content of a rock unless concentrations of potassium feldspar and micas are negligible (Cowan & Myers 1988; Hurst 1990). Clearly any change in clay mineral composition will most likely entail a change in radioactivity.

Chemical analyses of some typical sedimentary rocks show well the effects of the natural minerological variations outlined above. An arkose can contain over 5% K_2O while a silty clay contains only just over 3% K_2O (Cox *et al.* 1967). The log example (Fig. 4) illustrates this effect and shows a medium to coarse grained arkosic sandstone without clay but with up to 25% feldspar of which 10% is potassic and therefore radioactive, surrounded by a silty clay composed mainly of kaolinite with a very low natural radioactivity. The result is that the gamma-ray log radioactivity values of the silty clay and the sandstone are very similar. This is a serious complication when using gamma-ray log shapes.

Confusion between clay mineral and detrital mineral radioactivity is complicated even further by the fact that the non-quartz detrital components tend to vary in concentration with grain size just as the clay fraction does (a factor allowed for in the classification of sandstones as indicated previously, Fig. 3). Overall sandstone composition can, in fact, be expressed as a function of quartz grain size (Davies & Ethridge 1975). Orthoquartzites are generally fine to medium grained, shaly sandstones are fine grained and poly-detrital sands are coarse grained. In terms of natural radioactivity, this means that several regimes are in play against grain size and not just that of clay against quartz. The first example of this effect (Fig. 5) comes from the Middle Jurassic sandstones of the North Sea in which certain grain size classes contain up to 25% of micas (thin section count), naturally radioactive from their high potassium content (6–10%, Hurst 1990). The mica content is seen to vary with grain size changes while the gamma-ray reacts more to a combined function of clay and mica content. A second example shows a dramatic case in which an oil-bearing channel sandstone was considered from a petrophysical examination to have 'shaled out' (in Well 2, Fig. 6). Examination of cores showed that the channel continued but that in this particular well the lower part was dominated by large shale clasts and could be more properly called a shale conglomerate.

These examples clearly illustrate that a high gamma-ray value cannot be equated to high

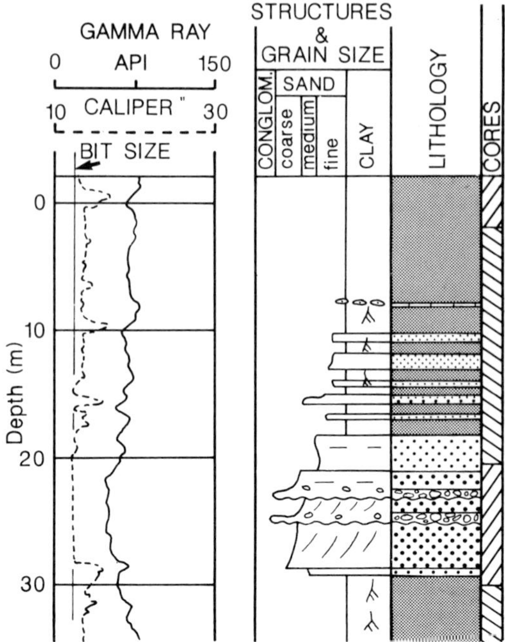

Fig. 4. Arkosic sandstone with an average 25% feldspar content (10% potassic) with a gamma-ray value similar to the surrounding silty shales.

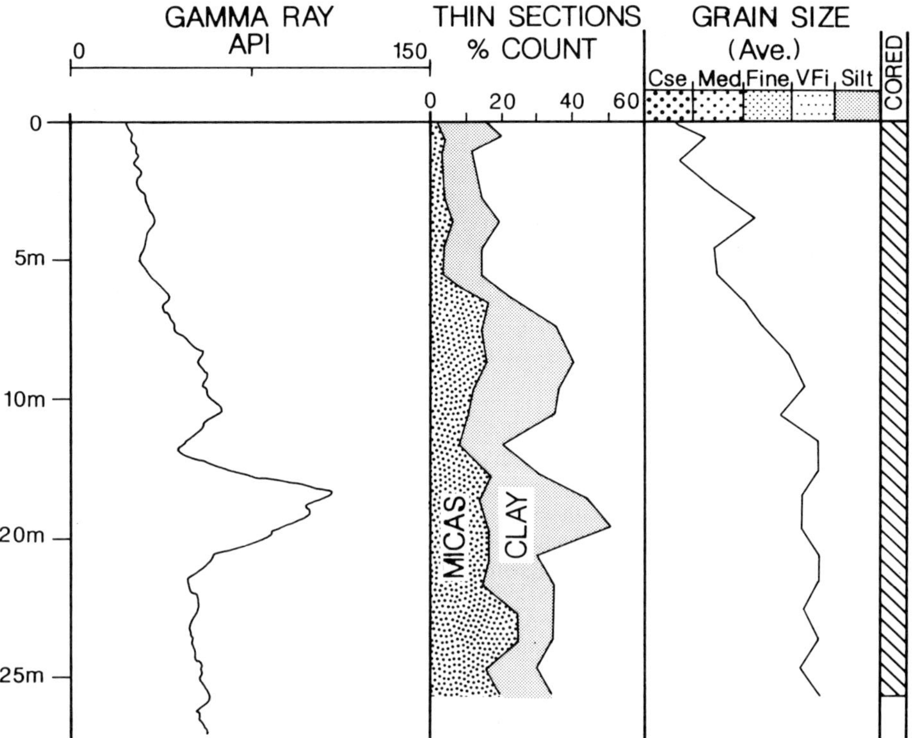

Fig. 5. Gamma-ray log in a sandstone with both radioactive grains (micas) and detrital clays. The micas vary more regularly with overall grain size than the clays.

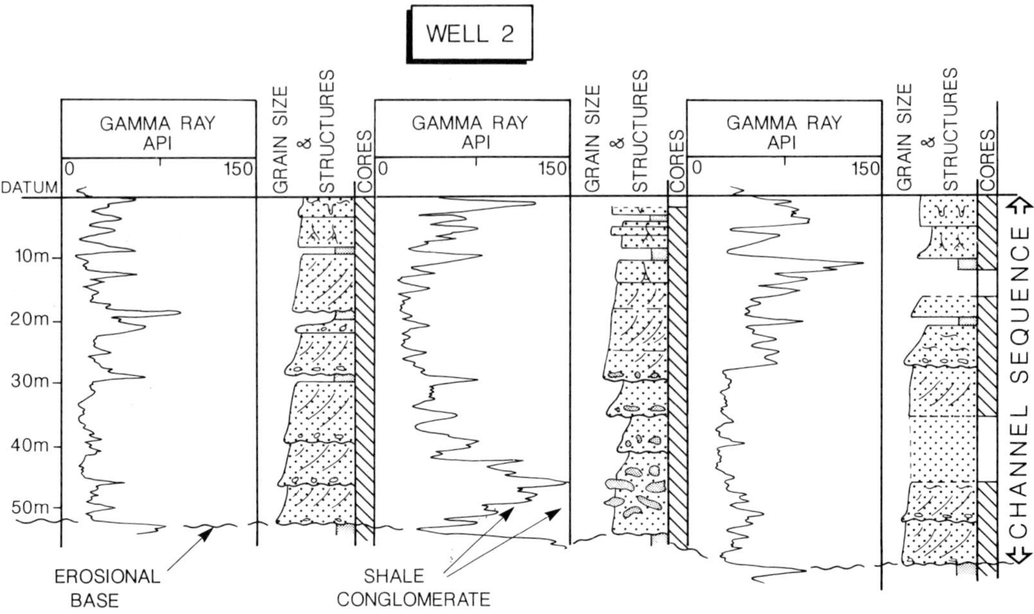

Fig. 6. Channel sandstone sequence across three wells with an interval of high gamma-ray values in Well 2 due to the presence of shale conglomerate.

clay content any more than a low value to a low clay content: the reality is much more complex. The gamma-ray log value of a sandstone is not necessarily related to its clay content.

The clay–grain size relationship

The relationship to be considered next concerns texture, being that between grain size and clay content. There is no doubt that a grain size to clay content correlation exists, but it is by no means the constant relationship required if the proposed log shape analysis methology is to work properly. The following examples illustrate the problems.

Sedimentary petrographers suggest that texture in sandstones can be related to depositional processes and, by inference, depositional environment (Visher 1969). Depositional environment is also believed to affect sandstone composition within, of course, the limits imposed by the grain size and mineralogical content of the sediment source. Some authors go so far as to claim that a depositional environment can be identified from composition (Davies & Ethridge 1975). Although sandstone texture and composition are identifiable as separate attributes, they are very closely related albeit in a variable manner. For example, a textural analysis of some German sandstones shows that clay content varies consistently with grain size in alluvial (Molasse) sandstones, but that in a well winnowed marine sandstone (Dogger) no such relationship is seen (Fig. 7). It appears that in rapidly deposited sands such as occur in the Molasse, the amount of clay is associated with the dominant grain size, but in environments where there is long duration winnowing, all clay is absent regardless of grain size: the clay content is no longer related to sand size fractions. These are just two cases.

Much modern work on primary textures and structures in sandstones aims to analyse the behaviour of various grain size classes under different hydraulic conditions. The mode of transport, of deposition or of erosion, is cross-linked to current velocity and expressed in the resulting deposit as a range in grain size (Allen 1985). The empirical differences in the relationship between clay content and grain size shown in the German sandstone examples (Fig. 7) have a theoretical basis. The variation is natural rather than exceptional and related to the dominant hydraulic processes in particular environments. Deposition from a turbidity current will create a different relationship between clay content and grain size to that created by marine winnowing or by deposition in a river current. Without attempting to resolve the intricacy of these inter-relationships, it is clear that clay to grain size correlations are as varied as are hydraulic conditions.

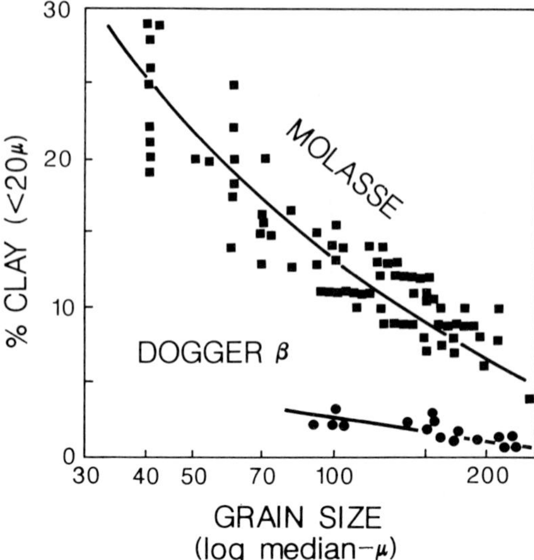

Fig. 7. The relationship between grain size and clay content in two German sandstones, the alluvial Molasse and the marine Dogger (from Pettijohn *et al.* 1972)

An additional complication which affects these observations on primary relationships is the frequent occurrence of clay mineral diagenesis in sandstones (e.g. Almon *et al.* 1976). Diagenesis may involve both clay mineral neoformation or destruction. Both change the clay to grain size relationships and, consequently, the relationship between the location of natural radioactivity in specific minerals and in grain size classes. (For example, dissolution of detrital potassium feldspar and neoformation of illite will change the location of potassium in minerals and thereby increase the radioactivity present in the clay-sized fraction.) Although certain diagenetic processes may be facies and grain size controlled (e.g. Hawkins 1978), the secondary changes can only make the primary clay to grain size relationships more complex and unpredictable. An added element of uncertainty in the understanding of gamma-ray log shapes.

Empirical gamma-ray log to grain size relationships

To put the preceeding somewhat theoretical aspects into context, it is necessary to examine empirical comparisons between cores and gamma-ray logs.

That there can be a close relationship between grain size and gamma-ray log values, is illustrated using core derived measurements (Fig. 8). A slightly different method of comparing core and log data shows a similar close relationship (Fig. 9A). Both these examples are from deltaic environments. Once a good gamma-ray log to grain size relationship has been demonstrated, it is a natural tendency to add a sedimentological dimension and an interpretation of depositional environment. Indeed, the most often quoted example of gamma-ray log shape is a fining upwards sequence from a fluvial channel or point bar with a classic bell-shaped gamma-ray log (Fig. 10). These convincing and valid examples can create a tendency to believe that a simple relationship between the gamma-ray log and sediment grain size is always present. Of course it is not.

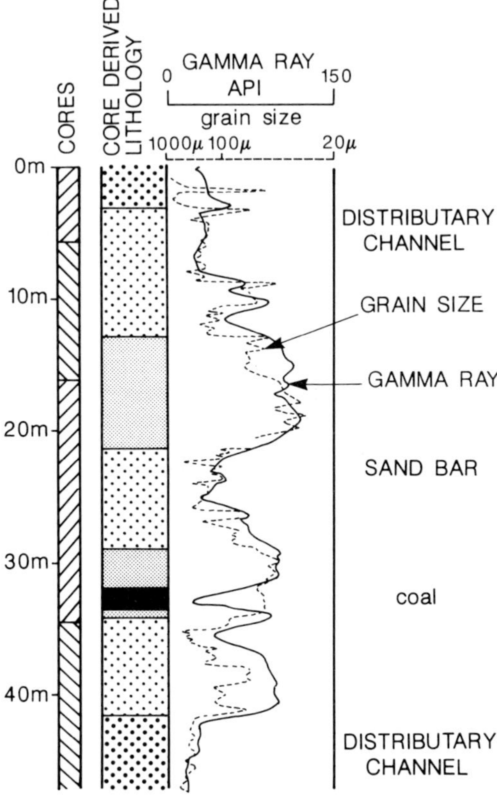

Fig. 8. Very close relationship between gamma-ray log values and core-measured sandstone grain size from a deltaic environment (from Simon-Brygoo 1980).

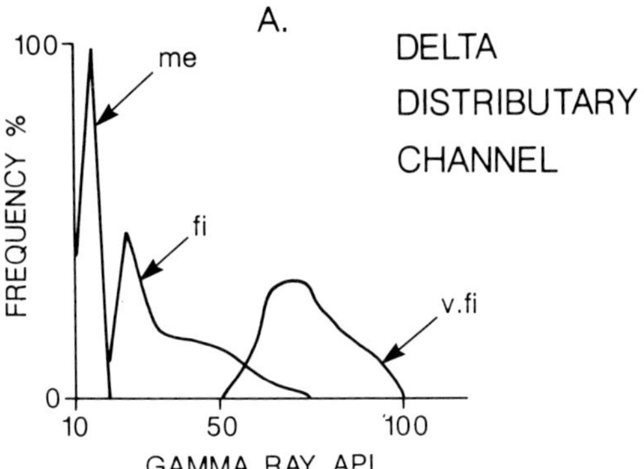

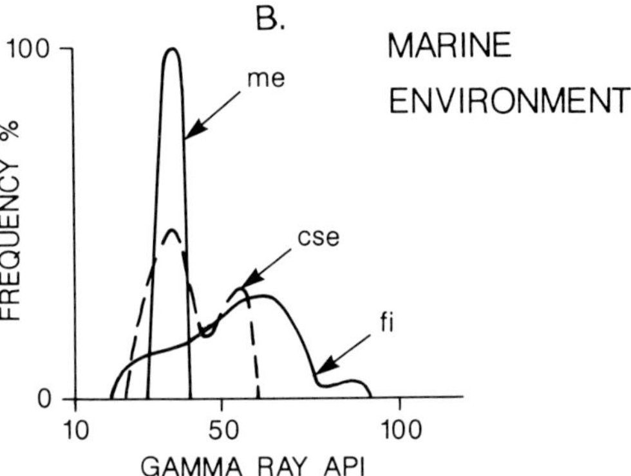

Fig. 9. Gamma-ray log values compared to sandstone grain size classes. A: Alluvial delta distributary sandstone. B: Winnowed marine sandstone. (from Simon-Brygoo 1980).

A sandstone deposited by turbidity currents (Fig. 11) illustrates that there is very little relationship between the texture seen in cores and the gamma-ray response. Log shapes give absolutely no indication of grain size changes. In this example, there appears to be a persistent clay-sized fraction regardless of the coarse grained fraction, be it sandstone or conglomerate. A similar lack of correlation is shown by a fluvial, fining upwards sequence identified from core (Fig. 12). It is not associated with the expected bell shape on the gamma-ray log but with a subdued funnel shape. Finally, in the general type of presentation, grain size classes are seen to have no relationship to gamma-ray values for the identified marine environment (Fig. 9B). These examples, in fact, represent the normal case rather than the unusual.

An aspect which adds complexity to the empirical comparisons is diagenesis. The effect is seen not just in the clay minerals, as has already been touched upon, but also in the detrital

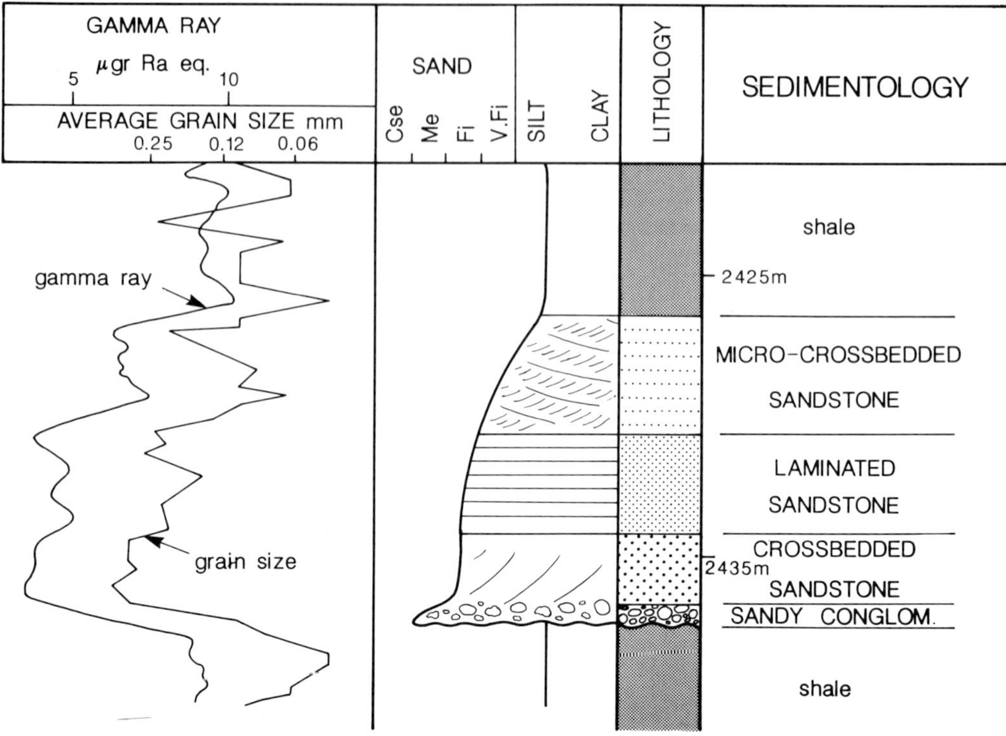

Fig. 10. Typical bell shaped gamma-ray log corresponding to a fluvial, fining upward point bar sandstone sequence (from Serra & Sulpice 1975).

minerals themselves, such as feldspars, which are chemically unstable in many sub-surface conditions. The effect that diagenetic changes have on radioactivity is very difficult to assess. For example on the shelf off mid-Norway, the depth-temperature controlled diagenetic change from kaolinite to illite requires an addition of potassium, the source for which may be pore water potassium derived from feldspar destruction (Bjørlykke *et al.* 1986). Formation of illite presumably concentrates potassium in the clay-size fraction of the rock at the expense of potassium feldspar present in the sand fraction. However, diagenetic clay minerals may be less radioactive than the detrital ones (A. Hurst, pers. comm.) so the overall effect on the gamma-ray log is difficult to evaluate. Diagenesis commonly causes changes in the distribution of the natural radioactivity in minerals, almost certainly confusing any simple relationship between grain size and gamma-ray log value.

Discussion

As the above examples show, there is a dilemma: sometimes the gamma-ray log shapes faithfully correlate with grain size variations, sometimes they do not.

As was stated at the outset, for the gamma-ray response to correlate with grain size variations two conditions must be met: the gamma ray must correlate with clay content, and the clay content must correlate with grain size. An inversion of this proposition is now more appropriate: when the gamma-ray log does correlate with grain size variations, then all the conditions necessary for the relationship to hold are met. These are that only an admixture of radioactive clay and non-radioactive quartz are involved and that the mode of deposition creates a close relationship between the grain size and the amount of dispersed clay. Empirically, it seems that it is especially in fluvio-deltaic environments (Figs 8 & 10) where these conditions are met (e.g. Flint *et al.* 1988). Moreover, in these cases it is not just the gamma-ray log shape which may be used; it is any variation in value, so that if core material is available, the grain size classes may be correlated in gross terms to actual gamma-ray log readings (Fig. 9A). But this does not create a general rule.

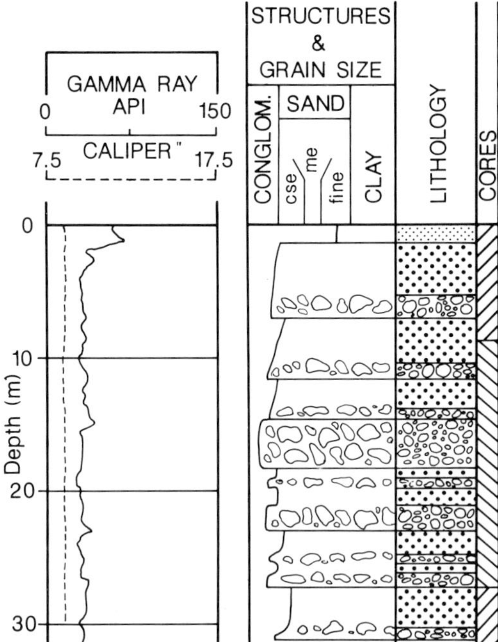

Fig. 11. Lack of correlation between gamma-ray log response and grain size variation in a very coarse grained sequence deposited by turbidity currents.

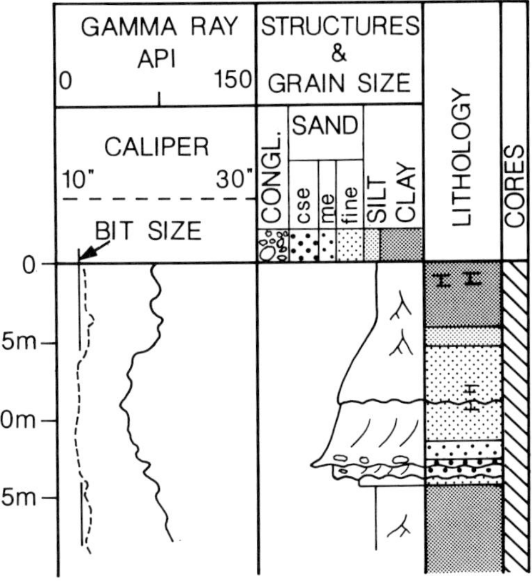

Fig. 12. Fining-upwards fluvial sequence with no bell shaped gamma-ray log.

There are two principal reasons for complications. Firstly, as has been illustrated (Figs 3–7), the ideal relationship between grain size and gamma-ray log response is disturbed by two primary factors, one compositional, the other textural. In terms of composition, it has been shown that it is quite normal in a sandstone for there to be non-clay detrital minerals which have a natural radioactivity. Their abundance is frequently related to grain size but it is quite independent of the clay fraction. A change in gamma-ray value simply indicates a change in the naturally radioactive detrital grains and not a grain size change with a concomitant variation in clay content. Secondly, the variation of the clay content has been shown not to always follow the grain size variation. Indeed the relationship is probably, in some way, related to the environment of deposition and is hence highly variable. Even when there are no naturally radioactive detrital grains present, gamma-ray variations do not reliably indicate grain size variations. This is the effect of texture. Added to this are the present but undefineable effects of diagenesis. Consequently, compositional, textural and diagenetic variability all cause the relationship between gamma-ray log shape or log value and grain size to break down.

It is quite clear that universal rules and a universal use of the gamma-ray log as an indicator of sandstone grain size variation is wrong. It may be correct in selected cases but this can only be confirmed after calibration with core material. The application benefits from an understanding of the principles involved.

This does not mean that there is no place for logs in the interpretation of depositional environments. Quite the contrary, it points to ways in which logs may be used. The gamma-ray log on its own has severe limitations. Furthermore, why should it be used on its own — because it is apparently simple? The spectral gamma-ray log could be used to refine the interpretation of the radioactivity of the simple gamma ray but most of the problems outlined above would still not be avoided. Also, why should only the coarse-grained intervals be analysed. In logging, unlike at outcrop, shales and silts are eminently 'visible'. Indeed, it is frequently the fine-grained intervals that have the most distinctive log signatures. The rule, therefore, is that analysing logs for facies information should include the whole log suite, not just the gamma-ray log, and the whole sediment sequence, reservoir and non-reservoir.

The final example illustrates this point (Fig. 13). It concerns a small fining-upward sequence surrounded by over-bank silty shales containing calcretes, and an upper, more complex, possibly

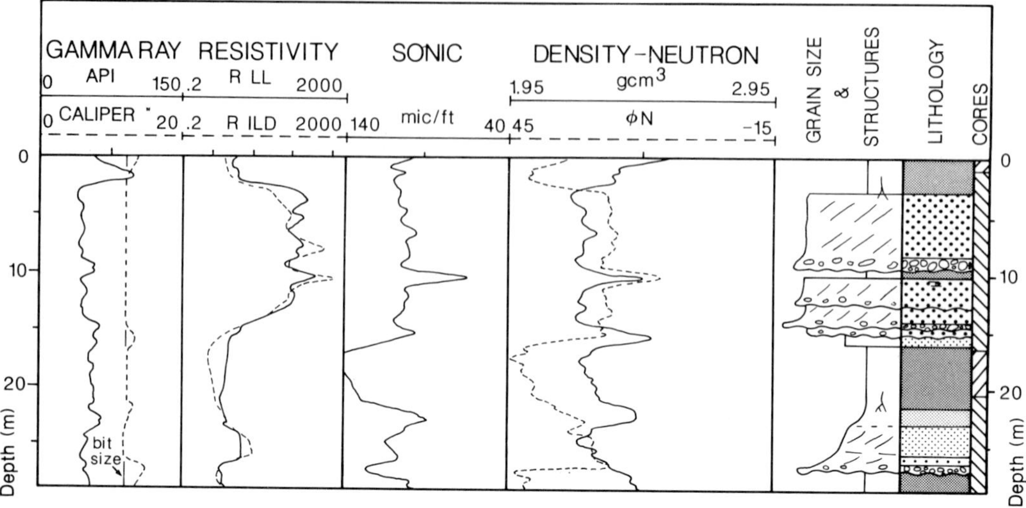

Fig. 13. Two sand bodies with a full set of corresponding logs. Lithological and textural information can be interpreted from each, especially the neutron density.

braided sand body. The shales are kaolinitic and have a low natural radioactivity. Conversely, the sands have a relatively high natural radioactivity because of abundant potassium feldspar, which has not been altered by diagenesis. The result is that the gamma-ray log is rather featureless and differentiates poorly between sand and shale. The neutron log, on the contrary, reacts to free and bound water content, thus allowing clear identification of the clay layers. The compact calcrete layers are seen on the density log but especially on the sonic log which is sensitive to horizontally continuous features. In the sandstone intervals the density log shows porosity variations which are related to textural changes. These are further linked in part to grain size variations and in part to diagenesis. The resistivity logs indicate textural changes in the sandstones in terms of differences in fluid displacements around the borehole. Each log, properly interpreted within its limitations, adds an element to the lithological, textural and eventually sedimentological description. The logs should lead us to a textural and compositional description, not a final interpretation of environment of deposition. Many other data are involved in that. In conclusion, therefore, ringing the virtues of a bell-shaped gamma-ray log is doing the science of geological log interpretation a gross disservice.

The author thanks the reviewers and editor for helpful comments and discussions which considerably improved the manuscript.

References

Allen, J. R. L. 1985. Loose-boundary hydraulics and fluid mechanics: selected advances since 1961. *In*: Brenchley, P. J. & Williams, B. P. J. (eds) *Sedimentology, Recent Advances and Applied Aspects*. Geological Society, London, Special Publication **10**, 7–28.

Almon, W. R., Fullerton, L. B. & Davies, D. K. 1976. Pore space reduction in Cretaceous sandstones through chemical precipitation of clay minerals. *Journal of Sedimentary Petrology* **46**, 89–96.

Bjølykke, K., Aagaard, A., Dypvik, H., Hastings, D. S. & Harper, A. S. 1986. Diagenesis and reservoir properties of Jurassic sandstones from the Haltenbanken Area, offshore Mid Norway. *In*: Spencer, A. M. *et al.* (eds) *Habitat of Hydrocarbons on the Norwegian Continental Shelf*. Graham & Trotman, London, 275–286.

Cowan, D. R. & Myers, K. S. 1988. Discussion-Surface gamma-ray logs: a correlation tool for frontier areas. *Bulletin of the American Association of Petroleum Geologists* **72**, 634–636.

Cox, K. G., Price, N. B. & Harte, B. 1967. *An Introduction to the Practical Study of Crystals, Minerals and Rocks*. McGraw-Hill, London.

Davies, D. K. & Ethridge, F. G. 1975. Sandstone composition and depositional environment. *Bulletin of the American Association of Petroleum Geologists*, **59**, 239–264

Dypvik, H. & Eriksen, D. Ø 1983. Natural radioactivity of clastic sediments and contributions of U, Th and K. *Journal of Petroleum Geology* **5**, 409–416.

FLINT, S., STEWART, T. H., GEVERS, E. C. A., DUBRULE, O. R. F. & VAN RIESSEN, E. D. 1988. Aspects of reservoir geology and production behaviour of Sirikit Oil Field, Thailand: an integrated study using well and 3-D seismic data. *Bulletin of the American Association of Petroleum Geologists* **72**, 1254–1269

FOLK, R. L. 1954. The distinction between grain size and mineral composition in sedimentary rock nomenclature. *Journal of Geology* **62**, 344–359

FONS, L. 1969. Geological Applications of Well Logs. *Transactions of the SPWLA 10th Annual Logging Symposium*, Paper AA.

HAWKINS, P. J. 1978. Relationship between diagenesis, porosity reduction and oil emplacement in Late Carboniferous sandstone reservoirs, Bothamsall Oilfield, E. Midlands. *Journal of the Geological Society* **135**, 7–24

HURST, A. 1990. Natural gamma-ray spectroscopy in hydrocarbon bearing sandstones from the Norwegian continental shelf. *In*: HURST, A., LOVELL, M. A. & MORTON, A. C. (eds) *Geological Applications of Wireline Logs* Geological Society, London, Special Publication **48**, 211–222.

PETTIJOHN, F. J., POTTER, P. E. & SIEVER, R. 1972. *Sand and Sandstone* Springer, New York.

REVERDY, X., ARGAUD, M. & WALGENWITZ, F. 1983. Minerological analysis required for log interpretation in complex lithologies. *Transactions of the SPWLA 8th European Symposium*, Paper H.

RIDER, M. H. 1986. *The geological interpretation of well logs*. Blackie, Glasgow.

RUSSELL, W. L. 1941. Well logging by radioactivity. *Bulletin of the American Association of Petroleum Geologists* **9**, 1768–1788

SELLEY, R. C. 1976. Subsurface environmental analysis of North Sea sediments. *Bulletin of the American Association of Petroleum Geologists*, **60**, 184–195

SERRA, O. & SULPICE, L. 1975. Sedimentological analysis of shale-sand series from well logs. *Transactions of the SPWLA 16th Annual Logging Symposium*, paper W.

SIMON-BRYGOO, C. 1980. *Analyse qualitative des diagraphies. essai de methodes d'interpretation*. Thesis, University of Bordeaux 1, France.

VISHER, G. S. 1969. Grain size distributions and depositional processes. *Journal of Sedimentary Petrology* **39**, 1074–1106.

The relevance of dip profiles from outcrops as reference for the interpretation of SHDT* dips

J. C. HERWEIJER[1], C. F. W. HÖCKER[1], H. WILLIAMS[1] & K. M. EASTWOOD[2]

[1]*Koninklijke Shell Exploratie en Produktie Laboratorium, Volmerlaan 6, 2288 GD Rijswijk ZH, Netherlands*

[2]*Shell UK Expro, Shell Mex House, Strand, London WC2R ODX, UK*

Abstract: The ability of SHDT processing techniques to produce detailed sedimentological information has been evaluated by comparing the data with 3000 ft of cores covering shallow marine, deltaic, lacustrine, fluvial and aeolian sedimentary environments. It has been shown previously that, by using side-by-side processing, realistic sedimentary dip sequences through sedimentary bodies can be obtained from SHDT data under normal conditions. However, optimum results require processing parameters to be chosen compatible with the scale of the sedimentary structures present. Furthermore, post-processing selection procedures are required to remove noise and to identify dips representing foreset inclination which can be interpreted in terms of flow direction and sand body orientation.

The sedimentological interpretation of downhole dips is often difficult. It may be improved by comparison with vertical dip sequences from outcrops. Sedimentary dip profiles have been measured as part of field work carried out in the Pocahontas Basin of Kentucky. Extensive roadcuts not only enable an unambiguous sedimentological interpretation, but also exhibit numerous relict blast-boreholes, in which vertical dip profiles can be measured under conditions similar to downhole logging. The data now available show that even contradictory dip sequences may be a proper representation of the reality. Systematic collection of sedimentary dip sequences will provide guidelines for a realistic interpretation of downhole records of sedimentary dips.

Processing of SHDT (Stratigraphic High Resolution Dipmeter Tool) resistivity curves can greatly support sedimentological interpretation. In particular, directional information derived from sedimentary dips is essential for detailed reservoir geological modelling. With the available recording and processing facilities, sedimentary dips down to 20 cm set thickness can be derived. However, dips computed on this small scale are commonly difficult to interpret: firstly the signal-to-noise ratio of dip computations is often unfavourable, and secondly the dips remaining after noise removal may show complex distributions. This problem can be most efficiently overcome by comparing downhole dips with measurements taken from outcrops (Steinmetz 1972, Cameron 1986).

Optimal processing parameters and a noise suppression scheme have been established by comparing SHDT and core data from a variety of sedimentary environments (Höcker *et al.* 1989). The use of different processing types and the effective removal of noise will be demonstrated here with an example well. To aid the interpretation of palaeo-transport directions, vertical dip sequences from fluvio-deltaic sand bodies are being evaluated systematically. An example of a dip profile from a fluvial channel fill of the Carboniferous in the Pocahontas Basin of Kentucky is described below.

*Mark of Schlumberger

Processing parameters

The key design feature of the SHDT is that two electrodes are mounted on each pad side-by-side with a spacing of only 3 cm. This specific arrangement enables the identification of foreset dips in small cross-bed sets.

In the following, we summarise the comprehensive survey of processing techniques and considerations for application given by Höcker *et al.* (1989). Only side-by-side processing, that is correlating curves from the electrodes of a certain pad and combining the results from adjacent pads to dip measurements, is capable of identifying the sedimentary dip in cross-bed sets thinner than 30 cm. To derive the dip of foreset laminae reliably the correlation window must not cross set boundaries. For most sedimentary environments (in particular for fluvial and coastal plain environments) the average cross-bed sets are around 15 cm thick. It was found that a correlation window of 10 cm (4 inch) length forms a practical compromise between technical limitations and the necessity to

From HURST, A., LOVELL, M. A. & MORTON, A. C. (eds), 1990, *Geological Applications of Wireline Logs* Geological Society Special Publication No. 48, pp. 39–43

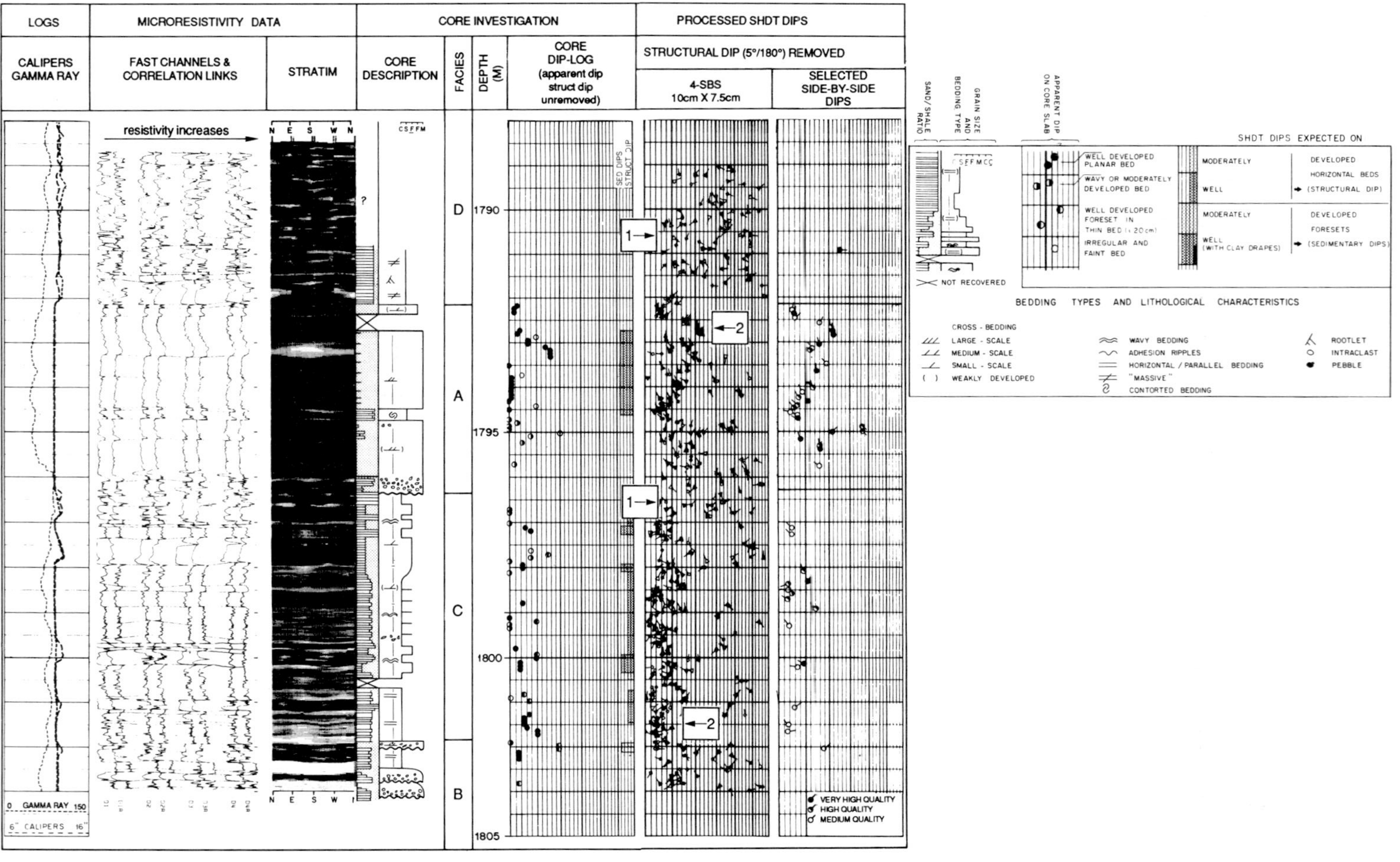

Fig. 1. Composite dipmeter data sheet with SHDT raw data; STRATIM resistivity image (resistivity intrapolated between raw curves) core data; SHDT processing results. The selected interval contains semi-arid fluvial/alluvial sediments with 4 facies: A, cross-bedded channel sands; B, crevasse sands with clay clasts and small-scale cross-bedding; C, silty overbank sands and shales with wavy and ripple bedding; D, rootletted floodplain shales.

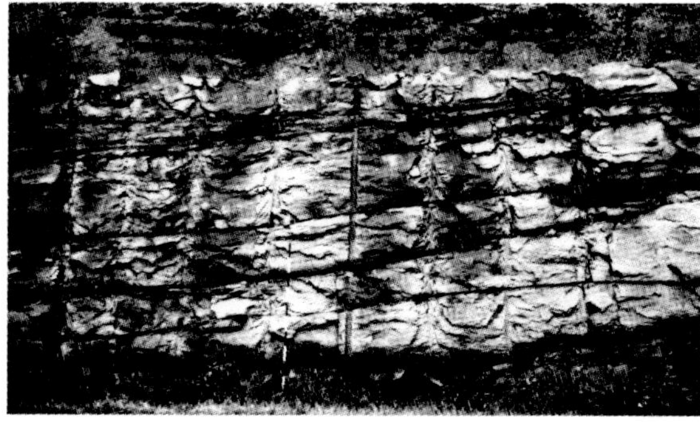

Fig. 2. Field measurements in Kentucky. (a) Heterolithic, laterally-accreted minor channel fill in the Carboniferous of the Pocahontas Basin, Kentucky. (b) Detail of A showing blast borehole used for dip measurements.

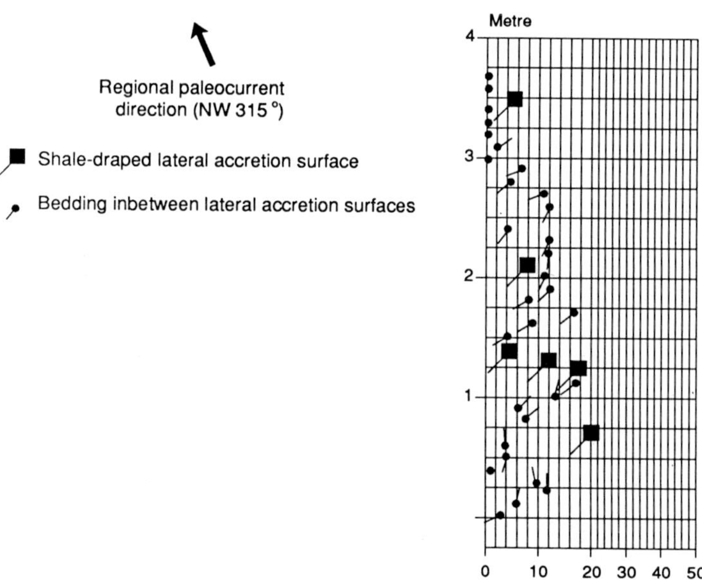

Fig. 3. Tadpole plot of the vertical dip sequence measured in the blast borehole in Fig. 2(b). Note the variable azimuth of bedding features between lateral accretion surfaces.

measure dip in as many and as small cross-bed sets as possible.

The output of 4-SBS* processing shows all (up to four) dips calculated at a certain depth (4-SBS column in Fig. 1). Highly consistent 4-SBS results (1) identify lamination that is present all around the borehole and that is planar at the borehole scale. High spread (2), on the other hand, may result from non-planar beds and also from processing artefacts, and the corresponding values are best discarded. Selection criteria have therefore been established on the basis of spread in 4-SBS results.

*Mark of Schlumberger

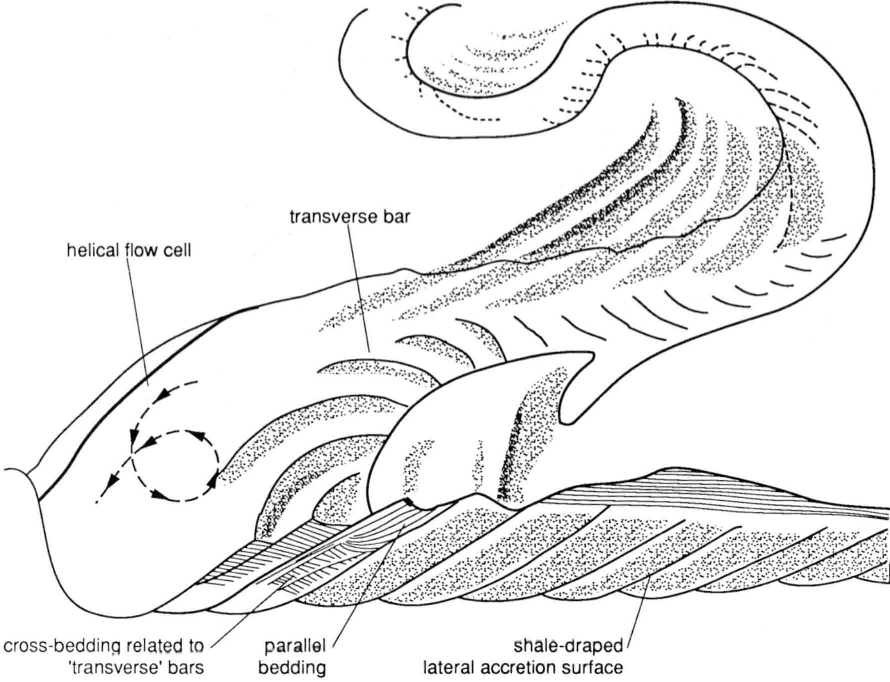

Fig. 4. Possible flow pattern and corresponding bedding features in a meander bend (after Smith 1987).

Inference of palaeotransport direction: an example

Figure 1 shows output from SHDT dip processing under conditions of 'average' resistivity contrast. The selected interval comprises semiarid alluvial sediments with cross-bedded channel sands (A), crevasse sands with clay clasts and small-scale cross-bedding (B), silty overbank sands and shales with wavy and ripple bedding (C) and rootletted floodplain shales (D).

The dip columns display core dips and the results of 4-SBS processing using optimum processing parameters. The right-hand column contains dips selected and averaged on the basis of 4-SBS spread. The number of tadpoles remaining and the distribution pattern compare favourably to core data. Note that the azimuth of core dips is not indicated and that their magnitude is underestimated as a result of measuring unoriented core slabs. A steepening-upward trend in the channel sands (1792–1793 m) is similar to dip sequences measured from outcrops in the vicinity of the well. Since the STRATIM display does not show any streaks of higher resistivity (lighter) in the conductive channel sand it is assumed that lateral accretion surfaces with a significantly different texture are absent. This dip profile is therefore interpreted to represent foresets within a low-sinuosity channel fill. The derived northwesterly channel direction compares well with outcrop observations.

Dip measurements on a fluvial channel outcrop

Unfortunately, vertical dip sequences are not always as simple as in the example well. During 1987, field work was carried out in Kentucky (USA) to investigate systematically dips in fluvial and deltaic sand bodies. Large road cuts enable three-dimensional investigation of sand bodies and direct measurement of palaeochannel directions. Accurate measurement of vertical dip sequences was possible due to the preservation of blast boreholes (Figs 2A & 2B). Using a sampling distance of 10 cm, one may compare the collected data with SHDT dips obtained by applying a 10 cm correlation interval in side-by-side processing.

Figure 2 shows an example of a sand body deposited by a high-sinuosity meandering river (laterally accreted point bar deposit) from which a rather complex dip sequence was derived

(Fig. 3). Dipping surfaces measured in the vertical profile include those of lateral accretion surfaces draped by shales, foresets of cross-beds and parallel beds.

Shale-draped lateral accretion surfaces throughout the vertical profile consistently dip toward the SW, whereas cross-bed foresets show a larger degree of variation.

(i) In the lower part of the sand body, dips of planar foreset are generally oriented at high angles to those of lateral accretion surfaces. These dips are attributed to transverse dune bedforms moving at an oblique angle across the point bar surface within a helical flow cell. Although bottom flow is directed downstream it also has a bottom flow component directed towards the inner bank (Jackson 1975; Collinson 1986) (Fig. 4).

(ii) In the upper part of the profile all dips are oriented to the SW, pointing down the point bar surface. This dip distribution is attributed to the presence of plane beds higher up on the point bar, which are deposited parallel to lateral accretion surfaces (Frazier & Osanik 1961).

In this example, dip readings from shale-draped lateral accretion surfaces indicate a direction of channel migration toward the SW, with the inference that the channel belt had a general NW–SE orientation. It is therefore essential that in interpreting dips obtained from SHDT data, we can distinguish those related to lateral accretion surfaces from those of trough cross-bed foresets. This distinction may be accomplished by correlating the dip profile with the resistivity data where the textural contrast between channel sands and relatively fine-grained, shaly lateral accretion surfaces is expressed in form of a deflection of the resistivity curves.

Conclusion

The downhole example shows that processing of SHDT data can provide accurate measurements of sedimentary dips. However, post-processing selection has to be applied to identify meaningful dips. An unambiguous dip sequence was derived in this example.

Interpretation of more complex dip sequences can only be done efficiently by comparing downhole data with outcrop measurements. Outcrop dips must be measured so that they are comparable to well data, i.e. as vertical sections. A statistically significant number of profiles is required from all common depositional environments to build up a reference set. This may be used to evaluate a number of alternative sedimentological models when interpreting dip sequences. By combining dipmeter data with other available information, sufficient data will be available to select the most likely solution.

In many cases, diagnostic features must be available to assign dip computations to specific sedimentary features, such as lateral accretion surfaces or foresets in the Kentucky example. Provided that a distinction can be made, an interpretation in terms of sand-body orientation still requires knowledge of the depositional process involved.

References

Cameron, G. I. F. 1986. Confidence and the identification of foresets in stratigraphic dipmeter surveys. *SPWLA, 10th European Formation Evaluation Symposium*, paper O.

Collinson, J. D. 1986. Alluvial sediments. *In*: Reading, H. G. (ed.) *Sedimentary Environments and Facies*. Blackwell, Oxford, 20–62.

Frazier, D. E. & Osanik, A. 1961. Point bar deposits, Old River Locksite, Louisiana. *Transactions of the Gulf Coast Association of Geological Societies*, **11**, 121–137.

Höcker, C. F. W., Eastwood, K. M., Herweijer, J. C. & Adams, J. T. 1989. Sedimentary dips from SHDT data. *Bulletin of the American Association of Petroleum Geologists*.

Jackson, R. G. 1975. Velocity — bed-form — texture patterns of meander bends in the lower Wabash River of Illinois and Indiana. *Bulletin of the Geological Society of America*, **86**, 1511–1522.

Smith, R. M. H. 1987. Morphology and depositional history of exhumed Permian point bars in the southwestern Karoo, South Africa. *Journal of Sedimentary Petrology*, **57**, 19–29.

Steinmetz, R. 1972. Sedimentation of an Arkansas river sand bar in Oklahoma: a cautionary note on dipmeter interpretation. *Shale Shaker*, **23/1**, 32–38.

Incorporating geological data in well log interpretation

D. E. KING

Schlumberger Well Services, Austin Systems Center, PO Box 200015, Austin, TX 78720, USA

Present address: Graduate School of Business, Stanford University, Rains Houses Box #568, Stanford, CA 94305, USA

Abstract: A multi-step technique is presented that permits a variety of quantitative and qualitative geological data to be incorporated in the log interpretation process to derive a detailed lithological analysis. The geological data are used to develop and refine explicit tool response and interpretation models. The models include synthetic wireline log data generated from input mineralogical and fluid data. These models are used in an interpretation system together with wireline data to provide a geologically reasonable lithological solution appropriate to the stratigraphic framework of a basin. An application of this technique is presented for a lithologically complex reservoir. A full suite of conventional logs was used together with core-derived data to provide an accurate level-by-level lithological analysis. The results contribute to the refined geological interpretation of the well.

Traditionally, geological data have not been used as direct input to the log interpretation process. Rather, these data have typically been used to check the quality of answers derived from various log interpretation programs. In this paper, a technique for incorporating geological data directly in the quantitative log interpretation process is presented. A wide variety of both quantitative and qualitative geological data can be included in the interpretation with this technique, including mineralogy, texture, sedimentary structures and contained fluids. An important feature of the technique is to provide a systematic method for inputting and processing geological data in wireline log lithological interpretation. In addition, the technique provides a routine procedure for lithological zonation that can be applied in a variety of geological settings using available geological data.

The technique has been implemented as a set of computer programs consisting of a log inpretation system and an allied lithology zonation program. The technique is a multi-step process centred on these two programs. The primary steps that comprise the technique include: (a) translating geological data (principally mineralogical and pore-type data) into a database of electrofacies using the interpretation system; (b) defining a set of formation models that reflect local geology for input to the interpretation system; (c) processing the wireline log data with the lithological zonation program and the data-base of electrofacies; (d) using the results of the lithology zonation to guide formation model combination in the log interpretation system.

The log interpretation system is the primary computational program. The principal application of the log interpretation system is to solve formation models. Formation models consist of explicit definitions of the geological parameters of a formation and of the wireline tools used in the solution of a log interpretation problem. The interpretation system can be used to calculate simultaneously the solutions of several formation models. Constraints developed from geological data are used to guide the selection of models relevant to a geological setting, and to determine how to uniquely combine the solutions of these models. Using input geological data, the log interpretation system provides a petrophysical solution appropriate to the geological characteristics of a well.

Applications of the interpretation system using mineralogical and fluid data are discussed. A method of using the interpretation system to develop a database of electrofacies from quantitative mineralogical data is presented. The use of the interpretation system to define and solve formation models together with geological constraints derived from electrofacies is also discussed. Finally, an example combining these various steps of the technique to solve a complex lithology log analysis problem is presented.

The log interpretation system

The log interpretation system is a set of computer programs that provides a framework for specifying and solving petrophysical models (Quirein *et al*. 1986; King & Quirein 1987). The principal components of a petrophysical model

From Hurst, A., Lovell, M. A. & Morton, A. C. (eds), 1990, *Geological Applications of Wireline Logs* Geological Society Special Publication No. 48, pp. 45–55

are wireline tool measurements, volumetric response parameters and formation volumes. Formation volumes are the specific mineralogical, textural and fluid components of a reservoir. Response parameters are empirically-derived responses of the volumes to the wireline tools (Ellis *et al.* 1988). For example, the formation volume quartz has a response parameter of 2.65 g cm^{-3} for the density tool. These three model elements — tools, parameters and volumes — are related to one another by response equations in the interpretation system. The response equations specify the relationship between the measured response of a wireline tool and the mineral and fluid volumes of the formation. A typical response equation for the density tool, for example, is given by:

$$\rho_b = \sum_{i=1}^{n} \rho_i V_i$$

where ρ_b is the measured response of the density tool in a formation consisting of n components, ρ_i is the density of formation component i, and V_i is the volume fraction of component i. The interpretation system allows the analyst to input any two of the three elements of a formation model and solves for the third element using response equations. This relationship is shown diagrammatically in Fig. 1.

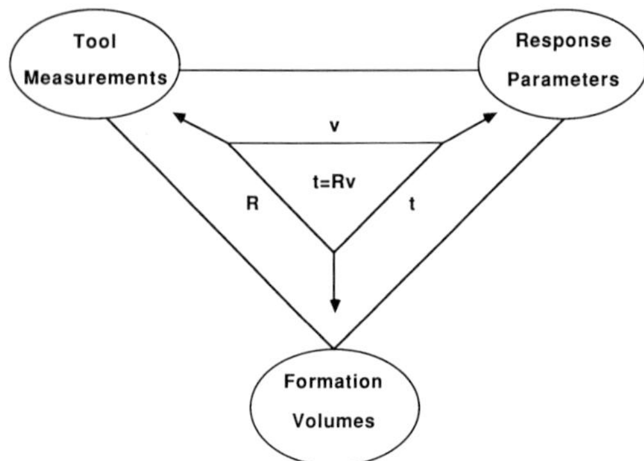

Fig. 1. Schematic representation of the principal elements of the log interpretation system. The interior triangle represents the solution method used by the system. Given any two elements, the system will solve for the third. For example, given R (response parameters) and v (formation volumes) on the interior triangle, the system calculates tool measurements on the exterior triangle. t = tool measurements; R = response parameters; v = formation volumes.

The most common mode of use of the interpretation system is to compute formation volumes of formation models, given wireline tool measurements and response parameters as input. The formation models are solved by the interpretation system using the response equations subject to user-specified geological constraints, such as mineralogical volume constraints and pore-type designations. The system is solving the standard log interpretation problem when used in this mode. The interpretation system can also generate simulated wireline log data. This is the forward modelling mode; formation volumes and response parameters are defined in a formation model and synthetic log responses are computed by the interpretation system.

The interpretation system can compute as many as five formation model solutions simultaneously when used to address the standard log interpretation problem. The ability to solve multiple models allows the analyst to create separate geological models with unique volumetric components that reflect the geological zonation of the well. This is particularly useful when the log analysis problem is underdetermined; that is, there are more formation volumes to be resolved than there are available logging tool measurements. The individual formation model solutions are calculated and combined to provide a level-by-level lithological solution of the formations in a well. The models are combined by the system using explicit logic derived from geological data input by the user. Specific methods of creating and combining the formation models are discussed in a following section.

Using the interpretation system to model geological formations

An important application of the interpretation system is to generate simulated wireline log measurements, or 'synthetic data', from input geologic models. This step of the technique accomplishes the important transition from 'real' geological data (e.g. mineral and fluid volumes) to wireline log representations of the same geological data. The synthetic data created in this process are used to define electrofacies. An electrofacies is a wireline log representation of a discrete geological section or rock type (Serra & Abbot, 1980). Using the interpretation system in this mode (i.e. the forward modelling mode), a database of electrofacies can be created from input geological data (King & Quirein 1986, 1987). The database of electrofacies is then used by a lithology zonation program to provide a systematic 'first-pass' lithological zonation of a well. The two steps involved in transforming the geological data using the interpretation system are discussed in detail below.

Creating rock models

The creation of an electrofacies begins with a geological model, called a rock model (King & Quirein 1987). Each rock model consists of the mineral and fluid components of the rock type of interest. The volumetric data used in the rock models are typically based on quantitative values obtained, for example, from X-ray diffraction data or thin section analysis. Alternatively, they may be derived from empirical representations including previous log analyses and visual core descriptions. The accuracy of the rock model and the electrofacies derived from it are, clearly, a function of the quality of this input data.

To define an electrofacies, the analyst first creates a porosity-free variant of the rock model containing only the mineral volumes. The rock model shown in Table 1 might be used to represent a zero-porosity anhydritic dolomite. Next, several porosity variants are created for each rock model by combining porosity data with the mineral data. Porosity is simply the sum of all fluid volumes and when added to the model all other mineral volumes are normalized

Table 1. *Volumetric rock model for 'Anhydritic Dolomite'*

Component	$\overline{X}$	σ	Range $\approx\overline{X} \pm 2\sigma$
Dolomite	67.5	5.0	57.5–77.5
Anhydrite	22.5	3.75	15.0–30.0
Calcite	5.0	2.5	0.0–10.0
Quartz	3.0	1.5	0.0–6.0
Illite	2.0	1.0	0.0–4.0

All values represent volumetric percentages.

Table 2. *Porosity-variant rock model for 'Anhydritic dolomite, 10–20 PU'*

Component	$\overline{X}$	σ	Range $\approx\overline{X} \pm 2\sigma$
Dolomite	64.0	4.7	54.7–73.8
Anhydrite	21.5	3.7	14.3–28.5
Calcite	4.75	2.4	0.0–9.5
Quartz	2.85	1.4	0.0–5.7
Illite	1.9	0.9	0.0–3.8
ϕ	5.0	2.5	0.0–10.0

All values represent volumetric percentages.

so that the sum of the means of all volumes is unity. Continuing with the above example, the analyst might create three porosity-variant rock models for an anhydritic dolomite based on economic porosity ranges for a particular reservoir (see Table 2) 'Anhydritic Dolomite, 0–10 PU', 'Anhydritic Dolomite, 10–20 PU,' and 'Anhydritic Dolomite, 20–30 PU'. For rock types that demonstrate a substantial non-linear variation in mineralogy and/or fluid content with changing porosity, the volumetric models can be adjusted to capture such trends. The volumetric rock models are in no way rigid or fixed; they are intended, instead, to be a statistical representation of the geological data from which they are derived.

Generating synthetic electrofacies

The rock models are transformed into electrofacies by the interpretation system through the application of the tool response equations to the mineral volumes and response parameters of the models. The interpretation system processes the rock models using a statistical procedure that assumes a normal distribution of the model volumes. Each volume is sampled randomly and the selected volumes are then normalized such that the sum of all volumes is unity. In a similar manner, the program adds 'noise' to each synthetic log by sampling statistically in the range of the tool's uncertainty. Tool uncertainty factors are known values that represent the statistical accuracy of the logging tool response. Typically, 500–1000 values of 'synthetic' log data are generated for each logging tool used in a rock model. These log data are used by the interpretation system to build the electrofacies definitions.

The electrofacies data are transferred automatically to a database where each electrofacies is represented by a minimum and maximum response for each log plus a matrix of correlation coefficients for each possible log pair. Table 3 is an example of a database electrofacies that was created using the anhydritic dolomite rock model from Table 2. Each electrofacies in a database represents a volumetric distribution in the multi-dimensional space defined by the input logs (typically six to eight different logs are used), and is, by definition, a multi-dimensional ellipsoid. The two-dimensional projection of each multi-dimensional ellipsoid is an ellipse. This is convenient in that one may examine crossplots of the ellipses for various log pairs as an aid in understanding the relationships of the electrofacies (Fig. 2). The relative differences

Table 3. *'Anhydritic Dolomite, 10–20 PU' electrofacies in database format. The first three rows contain the minimum wireline log value, the maximum wireline log value and the standard deviation of the mean of these two values for eight log measurements. The remaining rows contain the matrix of correlation coefficients for each possible combination of the eight log measurements. For example, the correlation coefficient of the neutron porosity log (NPHI) and the density log (RHOB), is −0.94. Units for each of the wireline logs are: RHOB, gm/cm^3; NPHI, porosity units; DT, μs/ft; GR, API units; PEF, barns/electron; THOR, ppm; URAN, ppm; POTA, wt%. Refer to Fig. 2 for log measurement acronyms.*

	RHOB	NPHI	DT	GR	PEF	THOR	URAN	POTA
Minimum	2.67	3.70	47.68	9.90	3.34	0.35	−0.64	0.06
Maximum	2.90	12.40	64.45	23.16	3.80	1.31	0.62	0.08
Std. dev.	0.038	0.032	3.21	0.98	0.19	1.63	2.25	0.004
Corr. coeff:								
RHOB	1.00							
NPHI	−0.94	1.00						
DT	−0.96	0.96	1.00					
GR	−0.07	−0.04	0.04	1.00				
PEF	0.41	−0.41	−0.39	−0.02	1.00			
THOR	0.03	0.02	0.02	0.55	−0.07	1.00		
URAN	−0.04	0.03	0.03	0.80	−0.03	0.00	1.00	
POTA	−0.27	−0.22	−0.22	0.56	0.07	0.72	0.00	1.00

in ellipsoidal volume between electrofacies reflect both geological changes among rock types (porosity, mineralogical variations, fluids, textural differences, etc.) and differences in the logging environment (condition of hole, contained fluids) (Delfiner *et al.* 1984).

Solving and combining multiple formation models with the interpretation system

A second important application of the interpretation system is to compute formation volumes from multiple formation models (i.e. the standard log interpretation problem). Two important considerations when using the system in this mode are the selection of formation models and the determination of the appropriate method to combine the formation model solutions. The interpretation system provides three methods for combining formation model solutions:

(i) via manual selection imposed by the analyst based on geological information,
(ii) via probabilities calculated internally by the interpretation system, which are a function of the volumetric solution of each formation model,
(iii) by external probability functions.

The technique presented here requires external probabilities derived from a lithology zonation program to combine formation models. Model selection and model combination methodologies are discussed in detail below.

Selecting formation models

The formation models reflect known geological properties in the reservoir of interest. In general, a formation model is created for each major lithological unit in a well. Formation models can also be created to solve geological problems related to texture, including secondary porosity and isolated porosity (Schlumberger 1986). The components of each model must be defined explicitly. When using the interpretation system to determine formation volumes, the formation models must contain explicit statements of the tool measurements and the response parameters of the minerals and fluids that are to be solved. Models may contain one or more of the same components but will generally have at least one unique component. As a general rule of thumb, one should include a volumetric component in a model only if it is known from external geological data to exist in the zone over which the model will be applied. If the analyst creates a shale model, for example, there is no need to include oil in the fluid response parameters of the formation model unless oil is known to occur in the shale. The number of volumetric components to be solved by a model is constrained to be equal to

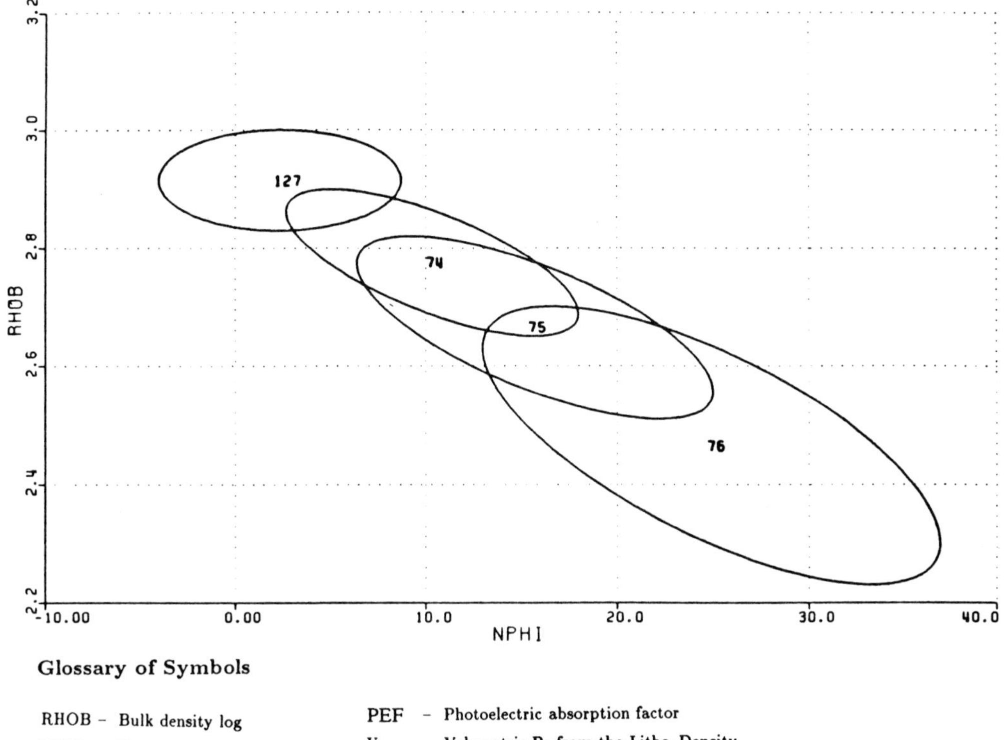

Glossary of Symbols

RHOB – Bulk density log
NPHI – Neutron porosity log
DT – Sonic compressional Δt log
VELC – Inverse of DT
GR – Gamma ray log
LDT – Litho-Density* tool
ϕ – Porosity
σ – Standard deviation
$\bar{x}$ – Mean volume %

* Mark of Schlumberger

PEF – Photoelectric absorption factor
U – Volumetric Pe from the Litho-Density
THOR – Thorium log
POTA – Potassium log
URAN – Uranium log
TPL – Electromagnetic propagation time log
Rxo – Resistivity of invaded zone
Rt – Resistivity of uninvaded zone
BHC – Borehole compensated sonic tool
PU – Porosity unit

Fig. 2. Two-dimensional projections of database electrofacies on RHOB against NPHI crossplot. 127 = 'Anhydrite'; 74 = 'Anhydritic dolomite, low ϕ'; 75 = 'Anhydritic dolomite, good ϕ'; 76 = 'Anhydritic dolomite, excellent ϕ'. RHOB axis units are g/cm^3 and NPHI axis units are porosity units.

or less than the number of wireline tools in the model.

Using external information to combine formation models

The formation models are input to a lithology zonation program that calculates the probability of each model. The lithology zonation program uses a modelled database of electrofacies, as described above, to calculate probabilities for each formation model at each level in a well. These formation model probabilities are then input to the interpretation system, where they are used to determine the appropriate combination of the formation models in the final formation evaluation result.

The lithology zonation program uses a Bayesian decision procedure and a database of electrofacies to assign the most appropriate lithology on a level-by-level basis (Delfiner *et al.* 1984; Serra *et al.* 1985). Following the application of user-specified constraints, such as minimum bed thickness and identification of bad hole and thin beds, the posterior probability of each electrofacies in the database is calculated, and the electrofacies with the highest posterior probability is assigned at each level. Most electrofacies in the database will have zero or near-zero probability, and generally only three to five electrofacies at each level will have significant probability. In addition, each electrofacies in the database is a member of a major lithology group (e.g., sandstone, limestone) and probabilities are calculated at each level for these groups. The probability of a major lithology group is computed by summing the probabilities of its member electrofacies.

Using the lithology zonation program, the analyst makes assignments of electrofacies to the formation models used by the interpretation system. The assignments are probability mappings. Each electrofacies in the database is mapped to one or more of the formation models. The lithology zonation program calculates the probability of each electrofacies at a level and maps the calculated probability to its assigned formation model(s). The total probability for a formation model at a given level is the sum of the computed probabilities of each electrofacies assigned to the model. For example, based on geological data the analyst might choose to create a 'sandstone' model, a 'limestone' model, a 'dolomite' model and a 'bad hole' model for a particular well. In this case, all electrofacies that were modelled originally as predominantly siliceous rock types would be assigned to the sandstone model, all electrofacies modelled as a majority of calcite would be assigned to the limestone model and so on. In an area where it is hoped to resolve more subtle mineralogical changes and a sufficient number of tools are available, the analyst might create a 'clean sand' model, a 'shaly sand' model and a 'carbonate sand' model. In this case a more innovative mapping of electrofacies to formation models might be undertaken; all electrofacies based on quartz-rich models might be assigned to the clean sand model, electrofacies modelled with a majority of quartz plus clay might be assigned to the shaly sand model and all electrofacies based on models containing a majority of quartz plus calcite plus dolomite might be mapped to the carbonate sand model.

When using external probabilities calculated by the lithology zonation program, the combined answer produced by the interpretation system is created either by selecting the formation model solution with the greatest probability at each level or by averaging the formation model solutions in proportion to their calculated probabilities. It is important to note that the mineral response parameters and tool response equations used to create the electrofacies database are identical to those used by the log interpretation system, ensuring consistency of all results.

Application of the technique to a complex lithology interpretation problem

The example comprises complex lithologies from a well in the Midland Basin of West Texas. The upper section of the well contains dense evaporite beds and is underlain by fine-grained shaly sandstones with frequent interlain tight dolostone beds. The electrofacies database contains approximately 125 electrofacies that cover a range of lithologies reported in the Permian Basin. The database contains modelled electrofacies based on X-ray diffraction and thin section analyses as well as empirically modelled electrofacies.

Results of the lithology zonation program over a 61 m (200 ft) section of the well, together with interpreted and depth-matched visual core descriptions, are shown in Fig. 3. The program results, on the right of the figure, demonstrate the strong contrasts in major facies over the displayed interval: evaporite and carbonate-rich rocks near the top followed by sands and shaly

Table 4. *Formation model assignments for complex lithology interpretation example*

Model	Description	Volumes	Input logs
1	Shaly Sand	Quartz, Calcite, Dolomite, Plag., Illite, Water, Oil	RHOB, NPHI, U, Rxo Rt, POTA, THOR, TPL
2	Carbonate	Calcite, Dolomite, Anhydrite, Illite, Water, Oil	RHOB, NPHI, U, Rxo Rt, POTA, THOR, TPL
3	Shale	Quartz, Dolomite, Plag., Illite, Chlorite, Water	RHOB, NPHI, U, POTA, THOR
4	Evaporite	Calcite, Dolomite, Halite, Anhydrite, Illite, Water	RHOB, NPHI, VELC, POTA, THOR

Refer to Fig. 2 for definitions of acronyms of input logs.

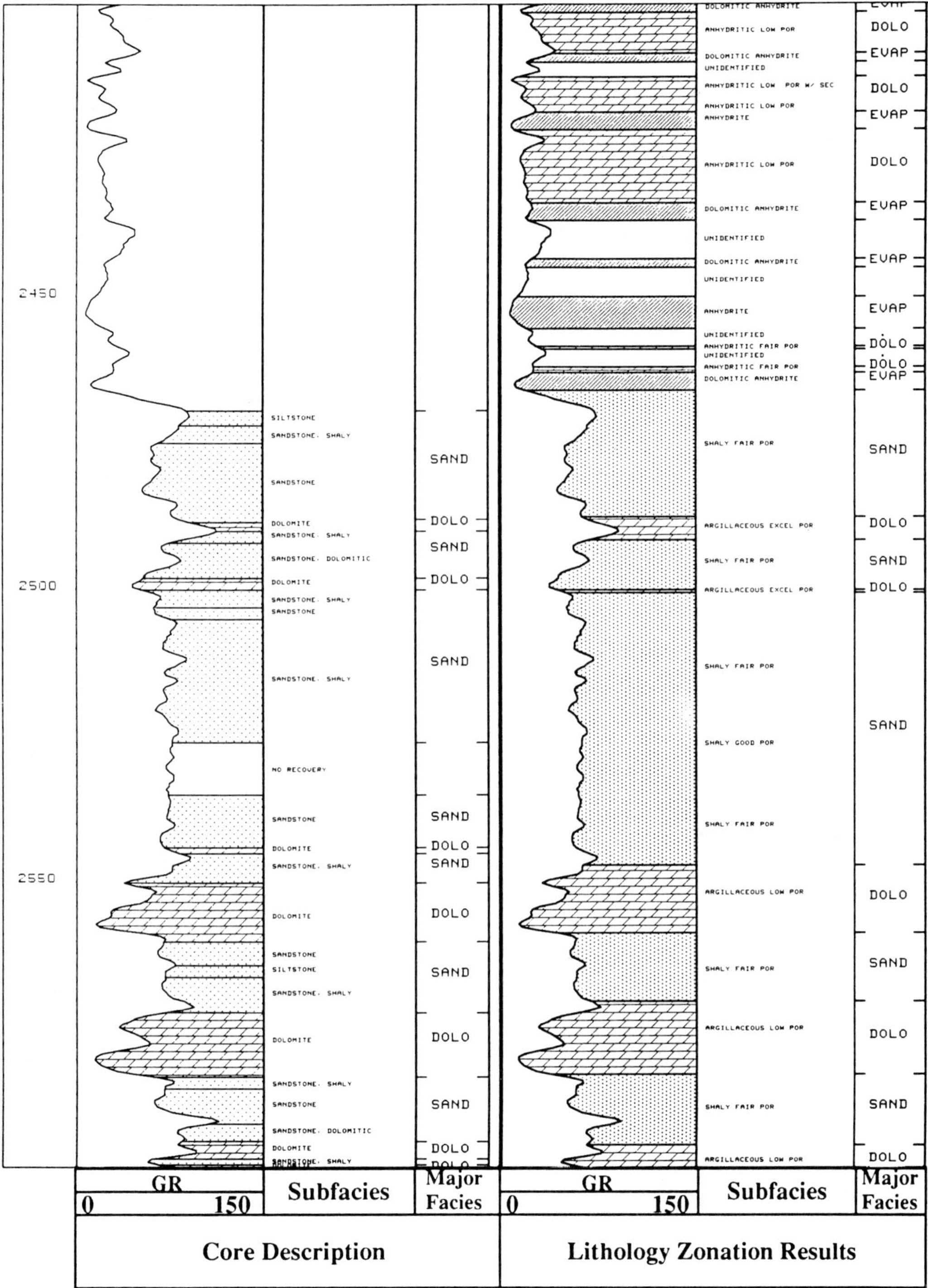

Fig. 3. Comparison of core results and automatic lithology zonation program results. Interpreted visual core descriptions are presented on the left side of the display, with subfacies descriptions in track 3 and major facies descriptions in track 4. Lithology zonation program results are shown on the right side of the display, with subfacies descriptions in track 6 and major facies descriptions in track 7. The lithological columns in tracks 2 and 5 are bounded by the same GR curve scaled in API units. Depth scale is in feet. For shadings, see Fig. 4.

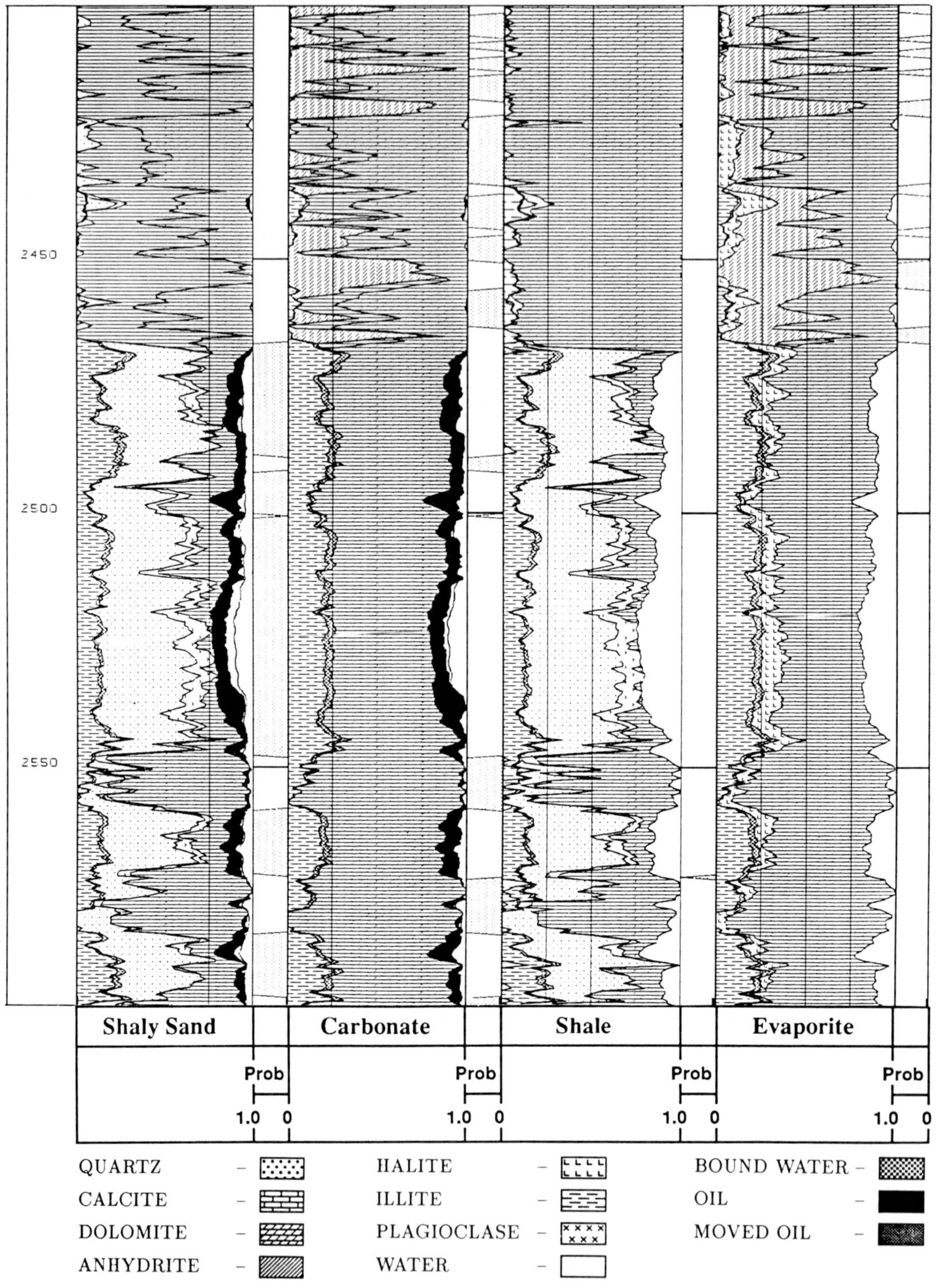

Fig. 4. Individual formation model solutions. Model definitions for each of these solutions are presented in Table 4. Each model is represented by a shaded track that presents the volumetric percentage of each component in the model. Cumulative volume percentages are scaled from 0 on the left edge of each display to 100 on the right edge of each display. Vertical lines in the shaded tracks indicate 25, 50, and 75 cumulative volume percentage. To the right of each shaded track is the model's probability track, scaled from 0 on the right edge to 1 on the left edge. Depth scale is in feet.

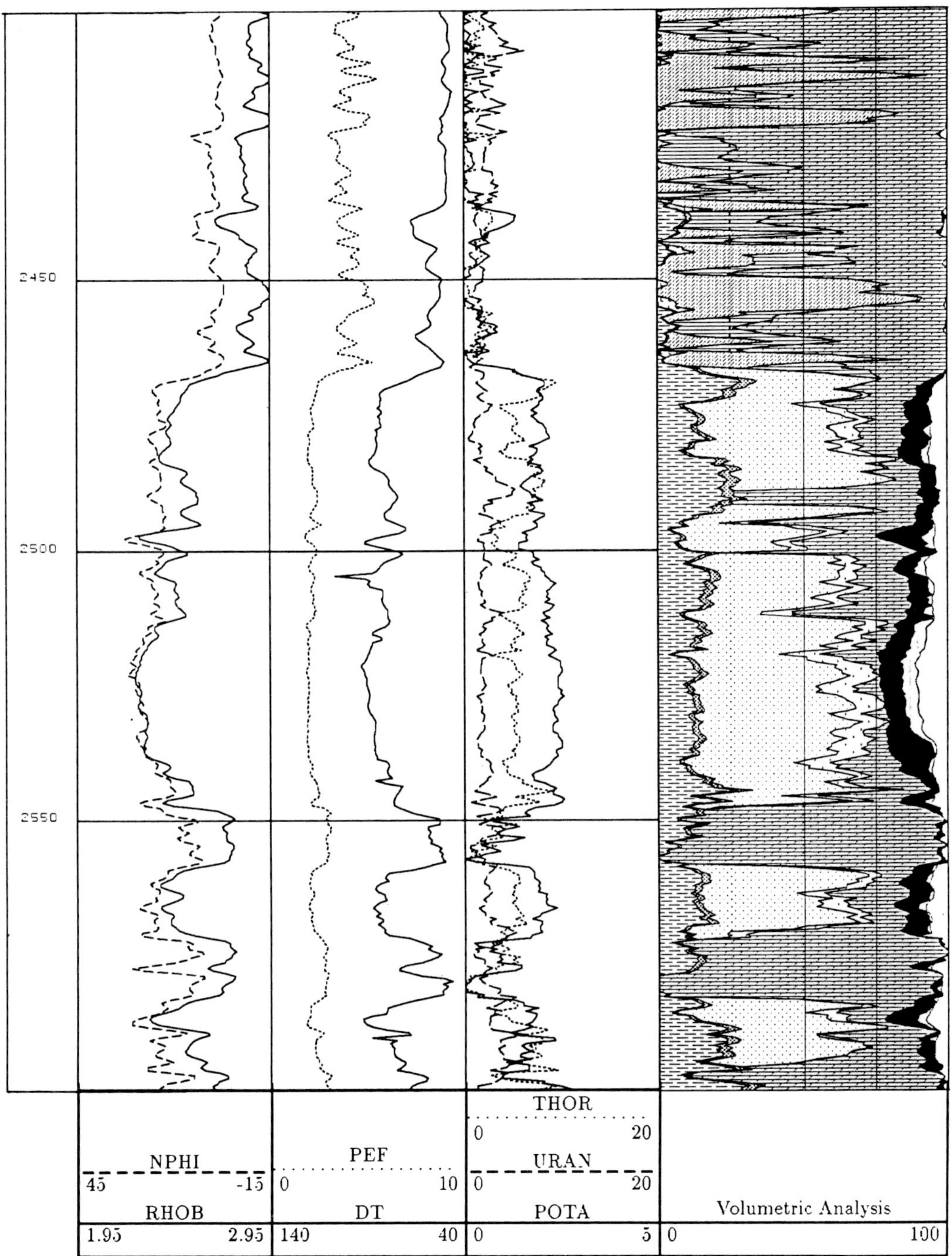

Fig. 5. Combined display of several input wireline log measurements (tracks 2–4) and the combined formation model solution (track 5). The combined solution is derived from the individual formation model solutions presented in Fig. 4. Units for log measurements are: NPHI, porosity units; RHOB, gm/cm^3; PEF, barns/electron; DT, μs/ft^{-1}; THOR, ppm; URAN, ppm; POTA, wt%. Depth scale is in feet. For shadings, see Fig. 4.

sands with dolomite interbeds in the remainder of the well. In general, the lithology zonation program has accurately identified the sand and dolomite intervals, as compared to core. In addition, there is some correlation between individual shaly sand electrofacies identified by the program and the shaly sands designated in the core description.

Four formation models were selected based on the results of the lithology zonation program: 'shaly sand,' 'carbonate,' 'shale' and 'evaporite' formation models. The results of a preliminary pass indicated that these four models would be adequate to resolve the lithologies present in this well. The formation model designations, together with the logging suite, allowed the mineral/fluid volumes and tool assignments shown in Table 4. Note that oil was not used in models three and four. This is because these models (shale and evaporite models, respectively) were not expected to contain hydrocarbons. Because of the close association of widely variable lithologies in this well, electrofacies mappings to formation models were based on major facies types. All sandstone electrofacies were assigned to the shaly sand formation model, all limestone and dolostone electrofacies were assigned to the carbonate formation model, all shale electrofacies were assigned to the shale formation model, and all evaporites were assigned to the evaporite formation model. The 'unidentified' electrofacies (i.e. no probable electrofacies identified at this level) was assigned to the shale formation model because it generally resulted from errant tool readings in bad hole due to sloughed shales.

The four formation model solutions calculated by the interpretation system and the probability of each solution over the same 61 m interval are shown in Fig. 4. In general, only one model at any given level has been assigned a probability by the lithology zonation program as indicated by the shaded curve to the right of each formation model solution in Fig. 4 (scaled from 100% probability on left edge to 0 probability on right edge of track). Over most of the well, there was no confusion in the major lithology assignments; all electrofacies that were assigned non-zero probability by the program at any level were from the same major facies. This is an indication of the confidence of the lithology assignments of the lithology zonation program and reflects the distinctive changes recorded by the wireline tools at major lithology boundaries. Note that the shale model was not assigned any probability, with the exception of a small interval around 785 m (2575 ft), indicating the general absence of shale lithologies and bad hole in this section of the well.

The combined interpretation solution from this interval together with several logs used in the solution is presented in Fig. 5. The formation model solutions were combined using the model at each level with the maximum probability assigned by the lithology zonation program. The maximum probability was used because the electrofacies were mapped to the formation models based on their major lithology type and the interpretation was designed to provide maximum discrimination between the major changes in lithology. The formation model probabilities generated by the lithology zonation program provided the log interpretation system with excellent control of the combination of the formation model solutions. The combined solution compares favourably with the core results and provides an accurate representation of the well, with evaporites and carbonates overlying sandstones, carbonates and occasional shaly sandstones.

Conclusion

A multi-step technique provides the capability of integrating geological data directly in the formation evaluation process. The technique relies on a model-based log interpretation system that processes both qualitative and quantitative geological data, and allows the analyst considerable flexibility when interpreting wireline logs. Using this technique, the analyst can systematically and routinely incorporate geological data when performing lithological analyses with wireline logs. The results confirm the potential and the utility of merging geological data with wireline data in the log interpretation process.

References

DELFINER, P., SERRA, O. & PEYRET, O. 1984. Automatic Determination of Lithology from Well Logs, *SPE Annual Technical Conference and Exhibition*, Paper SPE 13290.

ELLIS, D., HOWARD, J., FLAUM, C., MCKEON, D., SCOTT, H., SERRA, O. & SIMMONS, G. 1988. Mineral Logging Parameters: Nuclear and Acoustic. *The Technical Review* **36**.

KING, D. E. & QUIREIN, J. 1986. Wireline Log Modeling of Geological Data: Application to Automated Lithology Identification. *AAPG Bulletin* **70**, 607.

—— & ——. 1987. Systematic Incorporation of Geological Data in the Log Interpretation Process with Applications To Lithology Analysis. *Transactions of the Canadian Well Logging Society Eleventh Formation Evaluation Symposium*, Paper N.

QUIREIN, J., KIMMINAU, S., LAVIGNE, J., SINGER, J. & WENDEL, F. 1986. A Coherent Framework for Developing and Applying Multiple Formation Evaluation Models, *SPWLA 27th Annual Logging Symposium*, Paper DD.

SCHLUMBERGER 1986. *ELAN User's Guide*, Unpublished

SERRA, O. & ABBOTT, H. 1980. The Contribution of Logging Data to Sedimentology and Stratigraphy, *SPE Annual Technical Conference and Exhibition*, Paper 9270.

——, DELFINER, P. & LEVERT, J. C. 1985. Lithology Determination from Well-Logs: Case Studies, *SPWLA 26th Annual Logging Symposium*, Paper WW.

Stochastic reservoir description: a methodology

B. P. MOSS

Scientific software-intercomp (UK) Ltd, 5–8 The Sanctuary, Westminster, London, SWIP 3JS, UK

Abstract: Stochastic log interpretation methods are shown to provide more detailed information regarding the geology encountered in a complex formation than can be obtained by conventional interpretation methods. The next stage is to combine these data with qualitative data from other sources than wireline logs to provide information about the 'rock types' present. Such rock type definition provides the link between descriptive geological data and the likely hydrodynamic model of interest to reservoir engineers. The final stage in the reservoir description process is to relate the vertical sequence, observed and described in the well bore, to the distribution of the rock types in three dimensions within the reservoir body. This paper presents the methodologies behind each of these three stages in building a more complete reservoir description, and demonstrates their application on actual data from a producing field.

The essence of reservoir description is the characterization of heterogeneities in the formation comprising the reservoir under study. At a very basic level we are interested in the characteristics of the pore space, and of the pore throats that control the passage of fluids from one pore to its neighbours. In general, significant changes in pore and pore throat geometry occur at rock-type boundaries, both lateral and vertical. Within a particular formation the occurrence of distinct rock types that can be observed to transmit fluids differently is closely related to the various descriptive sedimentary facies that are present. Superimposed upon the distribution of facies within a reservoir are the larger-scale features of a structural nature, such as fault zones and unconformities. These may form barriers or conduits within reservoirs, and may be variably associated with formation fracturing, which in turn is controlled by facies on a smaller scale.

In reviewing the value of various data types in identifying and quantifying reservoir heterogeneities, Weber (1986) concludes that 'it is possible to identify and analyse almost every kind of heterogeneity provided one has suitable data and a proper data density'. The work reported in the present paper serves to extend Weber's encouraging conclusion by suggesting that not only should the appropriate data be collected, but the appropriate interpretation or analytical tools must also be available in order to make the best use of those data.

The following sections introduce three software tools that have been developed over recent years with which a log analyst and geologist can use wireline log data for more than just the normal calculation of porosity and saturation, especially when combined with other routinely available data. The collective aim of these tools is to assist in the building of models of the distribution of identifiable rock types which in turn will, it is hoped, lead to improved reservoir description. It is not the purpose of this paper to discuss in detail the workings of these programs; rather, it is to bring the broad methodologies that are involved before a wider audience and to illustrate the type of analysis work that is becoming possible. The three software tools deal each with one of the three steps that can be considered to comprise a full interpretation based upon available log and core data:

(i) detailed interpretation of the log data, in order to identify mineralogical variation that may be of importance in differentiating the reservoir units;

(ii) the identification and description of hydraulically distinct rock-types, generally related to the descriptive sedimentary facies present in the formation;

(iii) extending the well-based description away from the uni-dimensional view in the well vicinity and out into the reservoir volume between the wells.

By way of illustration of the techniques employed, the paper will summarize some of the results obtained during 1986 and 1987 (Moss 1986; Jelstad 1988; Myler & Haynes 1988) when using them to examine aspects of the Statfjord Formation in the Statfjord Field, Norwegian North Sea.

Log interpretation

The evaluation approach employed in the first of the software tools to be discussed — which has the specific aim of the interpretation of log data — involves the comparison of the observed

From HURST, A., LOVELL, M. A. & MORTON, A. C. (eds), 1990, *Geological Applications of Wireline Logs* Geological Society Special Publication No. 48, pp. 57–75

log response curve with a predicted curve generated by a user-defined petrophysical model, for each curve in the model. The program predicts log responses in terms of porosity, saturations, and mineral volumes, using selected log response equations. Thus the outputs from the program are: estimates of porosity, S_w, S_{xo}, mineralogy, the 95 per cent confidence intervals about these estimates and the log reconstructions for comparison with the actual curves.

The estimation process is based on minimizing an objective function which measures the differences between the model predictions and the actual logs. Such differences are expressed as a weighted sum to account for the variability of the different logs, and for their different units of measurement. The weights used are the variances of the logs which are calculated from the data by using an approximation to reality which holds that each log curve has a single value for its variance over a number of adjacent depths. In this manner the number of available degrees of freedom for use in calculating the variance terms is increased because we now use the log readings of many depth levels (typically 5 or 10) to estimate a single value for the variance of each log. Different logs, of course, have different variances for the same group of depths, and the analyst is able to alter the number of depths covered to suit particular circumstances in the formation, e.g. thinly layered formations should have fewer depths in each variance group. The solutions for the unknowns are found in the light of those variances, and are taken to be those values that are calculated when the objective function, that measures the differences between observed and modelled data, is at a minimum.

Treating variance in this manner obviates the requirement to pre-specify tool behaviour in order to determine their relative importance and accuracy, and, because they are calculated rather than *a priori* guesses, variance is used to produce confidence limits for the unknowns. Poor quality data are identified and automatically given lower weight in the analysis. Thus, the minimum of the objective function represents the best fit that can be achieved to the input data in the context of the selected petrophysical model definition. The analyst can override the assignment of variance if desired, although this should be a strategy that is used only when severe conflicts and ambiguities arise from the choice of model employed.

In general, the use of multiple log responses evaluated simultaneously in a stochastic analysis extends the amount of detailed mineralogical information that can be extracted from the logs. (However, there is an important caveat which is raised later.) By allowing the inclusion of 'error bars' around the input logs (i.e. recognizing the different log variances in the minimization function) log responses may be modelled to within a 'close fit' of the observed data as opposed to achieving the 'exact fit' required when the deterministic or 'error-free' analysis methods are used. In this way a log's contribution to the total model is assessed in respect of the consistency between the solution required to match that input log, and the solution required to match all other input logs. Thus, logs that 'agree with' each other reinforce their respective solutions and govern the final results obtained, whereas logs that do not 'agree' become outliers to the most likely solution. (The solutions referred to in log analysis comprise the volumes of minerals, of pore space, and of fluids contained therein).

The caveat alluded to above is simply that in order to differentiate the possible mineral volumes that may be present in the formation identified from core studies, the log analyst needs to have logs available that can resolve the required mineral sets. This constraint on the ability for log data to distinguish between the presence of particular minerals can be tested by use of confidence intervals calculated around the estimates of porosity, mineral volumes and saturations. Ambiguous, or poorly resolved, petrophysical models are associated with wider confidence intervals (less certainty) about their results, than are produced for more definitive petrophysical models. In part, the models' ambiguity is due to the fact that most log data are in fact highly correlated to each other, log to log (Moss & Seheult 1987). The number of discrete data items in a log data set is commonly fewer than the number of individual logs available, because of cross correlation between logs.

Multivariate statistical techniques can be used to gain insight into the 'effective' data contained within a log data set (Moss & Seheult 1987), but the principle can be demonstrated by using conventional crossplots of log data (Figs 1 and 2). The poor reservoir quality material (low spontaneous potential deflection) is associated with a wide range of gamma-ray readings (Fig. 1). This pattern is indicative of the presence both of non-radioactive pore filling cements such as calcite and/or kaolinite and of radioactive clay minerals (e.g. illite) that cause the reduction of reservoir quality measured on the spontaneous potential curve. The minerals will be distributed differently within the formation, owing to their different origins, and thus are likely to have differing impact on reservoir

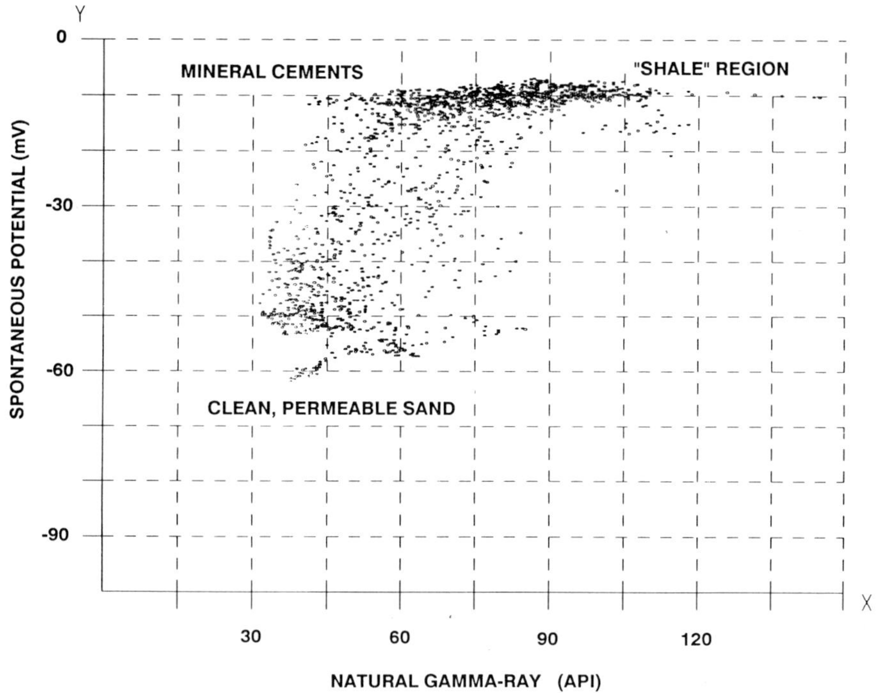

Fig. 1. Crossplot of spontaneous potential data against natural gamma-ray data, Eiriksson Member of Statfjord Formation, well 33/9–9. The low-quality formation (−10 to −15mV) has wide variation in gamma-ray readings: the permeability reduction is not due simply to 'shale'.

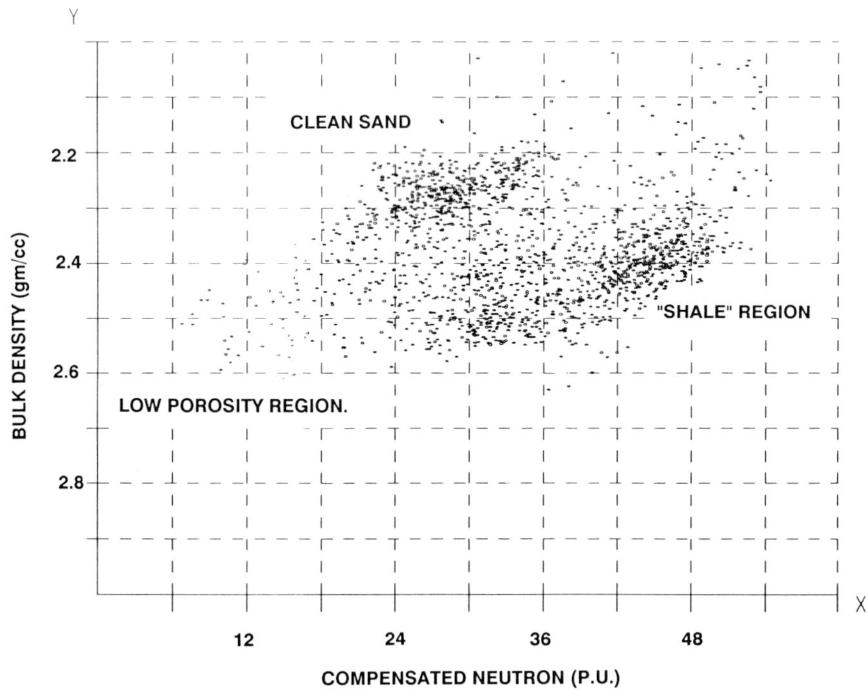

Fig. 2. Crossplot of bulk density against compensated neutron data, Eiriksson Member of Statfjord Formation, well 33/9–9. Trend of points from 'shale' region towards lower porosity indicates more complex mineralogy than simply quartz and 'shale'.

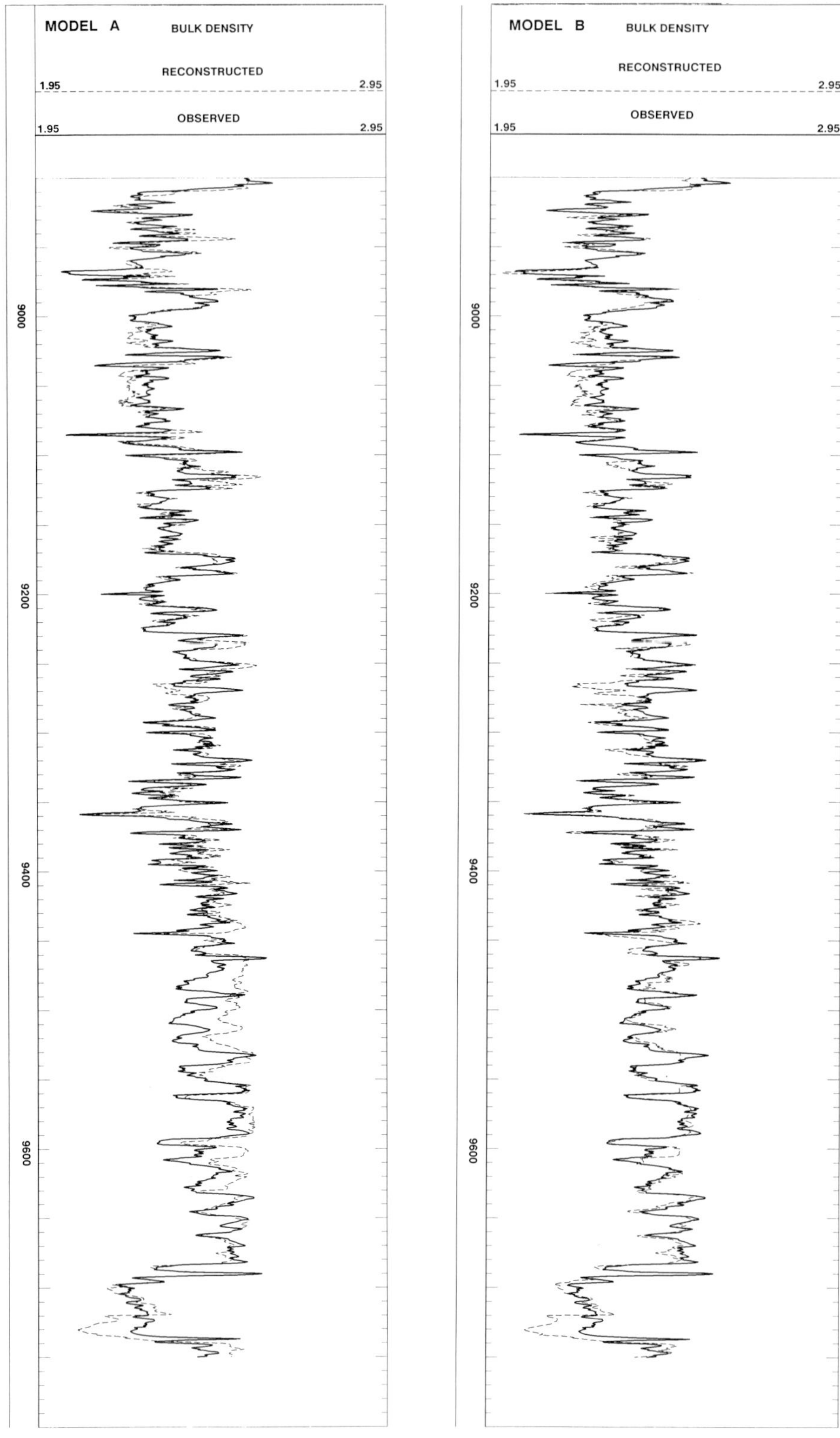

Fig. 3. Observed bulk density log data compared to predicted data arising from two different models: A, complex multi-mineral model; B, simple sand-'shale' model. Same data as Fig. 2.

quality within the reservoir as a whole. In a neutron-density crossplot of the same data interval (Fig. 2) the low-porosity data cloud (the lower edge of the data plotted) comprises a 'shale' response (higher neutron porosity, lower density) and a pore-filling cement response (lower neutron porosity, higher density). The differentiation between calcite and kaolinite cements is not possible on the basis of these crossplots and their constituent logs, alone. More data would be required.

The analyst is thus restricted to defining a mineral set for log interpretation work according to the mineralogical resolution of the data available. With only the basic curves present (bulk density, neutron porosity, natural gamma-ray, sonic transit time, spontaneous potential and resistivity data) detailed mineralogical investigations are rather problematic.

However, valuable information can be obtained from the results of a stochastic method by examining the patterns revealed in the residuals between the observed and the calculated, or predicted, data. Figure 3 is the trace plot of the density log curve from a single well overlain by the predicted density log curves resulting from two different petrophysical models. The poorer fit in Model A is obvious in the lower half of the data, but what of the upper half? In plots of the cumulative error between these density logs and their reconstruction built by the models used (Fig. 4), a vertical trend represents a good fit between the predicted and observed curves, whilst a sloping trend represents a poor fit. The magnitude of the horizontal axis is not important and is related to the units of measurement of the actual log. The fact that the vertical sections do not return to zero is not significant either; the cumulative effect of an increasing sum or a decreasing sum is what this diagram is designed to detect, as well as the points of change in the trend(s).

The character of the error cumulative summation (CUSUM) plots is different in each case. In Model A the upper half of the data are more or less on a vertical trend whilst the lower part of the data are on a sloping trend. We can infer that the fit is better in the upper section of Model A than in the lower. Model B, however, demonstrates the converse case. Both plots show the 'change of state' to occur at approximately 9240 ft. This depth coincides with the base of the Eiriksson Member, beneath which occurs the petrophysically distinct (on this evidence) Raude Member. (N.B. Model A includes a kaolinite cement term, whereas Model B does not.)

The study, of which this paper forms a summary discussion, (Moss 1986; Jelstad 1988; Myler & Haynes 1988) comprised log analyses for porosity, saturation and shale content for 13 wells in total from the Statfjord Formation of the Statfjord Field. The wells were all released to the public domain and so were completed prior to 1982. Their constituent log suites were restricted to the basic curves only, and so the scope for detailed mineralogical determination was limited. Reconstructed logs and other diagnostic indicators produced as part of

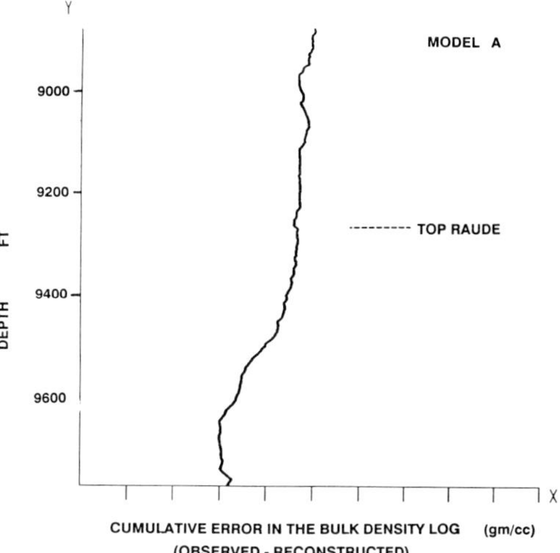

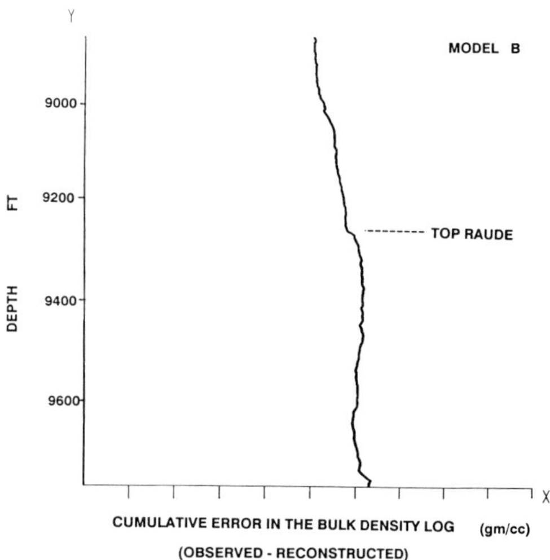

Fig. 4. Cumulative errors between observed and predicted bulk density data shown in the two cases of Fig. 3. A vertical trend indicates good fit, low error deviations from the vertical indicate poor fit, greater errors. (N.B. Absolute values are not important.) The top of the Raude Member is marked.

the method served to provide a good degree of confidence in the results for porosity and saturation. Conventional interpretation models do not routinely generate such diagnostic curves.

Extending the description to rock typing

That there is often a correlation between porosity and permeability in granular pore systems is indisputable, although there is probably never a direct causal relationship between these two attributes. This latter point is the reason for the frequently very poor correlation observed and for the lack of a universal relationship that can be found applicable in several different formations. Rather, the principal controls on permeability are in the variations in pore/pore throat geometry present and their consequent effects on pore wall surface area and flow path lengths (Pettijohn *et al.* 1973). The dominant factors controlling pore geometry often cannot be measured directly by current wireline log technology. Thus, their influence is difficult to account for when attempting to predict permeability from comparisons between wireline log data and measured core data. However, the properties of grain size, sorting, packing, pore type, degree of cementation and other textural information are routinely described in geological core analysis, and as such, form a large, albeit fairly subjective, potential source of information concerning variations in permeability. These subjective data combine to determine firstly the lithology present and, secondly, when available in some considerable detail, the descriptive sedimentary facies encountered in the well bore from which inferences concerning sedimentary architecture in the reservoir can be made. Thus, in reservoir description studies, it is desirable to combine wireline log data (which constitute widespread data) with core data (which provide highly detailed sedimentological information) to build a picture of the sedimentary facies present. These in turn, will have a major bearing on the reasons behind any observed fluctuations in permeability measurements.

To achieve this combination of diverse data types in an efficient manner computers are required in order to process the potentially large volumes of data available; the software to perform this task that has been used in this study has the following capabilities:

(a) allows for any input data, quantitative or qualitative, and selects the most effective data combinations;
(b) allows for any form of equation, linear or non-linear, to be constructed to examine correlations between data;
(c) uses graphics to enable the analyst to view the actual trends and scatter pictorially, to assist in deciding the best relationships to utilize;
(d) provides abundant feedback to the analyst to assess the statistical significance (and hence value) of one equation over another, or one 'group' definition versus an alternative.

The above list is not meant to be an exhaustive analysis of the requirements for such software, but serves merely to identify the more important characteristics needed, and perhaps also indicates why detailed rock-type investigation from combined log and core data is not attempted more frequently; more generalist programs often fall short in providing these facilities.

The software used in this study combines the above features and allows the analyst to investigate relationships between formation attributes and available log data using the principle of least squares multiple regression. The reader should refer to Davis (1986) and the extensive references and bibliography that he cites for more details on regression techniques. The least-squares procedure is a widely known and much researched mathematical method for predicting the values of one variable from combinations of other data available. The method is generally 'well behaved' providing certain basic assumptions hold true for the data being studied. The main assumptions are that the dependent variable values should be sampled from a single population with a single, common variance, that the observations should be independent of each other and that the independent (regressor) variables be uncorrelated with each other. A further assumption is that the model form to be fitted (linear, exponential, multiplicative etc.) is correct. The fact that these constraints are sometimes not achieved (often it is the independence of the regressors that is lacking) does not invalidate the process completely, but renders it difficult to be certain that the correct predictor model has been derived in such a case.

In the technique used for this paper many log curves can be studied at a single pass, with the algorithm able to point to the more effective terms to include. A central part of the approach used is to combine qualitative data with quantitative data into a single analysis to extend the level of detail attained. The qualitative data are used to partition the problem in a geologically relevant manner, so that individual relationships can be investigated for the sub-groups thus created. The aim is to define these sub-groups to represent different rock types in the formation.

For example, in the current study, the prediction of permeability has been shown to be improved by utilizing grain size data originally derived from core studies. These data, essentially subjective (not each grain is measured!), are used to subdivide the whole of the data set under study into different rock types, which has the effect of adhering to the assumption that the data to be predicted by a single equation should indeed come from a single population. In a heterogeneous formation with several rock types with distinct grain size distributions, the common procedure of lumping all the data into a single data set and searching for a single prediction probably violates this assumption. To be most effective, the grain size classes should be predicted in uncored sections by establishing correlations of the reported size distributions against a combination of log curves.

The program has been used first to discover which logs are most effective in providing information on the permeability distribution, and then, using similar logic, to establish a grain size predictor from the log curves. The permeability values studied are the horizontal air permeability measurements taken under simulated overburden conditions (5000 psi). Furthermore, it is assumed that transforming the permeability data by taking the logarithm at each point has the effect of converting the data to an approximately normal distribution.

Data preparation consists of generating core description curves from detailed lithological core descriptions and then depth matching them to the log data and conventional core data. A table of the rock characteristics reported in the cores is a useful construct to organise these data. The characteristics which enable a particular rock type to be classified are listed down the left-hand side of Table 1. The scales, although arbitrary, are chosen so that improved values of permeability are likely to be reflected in higher values of each characteristic. For example 'sorting' can have a quantitative representation ranging from 1 (very poorly sorted) to 7 (very well sorted), and reflect the fact that better sorting generally results in improved permeability.

Three cored wells (33/9–9, 33/9–A23 and 33/9–A42) had a detailed core description from which a complete set of core description curves could be generated. Wells A18 and A37 had only a lithology with just the grain size described.

For the analysis of these data 11 input curves were specified, and the program was set to indicate the six curves which combine to give the most information concerning permeability variation, i.e. the six-curve combination with the highest correlation coefficient with measured permeability.

The 11 curves that have been used as input for this analysis are: the environmentally corrected density log, the compensated neutron porosity log, the sonic transit time log, the computed effective porosity curve, the computed clay volume curve, the computed water saturation curve, and the following curves from core descriptions: grain size, grain roundness, grain sorting, and the degree of cementation.

A combinatorial method is employed to examine these curves for the combination of six which achieves the highest correlation with the variable to be predicted, in this case the horizontal air permeability. It is not within the scope of this paper to discuss the method of optimization used; the interested reader is referred to Raghavachari (1969) for further insight. Suffice it to say that the method is a computationally

Table 1. *Rock description classes*

Class Permeability:	1 Lower	2	3	4	5	6	7	8	9 Higher
Grain Size	Silt	V.fine	Fine	F-M	Med	M-C	Coarse	V.C	Granule
Sorting	V.poor	Poor	P-Mod	Mod	Mod-W	Well	V.well		
Roundness	V.ang	Ang.	Sub-A	Sa-Sr	Sub-R	Rnd	W.rnd		
Cementing	V.ind	Indur	Sl.Fr	Friable	V.fri	Uncon			
Vis Porosity	Nil	V.poor	Poor	Fair	F-G	Good	V.good	Excel	
Porosity Type	Poorly Conn.	Conn.	Inter. Gran.						
Fractures	Rare	Comm	Abund						
Mica Content	Abund	Comm	Rare						
Clay Content	Abund	Present	Trace						

efficient algorithm for accomplishing what is performed in a stepwise multiple regression model; in the latter, terms are included into the regression equation sequentially and the resultant combinations tested for improvements to the overall correlation coefficient, whereas in the more efficient former procedure, combinations of curves are examined simultaneously.

In this case the following six-curve combination was identified to have the best overall correlation with permeability: the density curve, the neutron curve, the effective porosity curve, the computed saturation curve, the computed clay volume curve, and the core grain size class curve. The algorithm looks at linear combinations of data and needs make no assumptions about the distributions of those data. If one assumes that the data are approximately normally distributed (AND), it is possible to use the Student's t-statistic to assess the relative significance of the individual curves in the suite by calculating the probability that their observed correlations are the product of chance or are 'real' (Davis 1986). On the basis of the t values thus calculated for this sample of 503 data points, three curves can be said to have strong correlations with the permeability data whilst the others are rather weaker and more prone to be pure chance. The strongest correlations are given by the density curve ($t = 5.09$), the water saturation curve ($t = 3.65$) and the grain size class curve ($t = 5.64$). Figure 5 is a crossplot of permeability versus the density log, to demonstrate the strong correlation that exists between these data.

Grouping was therefore set up to reflect the distribution apparent on this and other crossplots, and tested for significance with the software. The groups thus defined can be considered to represent rock-types apparently distinguishable from the available data.

1. Group 1 Grain size: silt to fine/med sand
2. Group 2 Grain size: fine/med sand to med/coarse sand
3. Group 3 Grain size: coarse sand to granule.

Group 2 was further subdivided into low water saturation points and moderate water saturation points.

The equation given below was tested in two ways by entering and processing it in the program. Firstly, without using the grouping system given above and, secondly, by subdividing the data on the basis of that grouping system.

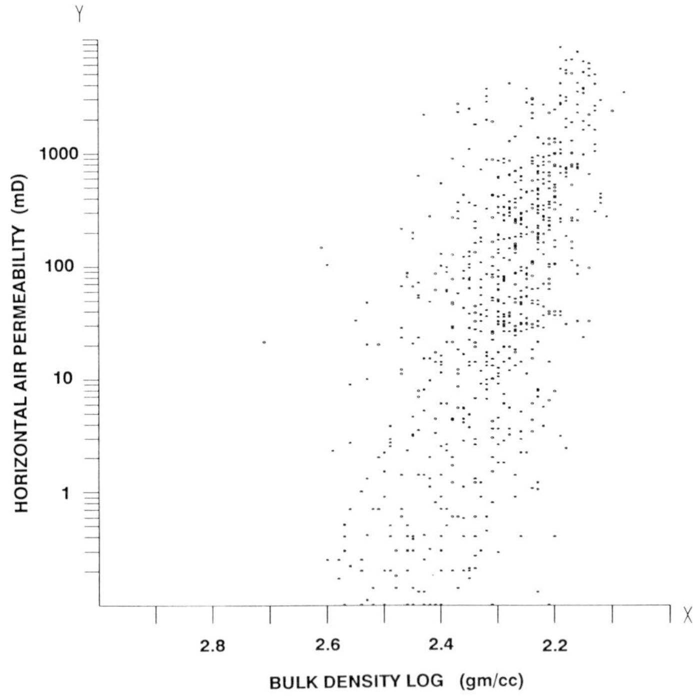

Fig. 5. Bulk density log against logarithmic scale horizontal air permeability, Eiriksson Member, Statfjord Formation, well 33/9–9.

$$\text{Log (permeability)} = P_1 + P_2\,\rho_c + P_3\,S_w \quad (1)$$

where ρ_c is the borehole corrected bulk density and P_1, P_2 and P_3 are coefficients calculated by the least squares algorithm in the program. These coefficients will have significantly different values in groups that are significantly different. The significance of each group was assessed using the *t*-value lookup table, with appropriate degrees of freedom. The output from the process run for the with-groups case shows an increase in the 'goodness of fit', compared to the without-group case (equation (1)):

without groups: r^2 (goodness of fit) = 0.47
with groups: r^2 (goodness of fit) = 0.58.

The detailed results breakdown suggests that water saturation variation between the groups is of least significance (Table 2: note the *t*-values in particular).

Given that recognition of grain size has been identified as improving the prediction of permeability, in order to extend the analysis to the uncored intervals it is necessary, first, to derive an equation from which it will be possible to predict the grain size in the non-cored zones. This will then provide a continuous trace which will be input to the permeability prediction to establish the groups for the input data.

A pass using the combinatorial algorithm was made to find out which combination of logs is most highly correlated to the grain size. This indicated that true resistivity (R_T), neutron porosity, effective porosity and gamma ray was the most effective combination.

The R_T has poor bed resolution but is still of use in grouping the data. The other three logs are well correlated with the grain size curve. On a projection of calculated porosity (ϕ) against true resistivity (Fig. 6) it is apparent that there is a split in the data at around 6 Ω m^{-1} in the moderate to high porosity points, which separates clean, water-bearing intervals from clean, oil-bearing intervals.

The data have been subdivided into broad groups on the basis of crossplot indications as follows:

1. Group 1. Clean water saturated points:
   ϕ = 0.1 to 0.4 fractional units
   R_T = 0 to 6 Ωm^{-1}
2. Group 2. Hydrocarbon saturated points:
   ϕ = 0.1 to 0.4 fractional units
   R_T = 6.1 to 1000 Ωm^{-1}

A generally satisfactory match of the measured grain size curve was achieved, once these groups were applied. In the data comprising this study, the high degree of correlation between permeability and the density log (Fig. 5) means that improvements in the prediction resulting from subdividing the data on the basis of grain size will be manifest in certain sections of the wells only. Rather more dramatic results have been seen in cases where only a poor correlation is seen initially, or where large scatter in the data exists. Such scatter often can be put down to variation in rock texture.

The aim of this stage of the analysis is to establish a subdivision of the section into layers that may have both geological and hydrodynamic significance. Thus, in this case, we are capable of subdividing coarser grained material from finer, through a prediction of grain size data that has involved the computed saturation curve. It is believed, therefore, that the sub-division thus derived has some geological and engineering value.

Table 2. *Detailed output; test for group significance*

Parameter	Log (permeability) = $P_1 + P_2\rho_c + P_3S_w$ Group 1 Fine Grain	Group 2 Med-C Grain Low S_w	 High S_w	Group 3 C-V. C Grain	All Data (No Groups)
Parameter Values					
P1	18.76	17.20	17.29	15.68	17.55
P2	−7.765	−5.552	−6.567	−5.899	−6.744
P3	0.384	−1.087	−1.814	−0.349	−0.856
No. of samples	206	59	126	260	740
T values					
P1	15.94	3.54	5.10	8.31	18.89
P2	9.46	2.66	4.76	7.05	15.74
P3	0.92*	3.38	1.68*	1.33*	4.90

T values indicate whether a correlation is significant or could occur purely by chance. Values marked * have 5–20% probability of occurrence purely by chance.

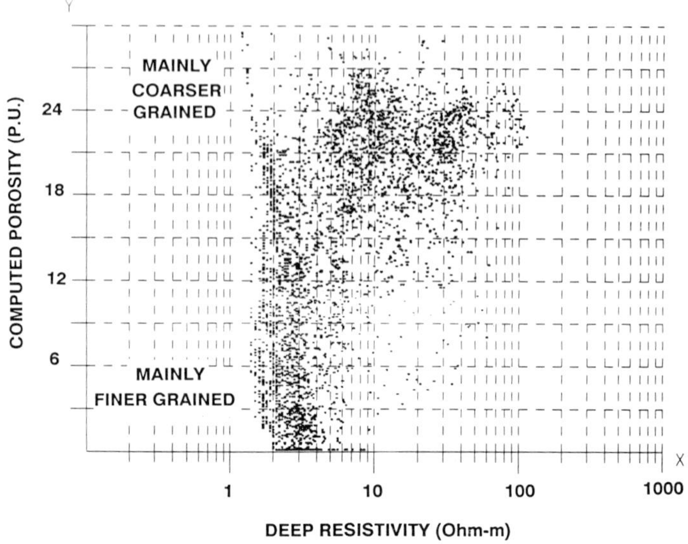

Fig. 6. Computed porosity against deep resistivity (R_T), Eiriksson Member, Statfjord Formation, well 33/9–9.

Prediction of reservoir architecture

In any discussion on the impact of reservoir heterogeneity on fluid movements within the formation, the issue of scale is a vital consideration. The data for the reservoir description are commonly derived from techniques which examine relatively small quantities of rock that are minute in comparison to the volume of reservoir depicted in a single grid cell used in a reservoir model. Thus, there exists a fundamental problem in scaling up the detailed reservoir description data to something representative of typical reservoir model grid block sizes.

To begin to approach a solution for this problem, the detailed reservoir architecture in three dimensions is required. From this, more accurate assessments of the size and shape of possible flow units can be made, and each flow unit can be ascribed a distribution of descriptive parameters that more closely reflects any spatial dependencies existing within that particular sedimentary facies. Haldorsen (1986) discusses in detail possible pitfalls and approaches in the process of relating the small scale descriptive data to the large-scale grid blocks typical of current simulator models.

The final stage of the current study was to utilize techniques designed to extrapolate 'rock types' or lithotypes into three dimensions away from the well bores. The distinguishable rock types have been identified and characterized into 'electrofacies' (Serra & Abbott 1980) by the integration of core and log data, both quantitative and qualitative, through the detailed processing that took place during the first two stages discussed above.

The technique that is used for this stage has been coded up into a computer program in view of the large amount of data to be handled. It is not intended that the algorithm predict the exact reservoir configuration, but rather that it should supply a number of reservoir descriptions from which the most likely can be postulated.

The reservoir is modelled as a three dimensional grid of cells, whose size and number are specified by the user. The size of the grid cells determines the scale of the description in each of the three axes. The reservoir grid cell model is filled from the bottom with the input lithotypes (rock types), in accordance with a set of model rules, and in accordance with the dimensions of the individual lithotypes as input by the user. The model's rules concern the manner in which a particular lithotype either erodes or is eroded by its neighbours and the most likely sequence of such lithotypes based upon information derived from the well data and entered as a transition matrix to the program. The input data are of the following general types.

Reservoir model dimensions.

Distribution of the thicknesses and areal dimensions of lithotypes.

Lithotype distribution in the basal layer.

Transition probability matrix to describe the vertical sequence as observed in each of the sample wells.

Deterministic sequences observed at those wells.

Average porosity and permeability for each lithotype.

The results from the model allow the user to view the reservoir along any x, y or z section, allowing both cross-sectional and plan views to be examined. Several other statistical results are also available from each construction.

The output data available from the program can be summarized as follows.

x, y or z direction sections.

'z' direction sections of connectivity representation.

Volumetric proportions of each lithotype as predicted from the algorithm for the entire grid of cells.

Average thickness of each lithotype within the entire grid of cells.

Average porosity and permeability of the reservoir as a whole built up of units predicted in their size and frequency by the algorithm.

Resultant transition matrix after reservoir construction.

Results of calculating the volume of connecting sands between well pairs as a proportion of the entire volume between such wells.

The Statfjord Formation of the Statfjord field has been well described by Roe & Steel (1985), who suggest a sedimentological model in terms of a coastal fan and fan delta complex on a low-energy coastline. They have described four facies associations, distinguished from the extensive cored intervals, which form an overall coarsening upwards sequence.

Subaerial fan deposits. Dominantly cross-stratified and plane bedded coarse grained sandstones deposited in two main regimes.

Braided stream deposits consisting mainly of coarse grained, poorly sorted sands and fine conglomerates.

Fan plain sediments consisting of coarse grained sands with mud clast conglomerates and thin mudstone laminations interpreted to result from flood controlled deposition.

Distal fan deposits. Mostly massive or biolurbated, fine to medium grained sandstones.

Lagoon bay deposits. Dark grey mudstones and siltstones containing fine, ripple laminated sands, interpreted to represent the pro-delta environments with periodic inundations.

Floodbasin bay deposits. Red and green bioturbated mudstones and silts occasionally with fine to very fine sandstones. Interpreted to represent embayments between fan lobes.

A diagrammatic representation of the sedimentological model devised for the Statfjord Formation (Fig. 7), shows a single lobe of a coalescing system of coastal fans which covered the Statfjord Field area during the late Triassic and early Jurassic. The modelling performed with the program during this study has been based on the model shown in the figure.

The Statfjord Formation is divided into three members, Raude, Eiriksson and Nansen. The lowermost, Raude Member, is the most continental of the three with the lowest sand to shale ratio, while the Eiriksson shows an increasing ratio upwards. The Nansen Member is interpreted to be a marine sand which covers much of the Statfjord Field area. The majority of the reservoir quality sandstones occur in the Eiriksson and Nansen Members while the Raude seldom contains much oil, due to its poor reservoir characteristics, and its low stratigraphic and, therefore, structural position.

The sand sequence can be divided into several deterministic sand layers which correlate across the whole field area. The major sand bodies represent lobes of the fan complex enclosed by the finer grained sediments.

The current implementation of the program requires that a datum be established at the base of the reservoir model, ideally a synchronous

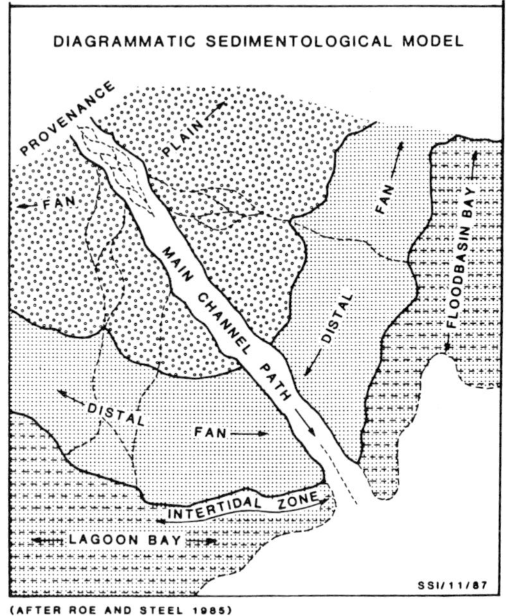

Fig. 7. Diagrammatic representation of the sedimentological model used as the basis for the three-demensional modelling (after Roe & Steel 1985).

horizon. This requirement is achieved by the selection of the layer that coincides with the base of the Eiriksson Member. A schematic map of the model area (Fig. 8) gives the approximate well locations and indicates the distribution of the oil column of the central Statfjord Field. This area was chosen for the modelling because it contains the cored wells and three of the remaining wells in the study data set.

In this case the grid is defined to comprise 46 cells in the x direction and 20 cells in the y direction, with each cell representing a surface of 500 feet square. In the vertical z direction, cells are defined as being 1.5 feet thick and there are 100 layers. The overall dimensions of the model are therefore 23 000 feet by 10 000 feet by 150 feet, and a representation of the relative size of the grid with respect to the well spacing is depicted in the top-left corner of Fig. 8. The orientation of the grid is rotated from north–south to allow a better representation of the sedimentological model. These input details are completely user definable.

The basic data for building the models were derived from the results of the previous studies. The assessment and prediction of permeability, conditioned by the use of grain size classes, has been used to define the well sequences in terms of lithotypes.

The rock-typing results for 33/9-A42 compare reasonably well with the core data available over the section of the data comprising the basal sand unit of the Eiriksson Member. The digitized grain size data from core are shown in track three (Fig. 9), together with the predicted grain size from the second stage of the analysis detailed above. The first track of the plot shows the lithotype groups determined, and plotted as the GROUP trace.

The prediction of the grain size and the subsequent groupings defined from that predicted curve constitute a series of electrofacies which are based on the log data used in the predictions. In this case the electrofacies are defined with the compensated neutron, the calculated effective porosity, the corrected gamma-ray and the true resistivity log (R_T), as these were the curves used in the prediction of grain size. The use of the R_T curve has unfortunately reduced the vertical resolution of the predicted grain size curve (track three, Fig. 9).

The group data thus defined were analysed and compared with the available core descriptions to establish the nature of the rock types. The following interpretations were made from these comparisons:

Group 0 was determined to be representative of the lagoon bay facies and to consist mainly of dark grey shales.

Group 1 correlated with the siltstones of the distal fan and floodbasin bay facies.

Group 2 was matched with the medium sands which are predominant in the fan plain facies.

Group 3 represents the coarse-grained channel sands which cut across the fan plain.

The groups were therefore renamed for the 3D model according to the following scheme:

Group 0 Shale
Group 1 Silt
Group 2 M_{sand}
Group 3 C_{sand}

The basis of the model is to define the observed well sequences as a Markov process and

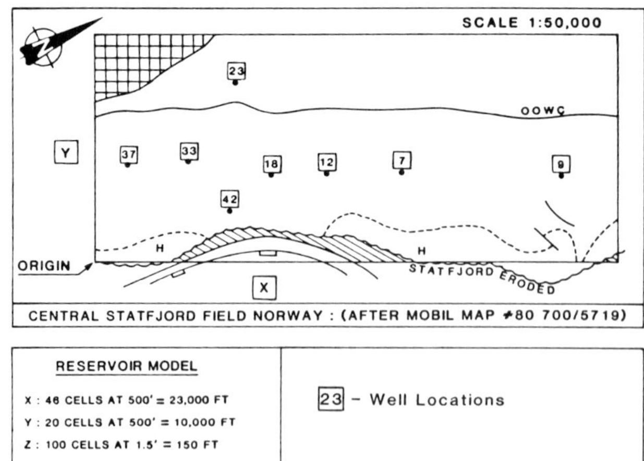

Fig. 8. Schematic index map.

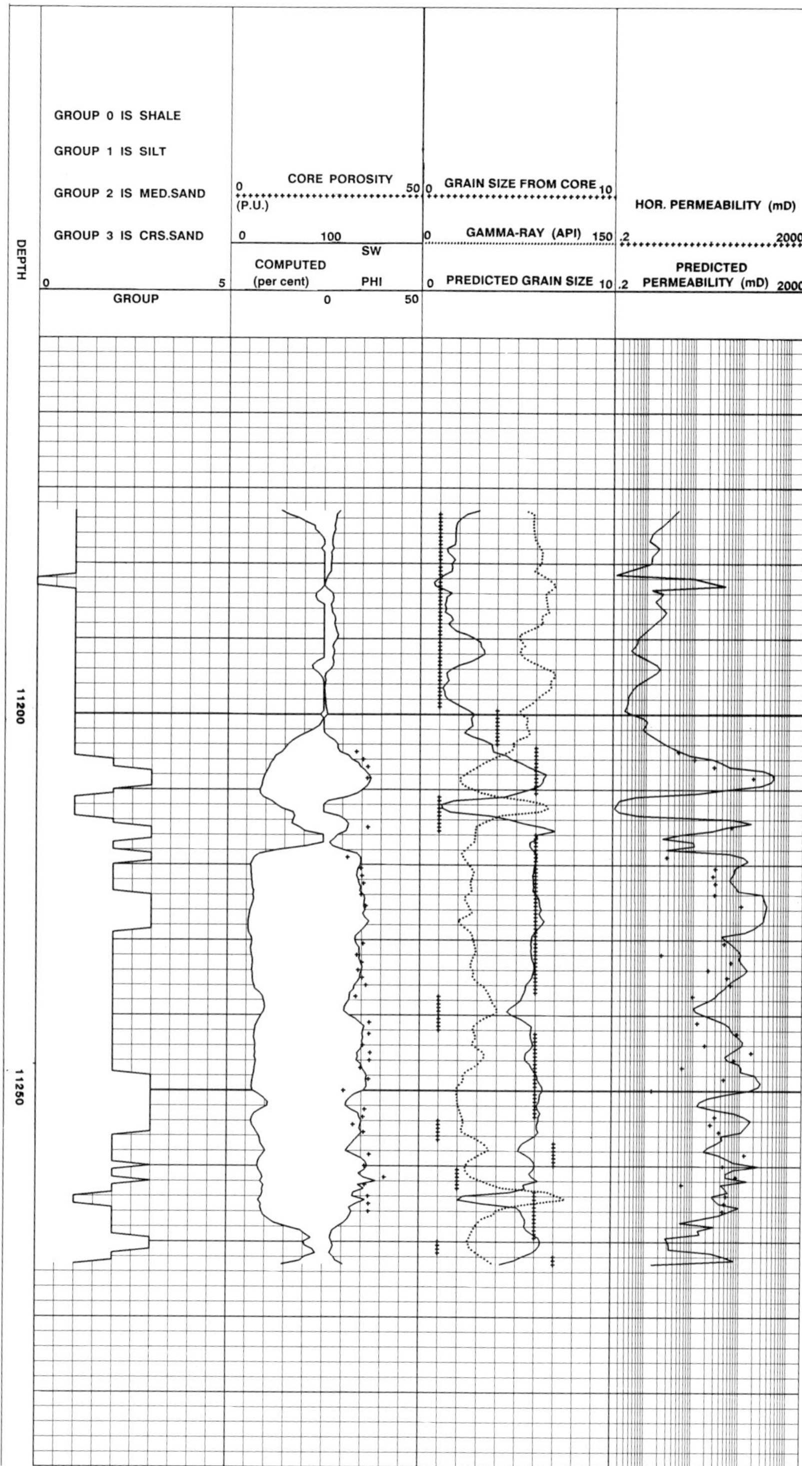

Fig. 9. Well 33/9-A42; results of lithotype grouping from log data, basal sand/silt unit of Eiriksson Member used in modelling exercise.

to build a matrix of counts of the transition from any one state into another. The transition counts are then converted to probabilities for use in the model. The derived probabilities are shown as the transition matrix in Table 3 for the four lithotypes defined. Note that the shale lithotype is only observed to occur below a silt. These probabilities are defined for unit volumes with the assumption that the population observed represents a good sample of the reservoir. If further data (e.g. seismic) were to indicate the presence of lithologies or rock types that were not observed in any of the wells, then this assumption obviously could not hold. In this case, however, we are in the happy situation whereby gross inter-well stratigraphic correlation is of high order and thus we can be confident that the wells do form a representative sample of the reservoir. We have assumed that the sequence of lithotypes is significantly different from a random distribution of the same volumetric proportions. In a pictorial representation of the transition probabilities (Fig. 10), the clear dependence between the silt and shale lithotypes can be observed.

In practice, the input probabilities do not match exactly to the probabilities calculated after the reservoir construction due to the random selection of the areal dimensions for each slab deposited during the building of the model, and in some cases the influence of the erosion rules specified by the user. The user has complete control over which lithotype erodes which, and in this specific case the Coarse Sand unit erodes all the others, the Medium Sand unit erodes the Silt and the Shale, whilst these latter two units do not erode any other unit. Within the algorithm the predicted transition matrix is calculated for comparison with the input data. Large discrepancies denote unstable model input and/or well conflict.

Table 3. *Initial 4 Lithotype model*

Transition counts

	C_{sand}	M_{sand}	Silt	Shale	(Total)
C_{sand}	0	27	2	0	(29)
M_{sand}	23	0	19	0	(42)
Silt	4	14	0	3	(21)
Shale	0	0	3	0	(3)
(Total)	(27)	(41)	(24)	(3)	(95)

Normalized probabilities

	C_{sand}	M_{sand}	Silt	Shale
C_{sand}	0	0.93	0.07	0
M_{sand}	0.55	0	0.45	0
Silt	0.19	0.67	0	0.14
Shale	0	0	1.0	0

The Silt and Shale lithotypes are interpreted to represent the Floodbasin Bay or intertidal facies associations.
The C_{sand} lithotype is interpreted to represent Channel sands.
The M_{sand} lithotype is interpreted to represent Fan Plain sands.

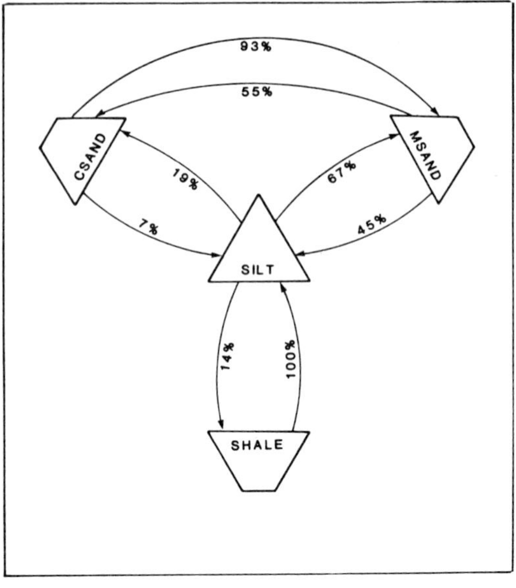

Fig. 10. Lithotype transition model, simple case.

The areal dimensions for each 'electrofacies' are chosen to reflect the concepts embodied in the sedimentological model (Fig. 8). The Silt and Shale lithotypes were modelled with the x and y dimensions having the same distributions while the channel sands (C_{sand}) and the fan plain sands (M_{sand}) were modelled with a cross-cutting relationship. The fan plain sands were modelled to be longer in the direction parallel to the x axis while the channel sands were longest parallel to the y axis. The thicknesses of the units are derived from the values sampled for the x and y dimensions, the larger the area the thicker the unit. Hence, measures of calculated unit thicknesses together with volumetric proportions offer insight into the accuracy of the model.

The model is built by repeated random sampling from the input transition probabilities and dimension distributions, such that 'slabs' of the input lithotypes are created within the grid framework of cells representing the reservoir. The algorithm detects empty grid cells and assigns a particular lithotype to each. That lithotype then has a certain size and thickness as governed by the dimension distributions provided as part of the input data, and thus placing it within the grid causes several cells to be assigned to the same lithotype. The algorithm next locates another empty cell, and repeats the sampling procedure to assign another — possibly the same — lithotype within the cell framework. Conflicts that arise in the event that different lithotypes are mapped to the same cell in subsequent passes are resolved by reference to the erosion rules that are input to the model. These will determine which lithotype 'prevails'. Well sequences can be honoured in the logic, or can be left to be estimated by the algorithm.

The present example compares two representations of the lithological sequence. One is a simplistic model with only four lithotypes and has proved far too simple. The other takes a different view and subdivides the sequence into two units each of which has different probabilities of transition between the finer and coarser-grained lithologies that comprise them. Diagrammatic representations of these two transition models can be seen in Figs 10 & 13.

The simpler transition matrix (Fig. 10) produces a rather inadequate representation of the reservoir (Figs 11 & 12). One vertical section (Fig. 11) is taken parallel to the depositional strike, whilst the other (Fig. 12) is orthogonal to the strike. Each shows that a preponderance of coarse sand is 'created' by this simple case, which is at odds with our expectations in that we are modelling a single, broadly fining-up unit corresponding to a single fan lobe. The fining upwards trend was completely missing from the model when based upon the simple transition case.

Therefore, a more complex transition matrix was generated (Fig. 13). Basically, the changes involved splitting the section into two layers, and creating a condition whereby once the model had passed from the lower lithotype selection to the upper lithotype selection, there was no return to the lower lithotype states . The main difference between the two sets of lithotypes was that a preponderance of finer grained material existed in the upper lithotype selection. This was designed to reflect the observed fining-upwards trend cause by the increasing occurrence of the delta-top sediments in the upper half of the section. The sections taken through

(vert. exag. approx 45 times.)

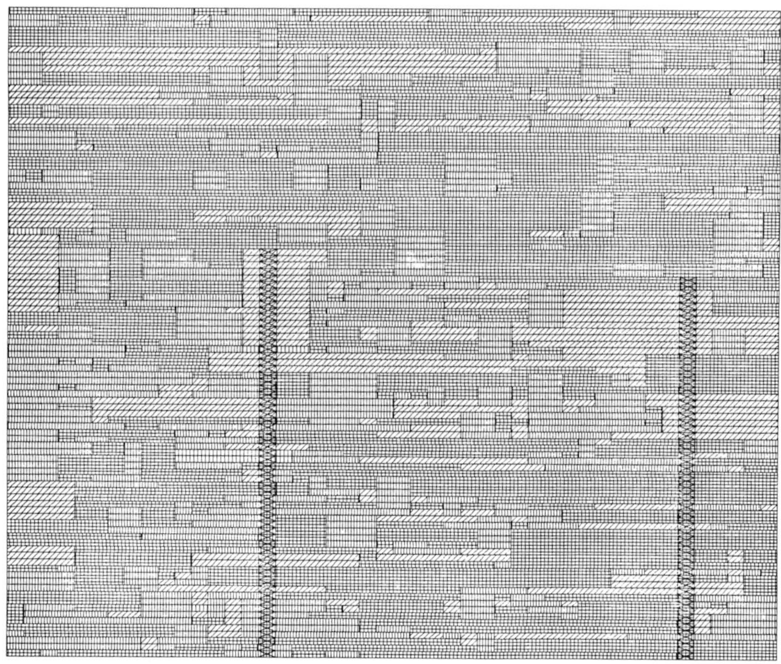

Fig. 11. Model output, simple case, cross section parallel to strike.

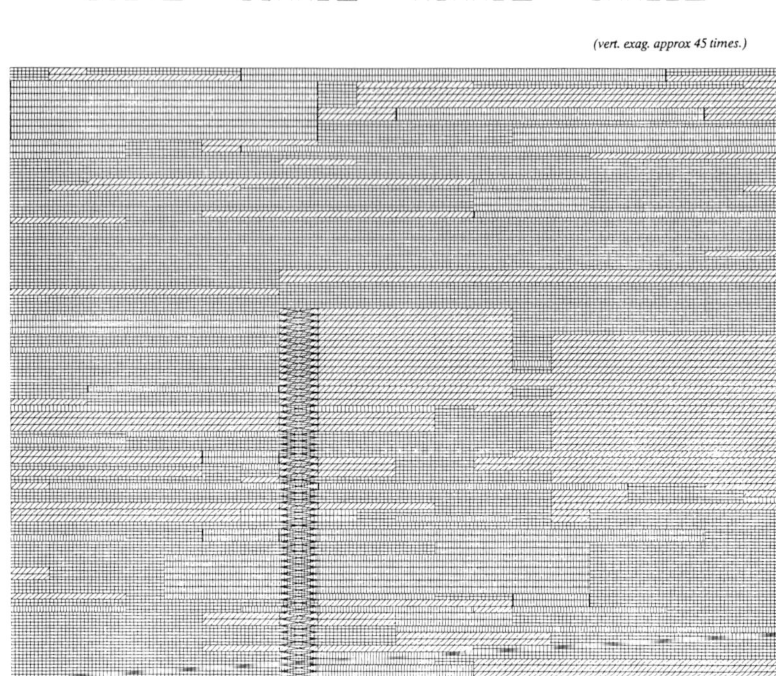

Fig. 12. Model output, simple case, cross section orthogonal to strike.

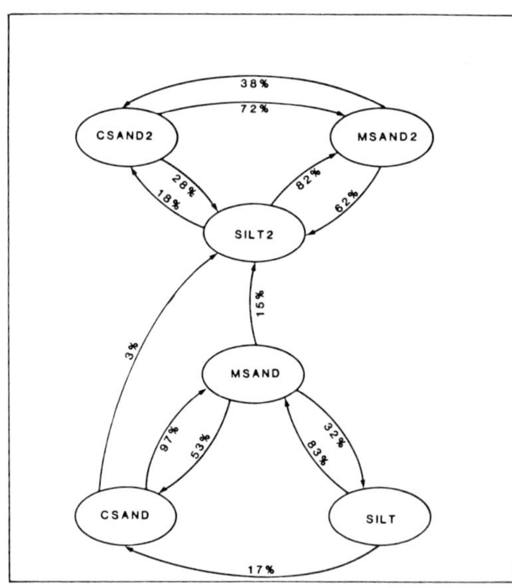

Fig. 13. Lithotype transition model, 6 lithotype case.

the model produced from using the more complex transition matrix (Figs 14 & 15) provide a more satisfactory representation of the observed fining-upwards sequence than had been obtained with the simpler model. However, the upper SILT2 unit average thickness is not matched well by the model (Table 4), and its definition should be re-examined.

The scale of the reservoir units that were being modelled by this method was sufficiently small such that conflicts between observed well sections and the three-dimensional distributions predicted from the statistics and by the erosion rules were not common. At present the program assumes that the lithotypes created from the well sequences are able to be eroded, with the exception of the vertical column that actually represents the well bore itself. Hence, it is possible for a lithotype to be 'eroded' with the exception of the well grid cell, leading to a conflict that is unresolved. Similar effects give rise to the model being dissimilar from one side of a well to the other. This is an artefact of the

(vert. exag. approx 45 times.)

Fig. 14. As Fig. 11, but with 6 lithotype case.

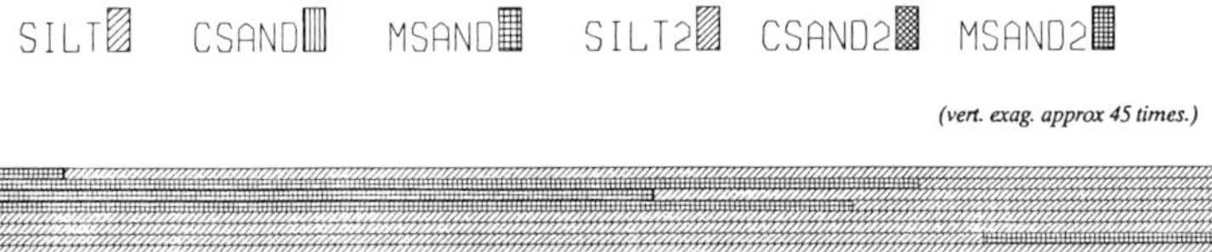

Fig. 15. As Fig. 12, but with 6 lithotype case.

Table 4. *Comparison of input and output statistics: 6 lithotype model*

	Transition matrix SILT	C_{sand}	M_{sand}	$SILT_2$	C_{sand2}	M_{sand2}
SILT	0	0.14 (0.17)	0.66 (0.83)	0.11 (0.0)	0.02 (0.0)	0.07 (0.0)
C_{sand}	0.06 (0.0)	0	0.66 (0.97)	0.2 (0.03)	0.01 (0.0)	0.06 (0.0)
M_{sand}	0.2 (0.32)	0.38 (0.53)	0	0.34 (0.15)	0.02 (0.0)	0.05 (0.0)
$SILT_2$	0.01 (0.0)	0.02 (0.0)	0.08 (0.0)	0	0.10 (0.18)	0.79 (0.82)
C_{sand2}	0.002 (0.0)	0.004 (0.0)	0.02 (0.0)	0.48 (0.28)	0	0.49 (0.72
M_{sand2}	0.005 (0.0)	0.02 (0.0)	0.04 (0.0)	0.67 (0.62)	0.27 (0.38)	0

Input probabilities in brackets.

Average thicknesses (ft)	From well sequences	3D model results	Well results
SILT	3.9	3.1	3.5
M_{sand}	4.2	5.1	4.6
C_{sand}	5.7	5.0	4.7
$SILT_2$	14.8	6.8	13.3
C_{sand2}	3.2	3.4	2.9
M_{sand2}	3.0	2.9	2.6

algorithm in its current formulation and can be avoided in future developments.

As noted earlier many passes are made with the algorithm, each building a complete reservoir model by the process of sampling from the reservoir units' dimension distributions, in order to select the most likely solution derived from the input data used. Selection is on the basis of comparison between the input transition characteristics and thickness properties and the output data produced after the runs (Table 4).

It must be stressed that these aspects of the work are incomplete at the time of writing this paper, as is the assignment of representative parameters to large grid blocks for use in reservoir simulation work. At present the model merely assigns volume-weighted averages of the required parameters, which is of course, a gross over-simplification. By virtue of the ability to identify separately the rock types that are present, it is hoped that even this simplistic approach is more meaningful than the averages produced without regard to the individual lithotype volumes and distributions. Some investigations into the effects of these "mega"-scale heterogeneities, such as rock-type variation, on fluid flow between wells have begun, with a view to performing better representation of the fine detail at the larger scale of the simulator grid block. However, beyond the rather glib statement that differences in modelled transmissibilities can be observed in the simulation output, arising from the suggested distributions of the flow units in the model, and that this distribution of transmissibility is considered to be of importance in explaining or modelling single well behaviour (e.g. well tests), it is too early to make definitive reports of the value of such extensions to the basic objective of predicting reservoir architecture based upon well data. We consider this to be a fertile area for continued study and will be pursuing our researches with enthusiasm.

Conclusions

Extending the information derived from log data beyond the calculation of porosity, water saturation and "shale" content requires that broad suites of log measurements be available that are capable of resolving between the minerals that may be present. This practice also requires an appropriate interpretation tool that can resolve between alternative ambiguous solutions, any of which may be supported by the data at first glance. One such possible methodology is outlined in the paper.

Estimating the permeability profile encountered at the well-bore from the log data available can be improved by the incorporation of qualitative data from geological descriptions into the analysis, to facilitate the subdivision of the formation into distinct rock-types. The resultant rock-type groups are facies dependent and should reflect different hydraulic properties which are of significance when simulating the movement of fluids within the reservoir.

The investigation of a reservoir's architecture based upon the observed sequence of sedimentary facies in the well-bores is facilitated by these first two analysis stages. The hydraulic significance of the facies variation can be examined once the particular rock-types have been identified. Of crucial importance to the applicability of the three dimensional models that can be built is the accuracy or representativeness of the dimensions ascribed to the various rock-types.

When utilizing the highly detailed geological models that can result from these procedures in further simulation modelling of the dynamic behaviour of the reservoir under production, the scaling-up to the generally large size of grid cell employed in the latter studies requires careful treatment to retain the impact of the detailed information. Further research is required to perfect suitable mathematical methods for use in this process.

The author is grateful to the many individuals and organisations who have been involved with the development of these three methodologies, in particular UK Dept. of Energy, British Gas Corporation, Enterprise Oil plc, Saga Petroleum A.S., and The Norwegian Institute for Energy Technology supported by the SPOR committee of the Norwegian Government. The work reported herein was carried out during 1986 and 1987 in cooperation with the Norwegian Institute of Energy Technology who were under contract to the SPOR committee of the Norwegian Government. This paper is drawn from the reports of that work: Moss (1986), Jelstad (1988), Myler & Haynes (1988). Log data used were processed and contributed by the Norwegian Petroleum Directorate. Co-workers on the project from The Norwegian Institute for Energy Technology were Mrs. Astrid Holstad and Mr. Oyvind Jelstad. The author is particularly grateful to these individuals and to several colleagues in Scientific Software Intercomp (UK) Ltd.: Peter Haynes, Nicholas Myler and Dr Nigel Goodwin; all of whom contributed greatly to the project work over the two-year period.

References

DAVIS, J. C. 1986 *Statistics and Data Analysis In Geology*. 2nd Edn, John Wiley, New York.

HALDORSEN, H. H. 1986. Simulator Parameter Assignment and the Problem of Scale in Reservoir Engineering. *In*: LAKE, L. W. & CARROLL, H. B. JR., (eds) *Reservoir Characterisation*, Academic Press, New York.

JELSTAD, O. 1988. '*Testing PERMPROBE and GEOPROBE, Reservoir Description of the Statfjord Formation*.' Institute for Energy Technology, Kjeller, Norway; Report IFE/KR/F-88/061

MOSS, B. P. 1986. *Reservoir Description of the Statfjord Sands, Block 33/9* Institute for Energy Technology, Kjeller, Norway, Report IFE/KR/F-86/071

—— & SEHEULT, A. 1987. Does Principal Components Analysis Have a Role in the Interpretation of Petrophysical Data? *Transactions of SPWLA 28th Annual Logging Symposium*, London 1987.

MYLER, N. C. & HAYNES, P. D. M. 1988. '*SPOR-OPT, Reservoir Description of the Statfjord Sands using PERMPROBE and GEOPROBE*.' Institute for Energy Technology, Kjeller, Norway, Report IFE/KR/F-88/035

PETTIJOHN, F. J., POTTER, P. E. & SIEVER, R. 1973. *Sand and Sandstone*. Springer, New York.

RAGHAVACHARI, M. 1969. 'On connection between zero-one integer programming and concave minimisation under linear constraints.' *Operation Research*, **17**, 680–684.

ROE, S. L. & STEEL, R. 1985. 'Sedimentation, Sea-Level Rise and Tectonics at the Triassic-Jurassic Boundary (Statfjord Formation), Tampen Spur, Northern North Sea.' *Journal of Petroleum Geology*, **8**.

SERRA, O. & ABBOTT, H. T. 1980. '*The Contribution of Logging Data to Sedimentology and Stratigraphy*.' SPE 9270, 55th Annual Fall Technical Conference, Dallas, Texas, Sept 21–24, 1980.

WEBER, K. J. 1986. 'How Heterogeneity Affects Oil Recovery.' *In*: LAKE, L. W. & CARROLL, H. G. JR., (eds) *Reservoir Characterisation*, Academic Press, New York.

The language of rocks: an example of the use of syntactic analysis in the interpretation of sedimentary environments from wireline logs

C. M. GRIFFITHS

University of Trondheim, The Norwegian Institute of Technology, Division of Petroleum Engineering and Applied Geophysics, The Petroleum Centre, N-7034 Trondheim-NTH, Norway

Abstract: The sedimentary sequence as recorded in vertical section, i.e. by boreholes and logged sections, is a very uneven representation of the passage of time. Thick sand sequences may be deposited instantaneously with respect to geological time, whereas thin shales may represent either millions or hundreds of years, depending on their environment of deposition. This uncertainty concerning the relative time values of sedimentary units and the breaks between them adds a certain spice to attempts to correlate rather than match well sequences. Consideration of Gressly's original facies concept, and the formal application of 'Walthers Law' suggests the basis of a sequence analysis tool that can be used to correlate adjacent wells. The syntactic pattern recognition technique demonstrated here is based on the development of a grammar of unit associations, and the parsing of log-derived lithology units. The development of stratigraphic syntax is shown for both wireline log and grain-size data, with example taken from the North Sea.

The aim of this paper is to present an approach to the interpretation of sedimentary environments from wireline log data. Although the approach presented here may possibly be of use in a geological interpretation system at some future date, the work was carried out largely with the aim of trying to understand how the types of rules that are subconsciously applied by geologists in an exploration context, using traditional data, can be formalized, and to examine the types of terminal symbol or language that may be of use in such rules when traditional data are not available.

The use of wireline data for environmental interpretation and stratigraphy has not been helped by the ambiguity inherent in translating petrophysical parameters to the more traditional hand-specimen classifications. Ambiguity is, however, not restricted to wireline data, and even if the ambiguity in translation could be successfully overcome, the ambiguity in the use of traditional data would still remain.

For example, in the field of clastic sedimentology, grain size, structure and texture are the most commonly used classification bases. Miall (1984, p 179) has shown, however, that if grain size alone is used in vertical profile then it would not be possible to make predictions of lithosome shape because similar coarsening and thickening upward profiles could be generated by a prograding alluvial fan, river-dominated delta, wave-dominated delta, barrier island, a storm-dominated shoreline, or a submarine fan. Thus, it is not only the descriptive unit itself but the context and sequence in which the unit is found that is of interpretative value.

Because of the human psychological necessity to reduce information to 'chunks' for efficient processing (Miller 1956), classification forms an essential role in the observational sciences, and yet the theoretical process of classification itself has not been subjected to as much debate in the geological literature as in, for example, geography. Despite this lack of attention, classification plays a central role in geological interpretation. The language of geological rules, the *lingua franca* of geological discourse, is composed largely of classes derived from sub-divisions of observations. As an example, take this description of a canyon-fill deposit:

> 'localised sand-prone intervals, possibly sand filled leveed feeder channels. . . . local areas of high amplitude, internally moundy, discontinuous reflections.'

A geologist or geophysicist would presumably feel relatively happy using this description, which contains at least eight classifications ('local', 'sand', 'possible', 'levee', 'feeder', 'channel', 'high', 'moundy', 'discontinuous'), and a mixture of genetic and descriptive, fuzzy and 'hard' classes. A machine-based interpreter would have severe problems with many of these classifications, as would a geologist attempting to put numbers on many of the characteristics for quantitative testing of hypotheses.

We must also be careful to choose appropriate classes for the questions that are being asked. For example, grain size is often used as a classi-

From HURST, A., LOVELL, M. A. & MORTON, A. C. (eds), 1990, *Geological Applications of Wireline Logs* Geological Society Special Publication No. 48, pp. 77–94

fier in clastic rocks for several good reasons: the modal grain size is easy to measure in the field, it is directly related to the energy of the depositional environment, it exerts a significant control on rock properties, and clastic rocks are by definition composed of grains bound together in one way or another. However, if we have seismic or petrophysical data and we wish to model the three-dimensional development of stratigraphic units and their relationships, then relating petrophysical data to modal grain size alone is not sufficient for the reasons stated above.

There are many problems involved in developing a geological terminology involving petrophysical data that can have predictive potential in modelling studies. This paper will examine some of the requirements for such a language, and make some suggestions for a systematic approach to the problem that can make use of both classical and petrophysical descriptors.

Languages, both natural and 'artificial', may usefully be considered as having two main components. The 'syntax' of a language describes its component symbols and the way they can be legally organized. The 'semantics' of a language, however, describe the possible ways of organizing legal sequences to give different meaning.

A semantic analysis can involve testing for logical inconsistencies and consequences within legal phrases as a guide to possible meaning. This paper will be concerned only with the possibility of testing for legal geological sequences, rather than the semantic analysis phase.

The need to provide fast and efficient compilers for computer languages, and the drive towards natural language computer communications has increased the amount of research into formal language structures. Computer systems at present are not very good at handling ambiguity, and computing languages generally have a limited vocabulary and well defined structure. A commonly used way of representing this structure, or 'syntax', is the Backus-Naur-Form (BNF) notation which was first developed to define the syntax of Algol 60 (Chomsky 1957).

An example of a syntax expressed using this notation would be:

```
<S> : := <A> <B> | <A> <C>
<A>: := <B> x | <B> Y
<B>: := d < C >
<C>: := s | t | u | v
```

This states that the non-terminal symbol 'S' is composed of a sequence of either 'A' followed by 'B', or 'A' followed by 'C'. Each of these symbols is itself composed of sequences of non-terminal and terminal symbols. The terminal symbols 'x', 'y', 'd', 's' etc. are the basic indivisible units of communication. In a geological context they could be facies of either genetic or descriptive origin. For a background to the geological use of syntax in stratigraphy the reader is referred to Griffiths (1988).

In a geological context, syntactic analysis can be used to examine the sufficiency of rules for the recognition of sedimentary environments and to formalise the prediction of facies sequences. The utility of syntactic analysis in this context is based on the ability to include both deep deterministic structures and stochastic models in an elegant rule base.

In addition to the rules of combination (the grammar), we need an interpreter of the rules. Such interpreters are called 'parsers', the use of which in a stratigraphic context are explained by Griffiths (1988).

The examples given in this paper were parsed using a simple context-free left-substitution parser with back-tracking. The aim here was to concentrate on how a geological syntax may be developed rather than on the parser itself.

We have seen how some classical geological terminal symbols are context dependent (for example, Miall's coarsening upward sequences), and this is directly equivalent to the use of the same word in different grammatical positions in English, or the same word groupings in different contexts. More sophisticated parsers are capable of considering several parallel interpretations during a single pass, and it may well be that these will also be more effective in a geological context.

A stratigraphic syntax

A syntactic rule set or grammar can be constructed from theory or derived inductively from the data, or a mixture of both. In the case of grammars derived from theoretical considerations, the parsing of real sequences provides a test of the theory. In the inductive case the syntax can help in the formulation of new principles, and provide insight into causal relationships.

Theoretical basis

In the fields of stratigraphy and basin analysis there are a few stated principles such as:

(i) Walthers Law of Succession of Facies,

which states that in a sequence with no breaks (i.e. faults or erosional unconformities) only those facies that are to be found beside each other at the present time can be superimposed in vertical section.

(ii) Law of Superposition of Strata, which states that in any succession of strata the beds were deposited in a simple vertical order from base to top.

(iii) Gressly's Facies concept which states that lateral changes in physical characteristics of rocks at a mappable scale are identified as discrete units.

Each of these stratigraphic principles has been the subject of much debate (Krumbein & Sloss 1963 p. 299– ; Miall 1984 p. 133 and p. 277; Davis 1983 p. 106– ; Reading 1979 p. 4– ; Visher 1984 p. 37–).

The facies concept has been subjected to various interpretations, but in the present paper it will be used in Gressley's original sense of being a collection of physical properties that are distinguishable from those of neighbouring rocks at a chosen scale (Krumbein & Sloss 1963 p. 317). In this sense it is close to Serra's electrofacies concept (Serra 1977).

Walthers Law has potential as a predictive tool if used carefully, and may be used to form the basis of a syntactic rule set. For any given depositional system at a given scale, there will be a set of facies relationships such that a traverse in a given direction in relation to the direction of flow of the transport medium would be expected to encounter a certain sequence of descriptive facies units with a certain probability.

In stratigraphy we thus have a set of terminal symbols (facies) that, if they are capable of being unambiguously mapped within a present-day depositional system, should also be capable of being identified, using the same syntax, from borehole sequences if there were no intervening periods of erosion.

An example of a syntax derived from theoretical considerations could use the Battery Point channel fill sequence (Walker & Cant 1984, p. 83) (Fig. 1).

Examination of the 'model' sequence might lead to the design of the following context-free grammar:

<Channel.floor> : := SS | A | <Channel.floor> A
<Bar.top> : := F | E | D
<VA> : := G
<Lower.channel> : := <Channel.floor> | <Lower.channel> C | <Lower.channel> B

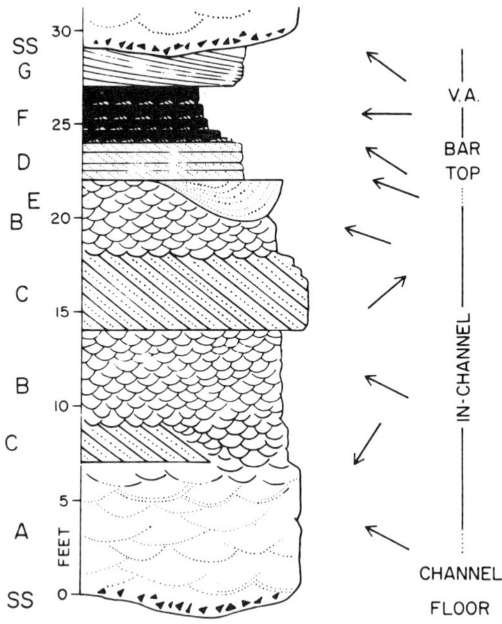

Fig. 1. Battery Point 'model' fluvial sequence (from Walker & Cant 1984 p.83).

<Upper.channel> : := <Bar.top> | <VA>
<Channel.fill> : := <Lower.channel> <Upper.channel>
<Channel.fill> : := <Channel.fill> <Upper.channel>
<Channel.sequence> : := <Channel.fill> <Channel.fill>
<Channel.sequence> : := <Channel.sequence> <Channel.fill>
<Channel.sequence> : := <Channel.sequence> <Upper.channel>

where SS = scoured surface, A = poorly defined trough cross-bedding, B = well defined trough cross-bedding, C = large planar tabular cross beds, D = small planar tabular cross beds, E = isolated scour fills, F = ripple cross laminated silts and muds, and G = low angle inclined stratification (Walker & Cant 1984 p. 82).

Formally speaking, this is an inverted BNF grammar. It is presented this way for two reasons. The first is that computationally it is more efficient, and secondly it is easier to follow the progressive interpretation of terminal to non-terminal symbols while reading. It is worth noting that grain size is included explicitly in only one terminal symbol (F). The remainder are a mixture of genetic and descriptive structural facies.

This grammar can be applied successfully to the full Battery Point section (Walker & Cant 1984, p 82), identifying the ten 'in-channel' sequences. In this case the grammar or syntax has been developed directly from the model and applied to the data. It is also possible to develop the rule set inductively from the data themselves.

Choice of terminal symbols in stratigraphy

Examination of the literature concerning choice of facies names and characteristics for quite limited environments reveals a state of terminal disarray. Even textbooks catering for undergraduate level stratigraphy and sedimentology disagree as to the basic building blocks of sequence description.

Facies terminology is often 'environment specific' which may mean that although an environmental interpretation using syntax analysis would be possible in each individual case, there is no overlap of facies units. This implies that choice of observational unit or descriptor must be made on the basis of a working hypothesis before attempting to develop or apply a syntax. The lack of a common facies is not only restrictive as far as the quantitative use of multiple working hypotheses in environmental interpretation is concerned, but it can lead to circular arguments. In a linguistic sense it is analogous to the comparison of syntax between German, English and Chinese. The basic unit of meaning ('sentence' in English) could not be investigated using syntactic analysis before the language was known. Therefore, syntactic analysis, and by implication the rule sets used in 'expert systems', using todays facies descriptors has its main function within rather than between environments. The principal environments as defined today have a similar role to that of national languages; there is little in the way of common terminology between them. Petrophysical measurement, however, is applied to vertical sections in many different environments and has the potential provide a common terminology for the testing of 'between environment' hypotheses.

Inductive development of a syntax

The formal application of syntactic analysis to sedimentary sequences faces two main problems. The first is the choice of terminal symbol, and the second is the consistent, formal, derivation of a syntax.

Choice of terminal symbol is often limited by the data available, for example grain-size and structural data from cores and sections, petrophysical (including dipmeter) data from oil wells, and impedance data or signal properties from seismic sections.

Examples will be shown using both grain-size and petrophysical symbols. The petrophysical terminal symbols will be based on the use of a numerical lithology derived from a subdivision of *M*-*N*-gamma space as discussed in Griffiths (1983). *M* and *N* are the slopes of the density-slowness, and density-neutron curves respectively (Burke *et al.* 1969); gamma is the formation natural gamma response. This is by no means the only possible type of numerical lithology, but it does satisfy some of the requirements of a repeatable petrophysical facies set that has an approximate relationship to mineralogy if not to grain size. By sub-dividing velocity-density space in a similar manner one can also produce a lithology that is porosity and mineralogy related. In the present case, *M*-*N*-gamma space is subdivided (Fig. 2) to give 125 different possible litho-states with 'hard' class edges. It would also be possible to use results from a standard log analysis package, classified on, for example, relative shale volume, matrix grain density and porosity etc. These are our sets of terminal symbols. We can then begin to derive a syntax from grain-size derived lithostates and then from petrophysical data.

Common to all facies relationships is the existence of at least two levels of oscillation. Beerbower (1964) (reported in Miall 1984, p 178) identified what he called 'autocyclic' and 'allocyclic' oscillations. Autocyclic oscillation can be considered to be due to the normal energy fluctuation along the borders of two neighbouring facies at any scale. Thus, facies

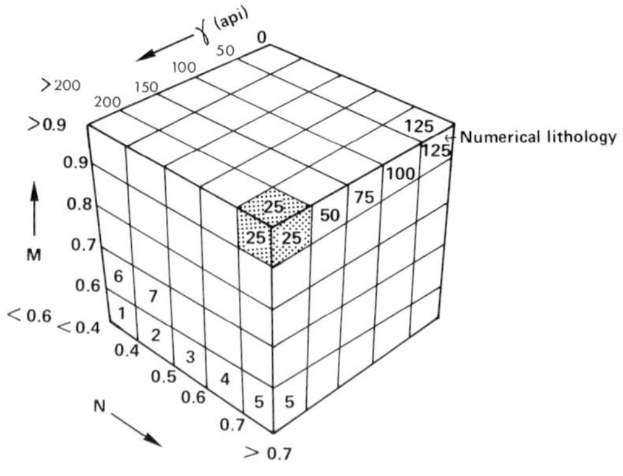

Fig. 2. Numerical lithology derivation in *M*–*N*–gamma space.

boundaries are not static but can oscillate about any point on the margin, and any vertical sequence would record these variations as a cyclicity over time. Allocyclic oscillations are controlled from outside the immediate environment, either by tectonic movement, climatic change, or eustatic sea level change.

Autocyclic oscillations may be considered as having two sub-types. The first would be gradational in nature and could be seen as an artefact of the classification process in that if one sets a boundary at an arbitrary place in a continuous sequence, such as a grain size and sorting distribution, one would anticipate neighbouring classes to occur as a function of random variation about the boundary. Examples of this type would be grain size class oscillations within a coarsening upward sequence, e.g. medium to coarse to medium to coarse. . . .etc. .

The second type would be chaotic in nature, resulting from instability in the sedimentation process. Such an oscillation would be typified by the association between basinal silt or muds and slumped sand from an unstable delta front. The association between abandoned river channels and overbank muds is also in this category. Prediction of sequences containing this type of oscillation depends on a knowledge of lateral controls of energy distribution in the environment and the probability associated with any given energy fluctuation.

Allocyclic oscillations are the most difficult to handle predictively on the scale of a few hundred metres, but are easier than autocyclic oscillations on the seismic and basin scale, and are therefore of more use in stratigraphic correlation.

From the syntactic point of view, the terminal symbols should be incorporated in a syntactic rule set in such a way as to express both main types of oscillation.

Autocyclic oscillations in a clastic depositional environment

Gradational. In a clastic sedimentary environment the energy of the transport medium varies in three dimensions so that a given flow strength oscillates about a given point in space. We should thus anticipate that the grain size transported by the energy at a given point will vary naturally with time. At the very lowest level our terminal symbols should reflect this grain size oscillation. We can examine this feature quantitatively by looking at a transition matrix where the classes are grain-size derived.

Table 1 shows the upward transition matrix (UTM) derived from the complete Battery Point fluvial sequence as published in Walker & Cant (1984, p. 82). The matrix shows autocyclic oscillations from scoured surface (SS) to well defined trough cross bedding (B), (B) to large planar tabular cross beds (C), and from (SS) to (A).

The oscillatory pairs and the degree of asymmetry are shown below:

States	Asymmetry
SS B	0 00
B C	0.14
SS A	0.31

where asymmetry is the directional bias in the oscillation (positive left to right i.e. SS to A in the example above).

The general syntax for this type of oscillation is shown below (as derived from the UTM).

1. <agl> : := x y | y x

or

2. <agl> : := x | y | <agl> <agl>

or

3. x <agl> y : := x y
   y <agl> x : := y x.

Case (2) is to be preferred in situations where stacked oscillations are possible. In cases (1) and (2) the states are *replaced* by the interpretation and are not available individually for higher order interpretations which must use the non-terminal symbol '<agl>'. In case (3), the individual states remain available.

The appropriate specific syntax derived from this is either:

1. <agl> : :=SS A
   <ag2> : :=SS B | B SS
   <ag3> : := B C | C B

Table 1. *The upward transition matrix (UTM) for the complete Battery Point fluvial sequence*

	SS	A	B	E	F	C	D	G	
SS	0	17	5	1	0	0	0	0	23
A	9	0	5	0	0	5	0	1	20
B	5	0	0	1	1	4	2	0	13
E	0	0	0	0	1	1	0	0	2
F	2	1	0	0	0	0	0	1	4
C	5	1	3	0	0	0	1	0	10
D	0	1	0	0	2	0	0	0	3
G	1	0	0	0	0	0	0	0	1

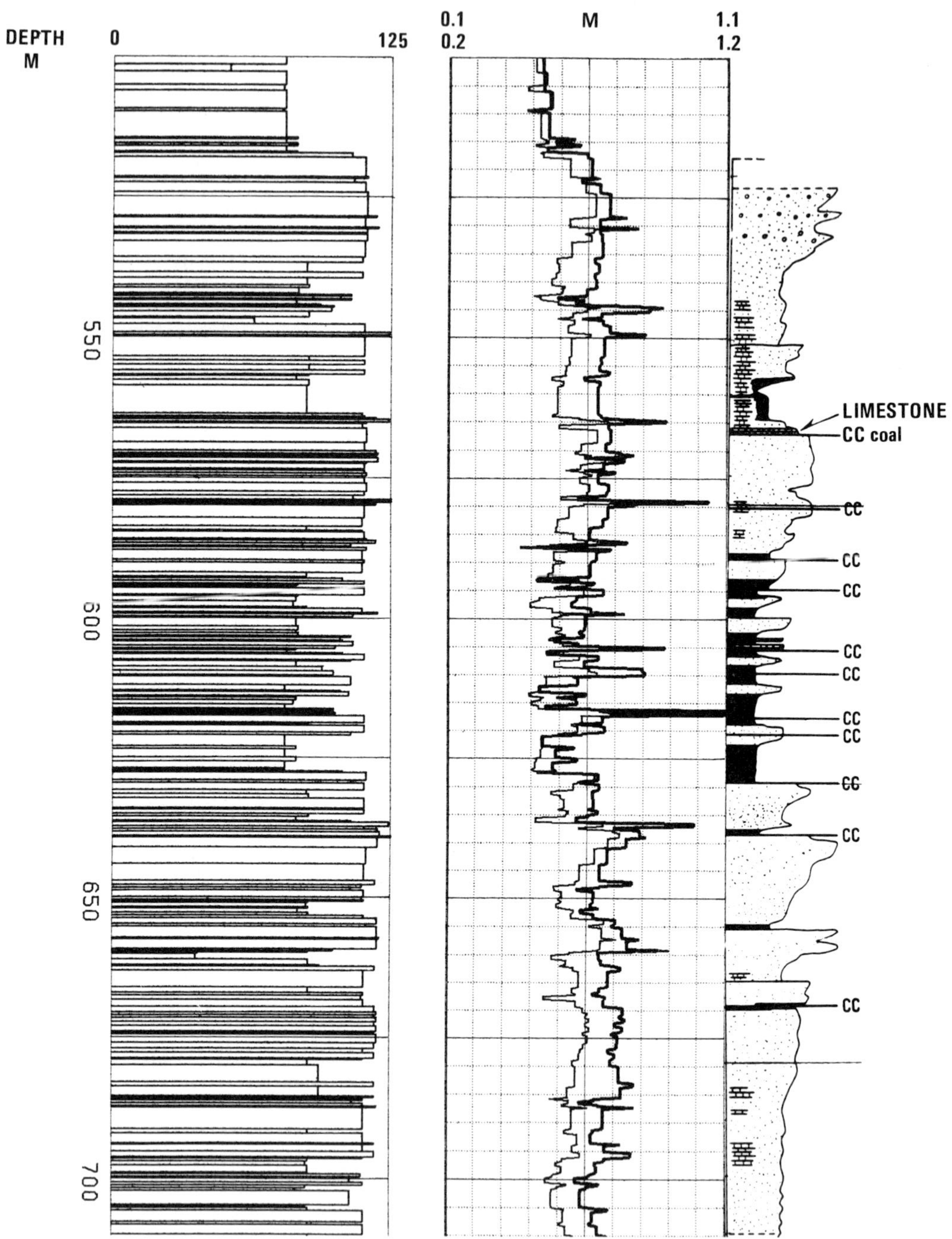

Fig. 3. Middle Jurassic deltaic sequence from a North Sea well. 'M' and 'N' values after Burke *et al.* (1969).

or

2. <ag1> : :=SS | A | <ag1> <ag1>
   <ag2> : :=<ag1> B | B <ag1>
   <ag3> : :=<ag2> C | C <ag2>

where <agx> is an autocyclic gradational unit.

The practical considerations of parsing a sequence such as 'SS B C' are such that case (2) is often more effective in deriving the most complete interpretation and allows nested oscillations.

Chaotic. Chaotic autocyclic oscillations would not necessarily be identifiable from the transition matrix alone, but in this case the 'substitutability' matrix may be of some help.

Substitutability analysis is discussed by Davis (1973, p 288) and Davis & Cocke (1972). Developed originally for the study of word substitution in written text, it identifies states that have the same relationship to other states lying either above or below.

It seems likely that the substitutability matrix can be of use in identifying chaotic autocyclic phenomena. High scores for state pairs on the downward substitutability matrix (DSM) (Table 2) identify consistent environmental oscillations that are possibly due to chaotic instability in the depositional process. In other words once a state is in existence its transition to one of two (or more) possible other states depends on the nature of an environmental trigger. Such oscillations, as discussed earlier, may be exemplified by the migration of distributary channels, or the erosion or silting up of a mouth bar complex.

Examination of the DSM reveals the following high-scoring pairs:

Unit A	Unit B	Score
SS	C	0.86
A	E	0.70
A	B	0.67
SS	G	0.67
E	D	0.63

The general syntax for the first pair is thus

<ac1> : := x SS | x C

where 'x' is any state that shows a significant transition to both SS *and* C on the UTM but is *not* involved in autocyclic gradational transitions. States 'A' and 'B' both have significant UTM transitions to both 'SS' and 'C', but only 'A C' can be considered as a valid candidate, leading to the specific syntax,

1. <ac1> : := A C

or

2. A <ac1> C: :=A C.

The non-terminal symbol <ac1> can be replaced by a genetic phrase if the cause of the relationship is known. In case (1) above, the possibility of including states 'A' and 'C' in higher order grammar has been removed, whereas in case (2) this possibility remains. Case (2) is generally preferred.

The complete autocyclic chaotic syntax would thus be:

A <ac1> C: :=A C
A <ac2> B: :=A B
A <ac3> G: :=A G
F <ac4> G: :=F G
SS <ac5> E: :=SS E
B <ac6> E: :=B E
B <ac7> D: :=B D
C <ac8> A: :=C A

where <acx> is an autocyclic chaotic boundary. An additional two pairs are of interest 'F to SS' and 'C to SS'; however, since 'SS' is the prime candidate for sequence base (see below) these transitions are upgraded to allocyclic.

Allocyclic. Lithologically, allocyclic events may be marked, for example, by the existence of a marine limestone or a volcanic ash, or a marine shale. In relation to transition matrices, such events only have significant probabilities in large data sets. They may possibly be found in the relationship between rankings on the upward and downward substitutability matrices. In vertical section, lithostates that have a lower than average score for upward substitutability (Table 3), combined with higher than average scores (row totals) on the downward substitutability matrix (Table 2) have some of the properties of sequence bases. That is to say that few other states have the same successor state, and many other states have similar underlying states.

Here, 'SS' is a clear candidate for the sequence base state and the resulting general syntax would be:

1. <sequ. base> : := x SS

or

2. x <sequ. base> SS : := x SS

where 'x' is any state that is not involved in an autocyclic gradational oscillation with 'SS' but shows significant transitions to 'SS' on the UTM. The specific syntax for this case would be:

F <sequ.base> SS: :=F SS

Table 2. *The downwards substitutability matrix (DSM) (high scoring states tend to have a common predecessor) derived from Table 1 data.*

	SS	A	B	E	F	C	D	G	
SS	1.00	0.04	0.67	0.30	0.18	0.86	0.58	0.67	4.29
A	0.04	1.00	0.67	0.70	0.05	0.00	0.03	0.04	2.52
B	0.67	0.67	1.00	0.46	0.00	0.50	0.17	0.46	3.94
E	0.30	0.70	0.46	1.00	0.29	0.44	0.63	0.00	3.82
F	0.18	0.05	0.00	0.29	1.00	0.31	0.37	0.00	2.19
C	0.86	0.00	0.50	0.44	0.31	1.00	0.55	0.55	4.21
D	0.58	0.03	0.17	0.63	0.37	0.55	1.00	0.00	3.33
G	0.67	0.04	0.46	0.00	0.00	0.55	0.00	1.00	2.71

Table 3. *The upwards substitutability matrix (USM) (high scoring states tend to have a common successor) derived from Table 1 data.*

	SS	A	B	E	F	C	D	G	
SS	1.00	0.12	0.01	0.00	0.39	0.30	0.43	0.00	2.25
A	0.12	1.00	0.83	0.31	0.68	0.87	0.00	0.78	4.58
B	0.01	0.83	1.00	0.52	0.60	0.66	0.13	0.73	4.46
E	0.00	0.31	0.52	1.00	0.00	0.00	0.63	0.00	2.46
F	0.39	0.68	0.60	0.00	1.00	0.75	0.18	0.82	4.41
C	0.30	0.87	0.66	0.00	0.75	1.00	0.07	0.83	4.48
D	0.43	0.00	0.13	0.63	0.18	0.07	1.00	0.00	2.45
G	0.00	0.78	0.73	0.00	0.82	0.83	0.00	1.00	4.16

C <sequ.base> SS: :=C SS
G <sequ.base> SS: :=G SS

Note that at this stage the grammar contains only 'gap' and 'cycle' information. The higher level grammar for sedimentological identification of intra-gap sequences must be supplied from theoretical considerations.

Examples

Brent Group, Northern North Sea

As a demonstration of some of these concepts a syntactic analysis of a Middle Jurassic Brent Group section is considered. This group of sands, shales, coals, occasional limestones and calcareous sands has been interpreted (Roberts *et al.* 1987, p. 332) as a deltaic sequence. It forms a productive reservoir unit in many parts of the North Sea and consists of several delta lobes with interspersed organic rich embayments and marine incursions. A section showing both core-logged lithologies and numerical lithologies is shown in Figure 3. The data sequence as logged from the core is shown in Table 4a.

A diagrammatic representation of the grain-unit relationships (Fig. 4) indicates the oscillatory units. For example 'mud' and 'cc' form a loop which can be broken by transition from 'cc' to the medium sand ('m'), or from 'mud' to 'vf' or 'f', but not to 's' or 'c'. Similarly the 'vc', 'c', and 'm' oscillations can be broken by transition from 'm' to 'f' or 'cc' but nothing else. This graph encapsulates the data from the UTM and substitutability matrices.

By examining the appropriate transition and substitution matrices (Tables 5, 6 & 7) and the graph, an appropriate syntax can be developed. The first step in developing the syntax is to look at possible sequence base candidates from a combination of the upward and downward substitutability matrices as explained above. Candidates for sequence bases are below ranked in order of likelihood (the lower the score the more likely):

Unit	Score
m	6
mud	6
c	6
s	7
vc	11
lst	12
cc	13
vf	13
f	16

Table 4a. *Well 35/8–1. Grain size sequence from the Brent Group.*

(base) s vf f m cc m f vc c vc mud vf f m c vc m cc s f m cc mud f
cc f mud cc mud vf mud cc mud vf mud cc 1st mud vf s f s vf s vf
cc vf f cc vf f c cc c f c m c m cc m 1st f s f m f vf f vf f vf
f m c vc c vc c vc (top)

Table 4b. *Well 35/8–1. Numerical Lithology sequence from the Brent Group*

(base) 82 107 112 87 113 108 113 112 87 88 113 112 107 113 112 107
112 113 118 113 108 113 112 118 93 118 93 88 113 118 119 118 119
118 119 118 113 114 112 113 118 113 88 37 49 99 119 113 108 113
108 113 118 113 114 119 120 119 113 124 107 82 87 113 112 113 114
107 82 107 113 108 113 114 107 106 107 106 107 106 107 82 107 82
99 97 99 100 82 107 113 108 107 108 107 82 107 118 112 106 107
108 113 88 113 112 87 112 114 112 118 113 112 113 108 113 114 113
119 118 113 114 112 114 124 118 113 112 87 112 113 112 113 112
113 123 113 88 113 124 118 107 113 108 113 112 113 114 119 113
118 119 114 113 114 (top)

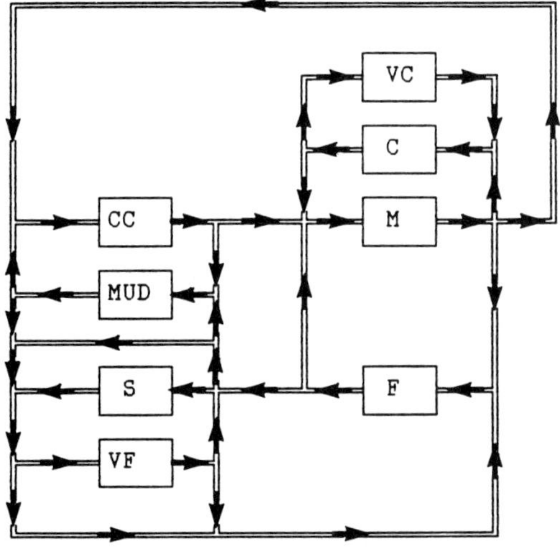

Fig. 4. Diagram of grain-size based facies relationships. Middle Jurassic deltaic sequence.

from which 'm', 'mud', 'c', and 's' appear equally likely. However, 's' is involved only in autocyclic gradational oscillations and can be dropped, leaving (by examination of the UTM):

1st <seq.bound> mud::=1st mud
vc <seq.bound> mud::=vc mud
f <seq.bound> c::=f c
vc <seq.bound> m::=vc m.

Next, we look for autocyclic gradational oscillations by examination of the UTM. Ranked candidates for autocyclic gradational units (are):

Unit A	Unit B	Asymmetry
vf	mud	−0.33
vc	c	−0.25
cc	mud	0
s	vf	0.2
m	c	0.2
s	f	0.2
m	cc	0.33
vf	f	0.4
f	m	0.43

giving a syntax as follows:

<cs>::=vc | c | <cs> <cs>
<org>::=cc | mud | <org> <org>
<fines>::=m | f | vf | s | <fines> <fines>.

Next are the autocyclic chaotic relationships. Ranked candidates for units involved in autocyclic chaotic oscillations are as follows (high scores indicate increased likelihood):

Unit A	Unit B	Score
s	mud	0.75
s	m	0.69
cc	c	0.60

Table 5. *Upward transition matrix for a Brent Group section, Middle Jurassic, North Sea*

	mud	s	vf	f	m	c	vc	cc	1st	row total
mud	0	0	4	1	0	0	0	3	0	8
silt	0	0	3	3	0	0	0	0	0	6
v.fine	2	2	0	7	0	0	0	1	0	12
fine	1	2	3	0	5	2	1	2	0	16
medium	0	0	0	2	0	3	0	4	1	10
coarse	0	0	0	1	2	0	5	1	0	9
v.coarse	1	0	0	0	1	3	0	0	0	5
coal	3	1	2	1	2	1	0	0	1	11
limest.	1	0	0	1	0	0	0	0	0	2

Table 6. *The downwards substitutability matrix (DSM) (high scoring states tend to have a common predecessor) derived from Table 5 data*

	s	vf	f	m	cc	vc	c	mud	1st	row total
s	1.00	0.43	0.62	0.69	0.36	0.13	0.35	0.75	0.24	4.56
vf	0.43	1.00	0.30	0.53	0.52	0.10	0.27	0.36	0.23	3.75
f	0.62	0.30	1.00	0.08	0.42	0.12	0.18	0.55	0.26	3.53
m	0.69	0.53	0.08	1.00	0.37	0.50	0.54	0.51	0.24	4.47
cc	0.36	0.52	0.42	0.37	1.00	0.25	0.60	0.18	0.51	4.21
vc	0.13	0.10	0.12	0.50	0.25	1.00	0.08	0.05	0.00	2.23
c	0.35	0.27	0.18	0.54	0.60	0.08	1.00	0.42	0.59	4.02
mud	0.75	0.36	0.55	0.51	0.18	0.05	0.42	1.00	0.53	4.36
1st	0.24	0.23	0.26	0.24	0.51	0.00	0.59	0.53	1.00	3.60

Table 7. *The upwards substitutability matrix (USM) (high scoring states tend to have a common successor) derived from table 5 data*

	s	vf	f	m	cc	vc	c	mud	1st	row total
s	1.00	0.65	0.31	0.26	0.46	0.00	0.13	0.69	0.50	4.00
vf	0.65	1.00	0.15	0.43	0.43	0.08	0.19	0.26	0.84	4.02
f	0.31	0.15	1.00	0.37	0.72	0.52	0.44	0.51	0.10	4.13
m	0.26	0.43	0.37	1.00	0.24	0.50	0.20	0.50	0.26	3.75
cc	0.46	0.43	0.72	0.24	1.00	0.53	0.20	0.39	0.62	4.58
vc	0.00	0.08	0.52	0.50	0.53	1.00	0.11	0.00	0.21	2.94
c	0.13	0.19	0.44	0.20	0.20	0.11	1.00	0.14	0.13	2.53
mud	0.69	0.26	0.51	0.50	0.39	0.00	0.14	1.00	0.14	3.63
1st	0.50	0.84	0.10	0.26	0.62	0.21	0.13	0.14	1.00	3.79

which, when combined with the autocyclic gradational results, can be summarized as:

<ac1> ::= <fines> <cs>
<ac2> ::= <org> <fines>.

Genetic names can be given to these relationships based on additional information, as follows:

<C.up>::= <fines> <cs> | <C.up> <C.up>
<Org.seq>::= <org> <fines> | <fines> <org>
<Org.seq>::= <Org.seq> <Org> | <Org.seq> <Org.seq>

and also to higher order rules:

<Delta.front.lobe>::= <cs> | <C.up> <Delta.front.lobe>
<Delta.front.lobe>::= <fines> <Delta.front.lobe>
<Embayment>::= <Org.seq> <fines>

and

<calc>::= 1st | <calc> <calc>

<Marine.incursion>::= <calc>.

Using the syntactic rule set developed above led to an interpretation as below:

(base)
<Embayment> <Delta.front.lobe>
<Delta.front.lobe>
<seq.bound> <Embayment>
<Delta.front.lobe>
<seq.bound> <Embayment> <Embayment>
<Embayment> <Org.seq>
<Marine.incursion>
<seq.bound> <Embayment> <Embayment>
<Embayment> <fines>
<seq.bound> <Delta.front.lobe>
<org> <Delta.front.lobe>
<fines>
<seq.bound> <Delta.front.lobe> <C.up>
<Embayment>
<Marine.incursion> <Delta.front.lobe>
<Delta.front.lobe> <Delta.front.lobe>
<Delta.front.lobe> <Delta.front.lobe>
(top)

Petrophysical lithostates

If we then take the same well, but look at the petrophysically derived terminal symbols only, then the matrices shown in Tables 8 to 10 are generated. The graphical representation of this structure is shown in Figure 5.

The same process that was used to derive a syntax from the grain-size symbols can be used to derive a syntax from the petrophysical data as shown in Table 11. The numerical lithology sequence for the same interval as the cored section above is shown in Table 4b.

The interpretation of the data based on the petrophysically derived syntactic rule set (shown in Table 11) is shown below:

(base)
<C.up> <c>
<seq.bound> <c> <C.up> <c> <ag1>
<seq.bound> <c> <C.up> <c> <C.up>
<Delta.front.lobe>
<Delta.front.lobe> <Delta.front.lobe>
<Delta.front.lobe>
<Embayment> <ag2> <c> <C.up>
<C.up> <Delta.front.lobe>
<ag5> <ag2> <Embayment> <C.up>
<seq.bound> <Delta.front.lobe> <ag1>
<seq.bound> <Delta.front.lobe>
<Embayment> <ag1>
<seq.bound> <c> <C.up> <c > <C.up>
<Delta.front.lobe>
<Marine.incursion> <Delta.front.lobe>
<Delta.front.lobe>
<Delta.front.lobe> <Delta.front.lobe>
<ag5> <Embayment>
<Embayment> <c> <ag1>
<seq.bound> <c> <C.up>

Table 8. *Upward transition matrix for the Brent Group section, Well 35/8−1. Petrophysically derived lithostates*

	82	107	112	87	113	108	88	118	93	119	114	37	49	99	120	124	106	97	100	123	row total
82	0	5	0	1	0	0	0	0	0	0	0	0	0	1	0	0	0	0	0	0	7
107	5	0	2	0	4	2	0	1	0	0	0	0	0	0	0	0	3	0	0	0	17
112	0	2	0	4	8	0	0	2	0	0	2	0	0	0	0	0	1	0	0	0	19
87	0	0	2	0	2	0	1	0	0	0	0	0	0	0	0	0	0	0	0	0	5
113	0	0	11	0	0	8	3	5	0	1	8	0	0	0	0	2	0	0	0	1	39
108	0	2	0	0	8	0	0	0	0	0	0	0	0	0	0	0	0	0	0	0	10
88	0	0	0	0	4	0	0	0	0	0	0	1	0	0	0	0	0	0	0	0	5
118	0	1	1	0	7	0	0	0	2	4	0	0	0	0	0	0	0	0	0	0	15
93	0	0	0	0	0	0	1	1	0	0	0	0	0	0	0	0	0	0	0	0	2
119	0	0	0	0	3	0	0	4	0	0	1	0	0	0	1	0	0	0	0	0	9
114	0	2	3	0	2	0	0	0	0	2	0	0	0	0	0	1	0	0	0	0	10
37	0	0	0	0	0	0	0	0	0	0	0	0	1	0	0	0	0	0	0	0	1
49	0	0	0	0	0	0	0	0	0	0	0	0	0	1	0	0	0	0	0	0	1
99	0	0	0	0	0	0	0	0	0	1	0	0	0	0	0	0	0	1	1	0	3
120	0	0	0	0	0	0	0	0	0	1	0	0	0	0	0	0	0	0	0	0	1
124	0	1	0	0	0	0	0	2	0	0	0	0	0	0	0	0	0	0	0	0	3
106	0	4	0	0	0	0	0	0	0	0	0	0	0	0	0	0	0	0	0	0	4
97	0	0	0	0	0	0	0	0	0	0	0	0	0	1	0	0	0	0	0	0	1
100	1	0	0	0	0	0	0	0	0	0	0	0	0	0	0	0	0	0	0	0	1
123	0	0	0	0	1	0	0	0	0	0	0	0	0	0	0	0	0	0	0	0	1

Table 9. *Upwards substitutability matrix derived from Table 8 data.*

	82	107	112	87	113	108	88	118	93	119	114	37	49	99	120	124	106	97	100	123	
82	1.00	0.00	0.28	0.00	0.00	0.23	0.00	0.11	0.00	0.00	0.41	0.00	0.19	0.00	0.00	0.43	0.96	0.19	0.00	0.00	3.81
107	0.00	1.00	0.50	0.52	0.33	0.51	0.51	0.46	0.09	0.40	0.39	0.00	0.00	0.00	0.00	0.12	0.00	0.00	0.65	0.52	5.99
112	0.28	0.50	1.00	0.55	0.16	0.86	0.80	0.71	0.15	0.68	0.44	0.00	0.00	0.00	0.00	0.28	0.21	0.00	0.00	0.83	7.45
87	0.00	0.52	0.55	1.00	0.49	0.65	0.65	0.63	0.24	0.38	0.71	0.00	0.00	0.00	0.00	0.00	0.00	0.00	0.00	0.67	6.49
113	0.00	0.33	0.16	0.49	1.00	0.00	0.00	0.10	0.33	0.32	0.46	0.00	0.00	0.03	0.06	0.26	0.00	0.00	0.00	0.00	3.55
108	0.23	0.51	0.86	0.65	0.00	1.00	0.94	0.83	0.00	0.56	0.52	0.00	0.00	0.00	0.00	0.11	0.24	0.00	0.00	0.97	7.41
88	0.00	0.51	0.80	0.65	0.00	0.94	1.00	0.81	0.00	0.56	0.41	0.00	0.00	0.00	0.00	0.00	0.00	0.00	0.00	0.97	6.65
118	0.11	0.46	0.71	0.63	0.10	0.83	0.81	1.00	0.00	0.48	0.68	0.00	0.00	0.27	0.47	0.05	0.12	0.00	0.00	0.83	7.58
93	0.00	0.09	0.15	0.24	0.33	0.00	0.00	0.00	1.00	0.54	0.00	0.00	0.00	0.00	0.00	0.63	0.00	0.00	0.00	0.00	2.98
119	0.00	0.40	0.68	0.38	0.32	0.56	0.56	0.48	0.54	1.00	0.25	0.00	0.00	0.00	0.00	0.69	0.00	0.00	0.00	0.58	6.44
114	0.41	0.39	0.44	0.71	0.46	0.52	0.41	0.68	0.00	0.25	1.00	0.00	0.00	0.25	0.43	0.19	0.43	0.00	0.00	0.43	6.99
37	0.00	0.00	0.00	0.00	0.00	0.00	0.00	0.00	0.00	0.00	0.00	1.00	0.00	0.00	0.00	0.00	0.00	0.00	0.00	0.00	1.00
49	0.19	0.00	0.00	0.00	0.00	0.00	0.00	0.00	0.00	0.00	0.00	0.00	1.00	0.00	0.00	0.00	0.00	1.00	0.00	0.00	2.19
99	0.00	0.00	0.00	0.00	0.03	0.00	0.00	0.27	0.00	0.00	0.25	0.00	0.00	1.00	0.58	0.00	0.00	0.00	0.00	0.00	2.13
120	0.00	0.00	0.00	0.00	0.06	0.00	0.00	0.47	0.00	0.00	0.43	0.00	0.00	0.58	1.00	0.00	0.00	0.00	0.00	0.00	2.54
124	0.43	0.12	0.28	0.00	0.26	0.11	0.00	0.05	0.63	0.69	0.19	0.00	0.00	0.00	0.00	1.00	0.45	0.00	0.00	0.00	4.21
106	0.96	0.00	0.21	0.00	0.00	0.24	0.00	0.12	0.00	0.00	0.43	0.00	0.00	0.00	0.00	0.45	1.00	0.00	0.00	0.00	3.40
97	0.19	0.00	0.00	0.00	0.00	0.00	0.00	0.00	0.00	0.00	0.00	0.00	1.00	0.00	0.00	0.00	0.00	1.00	0.00	0.00	2.19
100	0.00	0.65	0.00	0.00	0.00	0.00	0.00	0.00	0.00	0.00	0.00	0.00	0.00	0.00	0.00	0.00	0.00	0.00	1.00	0.00	1.65
123	0.00	0.52	0.83	0.67	0.00	0.97	0.97	0.83	0.00	0.58	0.43	0.00	0.00	0.00	0.00	0.00	0.00	0.00	0.00	1.00	6.79

Table 10. *Downwards substitutability matrix derived from Table 8 data.*

	82	107	112	87	113	108	88	118	93	119	114	37	49	99	120	124	106	97	100	123	
82	1.00	0.00	0.17	0.00	0.26	0.24	0.00	0.14	0.00	0.00	0.00	0.00	0.00	0.00	0.00	0.00	0.93	0.00	0.00	0.00	2.73
107	0.00	1.00	0.08	0.43	0.38	0.00	0.00	0.11	0.13	0.22	0.06	0.00	0.00	0.39	0.00	0.12	0.09	0.00	0.00	0.00	3.02
112	0.17	0.08	1.00	0.00	0.14	0.95	0.90	0.68	0.08	0.37	0.90	0.00	0.00	0.00	0.00	0.95	0.16	0.00	0.00	0.93	7.30
87	0.00	0.43	0.00	1.00	0.52	0.00	0.00	0.27	0.00	0.00	0.23	0.00	0.00	0.14	0.00	0.00	0.31	0.00	0.00	0.00	2.89
113	0.26	0.38	0.14	0.52	1.00	0.06	0.04	0.30	0.46	0.44	0.15	0.27	0.00	0.00	0.20	0.06	0.42	0.00	0.00	0.00	4.71
108	0.24	0.00	0.95	0.00	0.06	1.00	0.88	0.71	0.00	0.20	0.93	0.00	0.00	0.00	0.00	0.87	0.23	0.00	0.00	0.97	7.04
88	0.00	0.00	0.90	0.00	0.04	0.88	1.00	0.68	0.00	0.19	0.87	0.00	0.00	0.00	0.00	0.81	0.00	0.00	0.00	0.90	6.25
118	0.14	0.11	0.68	0.27	0.30	0.71	0.68	1.00	0.00	0.15	0.81	0.00	0.00	0.00	0.56	0.63	0.22	0.00	0.00	0.70	6.95
93	0.00	0.13	0.08	0.00	0.46	0.00	0.00	0.00	1.00	0.83	0.00	0.00	0.00	0.00	0.00	0.00	0.00	0.00	0.00	0.00	2.52
119	0.00	0.22	0.37	0.00	0.44	0.20	0.19	0.15	0.83	1.00	0.20	0.00	0.00	0.00	0.00	0.37	0.00	0.21	0.21	0.21	4.61
114	0.00	0.06	0.90	0.23	0.15	0.93	0.87	0.81	0.00	0.20	1.00	0.00	0.00	0.00	0.12	0.86	0.08	0.00	0.00	0.96	7.19
37	0.00	0.00	0.00	0.00	0.27	0.00	0.00	0.00	0.00	0.00	0.00	1.00	0.00	0.00	0.00	0.00	0.00	0.00	0.00	0.00	1.27
49	0.00	0.00	0.00	0.00	0.00	0.00	0.00	0.00	0.00	0.00	0.00	0.00	1.00	0.00	0.00	0.00	0.00	0.00	0.00	0.00	1.00
99	0.00	0.39	0.00	0.14	0.00	0.00	0.00	0.00	0.00	0.00	0.00	0.00	0.00	1.00	0.00	0.00	0.00	0.00	0.00	0.00	1.53
120	0.00	0.00	0.00	0.00	0.20	0.00	0.00	0.56	0.00	0.00	0.12	0.00	0.00	0.00	1.00	0.00	0.00	0.00	0.00	0.00	1.88
124	0.00	0.12	0.95	0.00	0.06	0.87	0.81	0.63	0.00	0.37	0.86	0.00	0.00	0.00	0.00	1.00	0.00	0.00	0.00	0.89	6.56
106	0.93	0.09	0.16	0.31	0.42	0.23	0.00	0.22	0.00	0.00	0.08	0.00	0.00	0.00	0.00	0.00	1.00	0.00	0.00	0.00	3.43
97	0.00	0.00	0.00	0.00	0.00	0.00	0.00	0.00	0.00	0.21	0.00	0.00	0.00	0.00	0.00	0.00	0.00	1.00	1.00	0.00	2.21
100	0.00	0.00	0.00	0.00	0.00	0.00	0.00	0.00	0.00	0.21	0.00	0.00	0.00	0.00	0.00	0.00	0.00	1.00	1.00	0.00	2.21
123	0.00	0.00	0.93	0.00	0.00	0.97	0.90	0.70	0.00	0.21	0.96	0.00	0.00	0.00	0.00	0.89	0.00	0.00	0.00	1.00	6.57

Table 11. *BNF syntax for well 35/8–1 Brent Group wireline log data*

87 <seq.bound> 106::=87 106
87 <seq.bound> 113::=87 113
107 <seq.bound> 113::=107 113

<ac97>::=97 | 100
<ac124>::=<ag4> <ag5> | >ag2> <ag5>
<ag1>::=107 | 106 | 82 | 108
<ag1>::=<ag1> <ag1>
<ag2>::=114 | 119
<ag2>::=<ag2> <ag2>
<ag3>::=118 | 93
<ag3>::=<ag3> <ag3>
<ag4>::=113 | 112 | 87 | 88 | 37
<ag4≤;:=<ag4> <ag4>
<ag5>::=124 | 123 | 49 | 120 | 99 | <ac97>
<ag5>::=<ag5> <ag5>
<c>::=<ag3> | <ag4> | <c> <c>
<C.up>::=<ag1> <c>

<Delta.front.lobe>::=<ag1> <Delta.front.lobe> | <c> <ag2>
<Delta.front.lobe> <ag2>
<Delta.front.lobe>::=<C.up> <ag2> | <c> <Delta.front.lobe>
<Embayment>::=<ag1> <ag5> | <Embayment> <ag5>
<ac124> | <c> <ag5>

<Delta.front.lobe> <Marine.incursion>::=<Delta.front.lobe> <c>
<ag1>

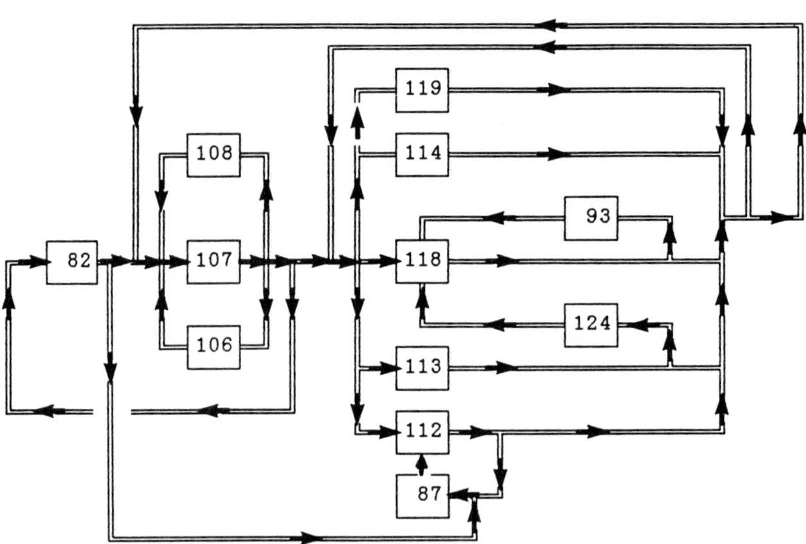

Fig. 5. Diagram of numerical lithology based facies relationships. Middle Jurassic deltaic sequence.

<Delta.front.lobe> <Delta.front.lobe>
<Delta.front.lobe> (top).

This interpretation uses a syntactic rule set derived from the transition and substitution matrices. The higher order (genetic) interpretations of 'Delta.front.lobe', 'Embayment', and 'Marine.incursion' are supplied by analogy from the core interpretation. However, the basic separation into autocyclic gradational, autocyclic chaotic and allocyclic structures could be performed by a machine using either grain-size or petrophysical data using the systematic approach described here.

Correlation between wells

Data

An additional study was carried out to examine the possibility of using the petrophysically-based syntax from one well for which an interpreted section existed, to interpret environmental sequences and sequence bases in other wells in which the same lithostratigraphic group had been identified. If successful this would offer the possibility of a stratigraphic correlation between wells using sequence boundaries derived from wireline data alone.

The first problem was that, in addition to the numerical lithology set that existed in well 35/8−1, the three other wells (33/9−4 and 33/9−5 from Statfjord, and 33/9−6 from Murchison) contained eight lithostates which were not among the twenty recorded from the 35/8−1 well. These additional lithostates were added to the grammar from 35/8−1 using relationships derived from the combined data matrices (sequence bases) and neighbour-connectedness in *M−N*-gamma space. The modified syntax was then applied to each of the four wells in turn. In the examples below, the following abbreviations are used:

R – top Rannoch (from published sources)
E – top Etive ::
N – top Ness ::
T – top Tarbert ::

Sx – sequence boundary (from petrophysical data alone)
EMB – embayment
MI – marine incursion
DFL – delta front lobe
C.up – coarsening upward sequence
c – calcareous lithostate

Results

Interpretation for well 33/9−4 (2725 to 2550 m).

Sequence number
S1 (base) <ag1>
S2 <seq.bound> <C.up> <C.up> *R* <c>
S3 <seq.bound> <C.up> <c>
S4 <seq.bound> <C.up> <C.up> <DFL> <MI> <c> *E*
S5 <seq.bound> <C.up> <C.up> <ag1>
S6 <seq.bound> <ag1> *N*
S7 <seq.bound> <ag1>
S8 <seq.bound> <ag1>
S9 <seq.bound> <C.up> *T*
S10 <seq.bound> <C.up> <C.up> <c>

Interpretation for well 35/8−1 Brent Group section (3680 to 3516 m).

Sequence number
S1 (base) <C.up> <c> <C.up> <c> <C.up> <c> <C.up> <c> <C.up> <c>
S2 <seq.bound> <c>
S3 <seq.bound> <DFL> <DFL> <DFL> <DFL> <c>
S4 <seq.bound> <EMB> <ag2> <c> <C.up> *E* <C.up> <DFL> <ag5> <ag260 <c>
S5 <seq.bound> <ag5> *N* <C.up> <DFL> <C.up> <DFL> <EMB> <C.up> <C.up> <c> <C.up> <DFL> <MI> <DFL> <DFL> <DFL> <DFL>
S6 <seq.bound> <ag5> <EMB> <c>
S7 <seq.bound> <ag5> <c> <C.up> >C.up> <DFL> <DFL> <DFL> (top)

Interpretation for well 33/9−5 Brent Group section (2781 to 2605 m).

Sequence number
S1 (base) <C.up> <C.up> <c> <ag1>
S2 <seq.bound> <ag5> <c> <ag1> *R*
S3 <seq.bound> <ag5> <DFL> <C.up> <ag5> *E* <c> <C.up> <ag1> <ag2> <ag1>
S4 <seq.bound> <ag1> <ag2> <C.up> <ag1> <ag2> <ag1> <c> *N* <C.up> <c> <C.up> <C.up> <C.up>
S5 <seq.bound> <C.up> <ag1> *T*

Interpretation for well 33/9−6 Brent Group (3140 to 2990 m).

Sequence number
S1 <EMB> <EMB> *R* <DFL> <DFL >DFL <MI> *E* <c> <C.up> *N* <c> <ag1> *T*

Interpretation summary. The inter-well correlation using these sequence numbers is shown in Fig. 6. Using the syntax from well 35/8−1 the system has managed to identify similar sequence patterns in three out of the four wells. The well from the Murchison field (33/9−6) has, however, a different sequence structure. The sequence base criteria does not function satisfactorily in this case although the environmental interpretations agree with the results from the other wells. The reasons for this are that the numerical lithostate relationships identified as being possible sequence bases in the whole data set did not occur in this well. This could point to log quality problems or systematic differences between the identified formations in Statfjord and Murchison areas.

The correlation is based on the nature of the interpreted units between the sequence boundaries. The formal lithostratigraphic formation, names are also shown for comparison. Occasionally, sequence boundaries, identified from the logs, and formation boundaries coincide as can be seen in the well interpretation, but usually there is a displacement of a few units.

The environmental interpretations are consistent, with sequences S3 and S4 showing the delta front lobe and coarsening upward units in the Etive, while the increased fines of the '<agl > <ag5>' oscillations are associated with the Tarbert and Rannoch. The Ness shows the oscillations between '<agl>' fines and coarsening upward units as would be expected.

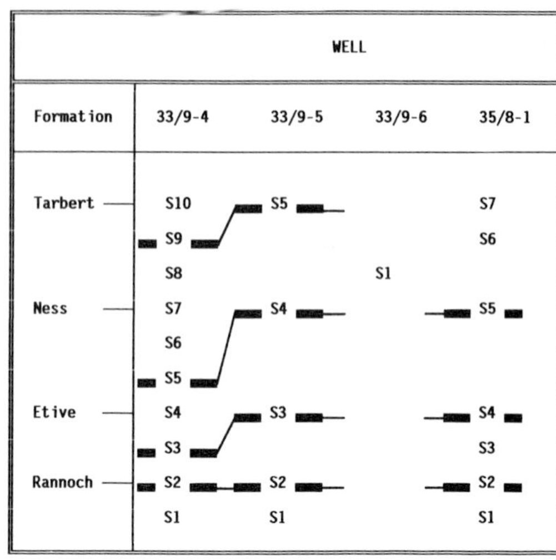

Fig. 6. Correlation diagram for the Brent sections of four wells.

Discussion

There is a wealth of literature on the subject of environmental interpretation and prediction from sub-surface data. Brief reviews by Cant (1984), and Serra (1985) point to many of the problems. Ambiguity exists in the relationship between a Gressly-type facies (or electrofacies), and an environment of deposition which may mean that it is necessary to study each facies as a part of a sequence rather than alone.

The use of syntactic pattern recognition techniques does not place any *a priori* constraints on the use of either descriptive- or genetically-defined sedimentary facies. A suitable analogy may be found in linguistics where, as stated earlier, syntactic rules describe the relationships between words which are themselves composed of letters arranged according to the rules of spelling. If descriptive, Gressly-type, facies are considered to be the letters from which the words of genetic facies are composed, then it may be that genetic facies can be used as terminal symbols in a syntax. This does not imply, however, that it is possible to avoid the quantitative definition of a descriptive facies as the basic building block of a rigorous interpretation system (i.e. to pursue the linguistic analogy, we must agree on the spelling of the words).

Griffiths (1988) has discussed the relative contribution that may be made by stochastic and syntactic analysis of sedimentary sequences. Fu (1982, p 3) pointed out that 'the syntactic approach to pattern recognition provides a capability for describing a large set of complex patterns by using small sets of simple pattern primitives and of grammatical rules (*sic*).' He also emphasises the fact that a syntactic approach allows elegant formulation of recursive structures. Stratigraphic data seems to be ideally suited to such a technique, being built up as it is of hierarchical sets of relatively straightforward observations.

This paper has hopefully emphasized the point that it is important to differentiate between the use of petrophysics as support data to traditional methods of interpretation, and the use of petrophysical data in a purely inductive way. Used inductively, petrophysics can make a positive contribution to interpretation and prediction.

The correlation exercise has shown that it may be possible to use a combination of wireline log data and syntactic pattern recognition to identify similarities in sequences in different wells. It should be noted that although the highest order genetic interpretation was provided by analogy with an interpreted core section, this need not be the case.

A failure to parse a given section may be due to many reasons. The facies may not be in a 'legal' sequence due to erosion or faulting. The syntax may be inadequate due to incomplete theoretical understanding or insufficient data for the induction process. Whatever the cause may be, the results should force the stratigrapher to reexamine the assumptions used in the analysis.

Conclusions

The results shown here have demonstrated:

1. A geological syntax can be developed on the basis of both core description and petrophysical states.
2. Sequences of litho-states can be parsed using such a syntax to give an environmental interpretation from both hand specimen and petrophysical lithostates. This provides one way of integrating core and log information.
3. A petrophysically based lithology used in conjunction with syntactic pattern recognition can provide a genetic component that is otherwise lacking from such measurements.
4. Resolving the ambiguity in the interpretation of petrophysically derived facies may need a semantic element in addition to the context-free grammar illustrated here.
5. This work indicates that quantitative tests of Walthers Law using petrophysical data are possible using various properties of the transition probability matrix.
6. If syntax analysis is to have a role in the interpretation of depositional environments, further quantitative work on a wide spectrum of environments is needed using both traditional and petrophysical data.
7. The use of syntax analysis with wireline log data may contribute to the stratigraphic correlation rather than matching of sub-surface data by providing a means of comparing interpretations rather than measurements.

This work was supported by the Norwegian Continental Shelf and Petroleum Technology Institute, and the Nordic Council.

References

BEERBOWER, J. R. 1964. Cyclothems and cyclic depositional mechanisms in alluvial plain sedimentation. *Bulletin of the Kansas Geological Survey*, **169**, 31–42.

BURKE, J. A., CAMPBELL, R. L. Jr. & SCHMIDT, A. W. 1969. The litho-porosity crossplot. *The Log Analyst* Nov-Dec.

CANT, D. J. 1984. Subsurface facies analysis: *In*: WALKER R. G. (ed.), *Facies Models*, 2nd. Ed. Geological Association of Canada., 297–310.

—— & WALKER, R. G. 1976. Development of a braided-fluvial facies model for the Devonian Battery Point sandstone, Quebec. *Canadian Journal of Earth Sciences*, **13**, 102–119.

CHOMSKY, N. 1957. *Syntactical Structures*. Mouton, The Hague.

DAVIS, J. C. 1973. *Statistics and data analysis in geology*. John Wiley, New York.

—— & COCKE, J. M. 1972. Interpretation of complex lithological successions by substitutability analysis. *In:* MERRIAM, D. F. (ed) *Mathematical models of sedimentary processes*. Plenum Press, New York 27–52.

DAVIS, R. A. 1983. *Depositional systems: a genetic approach to sedimentary geology*. Prentice-Hall, New Jersey.

FU, K. S. 1982. *Syntactic Pattern Recognition and Applications*. Prentice-Hall, New Jersey.

GAGLIO, S., MORASSO, P. & TAGLIASCO, V. 1983. Syntactic Techniques in scene analysis: *In:* ALEKSANDER, I. (ed.), *Artificial vision for robots*. Crane Russak and Co., New York.

GONZALES, R. C. & WINTZ, P. 1983. *Digital Image Processing*. Addison-Wesley Publishing Co., Reading, Massachusetts.

GRIFFITHS, C. M. 1983. An M-N-Gamma crossplot as the basis of a numerical lithostratigraphy with geotechnical implications., *Proceedings of the International Symposium on Recent Advances in Quantitative Stratigraphic Correlation*, Kharagpur, Dec. 12–13, 1983.

—— 1988. The nature of the geological representation language and consequent constraints on machine interpretation — Preliminary discussion and suggestions. *In*: SIMAAN, M. & AMINZADEH, F. (eds) *Advances in Geophysical Data Processing* Vol. 3, Artificial intelligence and Expert Systems in Petroleum Exploration. JAI Press Inc., Connecticut.

KRUMBEIN, W. C. & SLOSS, L. L. 1963. *Stratigraphy and sedimentation*. Freeman, San Francisco.

LEDLEY, R. S., 1964. High-speed automatic analysis of biomedical pictures. *Science*, **146**, 216–223.

—— *et al.*, 1965. FIDAC: Film input to digital automatic computer and associated syntax-directed pattern recognition programming system: *In*: J. TIPPET *et al.* (eds) *Optical and electro-optical information processing systems* MIT Press, Cambridge, Mass.

MIALL, A. D. (ed.), 1978. Fluvial sedimentology, *Canadian Society of Petroleum Geologists Mem.* **5**

—— 1984. *Principles of sedimentary basin analysis.* Springer, New York.

MILLER, G. A. 1956. The magical number seven, plus or minus two — some limits on our capacity for processing information, *Psychological Review* **63**, 81–97.

READING, H. G. (ed.), 1978. *Sedimentary environments and facies*. Blackwell, Oxford

ROBERTS, J. D., MATHIESON, A. S. & HAMPSON, J. M. 1987. Statfjord. *In:* SPENCER, A. M. *et al.* (eds). *Geology of the Norwegian oil and gas fields.* Graham & Trotman, London, 319–340.

SERRA, O. 1977. Methode rapide d'analyse faciologique par diagraphies differees. *Transactions of the 5th European Symposium of the SPWLA*, Paris, paper 9.

——, 1985. *Sedimentary environments from wire-line logs*. Schlumberger Publications.

VISHER, G. S. 1984. *Exploration Stratigraphy*. Pennwell Books, Tulsa.

WALKER, R. G. & CANT, D. J. 1984. Sandy fluvial systems, *In:* WALKER, R. G. (ed.) *Facies models*, 2nd edn. Geological Association of Canada.

WATANABE, S. 1985. *Pattern recognition: Human and mechanical*. John Wiley, New York.

WIRTH, N. 1976. *Algorithms + Data structures = Programs*. Prentice-Hall, New Jersey.

Heterogeneities in carbonate reservoirs: detection and analysis using borehole electrical imagery

R. NURMI[1], M. CHARARA[2], M. WATERHOUSE[3], & R. PARK[4]
Schlumberger Technical Services,
[1]P.O. Box 9261, Dubai, United Arab Emirates
[2]P.O. Box 4754, Abu Dhabi, United Arab Emirates
[3]P.O. Box 2836, Al Khobar - 31952, Saudi Arabia
[4]201 Competent House, New Delhi, 11001, India

Abstract: Porosity heterogeneities in carbonate reservoirs occur at all scales, from the size of a reservoir delineated by seismic data and/or well-to-well correlations, down to micro-scales which can only be revealed by scanning electron microscopy. The origin of porosity heterogeneities is varied; some result from the depositional system, while others are products of burial, diagenesis and tectonism. Defining the three-dimensional distribution of textures and facies in a carbonate reservoir is exceedingly difficult. At present, carbonate specialists rely on cores to provide an understanding of decimetre to micron-scale textures and fabrics in carbonate reservoirs. Cores are rarely available from every well in a reservoir and usually do not cover the entire reservoir interval. Thus, the characterization of carbonate textures, fabrics, pore types, and porosity distribution using wireline well logs in uncored wells provides additional valuable information. Such an approach requires the integration of all of the available data, including acoustic, nuclear, electric, and dielectric measurements for reliable geological and petrophysical analyses. The wide range of possibilities coupled with the complexity of many carbonate reservoirs, however, generally means that both core and logs have to be acquired, at least in a few key wells. The evaluation of borehole electrical imagery in carbonate reservoirs from different depositional settings with a variety of diagenetic histories, indicates that this technique is providing valuable, and sometimes new, information about porosity heterogeneities. Electrical imagery, which has a high resolution and provides three-dimensional data, usually reveals more of the complexity of pore distributions than standard well logs are capable of. Large individual pores, fractures, and vugs are often directly visible in these images, although microscopic pores are below the resolution of the technique. The porosity fabric, or decimetre-scale distribution of microscopic intergranular, intercrystalline and mouldic pores, is routinely defined, much as oil-saturated porosity in a core can be revealed by ultraviolet light. The correlation of electrical imagery with core samples, or drill cuttings in uncored intervals, can help to quantify the porosity fabric and also the geological interpretation of electrical texture and fabrics. On a decimetre-scale of examination, porosity is sometimes distributed homogeneously but is more commonly heterogeneous. In addition to vugs and fractures, there are four basic geometrical fabrics of decimetre-scale porosity. Layering is the most common, sometimes clearly associated with stratification. However, the thickness of porosity layers can be uniform or quite variable within the same formation. The next most common heterogeneous fabric is patchy porosity distribution. This fabric can either be patches of high porosity within low-porosity intervals or patches of low-porosity rock within porous reservoir zones. A common porosity fabric in Cretaceous and Tertiary carbonate shelf reservoirs is a three-dimensional convoluted continuous mixture of low- and high-porosity rock volumes. This porosity fabric is sometimes associated with isolated patches of either high- or low-porosity rock, although no obvious genetic transitions of these two fabrics have been observed in images, cores, or outcrops. The characteristics observed in electrical images must be summarized in a log format for large-scale correlation studies. Log summary plots of carbonate texture and fabric help to correlate long intervals of imagery with a variety of other data, including drill-cutting sample logs, other well-log data from the same well, and data from other wells. As specific porosity fabrics are generally associated with certain depositional or diagenetic facies, these log summaries also aid well-to-well correlation of facies. Many diagenetic and depositional facies with distinctive electrical fabrics are often not clearly defined by standard well logs because of the decimetre-scale averaging of rock properties by these logs.

From HURST, A., LOVELL, M. A. & MORTON, A. C. (eds), 1990, *Geological Applications of Wireline Logs* Geological Society Special Publication No. 48, pp. 95–111

Surprisingly, very few carbonate reservoirs, including the giant reservoirs of the Middle East (Nurmi 1986) are homogeneous. Even the relatively mono-mineralic low-energy shelf and basin facies are often not homogeneous (Fig. 1). This is because of the effect of stratification on the distribution of porosity. Moreover, thin shale beds and/or stylolites introduce important forms of reservoir heterogeneity to many carbonate rocks. Tectonism and subsurface movement of fluids are also important causes of reservoir heterogeneity. One of the most significant forms of reservoir heterogeneity is fracturing associated with folding and faulting. Both large-scale and micro-scale porosity heterogeneities have been shown to be associated with dolomitization (Longman 1981; Broomhall & Allan 1985). Moreover, the presence of hydrocarbons has helped to preserve reservoir porosity — the reservoirs themselves are large-scale porosity heterogeneities.

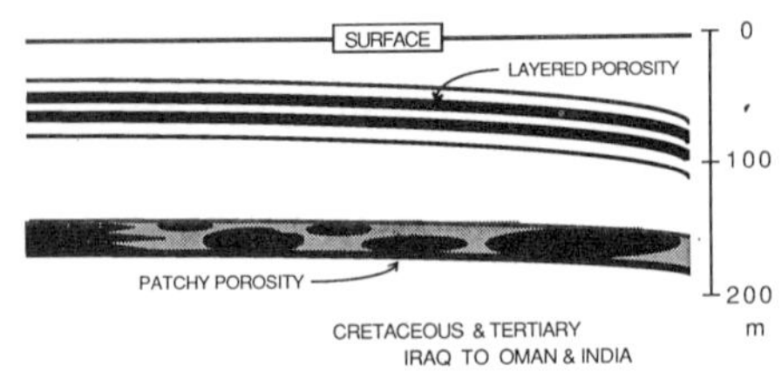

Fig. 1. Large-scale porosity heterogeneities — layered, patchy and interwoven — are common in carbonate reservoirs in the Middle East, North Sea, and offshore India.

'Heterogeneous porosity' is defined here as porosity which is either distributed in a non-uniform manner or composed of dissimilar elements or constituents within a specific volume or area. As the quantification of what constitutes heterogeneity is not well defined, the distinction between homogeneous and heterogeneous is usually relative, and may vary in each study. In most reservoir studies the distinction is ultimately made on the basis of economic considerations.

Many different sizes of heterogeneities are found in carbonate reservoirs. Post-mortems of water or enhanced oil recovery (EOR) flooding projects of a reservoir often reveal large-scale previously unrecognized heterogeneities. Some of these studies reveal that had the wells been drilled closer together, the recoverable reserves would have been increased (Barber *et al.* 1983). Many large-scale reservoir heterogeneities result from an irregular distribution of depositional facies and rock types. Reef reservoirs are large-scale porosity heterogeneities which generally began as depositional anomalies within a carbonate basin. Although reefs and associated facies are known to be depositionally heterogeneous at nearly every examinable scale (Jardine *et al.* 1977), so also are many carbonate grainstone shoal (Lewis 1988), tidal flat, shelf and slope deposits.

The size of porosity heterogeneities may vary from one reservoir zone to another in the same field, and from field-to-field in the same basin. Moreover, a review of numerous EOR projects has indicated that most reservoirs, both sandstone and carbonate, are more heterogeneous than initially recognized (DOE Panel 1982). In fact, the more data one collects, the more complex reservoirs appear to become. The evaluation of areas lying between wells within a reservoir and the integration of scale-dependent data continue to be major challenges for geoscientists. Although seismic and standard well log data show carbonate reservoirs to be homogeneous, cores and borehole electrical images commonly reveal heterogeneities. In such cases there are often differences between porosity measured on cores and that determined using standard well logs. Where porosity is distributed heterogeneously core porosity values should range both above and below the porosity values derived from well log measurements, because log values represent the average porosity of large volumes of rock. However, selective sampling of more porous rock is a common practice which results in core porosities being routinely higher than the carefully calibrated log porosities (Nurmi & Frisinger 1983). When sampling very porous and friable terrigenous sands, the reverse may be true: the core samples are often from the less porous material.

It is generally considered that at a micrometre or decimetre scale, carbonate pore systems are more heterogeneous than those of sandstones (Ham & Pray 1962; Choquette & Pray 1970). It was this heterogeneity of carbonate rocks that led to the introduction of 'whole core analyses' which use the entire width of core (Atkinson & Johnston 1949; Bynum & Koepf 1957) rather than the one-inch plugs routinely drilled out of sandstone cores. Unfortunately, the most permeable intervals, with a high volume of large secondary pores, are rarely recovered (Bulnes & Fitting 1945). Collins (1961), in a thorough study of core permeability, stressed that the analysis of porosity and permeability must consider the sample size, and introduced the concept of critial sample size. Schneider and Owens (1970) recommended that whole diameter cores be used in the study of displacement behaviour in heterogeneous carbonate rocks. However, in a study of relative permeability of vugular carbonates, Ehrlich (1971) concluded that cores that satisfactorily represent the reservoir are unobtainable.

As heterogeneities occur at all scales of measurement they affect all petroleum industry disciplines responsible for analysing reservoirs. In one of the earliest evaluations of heterogeneities, Alpay (1972) identified three scales of reservoir heterogeneity (megascopic, macroscopic, and microscopic) by analogy with groundwater studies (Hubbert 1940; Bear 1968). As these terms have not been applied consistently (Lewis 1988), they should either be avoided or redefined when used. The macroscopic-scale heterogeneities are the most difficult to evaluate as they are generally larger than the diameter of a core and smaller than the distances between wells. Haldorsen (1986), Lassete *et al.* (1986) and Lewis (1988) have recommended other terminology and hierarchical levels of heterogeneity scales.

Heterogeneity is a major factor influencing recovery efficiency in carbonate reservoirs (DOE Panel 1982). It is generally assumed, or hoped, that heterogeneities are randomly distributed. However, in an overview of carbonate reservoir heterogeneities Wardlaw (1978) concluded that heterogeneities in carbonates are commonly non-random. He further stated that although there is a great volume of geological literature concerning carbonate reservoirs, little illustrates how reservoir properties vary in three-dimensional space. Jardine *et al.* (1977) provides an overview of carbonate reservoir characteristics in western Canada. In a review of heterogeneities in genetically different sandstones, including Jurassic carbonate grainstones in southern England and the Paris Basin, Lewis (1988) questioned the representatives of porosity and permeability values of core plugs even for grainstone deposits. He concluded that pore systems in carbonate grainstones need to be examined at a variety of scales.

This paper addresses the detection of decimetre-scale heterogeneities within carbonate reservoir zones. The authors' experience with heterogeneities revealed by borehole electrical images prompted a systematic examination of these phenomena. During the examination of carbonate reservoirs in the Middle East, Egypt, India, Europe and the USA it was found that standard procedures and techniques used in core and well log evaluations of carbonate reservoirs often failed to reveal the heterogeneities which were seen by borehole electrical imagery. Furthermore, if some heterogeneities were recognized in the geological description of a cored interval, it was rarely related to the distribution of porosity. It was observed that neither routine core nor well log analyses include an assessment of porosity heterogeneity. Furthermore, even when porosity in small reservoir samples is homogeneous, the distribution of porosity in a larger volume may be heterogeneous. Thus, the detection of heterogeneities is often dependent on the manner of examination and on the technology used.

Borehole electrical imagery

Borehole electrical imagery is a relatively new well logging technology (Ekstrom *et al.* 1986) which underwent a short period of laboratory and field testing before its routine use in most petroleum basins. The development of electrical imagery grew from detailed evaluation of dipmeter data which revealed the potential of high-resolution resistivity measurements in the definition of the complexities of bedding, texture, fabric and heterogeneities in reservoir rocks. The physical design of the imagery tool and its electrical system is very similar to the high-resolution dipmeter (Ekstrom *et al.* 1986), with the major difference being the number of button electrodes on the four pads of each tool. The present Schlumberger dipmeter has two button electrodes on each of its four pads. Most other dipmeters have one button electrode on each pad, although they may have six rather than four pads. In order to produce images, many more button electrodes were placed on the pads of the imagery tool (54 on each of two pads on one version of the tool). The imagery electrodes are small circular buttons (0.5 cm) and the electrical current is measured at a very high sample rate (every 0.254 cm), which results in a resolution on the order of a centimetre. If a high electrical contrast exists between a very thin feature (less than the button electrode size) and the surrounding rock, it can be seen in the image, but the thickness of the feature is that of the button electrode rather than its true thickness.

The borehole imagery tool, like the dipmeter, is designed to focus small beams of electrical current into subsurface formations in boreholes (Ekstrom *et al.* 1986). It measures the ease with which this electrical current enters the formation. By placing numerous electrodes close together on the pads of this logging tool, and by sampling the variations of the electrical properties of the formation at a high rate, it is possible to produce realistic images of the electrical characteristics of the formation. The optimum use of the tool is in wellbores filled with conductive fluid (Ekstrom *et al.* 1986).

During the field testing of the prototype imagery tool in Jurassic carbonates of the Paris Basin, vugs were detected. Later, in a field test in a Conoco research well in Oklahoma, very fine (hairline) fractures were identified. Although many fabrics and textures on a decimetre scale were revealed, their geological origin was not always obvious. By comparing these images with core data, or other log measurements with a comparable vertical resolution, they could usually be interpreted in terms of lithology, porosity and fluid content. Comparisons of images with cores indicated that porosity heterogeneities larger than a centimetre can be detected if a resistivity contrast exists between the porosity and the non-porous rock. Individual pores less than a centimetre were generally not resolved.

In the plotting of electrical imagery data, it has become standard practise to designate the most conductive values as black and the most resistive values, or features, as white. Thus, shale beds and highly porous intervals appear black and the resistive low-porosity rocks white. The degrees of electrical resistivity between these two extremes are given appropriate shades of grey. With this shading, electrical imagery reveals the distribution of even very thin shales, which generally contain conductive clay minerals, and also the distribution of porosity, which is generally invaded by conductive drilling fluid.

The geological interpretation of electrical images is often straightforward if one is familiar with the geometry and electrical character of carbonate rock textures and fabrics in the logged rock sequence. However, the interpretation of electrical images in an unfamiliar rock sequence is best done after comparison with cores, just as the interpretation of infrared or ultraviolet photographs requires comparisons with the actual rock and a period of familiarization with such photography before their contribution to the recognition of porosity distribution can be realized.

There are some limitations to the use of this relatively new technology. Electrical imagery is generally restricted to wells drilled or filled with water-based muds. The tool is also sensitive to the rugosity of the borehole, with the quality of the image often degraded in 'washed-out' intervals. As the images only indirectly define porosity and/or shale content, it is important to correlate and calibrate them with core data or special high-resolution well log porosity and lithology data. However, even without such data, the images provide valuable input for designing pressure well-test programs and for indicating which reservoirs require higher resolution well log evaluations.

Decimetre-scale porosity heterogeneities

Most carbonate classifications of rocks, porosity, and fabrics require a genetic understanding of the rock features, or rely on the identification of microscopic particles and individual pores, many of which are shown by Ham (1962). When rock or individual pore types can be identified, the most appropriate carbonate terminology should be used, such as the vug and fracture terminology proposed by Choquette & Pray (1970). Individual microscopic pores (intergranular, intercrystalline, and mouldic pores) and grains which comprise most carbonate reservoir rocks are not resolvable in electrical images. However, the porosity fabric, or decimetre-scale distribution of microscopic pores, is revealed routinely by electrical imagery. This fabric is the spatial arrangement and orientation of the porosity distribution (Potter & Pettijohn 1963) at the scale defined by electrical imagery.

During the evaluation of porosity fabrics and textures revealed by electrical imagery, a geometrical description has generally been used, especially when the geological nature of the texture or fabric was not obvious. The observation that similar porosity fabrics occur in the same facies of different wells led to a systematic classification of the fabrics being revealed by electrical imagery. This fabric classification is actually independent of scale and can be applied to the analysis of data such as CAT, seismic data, pressure analyses and other types of borehole imagery. When individual pores are recognizable, the appropriate geological terminology is used. The fabric classification of decimetre-scale porosity does not replace existing genetic or geologic classifications of texture or fabric (Dunham 1962; Choquette & Pray 1970; Logan & Semeniuk 1976; Wanless 1979), but is used when geological interpretations of fabric are not certain or when the focus of a study is the distribution of porosity and flow characteristics.

Most classifications of centimetre- and decimetre-scale carbonate fabrics deal with the genesis of the fabric or fabric elements. Attention is rarely given to the decimetre-scale distribution and orientation of porosity, although the three-dimensional porosity fabric affects the flow of fluids and also subsurface measurements of porosity. As Longman (1981) pointed out, the 'bible' for pore type classification by Choquette & Pray (1970) fails to discuss the geometry and topology of carbonate pore systems.

It was recognized during our electrical imagery studies of carbonate reservoirs that most porosity fabrics could be defined as either 'homogeneous' or 'heterogeneous' in terms of their electrical properties. The thickness and nature of the boundaries of the homogeneous and heterogeneous intervals were examined and the geometry, or fabric, of the heterogeneities was analysed. The fabrics of the heterogeneous decimetre-scale porosity belong to one of four basic geometrical types (Fig. 2) — layered, interwoven, and two types of isolated. The four basic porosity fabric types are shown in the classification table as equal mixtures of homogeneous non-porous and porous rock. As equal mixtures, the four fabrics are placed half-way between the homogeneous end members to show the geometrical distinction between them. In nature, there is usually much variation in the relative volume of porous and non-porous rock fractions. Calculations of the percentage of the porous and non-porous fractions revealed by borehole imagery are aided using a computer workstation to determine whether they are layers, vugs, patches or interwoven porosity fabrics. However, the quantification of the actual porosity values of each fraction generally requires the geometrical analysis to be integrated with core and/or well log porosity data collected at a higher than normal sample rate.

As the geometrical distribution of porosity and impermeable barriers are factors which will determine the producibility of a reservoir, it is desirable to describe the geometrical distribution of porosity at every scale possible. Borehole electrical imagery allows one to define the decimetre-scale distribution of porosity and also the geometry of non-porous rock which may serve as a barrier to flow. Thus, if open fractures are detected, it is important to describe whether they intersect each other and whether the fractured rock is porous, or if the fracture system connects otherwise unconnected porosity.

If vugs are present, it is sometimes important

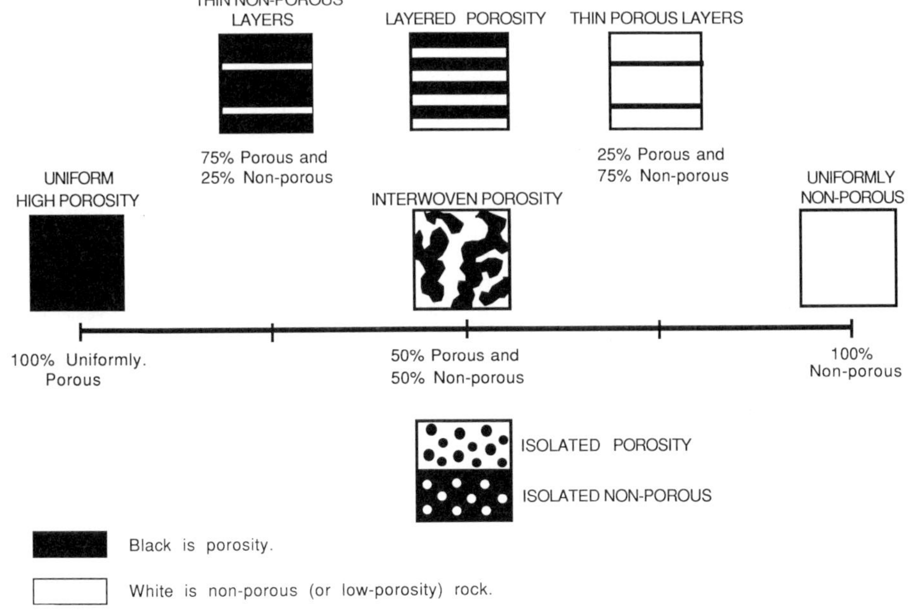

Fig. 2. Geometric classification of decimetre-scale porosity/resistivity fabrics present in carbonate rocks. The end members are relatively homogeneous porous and non-porous rock. Although four basic heterogeneous fabrics are shown as ideal equal mixtures of porous and non-porous rock, the relative proportions of porous and non-porous rock are usually not ideal. Unequal mixtures of the layered porosity fabric type are included to show how these mixtures vary.

to determine whether there is additional non-vuggy porosity, or if these vugs are connected. Vugs in thin porous dolomite intervals of the Arab Formation in Saudi Arabia, which formed from the leaching of *Cladocoropsis* debris, sometimes result in a very highly permeable layer. Centimetre size *Cladocoropsis* molds and vugs in the Arab-D member cores from the supergiant Ghawar Field have greater than Darcy permeabilities (Mitchell *et al.*, 1988). However, in the examination of cores or borehole electrical imagery, it is not obvious that these leached fossils have high connectivity over large areas of a reservoir. The slabbing of branching corals (*Porites*) by Ginsburg (James 1983) revealed how deceiving the analysis of three-dimensional forms can be when only viewing a planar surface cutting a complex skeletal form. A filling of a branching crustacean burrow from Florida Bay demonstrates how these could also be very complex and interconnected (Inden & Moore 1983). The geometrical classification of heterogeneous facies aids geological correlation and mapping, although the significance is enhanced when the facies can also be defined in geological terms. Even so, the systematic characterization of the geometrical characteristics of porosity is important to log analysts and petroleum engineers, providing insight into the presence of thin beds and permeability anisotropy. Each geometrical fabric has a distinct effect on the nature of fluid flow. The layered porosity fabrics detected by electrical imagery are likely to cause the flow characteristics to be anisotropic on the decimetre-scale, whereas patchy or isolated porosity may have poor permeability in any direction often with surprisingly low permeabilities even when bulk porosity is high. If the low-porosity fraction surrounding the isolated high porosity has some permeability, a dual (permeability) porosity system is present. The interwoven porosity type usually requires some additional analysis to define the direction in which the high-porous fraction is more continuous. Rocks with this porosity fabric usually have greater permeability than rocks with isolated fabrics. Each of these heterogeneities may result in poor hydrocarbon recovery. However, once the presence of the heterogeneities are recognized, it may be possible to enhance production by increasing the density of perforations or hydraulically fracturing the formation.

An accurate description of the geometrical distribution of porosity and permeability is important if appropriate expressions are to be derived for the flow characteristics of rocks (Havlena 1966). The layered porosity systems are treated as parallel permeability systems, whereas the rocks with isolated porosity fabrics are treated as series permeability systems. The rocks with interwoven fabrics can be either series or parallel, depending on the geometry of the fabric and its orientation with respect to flow. The geometry of porosity distribution determines how one should treat porosity/ permeability data (Havlena 1966). Electrical imagery that reveals decimetre-scale porosity fabrics identifies those carbonate reservoir zones which need to be evaluated using well logs with higher than normal sampling rates. The detection of reservoir heterogeneities also indicates which porous zones need to be evaluated critically using a larger scale of measurement, such as defined by transient pressure testing (Ayestaran *et al.* 1989).

Thin layers of porous and non-porous rock

Large-scale porosity layering is generally a function of the depositional history of the sequence. Numerous causes of thin and/or rhythmic bedding or cyclic sequences have been proposed, including orbital variations (Gilbert 1895; Schwarzacher & Fischer 1982), climate variations (Barron *et al.* 1985; Bottjer *et al.* 1986), and diagenesis (Simpson 1985; Ricken 1986). Recently, Bathurst (1987) summarized the layered fabrics caused by stratified cementation and selective compaction; these explain a variety of the layered porosity fabrics observed in electrical images. The decimetre- or larger-scale geometrical porosity distribution is typically layered, ranging in scale from stratigraphic formations or even groups, down to the microscopic concentric rings of an ooid. However, the internal geometry of porosity within the porosity layers defined by seismic data or well-to-well correlations, may be very patchy or interwoven. In other cases, reservoir zones that appear to be homogenous using well logs or seismic data may have obvious decimetre-scale layering in core samples and electrical images.

One of the most dramatic cases of this type of porosity distribution occurs in a Cretaceous reservoir zone of the Arabian Platform (Fig. 3), where electrical imagery revealed thin porosity layers which were smoothed out by the standard well logs. The well logs indicated a uniform, homogeneous porosity distribution within the reservoir. No shales or conductive minerals are present in the sequence and the resistivity layering is a function of porosity variations due to stratified cementation as described by Bathurst (1987). More commonly, large-scale highly

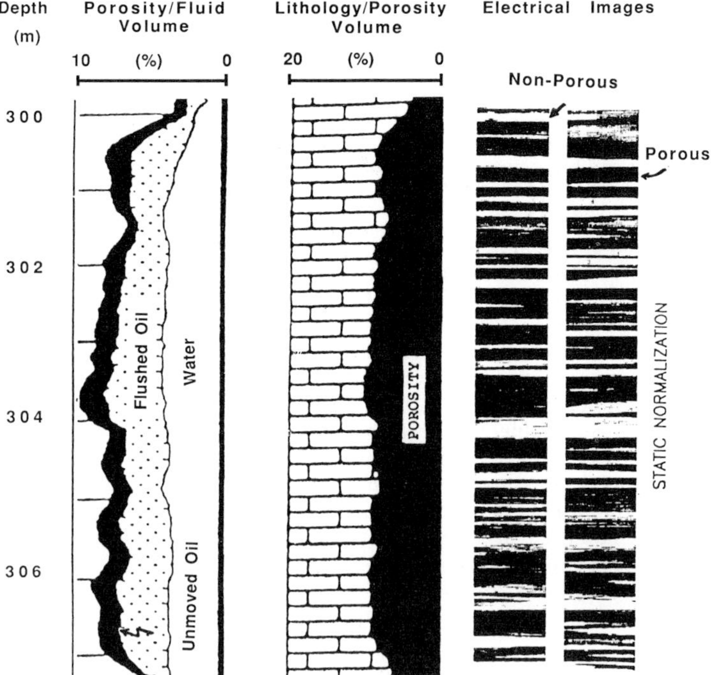

Fig. 3. A large-scale porosity layer comprised of smaller-scale (decimetre to centimetre size) porosity layers, is revealed by electrical images. The black layers in the images are clay-free, porous limestone, whereas the white layers are resistive cemented layers. The standard porosity logs (density, neutron, and sonic) were used in the quantification of porosity shown to the left. The vertical averaging of these measurements resulted in this layered Lower Cretaceous interval appearing to be a relatively homogeneous porous zone.

layered Cretaceous and Tertiary reservoirs contain decimetre-scale patchy or interwoven porosity fabrics. The Eocene reservoirs offshore Bombay (India) commonly have large-scale porosity layers with patchy and interwoven internal porosity fabrics. D'Heur (1984) indicated that only after a number of wells had been drilled was it realized that the chalk in the Greater Ekofisk area (North Sea) is an alternation of more and less porous beds, and that some layers could be traced from field to field.

The main reason for the porosity variations was thought to be related to early cementation. It was also found that the distribution of hydrocarbons had affected the later cementation of the chalk reservoirs resulting in porosity preservation on the high parts of the anticlinal structures.

In some reservoirs, porous layers are are much thinner than non-porous intervals. The importance of thin porous layers is probably greater than is generally realized, as such layers are often not recognized because of the averaging effect of standard well logs (Hartmann 1975). These thin porous layers are especially significant in low porosity gas reservoirs, such as the Khuff reservoirs around the Arabian platform. In the Khuff Formation, the presence of fractures may also be important in connecting the thin porosity layers (less than one foot thick), whose areal extent is probably not very great. Some Arab Formation reservoirs have thin porosity layers, such as Gaslia Field in Abu Dhabi.

Adjacent laminae within a cross-bedded grainstone sequence may also have great differences in porosity. Dipmeter studies and imagery detection of bedding laminae in cross-bedded Jurassic grainstones in the Paris Basin (Nurmi 1988) and in the Smackover Formation around the Gulf of Mexico (Nurmi & Frisinger 1983) revealed that the dipmeter detection of the cross-bedding was a result of centimetre-scale variations in resistivity. Core examination revealed that preferential cementation and compaction of laminae was the major cause of bedding resistivity contrasts.

Thin shales or clay seams can cause the permeability of seemingly high porosity carbonate intervals to be quite low, as with Miocene limestones of the Heera Field, offshore Bombay (India). An unpublished study demonstrated that micro-resistivity measurements of the dipmeter could reveal the presence of clay seams, thus improving the log estimates of permeability. In this case, very thin wisps of clay caused the permeability to be much lower than in intervals lacking such clay seams.

Stylolites can also be detected by the dipmeter and/or electrical images if there is an electrical contrast between the stylolite and the surrounding rock. If conductive material is concentrated along the stylolitic seams in a low-porosity rock interval, these seams appear as dark features (Fig. 4), much as they appear in black and white photographs of such rock. Less common are cases where the stylolite is more resistive than the surrounding rock, although this has been observed in the Thamama Field, Abu Dhabi. In electrical imagery, stylolites often have an appearance similar to thin shales but usually have less measurable radioactivity. In sequences with abundant stylolites, the concentration of insoluble material and the associated cementation cause the permeability to be much lower vertically than horizontally (Koepnick 1984; Burgess & Peter, 1985). However, in deep burial, stylolites may act as diagenetic pathways with leaching and secondary porosity associated with the stylolitization (Mazzullo 1981; Carozzi & Von Bergen 1987). Unfortunately, the resistivity difference between permeable and impermeable stylolites is insufficient to permit their distinction using electrical imagery.

The presence of thin shales and stylolitic seams can, however, affect the standard measurements of resistivity used in the calculation of water and hydrocarbon saturations. The conductivity of stylolites often causes low values of Archie's '*m*' exponent, as with open fractures (Watfa & Nurmi 1987) if the fractures are much more conductive than the surrounding rock.

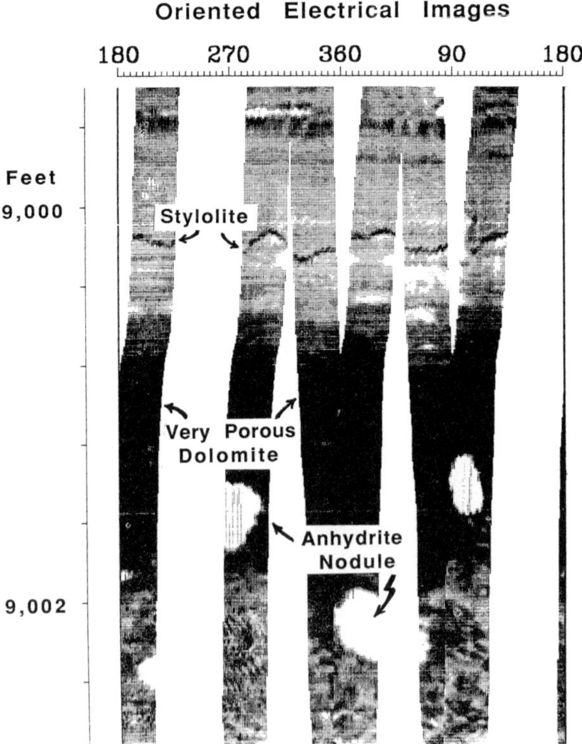

Fig. 4. Nodules of anhydrite — very resistive white features — are present in an Arab Formation reservoir zone. The dark interval is a porous dolomite with 5–12 micron-size intercrystalline pores. A stylolite seam is also visible because of concentrations of clay and traces of pyrite along the seam.

Patches of non-porous rock within a reservoir zone

Anhydrite nodules are one of the common forms of isolated non-porous rock within a porous zone in carbonate sequences. The high resistivity of these nodules is usually distinctive in electrical images. Their presence, if undetected, will generally cause standard log evaluations to be somewhat pessimistic, because of their high density and/or low porosity. Anhydrite nodules have been observed in the electrical images of porous carbonate facies of many formations, including the San Andres of the Texas Permian Basin, the Mission Canyon of the Williston Basin, the Smackover of the U.S. Gulf of Mexico region, and Arab and Khuff of the Arabian Platform (Fig. 4). One of the more dramatic examples is the highly porous Eocene dolomites of the Wafra Field in Kuwait, described by Danielli (1988). The recognition of these non-porous anhydrite nodules improved the evaluation of the producibility of this reservoir zone.

Patchy cement distribution is a common feature of many carbonate reservoirs, and is particularly abundant in the dolomitized pellet packstone facies of the San Andres reservoirs in West Texas and New Mexico. Al-Jallal (1987) documented the occurrence of patchy anhydrite cement distribution on a variety of scales in the Permian Khuff Formation of the supergiant Ghawar Field (Saudi Arabia). Sonic measurements can play an important role in the identification of anhydrite within a porous carbonate reservoir, providing no isolated pores are present, or if the volume of isolated pores can be determined independently of sonic measurements. Dielectric measurements combined with standard well log data can help to independently define the volume of isolated pores.

Patches of porosity

Electrical imagery often reveals decimetre- to centimetre-scale patches of porosity in carbonate reservoir rock. These patches may be remnants of intergranular or mouldic porosity that has been irregularly cemented, or may be unleached calcite patches after dolomitization.

Oomoulds in calcareous grainstones are well known isolated microscopic pores. They are usually recognized through core and thin section study. However, their small size, commonly less than 300 μm, precludes them being individually detected by either dipmeter or electrical imagery. Leaching processes often result in heterogenous oomouldic porosity fabrics.

Vugs, another common type of isolated pores, can be detected and recognized using electrical imagery and sometimes dipmeter microresistivity measurements, providing they are larger than one centimetre. If vugs or patches of high but isolated porosity are surrounded by impermeable rock their contribution to reservoir productivity would be very small. If, however, the rock surrounding the vugs or patches of high porosity is permeable, the vugs may contribute significantly to the storage capacity of the reservoir. The evaluation of the producibility of small vugs generally requires 'whole-core' analysis so that porosity determinations are geometrically representative of the reservoir. The evaluation of vugs too large to be analysed using cores is best carried out with transient pressure testing of zones defined by cores and/or electrical images. The very first porosity heterogeneities observed in an electrical image were vugs in a dolomitized Jurassic carbonate sequence in Schlumberger's test well in the Paris Basin. The vugs, which probably started with the dissolution of skeletal material, were of the order of a centimetre in diameter — just a little larger than the electrodes of the imaging tool used.

Large fossil moulds (rudists) and vugs in Jurassic Arab Formation biostromes and Cretaceous rudist build-ups (Mishrif & Shuaiba formations) of the Arabian Platform are easily detected using borehole imagery as they are very conductive features within more resistive carbonate rock (Nurmi *et al.* 1986a). Unfortunately, the interpretation of features in electrical images can sometimes be ambiguous. Very conductive material, such as pyrite, could produce features which could be confused with vugs in an electrical image. An indication of so-called 'sonic secondary porosity' in an interval with a vuglike appearance in the imagery can be used as support for a probable vug interpretation. Patchy porosity distribution and interwoven porosity fabrics can also indicate secondary porosity when none is actually present.

Interwoven porous and non-porous rock

A decimetre-scale mixture of porous and non-porous rock, both of which are physically continuous fractions of the rock, is referred to as interwoven porosity. Interwoven porosity fabrics have been observed in both electrical images and cores from a variety of reservoirs.

They are especially abundant in fine-grained Cretaceous and Tertiary carbonates, possibly because of the greater abundance of low-magnesium calcite in the shelf and basinal deposits from the Cretaceous onwards. In the central and eastern portions of the Arabian platform, such fabrics are found in low-energy shelf and basin facies, but are also associated with Lower Cretaceous (Thamama and Mishrif Formations) reef complexes in the United Arab Emirates. Large-scale interwoven porosity fabrics have been observed during porosity mapping using surface seismic data (Rubbens *et al.* 1983). Recent studies of Tertiary carbonate images from the western Arabian peninsula and offshore Bombay (India) have revealed similar interwoven porosity, as well as other types of heterogeneities. Hardman & Kennedy (1980) reported similar fabrics in North Sea Cretaceous chalks.

The interwoven porosity fabric in fine-grained rocks often appears to be a result of selectively preserved microporosity. Although Wilson (1985) attributed the microporosity in Arabian platform Cretaceous carbonates to the ubiquitous microporosity in fine-grained carbonate sediments, it is not clear why interwoven or isolated patches of porosity occur so often with very low-porosity and relatively impermeable rock. The reason why these decimetre-scale heterogeneities are so common remains unknown. One possible explanation is that these heterogeneities are formed by fresh water leaching, such as described by Friedman (1975) in his study of chalky microporosity in Tertiary carbonates of Indonesia. Harris *et al.* (1985), in studies of Cretaceous reservoir rocks on the Arabian platform, also suggested that the microporosity was due to leaching, perhaps associated with freshwater movements beneath interregional Cretaceous unconformities.

The few core observations made during the study of isolated porosity patches in Cretaceous rocks on the Arabian Peninsula revealed by electrical imagery are more consistent with observations of Moshier & Handford (1984). They proposed that the retained microporosity consisting of intercrystalline pores was likely to be preserved during burial by limited cementation, moderate mechanical compaction, and the common absence of chemical compaction. Our examination of interwoven porosity fabrics in Cretaceous and Tertiary limestones of the Arabian Platform and the Indian Ocean suggests that the preservation of the porous portion of the interwoven fabrics explains the presence of the porosity. However, there appear to be a variety of reasons for the presence of microporosity and much evidence for leaching of skeletal remains is observed in the upper parts of Cretaceous depositional sequences. These may have been sub-aerially exposed and/or subjected to fresh water leaching as a result of either sea level fluctuations or structural uplift.

A core sample of an interwoven porosity fabric from a Cretaceous reservoir on the Arabian Platform is shown in Fig. 5. The dark fraction of the core is oil-stained porous limestone, whereas the light-coloured fraction is recrystallized low-porosity limestone (Fig. 6). The porous and permeable rock is darker in the image because of the drilling fluid invading the rock. The less porous rock fraction is more resistive and thus very light in colour. Well log analysis of this interval indicated that the rock was a clay-free limestone with no other minerals present. Geochemical and mineral analyses of core samples confirmed that it was a limestone and showed no conductive minerals, such as clay or pyrite. Petrographic and SEM examination of the core revealed that the limestone contained micro-intercrystalline pores (Fig. 7), while the non-porous fraction contains evidence of recrystallization (Fig. 8).

A possible origin for the interwoven fabric and associated facies with isolated patches of

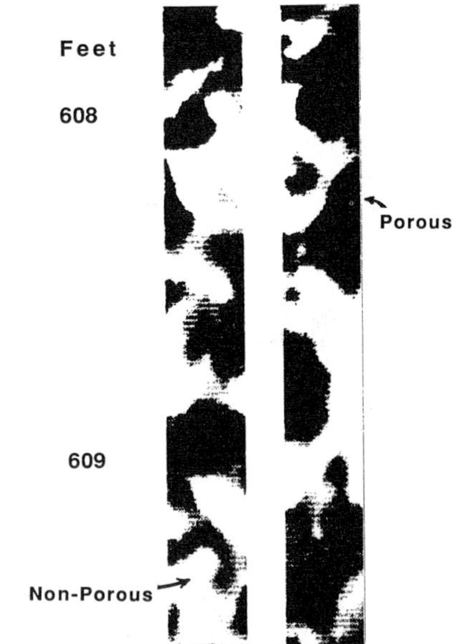

Fig. 5. A porosity fabric of interwoven non-porous and porous rock from a Middle Cretaceous reservoir interval in the United Arab Emirates is revealed by electrical imagery. The images displays are from positions 90° apart from other each other on the borehole wall.

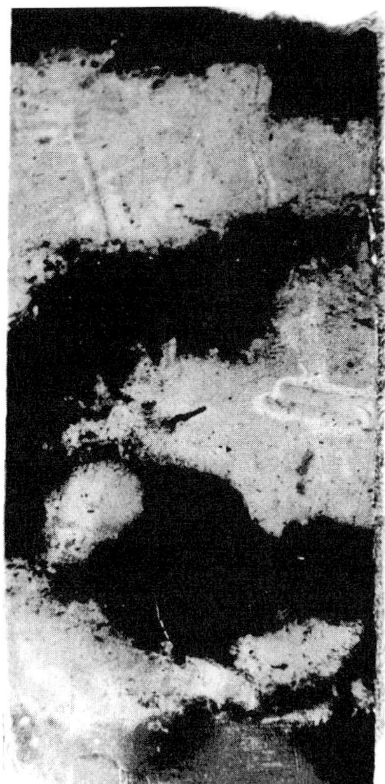

Fig. 6. An interwoven porosity fabric in a Middle Cretaceous reservoir is clearly visible because of oil staining of the microporous fraction of the rock. The light-colour fraction is relatively non-porous recrystallized calcite.

high porosity is burrowing, which has affected the subsequent diagenetic and compaction history. Some interwoven fabrics are similar to arthropod burrows, such as the large ramifying burrows of *Thalassinoides* (Fursich 1973). However, the three-dimensional examination of some interwoven fabrics precludes such a burrowing origin, as with porosity fabrics in North Sea Cretaceous chalks (Hardman & Kennedy 1980).

Various types of nodular limestones (Logan & Semeniuk 1976; Wanless 1979) may also show interwoven porosity fabrics. Whatever their origin, electrical images provide a means of detecting and evaluating the geometry. Interwoven fabrics cause many Cretaceous and Tertiary reservoirs to have dual porosity systems. In such reservoir zones it is important to determine whether the porous fraction of the rock is connected or isolated, and if connected, whether it is laterally and vertically connected. 'Whole-core' analysis is advised when such heterogeneities are present. If the scale of the heterogeneity is too large to be appropriately analysed using cores, one may need to rely on subsurface pressure tests to quantify the reservoir quality.

In some cases, both fractions of an interwoven fabric are porous with one fraction having a higher porosity than the other, as in the Simsima Formation outcropping in the United Arab Emirates (Fig. 9) and northern Oman. In this case, one of the more porous rocks, stained by reddish clay, comprises sand-sized skeletal

Fig. 7. Scanning electron microscopy reveals an abundance of microporosity in the oil-stained fraction of the core in Fig. 6 (magnification × 2610).

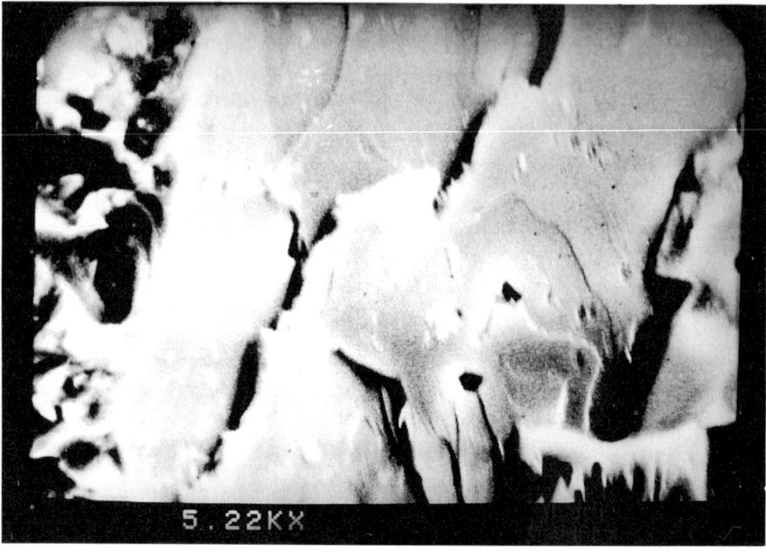

Fig. 8. SEM micrograph of the recrystallized calcite in the light coloured non-porous fraction of the core in Fig. 6 (magnification × 2610).

Fig. 9. Outcrop of the Maestrichtian Simsima Formation in northern Oman which contains an interwoven porosity fabric. The continuous dark fraction is porous nummulitic grainstone stained by a reddish clay. The less porous light coloured fraction is cemented rudist debris.

grains and the other fraction consists of less porous rudist debris.

The intimate mixture of two reservoir lithologies causes difficulty in evaluating this type of rock. In most cases, the reservoir properties of the two lithologies are distinctly different. The lower porosity rock fraction has a much higher irreducible water saturation or can be completely water saturated, while the higher porosity fraction is saturated with oil. Well log interpretation of reservoir potential is therefore difficult, especially when the interwoven fabrics are at the decimetre-scale.

Interwoven porosity fabrics are found in a variety of scales. The largest fabrics may be defined by seismic data, while the smallest may require a scanning electron microscope (Fig. 10).

Open and mineralized fractures

Fractures are an important type of reservoir heterogeneity. They commonly occur in the least porous intervals, although preferential fracturing of porous chalk has been observed. Open and permeable fractures appear as dark cracks in electrical imagery. Mineralized fractures also introduce heterogeneity to an otherwise porous interval and may behave as significant barriers to flow. If the fractures dip with respect to the axis of the borehole, they have a sinusoidal shape on electrical image plots which include the full circumference of the borehole. The dip and strike of such fractures are defined by using overlays containing cali-

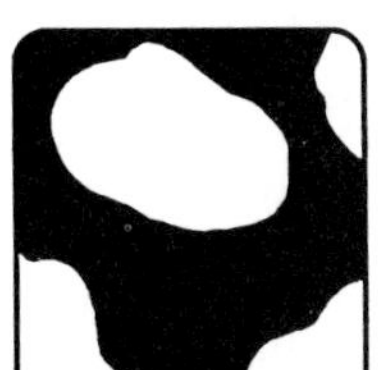

Fig. 10. Interwoven and patchy porosity (porosity is black portion) fabrics come in a number of sizes in Cretaceous and Tertiary carbonate reservoirs in the Middle East, North Sea and offshore Bombay, India.

brated sine waves which show dip angle and azimuthal orientations (Waterhouse *et al.* 1987). The electrical image characteristics of fractures in carbonate reservoir sequences have been described previously by Serra (1986), Nurmi *et al.* (1986b) and Dennis *et al.* (1987). The borehole images, when logged over long sequences and correlated with rock properties, such as mineralogy, clay volume, shale beds, porosity, curvature of folding, and faulting, provide insight into what controls the vertical distribution of fractures in carbonate reservoirs. The correlation of fractures and other formation characteristics is best made by summarizing the fracture intensity and correlating it with computer processed interpretations of the above-mentioned characteristics. Mineralized fractures appear as white fractures if the filling materials are resistive, such as calcite, dolomite, quartz, or anhydrite. However, clay-filled fractures could be confused with open fractures. Core data may indicate whether clay-filled fractures are to be expected in a particular area. The large-scale reservoir characteristics of fractured zones must be defined by some other approach, such as transient pressure testing.

Effects of heterogeneities on standard well logs

The presence of micrometre to decimetre-scale heterogeneities can dramatically affect the analysis of standard well logs, including the determination of lithology, total porosity, secondary porosity, and saturation. The effect of micrometre heterogeneities on reservoir analysis using logs has been observed in a number of carbonate reservoir rocks (Pittman 1971; Kieke & Hartmann 1973; Guillotte *et al.* 1979; Keith & Pittman 1983; Brie *et al.* 1985). The hydrocarbon recovery efficiency of different pore systems and the importance of micro-scale heterogeneities was summarized by Wardlaw (1980). However, there has been little mention of the effect of decimetre-scale heterogeneities on well log measurements or interpretation, with the exception of the presence of thin beds (Hartmann 1975). There are actually many log indicators of reservoir heterogeneity, although they generally do not reveal the type of heterogeneity that is present. The effects of decimetre-scale heterogeneities on logs are usually either misinterpreted as micro-scale heterogeneities or assumed to be caused by the presence of a layered porosity fabric.

In one of the first papers that addressed the log interpretation of heterogeneous carbonate reservoirs specifically, Winn (1957) stated clearly the common assumption made in log interpretation.

> 'In this paper a heterogeneous, carbonate reservoir is considered as a succession of layers of formations of radically different porosities, permeabilities, water saturations, and rock matrix types; i.e., the heterogeneity is vertical rather than radial to the well bore. Presumably each layer is more or less homogeneous.'

This misconception of carbonate heterogeneities still exists today, although it is generally implicit in the methodology of standard formation evaluation using well logs.

Heterogeneities of a decimetre-scale sometimes cause a mismatch of well log data from different runs if the logging tools, have a different orientation during each logging survey. Unfortunately, such a lack of repeatability is usually assumed to be a problem with the logging tool rather than being due to the presence of heterogeneities in the rock. A common indication of the presence of decimetre-scale heterogeneities is the occurrence of significant resistivity variations in the microresistivity measurements. These are not observed in the slightly deeper resistivity measurements which average out the resistivity character of a rock sequence.

The determination of porosity using sonic measurements can be adversely affected by the presence of decimetre-scale heterogeneities. In fact, some heterogeneously distributed porosity may not be detected as the acoustic wave travels along the fastest rock path. This is generally the lower porosity fraction of the rock. Thus, in carbonate rocks with either interwoven or patchy porosity fabrics, the sonic-derived porosity is less than the actual value. As total porosity defined by density-neutron logs is greater than porosity defined by sonic logs in rocks with such porosity fabrics, false indications of secondary porosity occur. Lithology estimates could likewise be in error if these velocities are used in the log interpretation of lithology.

Decimetre-scale heterogeneities may also cause the density/neutron determination of lithology and porosity to be in error, especially when the density and neutron tools measure different rock volumes, which thus have different properties. The neutron log measures a larger volume, and thus averages a greater volume than the density tool (Sherman & Locke 1975). Efforts to increase the vertical resolution by increasing the sampling rate of these tools improves the vertical resolution, but does not eliminate all of the anomalies due to non-layered heterogeneities. However, increasing the sampling rate and vertical resolution of logging measurements is an appropriate technique for sequences with thin alternations of lithology, porosity and other rock properties (Flaum *et al.* 1987) and also provides better well log data for quantification of some aspects of other possible heterogeneities. With careful examination of the deep and shallow measurements of each logging tool, such as the far and near detectors of the density and neutron logs, or the very shallow to deep resistivity measurements of the various electrical tools, it is possible to quantify the heterogeneities revealed by electrical imagery.

Conclusions

Porosity heterogeneities within carbonates are common and can cause the potential of a reservoir sequence to be incorrectly evaluated if standard to log and core evaluation approaches are followed. Electrical images only sometimes define directly a specific carbonate pore or lithology. Even so, they provide valuable information about decimetre-scale carbonate heterogeneities and porosity fabrics.

In geometrical terms, porosity heterogeneities of centimetre and decimetre scales can be classified into one of three basic fabric types — layered, isolated, and interwoven. As these fabrics are generally facies-related, they are valuable for geological correlation and mapping studies. An understanding of their geological origin usually requires careful examination of core data and correlation between the core and the electrical imagery.

Few of the reservoir heterogeneities observed have a chaotic or random distribution, but often they were not summarized in a standard characterization of a reservoir. In many of the reservoirs examined, it was found that when using data from cores and standard well logs, the presence of important heterogeneities went undetected. Porosity heterogeneities are often sensed by larger volume evaluations of transient pressure testing which can sometimes define quantitatively the volumes of mixed porosity types — the so-called dual porosity and dual permeability systems. However, it is difficult to distinguish what type of heterogeneity fabric caused the pressure response.

Layered and patchy porosity fabrics occur at nearly all scales of observation, although the type of porosity fabric that is observed is determined by the technology being used. Unfortunately, during the examination of a reservoir, it is not possible to study the porosity distribution at a variety of scales using the same data. This is due to the limitations of vertical resolution, data sampling rate, and methods of presentation. However, electrical images reveal layered porosity fabrics which are often missed by other reservoir examinations. Although the images are at the same scale as standard core photography, the electrical characterization often accentuates the porosity distribution which can be masked by other details and textures in the core.

The authors greatly appreciated the suggestions and comments of the reviewers of this paper: Carolyn Knight, A. C. Morton and M. W. Longman

References

AL-JALLAL, I. A. 1987. Diagenetic effects on reservoir properties of the Permian Khuff Formation in Eastern Saudi Arabia. *Proceedings of the 5th Middle East Oil Show*. SPE Paper **15745**.

ALPAY, O. A. 1972. A practical approach to defining reservoir heterogeneity. *Journal of Petroleum Technology* **24**, 841–848.

ATKINSON, B. F. & JOHNSTON, D. 1949. Core analysis of fractured dolomite in the Permian Basin. *Transactions of the AIME* **179**, 129–137

AYESTARAN, L., NURMI, R., SHEHAB, G. & SISI, S. 1989. Well test design and final interpretation improved by integrated well testing and geological efforts. *Proceedings of the 4th Middle East Oil Show*. SPE Paper **17945**, 167–180.

BARBER. A. H. Jr., GEORGE, C. J., STILES, L. H. & THOMPSON, B. B. 1983. Infill drilling to increase reserves — actual experience in nine fields in Texas, Oklahoma, and Illinois. *Journal of Petroleum Technology* **35**, 1530–1538

BARRON, E. J., ARTHUR, M. A. & KAUFFMAN, E. G. 1985. Cretaceous rhythmic bedding sequences: a plausible link between orbital variations and climate. *Earth & Planetary Science Letters* **72**, 327–340.

BATHURST, R. G. C. 1987. Diagenetically enhanced bedding in argillaceous platform limestones: stratified cementation and selective compaction. *Sedimentology* **34**, 749–778.

BEAR, J. 1968. *Dynamics of fluids in porous media*. Elsevier, New York.

BOTTJER, D. J., ARTHUR, M. A., DEAN, W. E., HATTIN, D. E. & SAVRDA, C. E. 1986. Rhythmic bedding produced in Cretaceous pelagic carbonate environments: sensitive recorders of climatic cycles. *Paleoceanography* **1**, 467–481.

BRIE, A., JOHNSON, D. L. & NURMI, R. D. 1985. Effect of spherical pores on sonic and resistivity measurements. *SPWLA 26th Annual Symposium Transactions Paper* **W**.

BROOMHALL, R. W. & ALLAN, J. R. 1985. Regional caprock-destroying dolomite on the Middle Jurassic to Early Cretaceous Arabian shelf. *Proceedings of the 4th Middle East Oil Show. SPE Paper* **13697**.

BULNES, A. C. & FITTING, R. U. 1945. An introductory discussion of the reservoir performance of limestone formations. *Transactions of the AIME*. **160**, 179–186.

BURGESS, C. J. & PETER, C. K. 1985. Formation, distribution, and prediction of stylolites as permeability barriers in the Thamama Group, Abu Dhabi. *Proceedings of the 4th Middle East Oil Show. SPE Paper* **13698**, 165–173.

BYNUM, A. S. & KOEPF, E. H. 1957. Whole-core analysis methods and interpretation of data from carbonate reservoirs. *Journal of Petroleum Technology* **15**, 11–15.

CAROZZI, A. V. & VON BERGEN, D. 1987. Stylolitic porosity in carbonates: a critical factor for deep hydrocarbon production. *Journal of Petroleum Geology* **10**, 267–282.

CHOQUETTE, P. W. & PRAY, L. C. 1970. Geological nomenclature and classification of porosity in sedimentary carbonates. *Bulletin of the American Association of Petroleum Geologists* **54**, 207–250.

COLLINS, R. E. 1961. *Flow of fluids through porous materials*. Reinhold, New York.

DANIELLI, H. M. C. 1988. The Eocene reservoirs of Wafra Field, Kuwait/Saudi Arabia Partitioned Neutral Zone. *In*: LOMANDO, A. J. & HARRIS P. M. (eds) *Giant Oil and Gas Fields*. SEPM Core Workshop **12**, 119–154.

DENNIS, B., STANDEN, E., GEORGI, D. & CALLOW, G. 1987. Fracture identification and productivity prediction in a carbonate reef complex. *62nd Annual SPE Meeting. Paper* **16808**.

D'HEUR, M. 1984. Porosity and hydrocarbon distribution in the North Sea chalk reservoirs. *Marine and Petroleum Geology*. **1**, 211–238.

DOE Panel 1982. *The Impact of Geologic Parameters on Enhanced Recovery*. U.S. Department of Energy. Report CONF-8110160, Bartlesville Energy Tech. Center.

DUNHAM, R. J. 1962. Classification of carbonate rocks according to depositional texture. *In*: HAM W. E. (ed). *Classification of Carbonate Rocks — A Symposium. Memoir of the American Association of Petroleum Geologists* **1**, 108–121.

EHRLICH, R. 1971. Relative permeability characteristics of vugular cores — their measurement and significance. *46th Annual SPE Meeting. Paper* **3553**.

EKSTROM, M. P., DAHAN, C. A., CHEN, M., LLOYD, P. & ROSSI, D. J. 1986. Formation imaging with micro-electrical scanning arrays. *SPWLA 27th Annual Symposium Transactions Paper* **BB**

FLAUM, C., GALFORD, J. E. & HASTINGS, A. 1987. Enhanced vertical resolution processing of dual detector gamma-gamma density logs. *SPWLA 28th Annual Logging Symposium Transactions Paper* **M**.

FRIEDMAN, G. M. 1975. The making and unmaking of limestones or the downs and ups of porosity. *Journal of Sedimentary Petrology* **45**, 379–398.

FURSICH, F. T. 1973. Thalassinoides and the origin of nodular limestone in the Corallian Beds (Upper Jurassic) of Southern England. *Neues Jahrbuch Fur Geologie und Palaontologie Monatshefte* **3**, 136–156.

GILBERT, G. K. 1895. Sedimentary measurements of geologic time. *Journal of Geology* **3**, 121–127.

GUILLOTTE, J. G., SCHRANK, J. & HUNT, E. 1979. Smack-over reservoir: interpretation case study of water saturation versus production. *Transactions of the Gulf Coast Association of Geological Societies* **29**, 121–126.

HALDORSEN, H. G. 1986. Simulation parameter assignment and the problem of scale in reservoir engineering. *In*: LAKE, L. W. & CARROLL, H. B. (eds) *Reservoir Characterization*. Academic, New York, 293–310.

HAM, W. E. 1962. *Classification of Carbonate Rocks — A Symposium. Memoir of the American Association of Petroleum Geologists* **1**.

—— & PRAY, L. C. 1962. Modern concepts and

classifications of carbonate rocks. *Classification of Carbonate Rocks – A Symposium*. W. E. Ham (ed). *Memoir of the American Association of Petroleum Geologists* **1**, 193–223.

Hardman, R. F. P. & Kennedy, W. J. 1980. Chalk reservoirs of the Hod Field, Norway. *In*: *The Sedimentation of the North Sea Reservoir Rocks*. Norsk Petroleums forening, Paper **XI**

Harris, P. M., Frost, S. H., Seiglie, G. A. & Schneiderman, N. 1985. Regional unconformities and depositional cycles, Cretaceous of the Arabian Peninsula. *In*: Schlee, J. S. (ed). *Interregional Unconformities and Hydrocarbon Accumulation. Memoir of the American Association of Petroleum Geologists* **36**, 67–80.

Hartmann, D. J. 1975. Effect of bed thickness and pore geometry on log response. *SPWLA 16th Annual Symposium Transactions Paper* **Y**.

Havlena, D. 1966. Interpretation, averaging and use of the basic geological-engineering data. *17th Annual Technical Meeting of the Petroleum Society of the Canadian Institute of Mining*, 236–247.

Hubbert, M. K. 1940. The theory of groundwater motion. *Journal of Geology*. **48**, 785–944.

Inden, R. F. & Moore, C. H. 1983. Beach environment. *In*: Scholle, P. A. Bebout, D. G. & Moore, C. H. (eds). *Carbonate Depositional Environments. Memoir of the American Association of Petroleum Geologists* **33**, 211–265.

James, N. P. 1983. Reef environment. *In*: Scholle, P. A., Bebout, D. G. & Moore, C. H. (eds) *Carbonate Depositional Environments Memoir of the American Association of Petroleum Geologists* **33**, 211–265.

Jardine, D., Andrews, D. P. Wishart, J. W. & Young, J. W. 1977. Distribution and continuity of carbonate reservoirs. *Journal of Petroleum Technology* **29**, 873–885.

Keith, B. D. & Pittman, E. D. 1983. Bimodal porosity in oolitic reservoir — effect on productivity and log response, Rodessa Limestone (Lower Cretaceous) East Texas basin. *Bulletin of the American Association of Petroleum Geologists* **67**, 1391–1399.

Kieke, E. M. & Hartmann, D. J. 1973. Scanning electron microscope application to formation evaluation. *Transactions of the Gulf Coast Association of Geological Societies* **23**, 60–67.

Koepnick, P. K. 1984. Distribution and vertical permeability of stylolites within a Lower Cretaceous carbonate reservoir, Abu Dhabi, United Arab Emirates. *In*: *Stylolites and Associated Phenomena – Relevance to Hydrocarbon Reservoirs*. Special Publication of the Abu Dhabi National Reservoir Research Foundation. 261–278.

Lasetter, T. J., Waggoner, R. J. & Lake, L. W. 1986. Reservoir heterogeneities and their influence on ultimate recovery *In*: Lake, L. W. & Carroll, H. B. (eds). *Reservoir Characterization*. Academic, New York.

Lewis, J. J. M. 1988. Outcrop-derived quantitative models of permeability heterogeneity for genetic different sand bodies. *63rd Annual SPE Meeting, Paper* **18153**.

Logan, B. W. & Semeniuk V. 1976. Dynamic metamorphism; process and products in Devonian carbonate rocks, Canning Basin, Western Australia. *Geological Society of Australia, Special Publication* **6**.

Longman, M. W. 1981. Carbonate diagnesis as a control on stratigraphic traps. *American Association of Petroleum Geologists Continuing Education Course Notes* **21**.

Mazzullo, S. J. 1981. Facies and burial diagnesis of a carbonate reservoir: Chapman Deep (Atoka) Field, Delaware Basin, Texas. *Bulletin of the American Association of Petroleum Geologists* **65**, 850–865.

Mitchell, J. C., Lehman, P. J., Cantrell, D. L., Al-Jallal, I. A. & Al-Thagafy, M. A. R. 1988. Lithofacies, diagenesis, and depositional sequence: Arab-D member, Ghawar Field, Saudi Arabia. *In*: Lomando, A. J. & Harris, P. M. (eds). *Giant Oil and Gas Fields*. SEPM Core Workshop **12**, 119–154.

Moshier, S. O. & Handford, C. R. 1984. Retention and enhancement of microporosity in micritic limestones – chalkification pathways. *SEPM Annual Midyear Meeting Abstracts* **1**. 57

Nurmi, R. D. 1986. Land of Giants. *Middle East Well Evaluation Review* **1** 6–21.

—— 1988. Geologic interpretation of well logs and seismic measurements in reservoirs associated with evaporites. *In*: Schreiber, B. C. (ed.) *Evaporites and Hydrocarbons*. Columbia University Press. New York, 403–459.

—— A. De la Cruz and Waterhouse, M. 1986a. Carbonate Close-Up. *Middle East Well Evaluation Review* **1**, 22–35.

—— & Frisinger, M. R. 1983. Synergy of core petrophysical measurements, log data, and rock examination in carbonate reservoir studies. *58th Annual SPE Meeting. Paper* **11969**.

—— Taha, M., Waterhouse, M. & Charara, M. 1986b. Fracture detection and geometrical analysis. *7th Biannual Petroleum Congress of Turkey* 203–213.

Pittman, E. D. 1971. Microporosity in carbonate rocks. *Bulletin of the American Association of Petroleum Geologists* **55**, 1873–1878.

Potter, P. E. & Pettijohn, E. J. 1963. *Paleocurrents and Basin Analysis*. Springer, Berlin.

Ricken, W. 1986. *Diagenetic Bedding*. Springer, Berlin.

Rubbens, I. B. H. M., Murat, R. C. & van Keullen, J. 1983. Seismic lateral prediction in chalky limestone reservoirs offshore Qatar. *SPE Middle East Oil Technical Conference*, Bahrain, SPE Paper **11451**.

Schneider, F. N. & Owens, W. W. 1970. Relative permeability studies of gas-water flow flowing solvent injection in carbonate rocks. *Journal of the Society of Petroleum Engineers* **16**, 23–29

Schwarzacher, W. & Fischer, A. G. 1982. Limestone-shale bedding and perturbations of the Earth's orbit. *In*: Einsele G. & Seilacher, A.

(eds) *Cyclic and Event Stratification*. Springer, Berlin. 72–95.

SERRA, O. 1986. *Advanced Interpretation of Wireline Logs*. Schlumberger.

SHERMAN, H. & LOCKE, S. 1975. Depth of investigation of neutron and density sondes for 35 percent-porosity sand. *SPWLA 16th Annual Symposium Transactions Paper* **Q**.

SIMPSON, J. 1985. Stylolite-controlled layering in an homogeneous limestone: pseudo-bedding produced by burial diagenesis. *Sedimentology* **32**, 495–505.

WANLESS, H. R. 1979. Limestone response to stress: pressure solution and dolomitization. *Journal of Sedimentary Petrology* **54**, 437–462.

WARDLAW, N. C. 1978. Geological studies. *In*: *A Treatise on Hydrocarbon Recovery from Carbonate Reservoirs*. Petroleum Recovery Institute. Research Report **RR-36**, 1–86.

—— 1980. The influence of pore structure in rocks on the entrapment of oil. *In*: *Facts and Principles of World Petroleum Occurrence*. Canadian Society of Petroleum Geologists, Memoir **6**, 193–208.

WATERHOUSE, M., CHARARA, M., NURMI, R. 1987. Focus on fractures. *Middle East Well Evaluation Review* **3**, 14–27.

WATFA, M. & NURMI, R. 1987. Calculation of saturation secondary porosity, and producibility in complex Middle East carbonate reservoirs. *SPWLA 28th Annual Symposium Transactions Paper* **NN**.

WILSON, J. L. 1985. Limestone and dolomite reservoirs. *In*: HOBSON, G. D. (ed). *Petroleum Geology — 2*. Applied Science Publishers Essex, 1–51.

WINN, R. H. 1957. Log interpretation in heterogeneous carbonate reservoirs. *SPE Permian Basin Oil Recovery Conference*. Paper **818-G**.

New techniques in lithofacies determination and permeability prediction in carbonates using well logs

M. H. DORFMAN, J.-J. NEWEY & G. R. COATES

Department of Petroleum Engineering, The University of Texas at Austin, Austin, TX 78712, USA

Abstract: Carbonate reservoirs pose many challenges to geologists and engineers. Lithofacies variations are often the key to the determination of commercial value and efficient reservoir exploitation. Thus, techniques which permit the identification of the involved lithofacies would be useful. This paper illustrates a successful application of multi-dimension techniques in two-dimensional space, by using cross plots of a variety of well logs that have a principle component sensitivity to the lithofacies. This paper also examines the appropriateness of techniques which relate log data to core-derived permeability. These approaches have previously been associated with the intergranular shaly sand rocks and not carbonates. The technique shown approaches this from a Kozeny-core correlation which is used to predict the surface area parameter. This is then followed by development of a link to log-derived bulk water values. The log derived variables can be made to work providing irreducible conditions are present or can be established.

Identification of carbonate rock type and texture from well-log response can serve as a valuable aid in geological characterization of reservoirs and in development of depositional models for both exploration and production. Such identification can also be used to trace lateral extent and continuity of individual reservoir zones within complex reservoir systems, in many cases optimizing development well locations for improved primary drainage patterns and for efficiency in subsequent enhanced recovery operations.

Geologists have traditionally relied on core data for sedimentological interpretation, but complete coring in all field wells is rare because of high cost; total core recoveries are also rare because of mechanical problems. As a result, the technique of matching limited core facies data with well-log response, and subsequent facies identification from well logs, can often lead to complete facies discrimination and refinement of depositional models. Since open-hole logging suites are routinely run in all wells, they provide complete data throughout both the reservoir and adjoining rock systems.

The technique, which has been successfully in several fields, involves correlating facies from core description with log response and crossplotting various log parameters of individual facies by a variety of methods. The key to selection of crossplots which distinguish sedimentary facies is for the facies to have textural and mineralogical characteristics which affect log measurements. Carbonate systems consist typically of a complex series of vertically sequenced rocks that may be classified as clean grainstones and packstones, grading to mud-supported wackestones and mudstones. Porosity within reservoir systems may be intergranular, intraparticle, vuggy or moldic, and sometimes includes fractures or bimodal porosity systems. Regardless of the classification scheme employed, characteristic well-log response patterns are often discernable within individual facies and can be used for both geological characterization and engineering quantification of reservoir performance.

Using logs to identify sedimentary facies is also expected to help predict permeabilities. Such correlations are approachable by using permeability information from cores of the facies studied, and a model, like Kozeny's, which links such parameters to permeability.

Current studies in lithofacies determination

Well-log crossplots have long been in general use for determination of mineralogy and porosity (Raymer & Biggs 1963; Burke *et al.* 1969) and also more specifically for discrimination of carbonate facies (Asquith 1979). Recently Basham & Dorfman (1983), Keith & Pittman (1983), Dorfman & Dupree (1985) and Kamon & Dorfman (1985) successfully distinguished facies and associated reservoir characteristics using primarily well-log responses correlated with limited core data. Dorfman & Dupree (1985) gave an outline of a method that might be used for permeability prediction in carbonates which is developed in this paper.

Variation in sonic and neutron porosity was used to differentiate three major facies (grainstone, wackestone and lime mudstone) in the

From HURST, A., LOVELL, M. A. & MORTON, A. C. (eds), 1990, *Geological Applications of Wireline Logs* Geological Society Special Publication No. 48, pp. 113–120

Sligo Formation (Lower Cretaceous) in South Texas (Basham & Dorfman 1983), where porosity is above 5% (Fig. 1). The formation has intergranular porosity in the grainstone reservoir facies with limited diagenesis in the supratidal facies, and is comparatively uncomplicated. Distinction of facies depends upon the difference between total porosity as recorded by the neutron log and porosity calculated from the sonic log. This difference, caused by low porosity values from the sonic log due to poorly connected porosity, is usually referred to as 'secondary porosity' by the log analyst.

Keith & Pittman (1983) examined a more complex system involving both a unimodal and bimodal pore system within a grainstone facies in the Rodessa Formation (Lower Cretaceous), Running Duke Field, Texas. In this gas field, water saturation calculations indicated similar saturation values in both pore systems, but water-free gas was produced from only the bimodal pore system. The two systems were differentiated by use of an R_t against R_{xo}/R_{mf} crossplot. Differences in pore geometry often create differences in the depth of mud filtrate invasion or the efficiency with which mud filtrate displaces hydrocarbons and formation water. Depending upon saturations and the values of R_{mf} and R_w, differences in filtrate invasion characteristics here resulted in a unique contrast between formation resistivity in the undisturbed portion of the reservoir, R_t, and resistivity in the region flushed by mud filtrate, R_{xo}.

The work of Kamon & Dorfman (1985) involved a complex grainstone/mudstone sequence in the San Andres Formation (Permian), West Seminole Field, Texas. In this oil field, extreme diagenesis is found in the dolomitic rocks, with both moldic and intraparticle porosity present throughout the rock sequence, and interparticle porosity present in the grainstone facies. The two facies were distinguished by use of an $R_t/(R_{xo}/R_{mf})$ against $R_t/(R_s/R_z)$ crossplot (Fig. 2). As in the case of Keith & Pittman (1983), the crossplot works because of the differences in mud filtrate invasion.

Dorfman & Dupree (1985) examined the San Andres Dolomite (Permian), Hanford Field, Texas, a series of shoaling upward sequences with coarser grainstones deposited on the top of the shoals; mudstones and wackestones, representing lower depositional energy, were deposited away from the shoal and along its flanks (Kumar & Foster 1982). The depositional cycle comprises approximately 1000 feet of cyclic sediments, capped by supratidal deposits of anhydrite and terrigenous shales.

Only two of the 35 field wells were cored, but geological descriptions from the cores were sufficient to provide facies information for log studies. Digitized logging parameters for each

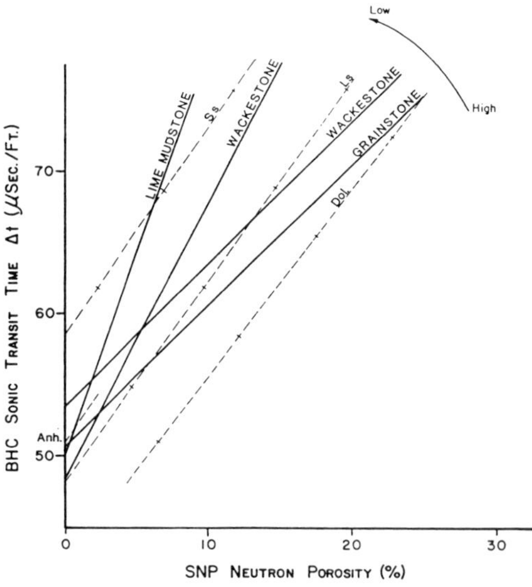

Fig. 1. Cumulative sonic-neutron crossplot showing combination of observed linear data trends. Note transition from high energy to low energy facies deposition.

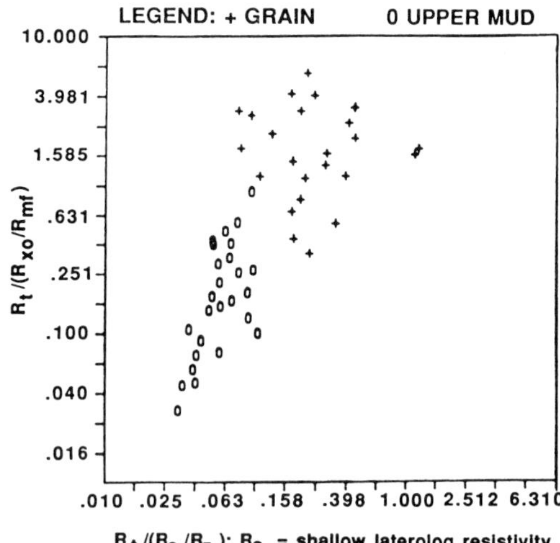

Fig. 2. Resistivity crossplot from logs from the West Seminole Field where R_t is the true resistivity, R_{xo} is the resistivity in the flushed zone, R_{mf} is the mud filtrate resistivity, R_s is the shallow laterolog resistivity and R_z is the estimated resistivity of mud filtrate and formation water in the partially flushed zone.

facies were then crossplotted following the method of Fertl (1981), and it was found that a plot of R_t against ϕ_ρ (Fig. 3) as well as a plot of R_t against ϕ_N, could be used to distinguish the various facies. Since all R_t values were obtained within the oil column at irreducible S_w, the principal variant that differentiates facies is presumed to be the cementation exponent, m, in the Archie equation. The value of m depends upon the tortuosity of the electrically conductive paths provided by interstitial water. In general, higher values of m are associated with more tortuous pore geometries. In this case, successively finer grained sediments (e.g., lime mudstones) exhibit more tortuous pore geometries than the grainstones.

We conclude that, generally, crossplots involving functions of porosity against resistivity differentiate facies due to variations in the cementation exponent, m while crossplots of resistivity values, e.g. R_t against R_{xo}/R_{mf}, represent changes in the characteristics of mud filtrate invasion when plotted in zones at or near irreducible water saturation conditions providing the grain size characteristic is uniform. A more detailed study of m value relationships in carbonates is found in Lucia (1983).

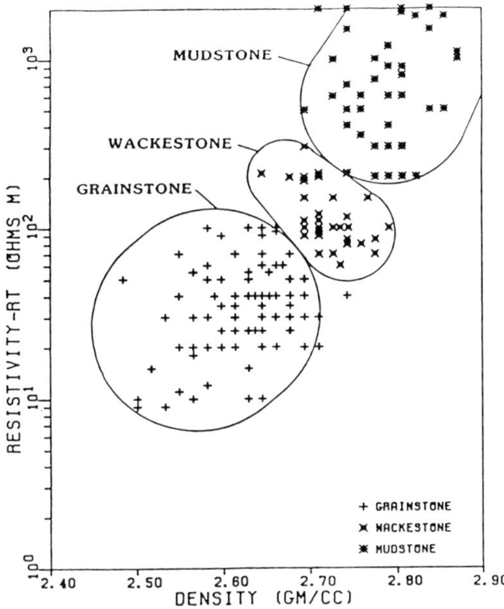

Fig. 3. Crossplot of R_t against bulk density, Hanford Field.

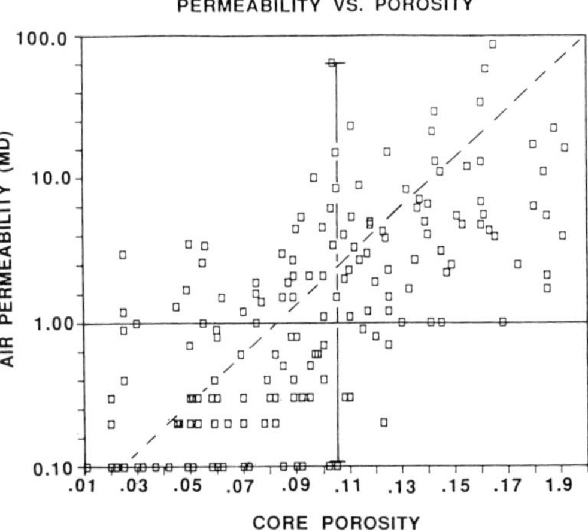

Fig. 4. Plot of core permeability versus core porosity for the Boecker 215–1 well, illustrating the poor linear trend. Note the variation in permeabilities for the 10–11% porosity range.

Prediction of permeability in carbonates

Background

Estimation of permeability from core data or from well logs is not new. In fact, the process most commonly used is to estimate permeability from core data using a semilog plot of K against ϕ. While this technique is often useful in fairly homogeneous sandstones, it is often impossible to apply in heterogeneous carbonates. Figure 4 shows a K against ϕ crossplot in the San Andres Dolomite, Hanford Field, Texas, from cores of one well. As a result, other methods must be applied to attempt to estimate permeability. The concept used in this work was first presented by Wyllie & Rose (1950). Their work proposed use of the Kozeny equation in estimating the permeability of consolidated porous media having regular pore geometry. A later paper by Wyllie & Spangler (1952) first implied that permeability could be estimated using electrical resistivity measurements and a modified form of the Kozeny equation.

The problem in applying the Kozeny equation to complex porous media has been obtaining a direct measure of specific surface area and a parameter called the Kozeny constant. Wyllie & Spangler showed that values of specific surface area, obtained using different techniques of measurement on the same porous medium, varied greatly. However, specific surface area of the pore surface is a controlling factor in the permeability of carbonates. A log–log plot (Fig. 5) shows the effect of particle size on the permeability of uniformly cemented, non-vuggy

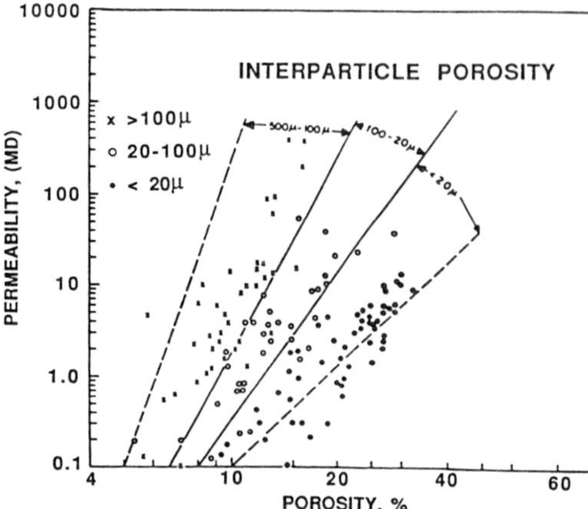

Fig. 5. Relationship between porosity and permeability for various particle size groups in uniformly cemented, non-vuggy carbonate rocks (after Lucia 1983).

carbonates from which Lucia (1983) concluded that the size and amount of intra-particle porosity were major factors controlling permeability in carbonates. Brooks & Purcell (1952) presented a laboratory method for measuring specific surface area using gas adsorption. Their results, presented for a variety of sandstones and limestones, showed good repeatability of measurement. Unfortunately, measurement of specific surface area by gas adsorption requires elaborate apparatus which is not standard equipment in a typical core laboratory.

The difficulty in direct measurement of specific surface area in the laboratory has been circumvented in this study. Incorporating a modified form of the Kozeny equation, values for core permeability and porosity are plugged into the equation, and specific surface area values are then back-calculated. The method avoids the errors in measuring specific surface area using a visual method (Wyllie & Spangler 1952). Furthermore, values obtained with the proposed method compare well in range and magnitude, to values obtained for other carbonate rocks using the gas adsorption techniques.

The other problem in applying the Kozeny equation to complex porous media had been measurement of the tortuosity parameter. Wyllie & Rose (1950) suggested that tortuousity could be estimated by measurement of the formation resistivity factor. Winsauer (1952) presented theory and laboratory measurements that substantiated the earlier premise of Wyllie and Rose. They found that for clean, brine-saturated sandstones, tortuosity is related to the formation resistivity factor and the porosity of the rock. Both of these parameters can be directly measured using well logs in water-bearing formations, and estimated in oil-bearing formations from secondary correlations.

In theory then, tortuosity could also be estimated in clean formations using well-log measurements. Furthermore, if a correlation could be obtained between log-measured parametres and specific surface area, permeability could be estimated in wells without core, using the Kozeny equation and well logs. However, in dealing with logs through oil-saturated zones, it is important to recognize that a correlation should exist between specific surface area and irreducible water saturation (Archie 1952); this correlation is elaborated on in this work.

Application of the Kozeny Model

The Kozeny equation is an analytical solution for fluid flow through porous media and is obtained by combining the Hagen–Poiseuille law with Darcy's law. The general Kozeny formula is

$$K = \frac{\phi^3 \times 10^8}{2\tau^2 S_v^2(1-\phi)^2} \tag{1}$$

where K is the absolute permeability of the porous media (in darcies), ϕ is the porosity as a fraction of bulk volume, S_v is the specific surface area of mineral constituents in cm^{-1} and τ is the tortuosity factor (dimensionless) which has been shown (Winsauer 1952) to be related to formation factor and porosity in the following manner:

$$\tau^2 = F\phi^{-2}. \tag{2}$$

The formation factor commonly used for carbonates is

$$F = \phi^{-2}. \tag{3}$$

Combining (2) and (3) into (1), gives

$$K = \frac{\phi^4 \times 10^8}{2S_v^2(1-\phi)^2}. \tag{4}$$

The Kozeny equation assumes laminar incompressible flow with end effects ignored. This application assumed no end effects in laboratory measurements of permeability, which are normally minimized. Another assumption is lack of channeling in the porous media, a factor which might be violated if fractures or large vugs are present. Since the predominant porosity in this San Andres Dolomite is intergranular, this assumption is probably not violated. The final assumption is a cementation exponent of 2, which cannot be validated in our case, although similar formations in nearby fields have an average cementation exponent closely approximating 2.

In order to proceed with estimation of permeability, core porosities from one well (Boecker No. 215–1) were compared with log porosity obtained by density/neutron crossplot porosity (Fig. 6). Some deviations occur in extremely low porosity ranges, but in general there is good agreement between core-and log-derived porosity. The modified Kozeny equation (4) was then rearranged as follows:

$$S_v = \left(\frac{\phi^4 \times 10^8}{2K(1-\phi)^2}\right)^{1/2} \qquad (5)$$

and core porosity and permeability for 336 core samples from this well were used to obtain values for S_v (Fig. 7). A distribution of S_v values is shown in Fig. 8, with values ranging from 100 cm^2/cm^3 to just under 7000 cm^2/cm^3. Modal values of 1000 to 2000 cm^2/cm^3 (Fig. 7) are similar to gas adsorption measurements made on some Ordovician limestones (Brooks & Purcell 1952). We also note that S_v increases with a decrease in grain size, a finding that is in agreement with Lucia (1983). However, since pore size, pore geometry, and roughness of pore walls are controlling factors in formation permeability, it is beneficial to think of specific surface area in terms of pore size and geometry rather than grain size.

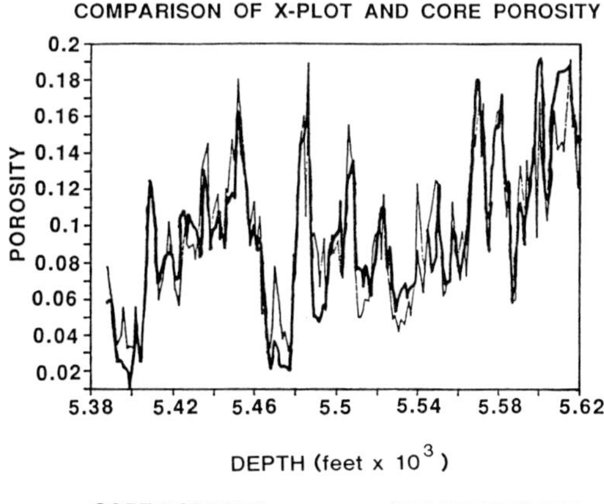

Fig. 6. Comparison of the (CNL/FDC) crossplot porosity curve with the core porosity curve for the Boecker 215–1 well. Note the division of the two curves at low porosity ranges.

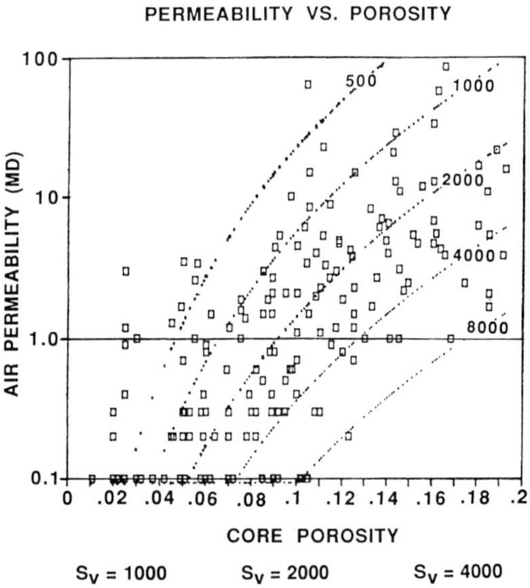

Fig. 7. Core permeability plotted against core porosity for Boecker 215–1. Curved lines represent solutions to the Kozeny equation for constant values of specific surface area.

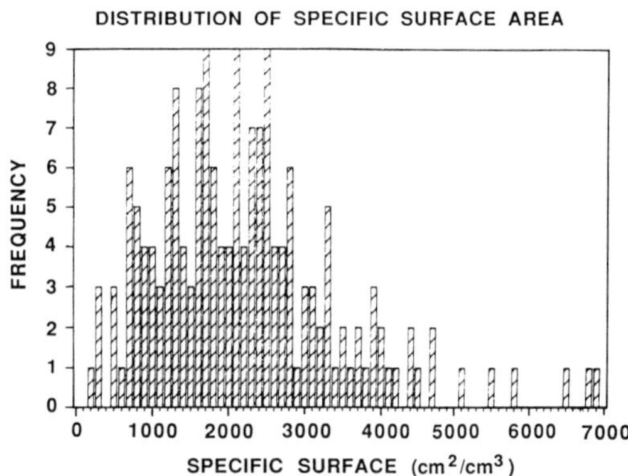

Fig. 8. Frequency distribution of calculated specific surface areas, from the San Andres producing horizon in Boecker 215–1, Hanford Field (Gaines County, Texas).

Petrophysical studies have shown that, in a water-wet reservoir, the irreducible or connate water saturation is a function of pore size. In general, a rock with smaller pores will have a higher irreducible water saturation than a rock with larger pores. This phenomenon is caused by the interfacial forces that exist between the oil-and water-wet phases in the reservoir and the specific surface area. These forces are inversely proportional to the radii of the pores in which the fluids interact. In the smaller pores, higher capillary forces are exhibited. As a result, water tends to reside preferentially in the smaller pores. In the oil column of a carbonate reservoir, the irreducible water saturation will tend to vary according to the predominant pore size. In addition, if water is exposed to highly oil-saturated rock in the oil column, it will tend preferentially to invade the portion of the reservoir having smaller pores.

Since specific surface area is directly related to pore size, we focused on correlating specific surface area to log-derived values of water saturation. In addition, since the wells at Hanford Field were drilled with a saltwater-based mud, the possibility of an imbibition effect was also examined. Theoretically, if invasion of mud filtrate did occur in the oil column, it would invade preferentially at levels with predominantly smaller pores. Reservoir sections having larger pores would have little or no invasion.

Specific surface area was plotted against various log-derived saturation values. A correlation was observed between specific surface area and (1) ϕS_w, the bulk volume fraction of water in the virgin zone, and (2) ϕS_{xo}, the bulk volume fraction of mud filtrate in the invaded zone.

The bulk volume fraction of water in the virgin zone is calculated using the following formula:

$$\phi S_w = (R_w/R_t)^{1/2} \quad (6)$$

where S_w is the water saturation in the virgin zone, R_w is the resistivity of the formation water (in ohm m) and R_t is the true resistivity of the virgin zone (in ohm m). The bulk volume fraction of filtrate in the invaded zone is calculated in a similar manner:

$$\phi S_{xo} = (R_{mf}/R_{xo})^{1/2} \quad (7)$$

where S_{xo} is the mud filtrate saturation in the invaded zone, R_{mf} is the resistivity of the mud filtrate (in ohm m) and R_{xo} is the true resistivity of the invaded zone (in ohm m). Both (6) and (7) assume a cementation exponent of 2 and a saturation exponent of 2. As discussed earlier, the assumption of 2 as a cementation exponent is justified. The saturation exponent is arbitrarily taken as 2 also.

The relationship between the quantities expressed in (6) and (7) and specific surface area can be seen in a plot of ϕS_{xo} against ϕS_w (Fig. 9). Note that $\phi S_{xo} < \phi S_w$ occurs; this is probably caused by incomplete flushing by the invasion process. The plot has specific surface area, divided by 1000, labelled as the Z axis. The general trend observed is that low values of specific surface area, located in the lower left hand corner of the plot, correspond to low values of both ϕS_{xo} and ϕS_w, and high values of specific surface area correspond to higher ϕS_{xo} and ϕS_w amounts in the upper right-hand corner of the plot. This is further illustrated by adding the $\phi S_{xo} \times \phi S_w$ = constant hyperbolae for 0.0001, 0.0002, 0.0008 and 0.0016. These observations are reasonable since high specific surface areas mean smaller pores. Smaller pores result in higher irreducible water saturations in the virgin zone, explaining the increased ϕS_w. Furthermore, mud filtrate tends to invade zones preferentially where the predominant pore size is smaller. This phenomenon explains the high ϕS_{xo} estimates for higher specific surface areas.

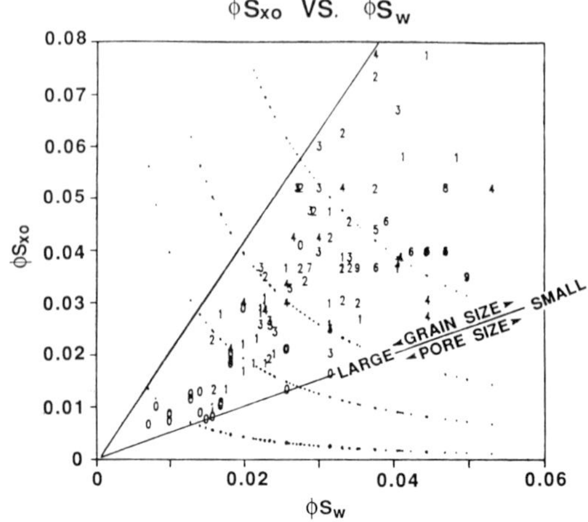

Fig. 9. Well-log derived values of ϕS_{xo} against ϕS_w for core data in the oil column of Boecker 215–1. Points are labelled with specific surface areas/1000.

A more obvious correlation is observed when specific surface area is plotted against the product of ϕS_{xo} and ϕS_w (Fig. 10). The plot shows that an increase in specific surface area corresponds to an increase in both ϕS_{xo} and ϕS_w. By performing a linear regression fit through the data we now have a model for predicting specific surface area. By calculating ϕS_{xo} and ϕS_w using resistivity logs, we can obtain a predicted value of specific surface area in wells that have no core. Then, an estimate of permeability is obtained by plugging the predicted specific surface area value and a log-derived estimate of porosity into (4).

It is important to stress that this correlation, based on an irreducible water saturation, applies only to the oil column of the reservoir. Some other method must be used in zones below the oil/water contact or if hydrocarbon changes from oil to gas. Since the zone of interest is normally

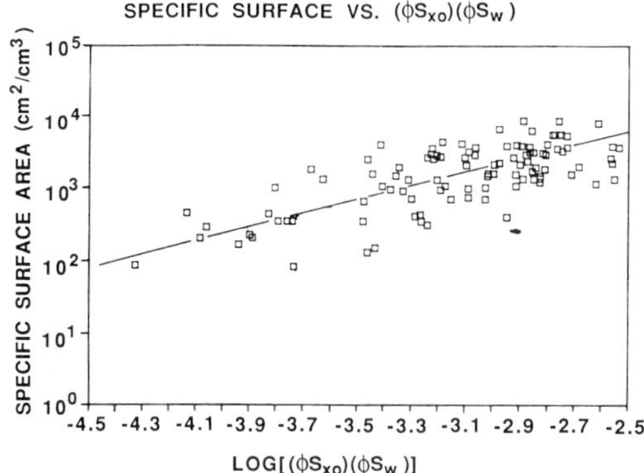

Fig. 10. Correlation between specific surface area and $S_{xo}\phi S_w$ in the oil column of Boecker 215−1.

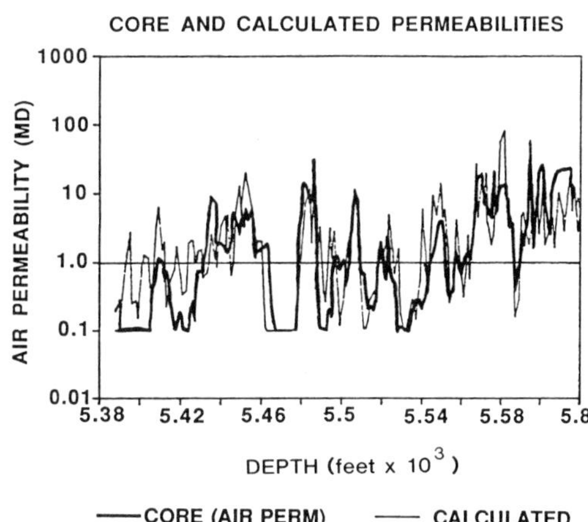

Fig. 11. Comparison of calculated and core-derived permeabilities in Boecker 215−1.

above the oil/water contact, this should not be a major problem.

Results

The ultimate test of this Kozeny model is comparison of permeability estimates with core-derived permeabilities (Fig. 11). Differences occur primarily in the very low permeability ranges and occasionally in the very high permeability ranges. The latter may result from the few large vugs that occur in the San Andres and are mentioned in the core description. Another probable cause of error is in sampling of cores. Since carbonate cores often have extreme changes in permeabilities over small intervals, it is common for physical sampling to include the best portion of the core for analysis. This may not, in some cases, be representative of the entire interval. In general, the technique appears to give permeabilities based on well-log response that compare favourably to those obtained by laboratory measurement of cores.

In the Hanford Field, all wells have been subjected to permeability estimation by the technique described herein. A cross section of wells across the field shows that the various zones have characteristic permeability profiles

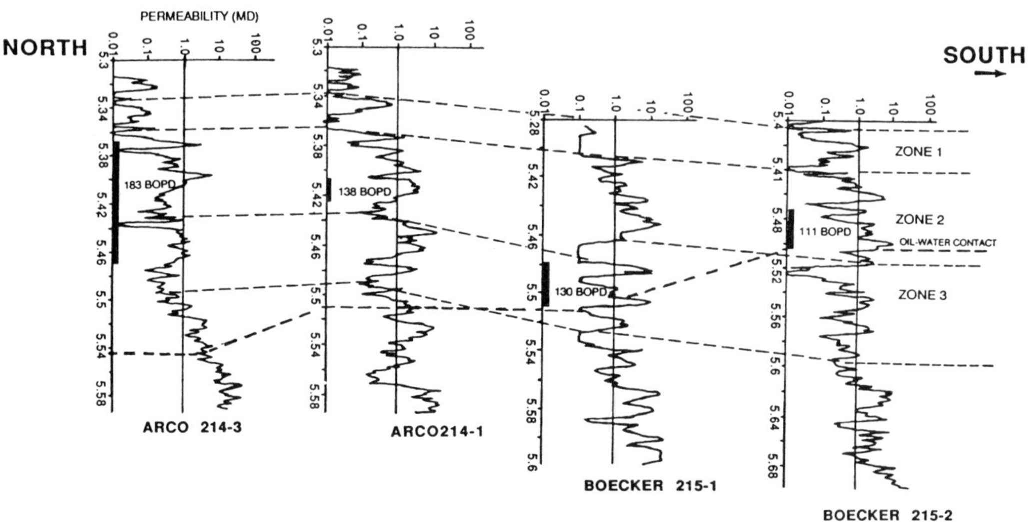

Fig. 12. North−south cross section through the Hanford Field, Gaines County, Texas. Curves represent estimated permeabilities using the Kozeny equation. Depths in thousands of feet.

that can be correlated field-wide (Fig. 12). We can also see that not all wells are completed in the same permeable intervals, and that in some cases, other zones in individual wells contain permeabilities as good as, or better than, the completion interval. Since only two wells in this field were cored, the log-derived permeabilities appear to offer an alternative method for correlating individual permeable zones and of estimating the permeability in each; they also provide a method for consistent well completions in comparable zones for effective drainage and for more efficient EOR operations.

References

ARCHIE, G. E. 1952. Classification of Carbonate Reservoir Rocks and Petrophysical Considerations, *Bulletin of the American Association of Petroleum Geologists*, **36**, 278–298.

ASQUITH, G. D. 1979. *Subsurface Carbonate Depositional Models: A Concise Review*, The Petroleum Publishing Company, Tulsa, Oklahoma.

BASHAM, M. & DORFMAN, M. 1983. Facies Characterization of Carbonates by Use of Well Logs, Sligo Formation (Lower Cretaceous) South Texas, *Transactions, SPWLA* **2**, paper SS.

BROOKS, C. S. & PURCELL, W. R. 1952. Surface Area Measurements Made on Sedimentary Rocks, *Transactions, American Institute of Mining and Metallurgical Engineers, Petroleum Development and Technology*, **195**, 289–296.

BURKE, J. A., SCHMIDT, A. W. & CAMPBELL, R. L. 1969. The Litho-porosity Crossplot, *The Log Analyst*, **10**, 25–43.

COATES, G. R. & DUMANOIR, J. L. 1973. A New Approach to Improved Log Derived Permeability, *Transactions, SPWLA*, Paper R.

DORFMAN, M. & DUPREE, J. 1985. New Techniques for Facies Determination in Carbonate Reservoirs Using Well Logs, *Proceedings, Third European Oil Recovery Conference, Rome*, **1**, 13–23.

FERTL, W. 1981. Open Hole Crossplot Concepts: A Powerful Technique in Well Log Analysis, *Journal of Petroleum Technology* (March), 535–549.

KAMON, K. & DORFMAN, M. 1985. Identification of Carbonate Depositional Texture from Well Log Response: West Seminole Field, Gaines County, Texas, *Proceedings, Permian Basin Research Conference*, Society of Economic Paleontologists and Mineralogists, **86–26**, 75–78.

KEITH, B. D. & PITTMAN, E. D. 1983. Bimodal Porosity in an Oolitic Reservoir: Effect on Productivity and Log Response, Rodessa Limestone (Lower Cretaceous) East Texas, *Bulletin of the American Association of Petroleum Geologists*, **67**, 1291–1399.

KUMAR, N. & FOSTER, J. D. 1982. Effect of an Underlying Bioherm on the San Andres (Permian) Reservoir and Trap, Handford Field, Gaines County, Texas, *Bulletin of the American Association of Petroleum Geologists*, **66**, 2571–2583.

LUCIA, J. 1983. Petrophysical Parameters Estimated from Visual Descriptions of Carbonate Rocks: A Field Classification of Carbonate Pore Space, *Journal of Petroleum Technology*, (March), 629–637.

RAMER, L. & BIGGS, W. P. 1963. Matrix Characteristics Defined by Porosity Computations, *Transactions, SPWLA*, Paper X.

WINSAUER, W. O. 1952. Resistivity of Brine Saturated Sands in Relation to Pore Geometry. *Bulletin of the American Association of Petroleum Geologists*, **36**, 253–277.

WYLLIE, M. R. J. & ROSE, W. D. 1950. Some Theoretical Considerations Related to the Quantitative Evaluation of the Physical Characteristics of Reservoir Rock from Electrical Log Data. *Journal of Petroleum Technology*, **189**, 105–118.

—— & SPANGLER, M. B. 1952. Application of Electrical Resistivity Measurements to the Problem of Fluid Flow in Porous Media, *Bulletin of the American Association of Petroleum Geologists*, **36**, 359–403.

Stratigraphic correlation

Sediment cyclicity from well logs

P. F. WORTHINGTON

BP Research, Sunbury Research Centre, Chertsey Rd, Sunbury-on-Thames, Middlesex, TW16 7LN, UK

Abstract: The application of wireline well logs to the recognition and interwell correlation of sediment cyclicity is governed by the resolving powers of the different logging tools relative to measured sedimentation rates and cyclic periodicities. These variables, which are not always well defined, give rise to diverse capabilities of well logs in recording and correlating cyclic events. It is therefore important that criteria be established for the successful application of wireline logs in cyclostratigraphy, especially when a succession reflects several lithoperiodicities due to climate forcing. Quantification of these criteria leads to the introduction of a log association parameter (LAP) to compare the effective cyclic resolutions of different logging tools in diverse geological environments. This parameter relates sedimentation rates and tool vertical resolutions to the targeted time-periods of cyclicity. It is tool- and site-specific but is independent of formation lithology and chemistry. High LAP values, above the critical level of unity, suggest a strong cyclic resolution. Similar supracritical LAP values for (different) logs in different boreholes imply a high degree of potential interwell correlation. The benefits of a LAP-based approach are evident from illustrative examples, even at this preliminary stage. Field application ideally requires the conjunctive use of log and core information, a strategy which reflects the current trend in earth science towards a more meaningful integration of data measured at different scales.

Cyclic sedimentation involves a series of lithological elements (A, B, etc.) repeated through a succession. These elements may be grouped together (AB) and described as a cycle. The simplest form of cyclicity is the AB AB type. More complex forms, e.g. ABC ABC, may contain AB AB components of different length and period, perhaps due to climate forcing (Arthur & Garrison 1986). A standard text on cyclic sedimentation is that of Duff *et al.* (1967).

The recognition of sediment cyclicity from wireline well logs that respond predominantly to mineralogy and porosity constitutes an important potential facility for stratigraphic correlation. There are many examples of cyclicity in the sedimentary record (e.g. Gradstein *et al.* 1985; Fischer 1986). Naturally, these show varying periodicities, regional correlatabilities, and degrees of completeness. Such diversity requires a full and representative sampling of the relevant sedimentary interval if cyclic phenomena are to be investigated meaningfully. The essential requirements for effective sampling are continuity, resolution and characterization at in situ conditions. Core data alone are unlikely to satisfy the criterion of continuity and may not approximate in situ conditions. Wireline log data do constitute continuous records at in situ conditions but they may not possess the resolving power to indicate cyclic changes on a fine scale. The optimum approach is to use core data and well logs conjunctively, a strategy that requires a knowledge of the limitations and capabilities of well logs in recording cyclic phenomena.

This paper evaluates quantitatively the effective resolving powers of different logging tools over intervals which show Milankovitch cyclicity. The scope of the study includes, but extends beyond, an evaluation of the recognition through logs of individual litho-elements within a cyclic succession. The aim is to provide an initial quantitative foundation for the eventual interwell correlation of log-derived sediment cyclicity. It is hoped that the criteria established will contribute to a more effective application of wireline logs in cyclo-stratigraphy as a whole, not just within formation or reservoir units but perhaps more extensively in intra- and inter-basin studies.

Log applications to sediment cyclicity

Many investigators have used well logs to study cyclic variations in sedimentary deposition. Some recent examples are cited in Table 1: three of these examples relate specifically to Milankovitch cyclicity. Well logs have been used to study sediment cyclicity over an age range from Devonian to Tertiary in situations which reflect various causative factors. Furthermore, the use of well logs has allowed cyclic correlation both within and between individual basins. However, the Lamont–Doherty Borehole Research Group (1987) have demonstrated that

From Hurst, A., Lovell, M. A. & Morton, A. C. (eds), 1990, *Geological Applications of Wireline Logs* Geological Society Special Publication No. 48, pp. 123–132

Table 1. *Examples of log applications to sediment cyclicity*

Citation	Formation/location	Age	Principal logs	
Watney (1980)	Lansing & Kansas City Groups, NW Kansas	Carboniferous	Gamma ray	Neutron
Lott & Strong (1982)	Sherwood Sandstone, Southern England	Triassic	Gamma ray	Sonic
Whittaker *et al.* (1985)	Old Red Sandstone, Wales	Devonian	Gamma ray	Sonic
Fracasso & Hovorka (1986)	San Andres, Texas	Permian	Gamma ray	
Penn *et al.* (1986)	Callovian–Kimmeridgian, Eastern England Shelf	Jurassic	Gamma ray Density	 Sonic
* Golovchenko *et al.* (1987)	Hemipelagic Muds, Weddell Sea	Miocene/Pliocene	Induction Gamma ray	
* Imbrie *et al.* (1987)	Niobrara, Wyoming Chalk, Northern Germany	Cretaceous	Microlog Normal resistivity	Neutron
* Lamont–Doherty Borehole Research Group (1987)	Marine sediments, Labrador Sea	Pliocene	Induction Geochemical	Sonic

* Applications to Milankovitch cyclicity.

well log resolution can be inadequate to resolve cyclicity at lower sedimentation rates. The quantification of log cyclic resolution, in the context of Milankovitch but with a general application, forms the substance of this paper.

The cyclic model

In order to study the log resolution of cyclic events, it has been useful to generate a simplified cyclic model which is internally consistent from a petrophysical standpoint, i.e. the assigned petrophysical parameters are physically compatible. The cyclic model has been specified to be a composite of three discrete periodicities (20, 40, 100 ka), of equal amplitude, which approximate (Table 2) Milankovitch climate-forcing orbital cycles (Arthur & Garrison 1986). The approximation permits a composite cyclicity which repeats every 200 ka (Fig. 1). Although this model does not allow amplitude variations within its cyclic components, and assigns to each of the latter a single frequency rather than a frequency band, it is considered sufficiently realistic for an initial study of log cyclic resolution to be meaningful.

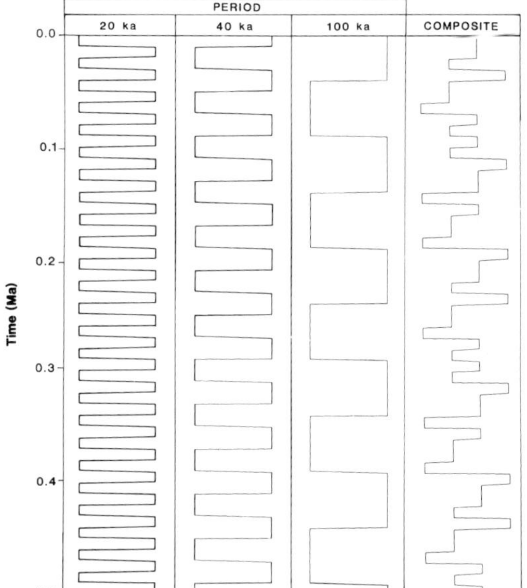

Fig. 1. Cyclic model (fourth track) with three constituents that approximate astronomically-driven palaeoclimatic cycles.

Table 2. *Cyclic periods*

Cause	Milankovitch (ka)	Model approximation (ka)
Precession	19 23	20
Obliquity	41	40
Eccentricity	95 413	100

The detail of the composite model is determined by the shortest constituent period. For a given sedimentation rate that is intuitively 'measured' post-compaction and presumed constant over model intervals, temporal cycles can be transformed directly to spatial cycles. In our composite model (Fig. 1), the half-length of the

20 ka cycle corresponds to the thickness of the smallest element. The thickness of the turreted (or blocked) segments can be varied, an exercise which requires a proportionate change in the implied sedimentation rate according to the relationship.

$$\text{half-period (ka)} = \frac{\text{segment thickness (m)}}{\text{sedimentation rate (m/Ma)}} \times 10^3. \quad (1)$$

A key issue is the effectiveness of logging tools in resolving segments of different thickness, for any inferred spatial resolutions can subsequently be transformed to temporal resolutions through equation (1). This aspect will be explored for three different logging tools, the natural gamma ray, the density and the induction (resistivity) devices. Although targeted on Milankovitch cyclicity, the study does provide useful guidelines for well-log correlations for any cyclic sequences.

The resolving power of logging tools in a cyclic succession can be investigated using the model of Fig. 1. This requires that realistic values of 'measurable' petrophysical parameters be pre-assigned to the various turreted segments which, through the mathematics of summation, can assume one of only four levels. This assignment has been effected in accordance with the following petrophysical specifications.

(i) The model formation is a brine-saturated 'shaly' sand, i.e. one containing uniformly dispersed but unspecified clay minerals, in which the shaly constituents have the same density as silica (2.65 g cm^{-3}): the four permissible bulk densities are 2.2, 2.3, 2.4, and 2.5 g cm^{-3}.

(ii) These densities correspond to a porosity range of 10–30%: over this range changes in porosity are attributable solely to changes in the degree of (uniformly dispersed) shaliness.

(iii) The gamma-ray log is a faithful linear shale indicator, i.e. the log response is directly proportional to clay-mineral content: the four permissible gamma-ray levels are 20, 50, 80 and 110 API units.

(iv) These gamma-ray levels correspond to a shale volume fraction range of 0–60%: over this range shale volume fraction is related to formation resistivity through the Simandoux (1963) equation.

(v) The four permissible formation resistivities are 0.7, 1.0, 1.5 and 2.5 Ω m: these are broadly consistent with the bulk densities through the Simandoux equation, the standard Archie (1942) equation with shale-corrected input, and the mixing relationship between porosity and density.

These petrophysical specifications allow the cyclic model of Fig. 1 to be represented in three petrophysical successions of gamma-ray, density and induction resistivity (Fig. 2). The successions can be described as idealized well logs, for these data distributions are what the corresponding logging tools would record if they were capable of perfect vertical resolution and if exact corrections could be applied for the effect of the borehole environment. The model of Fig. 2 forms a basis for the following considerations of well log response to cyclicity.

Cyclic resolution of well logs

The effectiveness of well logs as indicators of sediment cyclicity has been investigated by generating synthetic logs which relate to the petrophysical successions of Fig. 2. This has been achieved using a programmed well log simulator which utilizes determined response functions for generic gamma ray, density and deep induction logs. The generic tool specifications are closely representative of commercial logging tools. However, although realistic, they are not claimed to describe the specific response characteristics of any given service tool. Furthermore,

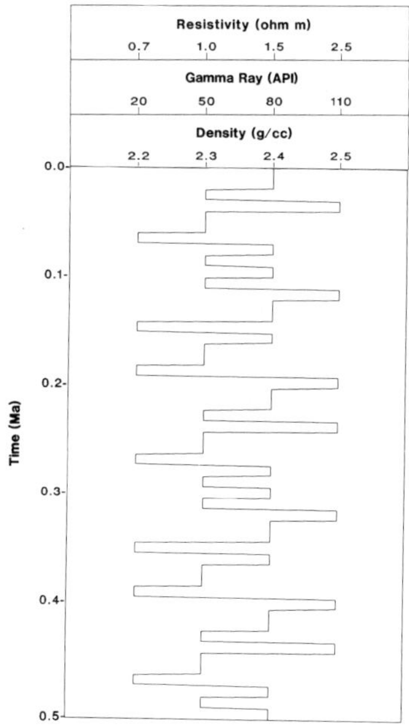

Fig. 2. Assignment of petrophysical parameters to the cyclic model.

the simulator takes no account of log environmental effects due to the borehole and to invasion since these are not of prime interest in this paper. Yet again, the system is presumed to be noise-free.

The response of the generic density tool to the model petrophysical succession is shown in Fig. 3 for three different bed thicknesses and with a log sampling interval of 3 cm. As mentioned earlier, in our model the term 'bed thickness' relates to the half-length of the 20 ka cycle, i.e. it is the thickness of the shortest element in the composite model of Figs 1 and 2. The use of a common timescale in Fig. 3 implies three different sedimentation rates which can be calculated from equation (1). It can be seen from the left track of Fig. 3 that with a bed thickness of 0.15 m, the density log is incapable of following the turreted features of the model. In the centre track the bed thickness has been expanded to 0.3 m, the common timescale implying a higher sedimentation rate, and the density log now follows, but does not yet faithfully represent, the density values used to construct the model. Finally, in the right-hand track, where the bed thickness has been further expanded to 0.46 m, there is close agreement with the parametric levels of the model. In

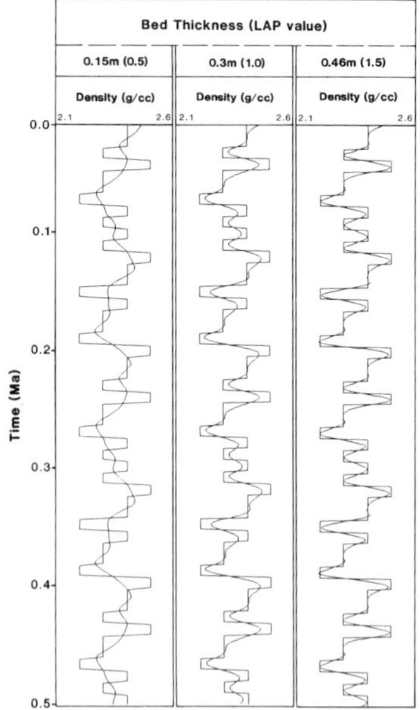

Fig. 3. Simulated density logs of the cyclic petrophysical model for three different bed thicknesses: bed thickness is defined as the half-cycle length of the 20 ka period. The corresponding LAP values (for the 20 ka period) are dimensionless ratios of bed thickness to log sensitivity (cf. Table 3).

essence, the vertical resolution of the generic density tool is adequate to identify the correct density values within our model when the bed thickness is uniformly 0.46 m but not when it is 0.3 m: at 0.15 m the trends of the model are no longer reflected in the log.

The next question concerns the effectiveness of the density tool in resolving the cyclicities (20, 40, 100 ka) that were assumed in constructing the model. The answer is to be found in Fig. 4 which compares Fourier power spectra for the three examples of Fig. 3. All three cyclicities are strongly evident in Fig. 4c which corresponds to a bed thickness of 0.46 m. When the bed thickness is reduced to 0.3 m all three cyclicities are again evident (Fig. 4b) but the manifestation of the 20 ka cycle is becoming marginal. The 20 ka cycle is not resolved for a bed thickness of 0.15 m at which the resolution of the 40 ka cycle is becoming marginal (Fig. 4a). Recalling that we have specified 'bed thickness' to be one half of the length of the 20 ka cycle, it is possible to infer the 20 ka cyclic resolution limits for the density tool in terms of bed thickness. This has required further calculations for a range of bed thicknesses. The concept has been extended to the longer periodicities but in these cases the cyclic resolution limits have had to be expressed in terms of half-cycle lengths which cannot correspond explicitly to the smallest bed thickness used in the model, for the latter is determined by the shortest consistent period. The inclusion of longer periods provides an additional test of log cyclic resolution. The entire computational process has been repeated for the gamma-ray and induction tools.

The results of these computations are summarized in Table 3. Here, the tool intrinsic resolutions relate to the actual design and dimensions of commercial tools. The term 'log sensitivity' describes a minimum half-cycle length that is necessary for the recognition or resolution by the appropriate log of that cyclicity within a power spectrum. The term 'log parametric resolution' describes a minimum bed thickness that is needed for the parametric levels of the petrophysical model to be correctly attained. The data of Table 3 are influenced by the assumptions of no noise and no environmental effects, by the need to consider a generic tool, and by the ranges and distribution of petrophysical parameters within the cyclic model. Despite these constraints, Table 3 is considered to provide a generally useful guide to logging tool resolution capability for detecting and resolving cyclicity. In cases where logs are used solely to indicate the composite litho-

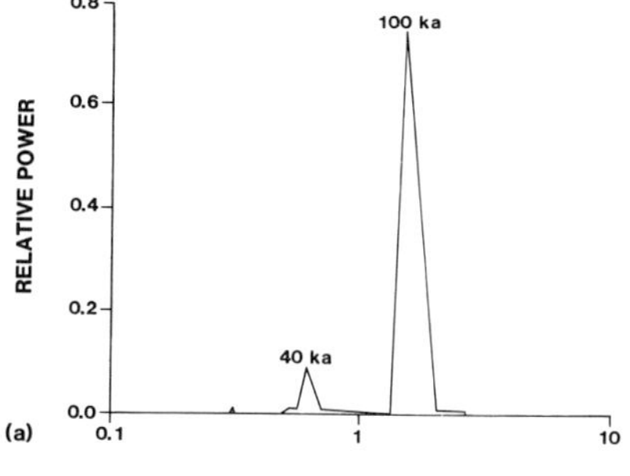

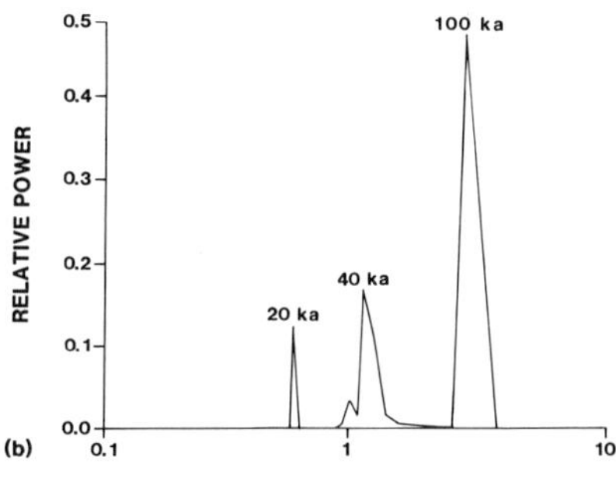

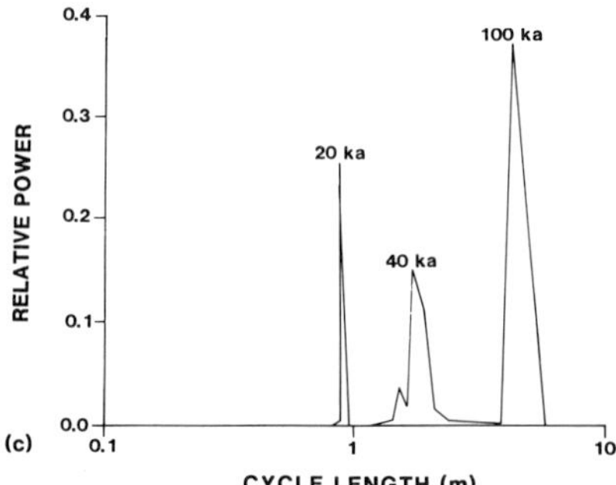

Fig. 4. Fourier power spectra of simulated density logs of the cyclic petrophysical model for bed thicknesses of (a) 0.15 m, (b) 0.3 m, (c) 0.46 m.

cyclic nature of a succession, the log sensitivity identifies the minimum bed thickness that can be recognized by the logging tool: thinner beds that conform to the parametric ranges considered here will not be seen.

The concept of log sensitivity as introduced above can be extended to provide information on the minimum sedimentation rates that are needed for the effective resolution of the three cyclicities assumed in our model (Table 4). These estimates will have to be increased where there are pronounced departures from our assumption of accurate measurements which are not affected by noise and environmental factors. However, real-life situations generally approximate the ideal sufficiently to render these figures useful. Recognition of the 100 ka cycle by the density log requires a sedimentation rate of at least 6 m Ma^{-1}. At the other extreme, the induction log requires a sedimentation rate of at least 91 m Ma^{-1} before the 20 ka cycle can be satisfactorily resolved. With these widely varying requirements, it is desirable to have some means of assessing log effectiveness in cyclic resolution, as a basis for the correlation of cyclic phenomena between well sites.

Log association parameter (LAP)

The cyclic resolving power of well logs can be quantified through a log association parameter (LAP) defined as follows:

$$\mathrm{LAP} = \frac{s\,\tau}{2\,h} \times 10^{-6} \qquad (2)$$

where s is the sedimentation rate (m Ma^{-1}), τ is the cyclic period (a), and h is the log sensitivity (m). This parameter reduces to the ratio of half-cycle length to the log sensitivity (here determined for idealized conditions). In the particular case where (constant) bed thickness is equal to

Table 3. *Well log resolution*

Log	Tool intrinsic resolution* (m)	Log sensitivity† (m)	Log parametric resolution† (m)	
Gamma ray	0.2–0.3	0.3 –0.4‡	0.6 –0.75	
Density	0.4	0.25–0.3‡	0.45–0.6	
Induction	2.1–2.4	0.75–0.9‡	1.2 –1.5	(Conductive beds)
			10	(Resistive beds)

* After Allen *et al.* (1988).
† Assumes no environmental effects, no noise.
‡ Values adopted in this study.

Table 4. *Minimum sedimentation rates for log cyclic resolution (no environmental effects, no noise)*

Tool	Sedimentation rate (m/Ma)		
	20 ka	40 ka	100 ka
Gamma ray	38	19	8
Density	30	15	6
Induction	91	46	18

half-cycle length, LAP is the ratio of bed thickness to log sensitivity (cf. Fig. 3). Where it is desired solely to identify through logs the lithological elements of a cyclic succession, the key ratio is that of the minimum bed thickness to log sensitivity (Table 3), and it is not necessary to draw upon the concepts of equation (2). LAP is dimensionless.

A consideration of equation (2) reveals intuitively the following properties for idealized conditions:

LAP > 1 implies good cyclic resolution of the targeted periodicity;
LAP = 1 implies critical resolution;
LAP < 1 implies unresolved cyclicity.

The log association parameter is period-specific, e.g. the same log at the same site will show higher LAP values for longer periods of cyclicity. LAP is site-specific in that it depends on the sedimentation rate at a well site. It is also tool specific since it depends on log sensitivity: the sharper the sensitivity, the higher the LAP value for a specific period and a given sedimentation rate. In other respects, LAP is largely independent of formation lithology and chemistry. For example, a variable mineralogy would only impact on the LAP concept if it gave rise to inter-bed petrophysical contrasts far in excess of those considered here. Even so, such departures can easily be accommodated where necessary by using a range of appropriate petrophysical models to verify log sensitivities for these new situations.

Model examples

As an illustration of the use of LAP, Fig. 5 shows three simulated logs for a uniform bed thickness of 0.15 m within our petrophysical model. The targeted period of cyclicity is 40 ka. The LAP values range from 0.33 for the induction log to 1.0 for the density log. The amount of detail in the logs increases with increasing LAP. If these logs related to different sites, the

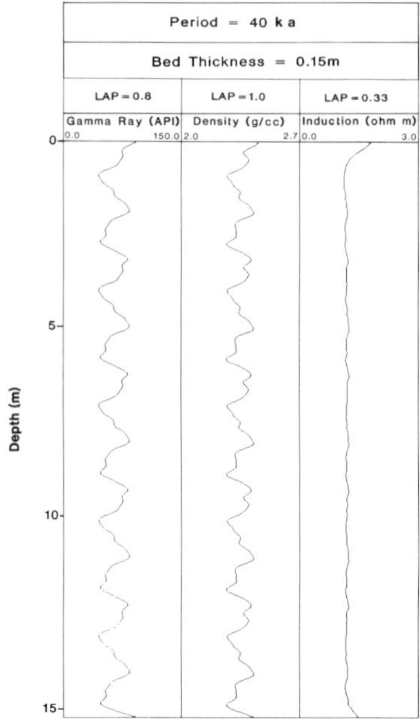

Fig. 5. Simulated logs of the cyclic petrophysical model for a bed thickness of 0.15 m. LAP values (for targeted period of 40 ka) are dissimilar.

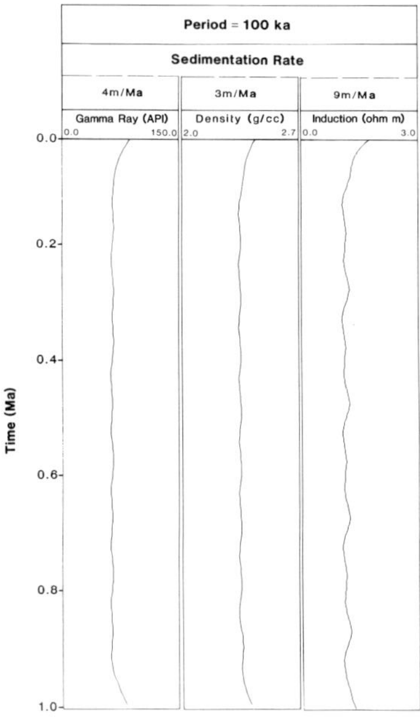

Fig. 6. Simulated logs of the cyclic petrophysical model for a specified subcritical LAP of 0.5 (for a targeted period of 100 ka). The calculated sedimentation rates are tool-specific and intuitively relate to different 'sites'.

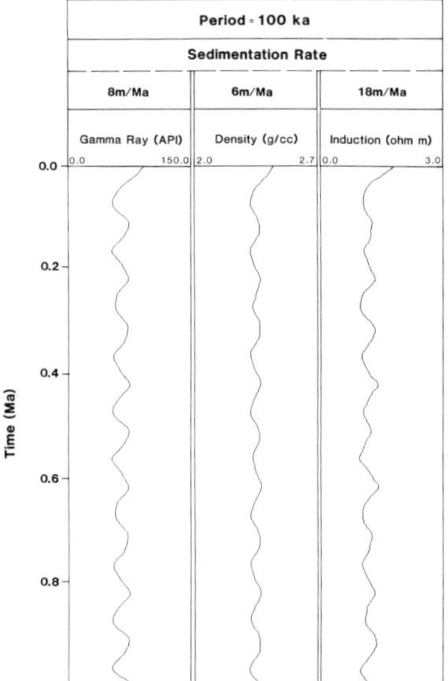

Fig. 7. Simulated logs of the cyclic petrophysical model for a specified critical LAP of 1.0 (for a targeted period of 100 ka).

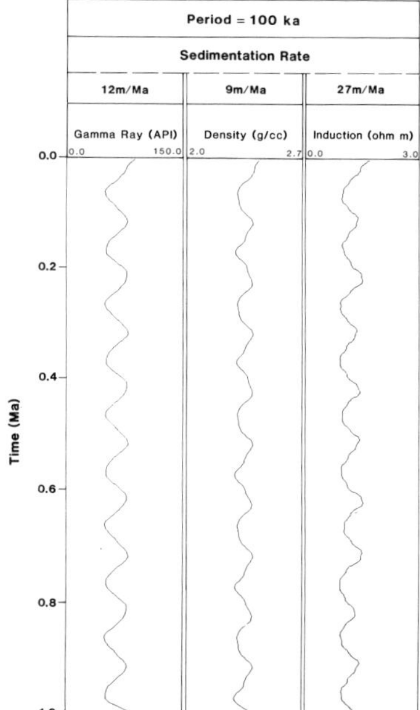

Fig. 8. Simulated logs of the cyclic petrophysical model for a specified supracritical LAP of 1.5 (for a targeted period of 100 ka).

poor resolution of the induction log would preclude interwell correlation of all three.

Figure 6 shows three simulated logs corresponding to different sedimentation rates chosen so that LAP = 0.5 for a targeted period of 100 ka. The logs reveal very little detail and they therefore cannot be correlated meaningfully.

Figure 7 shows three simulated logs corresponding to different sedimentation rates chosen so that LAP = 1.0, again for a targeted period of 100 ka. The logs contain more detail and are evidently correlatable.

Figure 8 repeats the exercise for sedimentation rates chosen so that LAP = 1.5. The detail contained within the logs is now greater still. The similar shapes of the logging curves indicate a high degree of correlatability.

These examples lead to the following ground rules for cyclostratigraphic correlation using well logs:

(i) High LAP values (> 1) are needed to resolve cyclicity.
(ii) Similar and high LAP values indicate strong inter-log correlation of tool responses even between different logs, e.g. density and induction.
(iii) Low LAP values (< 1) will not resolve cyclicity.
(iv) Dissimilar and low LAP values indicate weak inter-log correlation even between the same logs, e.g. density logs at different sites.

These ground rules presuppose that the targeted cyclicity is manifested through all those physical properties which the tools purport to measure, in this case natural radioactivity, electron density, and electrical conductivity. Where the cyclicity does not correspond to changes in a given physical property, the corresponding wireline log will not be useful in studies of sediment cyclicity, regardless of tool resolution capability.

It should be noted that LAP = 1.0 is critical for noise-free, environmentally corrected situations. Field data will not be so ideal and, if log sensitivities are based on idealized model studies, it is to be expected that much higher LAP values will be needed for effective resolution of cyclicity in field situations. These aspects will now be explored.

Field examples

The messages of the previous section are illustrated through logging data from site 646 B in the Labrador Sea, drilled during Leg 105 of the Ocean Drilling Programme (ODP) (Srivastava *et al.* 1987). The logging practices of

ODP have been described previously (Worthington *et al.* 1987).

Selected logs from this site (Fig. 9) reveal a sharp increase in sedimentation rate around 3795 m, reportedly from 52 m Ma^{-1} in the older sediments to 100 m Ma^{-1} in the younger (Imbrie *et al.* 1987). Reference to Table 4 suggests that for a noise-free, environmentally-unaffected situation, these sedimentation rates would be sufficient for the 40 and 100 ka cycles to be detected by the induction tool but that the 20 ka cycle would be unresolved over the interval of lower sedimentation rate. However, in reality, this is not the case.

Figure 10 shows a Fourier power spectrum for the deep induction log over the lower interval. The LAP values have been calculated, for idealized conditions, for the Milankovitch climatic cycles of periods 23, 41 and 95 ka. These values suggest that the 23 ka cycle is not resolved, the 41 ka cycle is critically resolved, and the 95 ka cycle is strongly resolved. In fact, the 41 ka peak is barely discernible and it is therefore likely that LAP values somewhat higher than unity might more realistically define the lower limit of cyclic resolution in field situations which depart significantly from the idealized conditions assumed in generating the estimates of log sensitivity. These departures can be caused by borehole rugosity, washouts and drilling mud composition, in the borehole environment, and by discontinuities and variable sedimentation rates, in the sedimentary record.

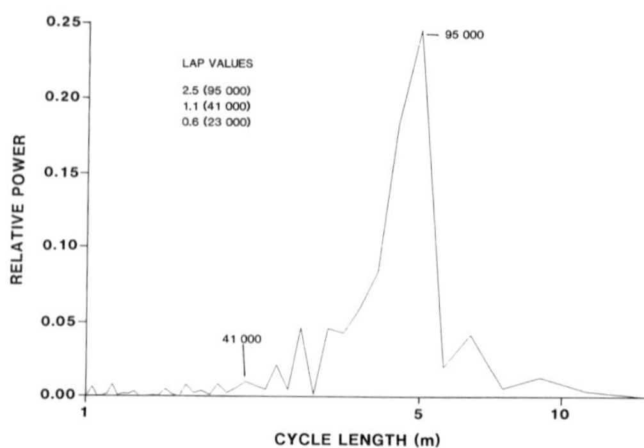

Fig. 10. Fourier power spectrum for the deep induction log over the depth interval 3795 – 3840 m at ODP site 646 B. The LAP values for the reported sedimentation rate of 52 m/Ma suggest strong resolution only of the 95 ka cycle.

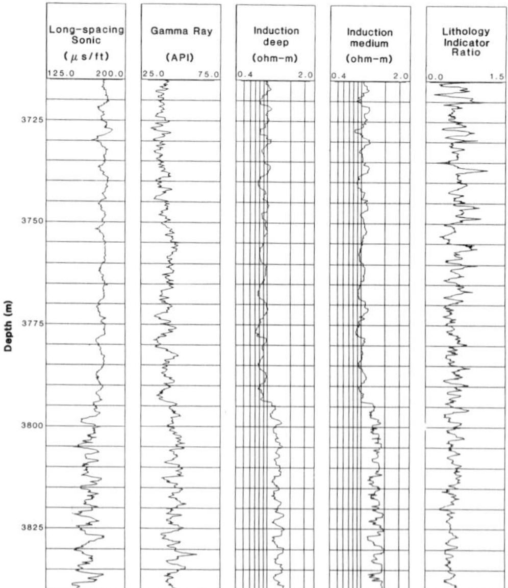

Fig. 9. Well logs from ODP site 646 B in the Labrador Sea including the Lithology Indicator Ratio (Si/(Si + Ca)) from the Geochemical Logging Tool.

Figure 11 shows a power spectrum for the deep induction log over the interval of higher sedimentation rate. The calculated LAP values suggest that the 23 ka cycle is critically resolved, and the 41 ka and 95 ka cycles are strongly resolved. In fact, the 23 ka event is lost in background and there is therefore no log evidence for the 23 ka cycle over both lower and upper intervals. There is also a strongly resolved peak at 70 ka, perhaps attributable to local or regional phenomena or possibly to an unrecognized change in sedimentation rate within part of the analysed interval.

These examples illustrate that LAP values of

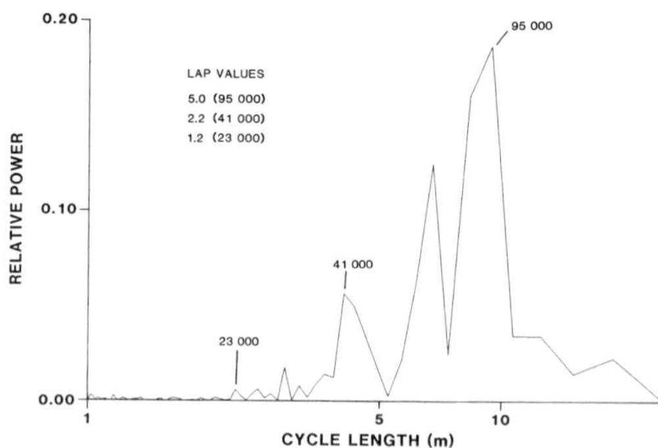

Fig. 11 Fourier power spectrum for the deep induction log over the depth interval 3715–3790 m at ODP site 646 B. The LAP values for the reported sedimentation rate of 100 m/Ma suggest strong resolution of the 95 and 41 ka cycles.

the order of two may be needed for effective resolution of cyclicity in field situations where the log sensitivity used to calculate LAP has necessarily been derived from idealized models. This is equivalent to saying that the minimum sedimentation rates of Table 4 should be approximately doubled for effective resolution in field situations. Further case studies should refine this observation. Calculated LAP values of unity suggest no more than a marginally discernible event within the background of a Fourier power spectrum. Still lower LAP values result in no discernible events.

Conclusions

Log sensitivities to cyclic changes in sediments have been identified for noise-free, environmentally unaffected gamma ray, density and induction logs. These sensitivities are partly governed by the design of the petrophysical model used for log simulation, in particular the attainable levels of, and contrasts between, the various input parameters. Log applications solely to identify the composite litho-cyclic nature of the succession require that log sensitivity be greater than the minimum bed thickness. For Milankovitch applications, the identification of each constituent periodicity requires that log sensitivity be greater than the corresponding half-length of the cycle. Minimum sedimentation rates needed for the recognition of Milankovitch cycles have been identified from the characteristics of these idealized logs.

The ability of a log to recognize a given cyclicity at a particular site has been quantified for ideal situations through a log association parameter (LAP). Similar LAP values above the sensitivity threshold imply a high degree of correlatability between different logs and separate sites. Although the threshold criterion appears to relate also to (non-idealized) field situations, higher LAP values of around two seem to be needed for effective cycle recognition against natural geophysical background. Many more case histories, supported by more extensive model studies, would allow this conclusion to be developed further.

These observations relate to sections over which the sedimentation rate is fairly constant and within which there are no discontinuities. If these conditions are not satisfied, artefacts will be introduced into the power spectra which might not allow the target cyclicities to be recognized. This means that a power spectrum alone cannot confirm cyclicity but merely indicate its occurrence. Core calibration of log derived cyclicity is therefore essential both to confirm log cyclicity and to identify the climatic or mineralogical source of the log cycles. It is, of course, recognized that this calibration should be restricted to intervals over which both core and log data are comprehensive.

Geophysical noise, variable sedimentation rates and discontinuities all reinforce the requirement for core calibration of log responses with appropriate scale reconciliation. The meaningful integration of data measured at different scales is a prerequisite for the most effective implementation of cyclostratigraphy in pure and applied earth science.

The author thanks Mr T. J. Skinner for his assistance with computer applications and Dr R. D. Jarrard, Ms X. Golovchenko and Dr J. R. W. Bellamy for helpful suggestions. The British Petroleum Company plc has sanctioned publication of this work.

References

ALLEN, D., ANDERSON, B., BARBER, T., DES LIGNERIS, S., EVERETT, B., FLAUM, C., HEMINGWAY, J. & MORRISS, C. 1988. Advances in high resolution logging. *The Technical Review* **36**(2), 4–14.

ARCHIE, G. E. 1942. The electrical resistivity log as an aid in determining some reservoir characteristics. *Transactions of the American Institute of Mining and Metallurgical Engineers*, **146**, 54–62.

ARTHUR, M. A. & GARRISON, R. E. (eds) 1986. Milankovitch cycles through geologic time. *Paleoceanography* **1**, 369–586.

DUFF, P. McL. D., HALLAM, A. & WALTON, E. K., 1967. Cyclic Sedimentation. *Developments in Sedimentology* **10**, Elsevier, Amsterdam.

FISCHER, A. G. 1986. Climatic rhythms recorded in strata. *Annual Reviews of Earth and Planetary Sciences*, **14**, 351–376.

FRACASSO, M. A. & HOVORKA, S. D. 1986. Cyclicity in the Middle Permian San Andres Formation, Palo Duro Basin, Texas Panhandle. *Reports of Investigations of the Bureau of Economic Geology* **156**, University of Texas.

GOLOVCHENKO, X. *et al.*, 1987. Milankovitch cycles detected from wireline logs: results from ODP Site 693, Antarctic continental margin. Abstracts prog. *Geological Society of America*. **19**, 679.

GRADSTEIN, F. M., AGTERBERG, F. P., BROUER, J. C. & SCHWARZACKER, W. S. (eds.) 1985. *Quantitative Stratigraphy*, D. Reidel, Dordrecht.

IMBRIE, J. *et al.* 1987. Scientific goals of an Ocean Drilling Programme designed to investigate changes in the global environment. *Report of the Second Conference on Scientific Ocean Drilling (COSOD II), Strasbourg, France, 6–8 July*, 15–46.

LAMONT-DOHERTY BOREHOLE RESEARCH GROUP 1987. *Wireline Logging Manual*, Vol 2, Lamont-Doherty Geological Observatory, Palisades, NY.

LOTT, G. K. & STRONG, G. E. 1982. The petrology and petrography of the Sherwood Sandstone (?Middle Triassic) of the Winterborne Kingston Borehole, Dorset. *UK. Institute of Geological Sciences Report* **81/3**, 135–142.

PENN, I. E., COX, B. M. & GALLOIS, R. W. 1986. Towards precision in stratigraphy: geophysical log correlation of Upper Jurassic (including Callovian) strata of the Eastern England Shelf. *Journal of the Geological Society, London,* **143**, 381–410.

SIMANDOUX, P. 1963. Dielectric measurements on porous media: application to the measurement of water saturations: study of the behaviour of argillaceous formations. *Revue Institut Francais du Petroles* **18**, supplementary issue, 193–215. (Translated text in Shaly Sand Reprint Volume, SPWLA, Houston, pp IV 97–124).

SRIVASTAVA, S. P. *et al.* 1987. Baffin Bay and Labrador Sea. *Proceedings* ODP, Initial Reports (Pt. A.) **105**.

WATNEY, W. L. 1980. Cyclic sedimentation of the Lansing-Kansas City groups in northwestern Kansas and southwestern Nebraska. *Bulletin* **220**, Kansas Geological Survey.

WHITTAKER, A., HOLLIDAY, D. W. & PENN, I. E. 1985. *Geophysical logs in British stratigraphy*. Geological Society, London, Special Report **18**.

WORTHINGTON, P. F., ANDERSON, R. N., BELL, J. S., BECKER, K., JARRARD, R. D., SALISBURY, M. H. & STEPHEN, R. A. 1987. Downhole measurements for the Ocean Drilling Programme. *Report of the Second Conference on Scientific Ocean Drilling (COSOD II), Strasbourg, France, 6–8 July*, 131–136.

Interwell matching using a combination of petrophysically derived numerical lithologies and gene-typing techniques

C. M. GRIFFITHS[1] & S. BAKKE[2]

[1]*University of Trondheim, Division of Petroleum Engineering and Applied Geophysics, Trondheim, Norway*

[2]*Continental Shelf and Petroleum Technology Institute, Trondheim, Norway*

Abstract: The use of gene-typing techniques in stratigraphic matching problems has been steadily developing since Smith & Waterman published an example in 1980. The regular use of these techniques to assist in stratigraphic correlations using wireline log data has, however, been hindered by the lack of suitable precisely defined lithological units. In an oil-field production or exploration environment, the porosity logs and a natural gamma log can be used to derive various types of numerical lithologies. These lithologies provide a convenient starting point for the use of gene-typing algorithms since they comprise a relatively small set of precisely defined reproducible units. The advantage of using a gene-typing technique is that it enables multiple matching hypotheses to be produced from sequences which contain gaps, stretched zones, and duplicated sections. These matches can assist the production of stratigraphic correlations. Examples of matching are presented from reservoir sections in the Statfjord and Murchison fields.

Stratigraphic correlation plays an important role in many aspects of geological interpretation and prediction. At both reservoir and basin scales, the identification of sedimentary units of similar age is an essential prerequisite to palaeogeographical analysis and prediction.

Schwarzacher (1980) has discussed the difference between matching and correlation. He points out that stratigraphic correlation, or the hypothesis that two or more sedimentary units are contemporaneous, can be only right or wrong; there are no half measures. Most available techniques using wireline log data are concerned with matching hypotheses rather than correlation. Matching can be, and usually is, purely expedient. Matching is the degree to which two or more units are similar on the basis of one or several measurements. The choice of measurement(s) is often governed by practical considerations. Examples would be the matching of flow units in reservoir studies using core analyses or wireline logs, or the matching of weak zones in site investigation holes using drill rate data. A high degree of similarity only occasionally implies stratigraphic correlation (for example when comparing volcanic ash layers). Matching, as Schwarzacher points out, is an important tool in correlation studies, but should be distinguished from stratigraphic correlation.

The use of lithostratigraphic data in true stratigraphic correlation is subject to many problems. Lithostates, unless defined in terms of fossil content and/or radionucleotides, tend to be repetitive throughout time. Erosion can remove parts of a sequence. Differential sedimentation rates can produce thickness changes in sedimentary units over short distances, and the distribution of energy in many sedimentary environments produces lateral changes of lithology within chronostratigraphic units.

Within stratigraphic nomenclature committees this must cause a great deal of confusion. A formal lithostratigraphic unit is defined according to its physical properties rather than its age (Hedberg 1976), and yet it is very common to read a formation definition that includes details of lateral change in composition. This implies that even formally defined lithostratigraphic units are often used chronostratigraphically in practice.

Within the field of wireline log correlation there have been a series of developments that recognize these problems. Startzman & Kuo (1987), and Wu & Nyland (1987) both address the problem of lateral facies change in isochronous units. They both reduce the log curve(s) to symbol form and examine the sequential relationships between symbols. Wu & Nyland go on to relate the symbol string to sedimentary sequence, and then match the sequences between wells.

Smith & Waterman (1980), Howell (1983) and Waterman & Raymond (1987) suggested the possibility of using gene sequence matching algorithms in the analysis of sedimentary sequences. These techniques, originally developed by molecular biologists and linguists,

From HURST, A., LOVELL, M. A. & MORTON, A. C. (eds), 1990, *Geological Applications of Wireline Logs* Geological Society Special Publication No. 48, pp. 133–151

are attractive because they offer a means of handling:

a. gaps in both sequences under comparison;
b. repetition of parts of the sequence;
c. stretching of parts of the sequence;
d. knowledge concerning permitted matches;
e. the presentation of multiple, ranked, matching hypotheses

These points address some of the problems associated with the use of lithology as a correlation tool.

One important requirement for the reliable use of gene sequence analysis algorithms is the existence of a finite, reproducible, alphabet. This is not the case for, for example, cross-correlation of log curve shapes or grain-size curves, where interval or ratio scale data can be compared. If however the sequence consists of ordinal or nominal scale data (e.g. 'sandstone', 'shale'), which is true of many geological classifications, then gene sequence algorithms can be used provided that the definitions of the units are rigorously adhered to. In a wireline log context this means that the log values should be converted to a finite, exogenous, set of states or classes. An exogenous class (Evans 1977) is one for which the class boundaries are fixed before data is examined, such as the Wentworth grain-size classification.

Other algorithms that take gaps, stretching and duplication into consideration have been published by Gordon & Reyment (1979) and Ghosh (1984). Gordon and Reyment used a dynamic programming algorithm that is very similar to the algorithm used here, but they applied it to raw log traces rather than derived units. Given the problem of standardizing log data it may be that the approaches detailed by Wu & Nyland (1987) and in this paper, are less disturbed by logging error.

This paper demonstrates the use of gene-typing algorithms in primarily a matching role, but with suggestions as to extension to true stratigraphic correlation.

Gene-typing algorithms

The algorithms under discussion arose from sequence comparison work by several authors in the 1960s in the fields of speech analysis and molecular biology. The particular dynamic programming algorithm used here can be traced back to work published by Levenshtein (1966). A good overview of the subject can be obtained from Sankoff & Kruskal (1983). The first published applications in geology were by Gordon & Reyment (1979) and Smith & Waterman (1980). Wireline logging applications have been described by Wu & Nyland (1987) and Griffiths & Bakke (1988).

One of the basic analysis tools in this field has been a 'dynamic programming method' used in conjunction with the Levenshtein (1966) distance. Levenshtein (1966) introduced a concept of distance between two sequences as being the minimum number of insertions, deletions or substitutions necessary to convert one sequence into the other (Sankoff & Kruskal 1983, p. 18).

Descriptions of the basic method can be found in Smith & Waterman (1980), Sankoff & Kruskal (1983, p. 25–29), Howell (1983), Waterman & Raymond (1987), Wu & Nyland (1987) and Griffiths & Bakke (1988). Reference should be made to these papers for details of the algorithm. A Fortran program implementing the technique is presented in Howell's paper. Although the basic recursive algorithm is used in all these applications, there is a wide variation in choice of both lithostates and the way the matching costs are handled.

Because geological units are not deposited totally at random, and we have at least some idea of the possible lateral extent of different lithologies, it is possible to include this knowledge in the cost or distance function. Howell (1983) discussed the use of a cost matrix in which (using grain-size based classes) she allowed lower cost substitutions of neighbouring classes, and a certain amount of geological knowledge was thus built in. Waterman & Raymond (1987 p. 121) discuss the use of major lithological class, minor class, modifying adjective, and strata thickness as contributory factors in building a distance matrix. Typical additional geological considerations would be: likely lateral extent of a given facies, grain size distribution, probability of occurrence in a given environment, palaeocurrent direction etc.

The cost structure used in the present paper differs from that used by previous workers. Instead of assigning a range of cost values to different substitution possibilities (such as a sandstone by a shale, or a coal by a limestone), we keep the cost of an incorrect substitution constant, but provide a range of conditions under which a zero-cost substitution is allowed. The cost values we used are described in the section headed 'Examples'.

Numerical lithology

The use of the sequence substitution algorithms in practice depends on the existence of a precise, finite alphabet or set of terminal symbols. In gene-typing applications these symbols are well defined, i.e. the genetic code consisting of five bases A, G, T, C, and U.

In a geological application one has a wide choice of possible units, with varying qualities of definition. The Los Alamos work published by Smith, Waterman and Howell has concentrated on using the sort of data that is available from cores or hand specimens, i.e. grain size, unit thickness, and genetic classes. These units have the advantage of being familiar to most geologists, and of being related to the energy of the depositional system. In an exposed section or a well with full core recovery then such a choice may be adequate. However, in most boreholes the only reproducible data for much of the drilled interval are the wireline logs. In such a case one is faced with the choice of either using the log data as valid lithology descriptors on their own, or of converting the log readings to a more traditional petrography. Wu and Nyland's application used petrophysical data from gamma, porosity and resistivity logs and converted them into genetic and grain-size classes before using the string comparison algorithm. The conversion of log values to traditional sedimentary descriptors such as 'sandstone' and 'shale' before further analysis introduces more uncertainty than is necessary. Put another way, if the conversion from logs to lithology is a hundred per cent certain, then the raw log values are as easy to use as the names given to them, and if the conversion is less than one hundred per cent certain then the uncertainty in the interpretation cannot be less than the uncertainty in the log values.

The translation of wireline log data to grain-size data is, and has been, the subject of much research for many years. The problem is largely unresolvable due to inherent ambiguities in the relationship between assemblages of clastic grains of varying mineralogy and grain-size distributions, and the physical response of the assemblage. Thus for any single petrophysical measurement, the same response can be achieved by many different combinations of mineralogy, grain-size and porosity. With measurements on several variables this ambiguity may be reduced, but in practice, with an economically defensible log suite, not removed. An alternative approach is to use the log response itself, suitably classified, as the basic symbol to be matched. The 'electrofacies' concept (Serra 1977) would provide a suitable basis for sequence comparison if the facies units could be grouped into a finite exogenous set of classes.

The numerical lithologies that will be used in the following examples have been discussed by Griffiths (1983a,b). These lithostates are derived from a subdivision of *M*-*N*-gamma space as shown in Fig. 1.

The *M*-*N*-gamma space has been subdivided to give a total of 125 numerical lithologies. '*M*' and '*N*' are the matrix mineralogy related values proposed by Burke *et al.* (1969). '*M*' is the slope of the density-slowness plot, and '*N*' is the slope of the density–neutron porosity plot for any given mineral combination, as porosity changes. Gamma is the natural gamma response.

There is considerable ambiguity in interpretation of any individual *M* or *N* value in mineralogical terms, and the *M* value assumes a linear porosity-acoustic response which is not strictly valid. Nevertheless, within limited field or basin areas, experience has shown that the values are reasonably stable and offer a means of summarizing multivariate log data in a way that makes them amenable for use in the matching algorithms described here. With the class boundaries as shown, most shales group in the region with lithostate numbers below 83 (i.e. 76 to 83, 51 to 58, 26 to 33, and 1 to 8). Organic-rich shales often have values between 2 and 35. Sands often correspond to states 114, 119, 89, 94, and calcareous units (limestones and cemented silts and sandstones) 113, 112, 118, 87, 88, and 93. Generally speaking, increasing gamma response reduces the lithostate number in jumps of 25, increasing shalyness or siltyness reduces the number in jumps of 5 (class change in the *M* direction), whereas increasing carbonate content reduces the state number by ones and twos (class change in the *N* direction).

These translations into mineral content are however only approximations for the convenience of comparison with hand specimen descriptions. We are interested here in examining the possibility of using the petrophysically

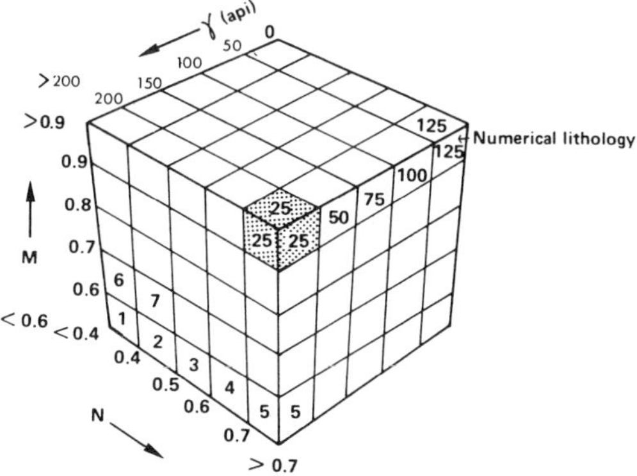

Fig. 1. The relation between *M*-*N*-gamma values and the numerical lithologies used in this example.

defined lithostates in their own right as symbols in the matching process. Grain size is only very indirectly related to this numerical lithology system.

An additional processing step is necessary before the log values are assigned numerical lithology values. As Wu & Nyland (1987) point out, it is necessary to block, segment or zone the logs before the values within the segment boundaries can be compared. There have been many segmentation techniques proposed for wireline logs. These have been reviewed in various articles (e.g. Hawkins & ten Krooden 1979, Hawkins & Merriam 1975, Shaw & Cubitt 1979, Griffiths 1982b).

The approach used here is, rather than attempt to produce large 'homogeneous' zones (which will always depend on a rather arbitrary definition of 'homogeneity'), to concentrate on reducing the effect of the difference between the logging tool measurement averaging distance and the tool sampling interval. A logging tool usually has a longer source-to-receiver distance than the sampling interval, and this means that a single, sharp, lithological contact or bed boundary in the rock sequence is transformed into a series of three or four intermediate values as the logging tool passes the contact. If the intermediate values are retained they will introduce spurious lithological units that complicate the substitution process. The intermediate values contain sequence information in that they are a function of the fact that one unit overlies another, but in sections that may have been subject to erosion they are best removed. The algorithm used (given the difficulty of obtaining tool response equations from major logging companies) is one published by Griffiths (1982a). This is based on a runs-test modification to the Kulinkovich segmentation technique (Kulinkovich *et al.* 1966). The result is a set of blocked log traces that are combined to give a set of numerical lithologies. The numerical lithologies described here, and particularly the class subdivision values are of course by no means the only ones available. The author has also experimented with velocity–density based numerical lithologies for matching in seismic and porosity contexts. The important point is that a precisely defined lithology (even though arbitrary) can be designed for a specific purpose, and one should not expect that one lithology classification is sufficient for all needs. There is no 'lithology for all seasons'. The ideal chronostratigraphic lithology of course would be defined in terms of isotope content and permit direct chronostratigraphic correlation. This may be somewhat of a long-term goal however.

Correlation by cluster analysis would be the antithesis of the present approach for several reasons. First of all the clusters are unstable in the presence of changing data sets, (this would be like redefining 'sandstone' each time one examined a sedimentary section). Secondly a Q-mode cluster analysis would identify similar rock units in two or more wells, and this is exactly what we are trying to avoid with the geological gene-typing algorithms. Stratigraphic correlation must try to use the knowledge concerning anticipated lateral variation in lithology, not produce a matching which is probably diachronous. Thirdly the interpretation of dissimilarity matrices from cluster analyses is notoriously subjective (see Griffiths 1982b) and thus the number of clusters to be included in the matching will depend to a large extent on the analysts expectations rather than on the geology.

The numerical lithologies described here provide finite, precisely defined sets of petrophysical symbols that can be compared using the dynamic programming method, and offer the possibility of another geological application of wireline logs. The electrofacies described here is just one of an infinite number of possible subdivisions for specific purposes. Other classifications based on gamma spectrometer data have been suggested by Collins & Doveton (1988).

Examples

As an example of the use of the method we have chosen a set of three North Sea wells. The wells are from the Statfjord and Murchison fields and are several kilometers apart. The stratigraphic units of interest are the Lower Jurassic Statfjord Formation and Dunlin Group, and the Middle Jurassic Brent Group. The stratigraphy of these units has been described in several publications over the past ten years and has been summarized in Spencer *et al.* (1987).

The Statfjord Formation includes alluvial sands, silts and shales. The Dunlin Group was deposited in a marine environment and consists mostly of clay and shale with minor silts and sand (Spencer *et al.* 1987, p. 345). The Brent Group consists of a series of sands, shales, carbonates and coals that have been interpreted as a laterally extensive deltaic series (Spencer *et al.* 1987, p. 332). It is perhaps worth noting however that Blackbourn (1984) comments on some of the problems in developing sedimentological and diagenetic models for the Group, in particular the shortage of definitive sedimentary structures.

The Brent Group has been subdivided into a series of formations, the Broom (eldest, not discussed further here), Rannoch, Etive, Ness and Tarbert. The Etive consists of massive sands, whereas the Ness and Tarbert formations have more thin sands and coals. The Rannoch Formation can be relatively silty.

These sections illustrate some of the differences between lithostratigraphic matching and stratigraphic correlation. Whereas use of the numerical lithology in the Statfjord Formation allows matching of petrophysical states with similar petrophysical properties, this will only have stratigraphic meaning if the particular numerical lithology is known to have a significant lateral extent, at least to the order of magnitude of the well spacing.

The numerical lithology sequences are shown in Fig. 2. Wells 33/9−4 and 33/9−5 are on the Statfjord field, while 33/9−6 is on the Murchison field. The basic lithostratigraphic subdivision is as follows, with reference to Fig. 2.

L	Viking Group
K	Viking Group shales
J	Tarbert Formation
I	Ness Formation
H	Etive Formation
G	Rannoch Formation
F	Drake Formation
E	Cook Formation
D	Burton Formation
C	Amundsen Formation
B	Eirikson Member, Statfjord Formation
A	Raude Member, Statfjord Formation

Both the numerical lithology curve and the continuous $M-N$ plot show marked qualitative changes with depth and it would seem reasonable to label formations in sequence above and below the top Etive. However, away from this distinct marker the nature of the lithological units varies. For example, to what extent are we justified in calling the unit overlying the Etive the Ness Formation in both wells? Apart from the fact that it overlies the Etive, do the two units have anything descriptive in common (and common description is supposed to be the basis of lithostratigraphy). If they are similar, how much of the depth interval is equivalent, and how much is not? The use of gene-typing algorithms can help to answer these questions.

In the case of Well 33/9−6 the algorithm will be used to suggest the location of a Brent equivalent from the whole well. This can be done 'blind'; sections of the well can be compared to Well 33/9−4 and Well 33/9−5 and the likely lithostratigraphic equivalents scored. The comparison giving the minimum cost of substitution is the closest formation equivalent. The existence of a lithostratigraphic equivalent does not, of course, imply age equivalence in the absence of a palaeogeographic model.

The cost rules used in this example are as specified below:

(1) $w(a_i, 0) = 1.0$ i.e. a gap in sequence a
(2) $w(0, b_j) = 1.0$ i.e. a gap in sequence b
(3) $w(a_i, b_j) = 0.0$ for $a_i = b_j$
according to different levels of substitution.
(This means that all substitutable lithostates, as defined in the following section, have a zero cost of substitution.)
(4) $w(a_i, b_j) = 4.0$ for $a_i \neq b_j$
according to any rule of substitution.
This means that lithostates from sequence a or b that cannot be substituted according to the specified rules are heavily penalized.

Substitution levels

The substitution levels are related to position in M-N-gamma space. The substitution level controls the condition under which rule (3) operates, i.e. substitution at zero cost.

As mentioned above, this is an alternative to having different cost values for different substitutions. It would be possible to operate both schemes simultaneously, i.e. varying rule (4) cost values as well as zero-cost substitution rules but this has not been done here.

Level 0 allows no substitution of numerical lithologies at zero cost. i.e. only exactly equal lithostates are allowed. This will generally not produce stratigraphic correlations, but could give information concerning flow unit equivalence for example.

Level 1 allows the substitution of one class width in the gamma direction only.

Level 2 allows for substitution of one class width in the M direction only.

Level 3 allows for substitution of one class width in both the N and gamma directions.

Level 4 allows for substitution of one class width in all three M, N and gamma directions.

The geological consequences of these substitution levels is that level 0 does not allow for lateral lithostate change at any stratigraphic level. This condition is rarely successful in giving matches that can be compared with biostratigraphic correlations.

Level 1 allows for lateral changes in radioactivity, (commonly shalyness but also mica content). It also allows for autocyclic oscillation in gamma values.

Level 2 allows for changes in sand, silt and shale content (since one cause of variation in

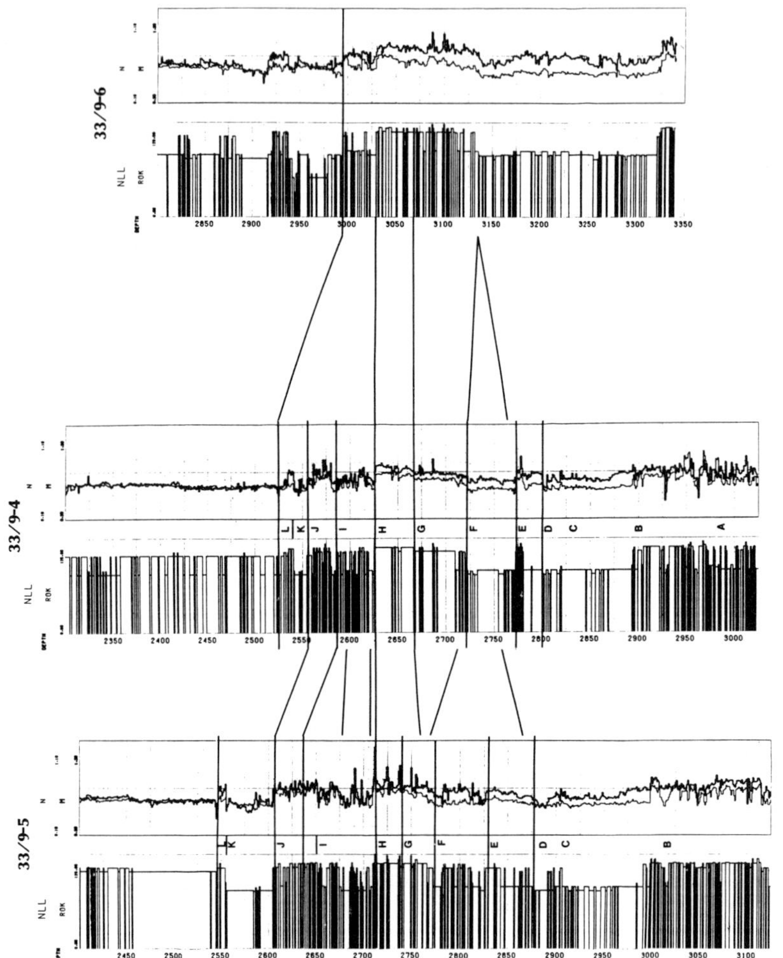

Fig. 2. The three wells used in the study. Bold letters indicate formation correlations derived from cuttings. Lines between wells depict correlations derived from matching of numerical lithologies.

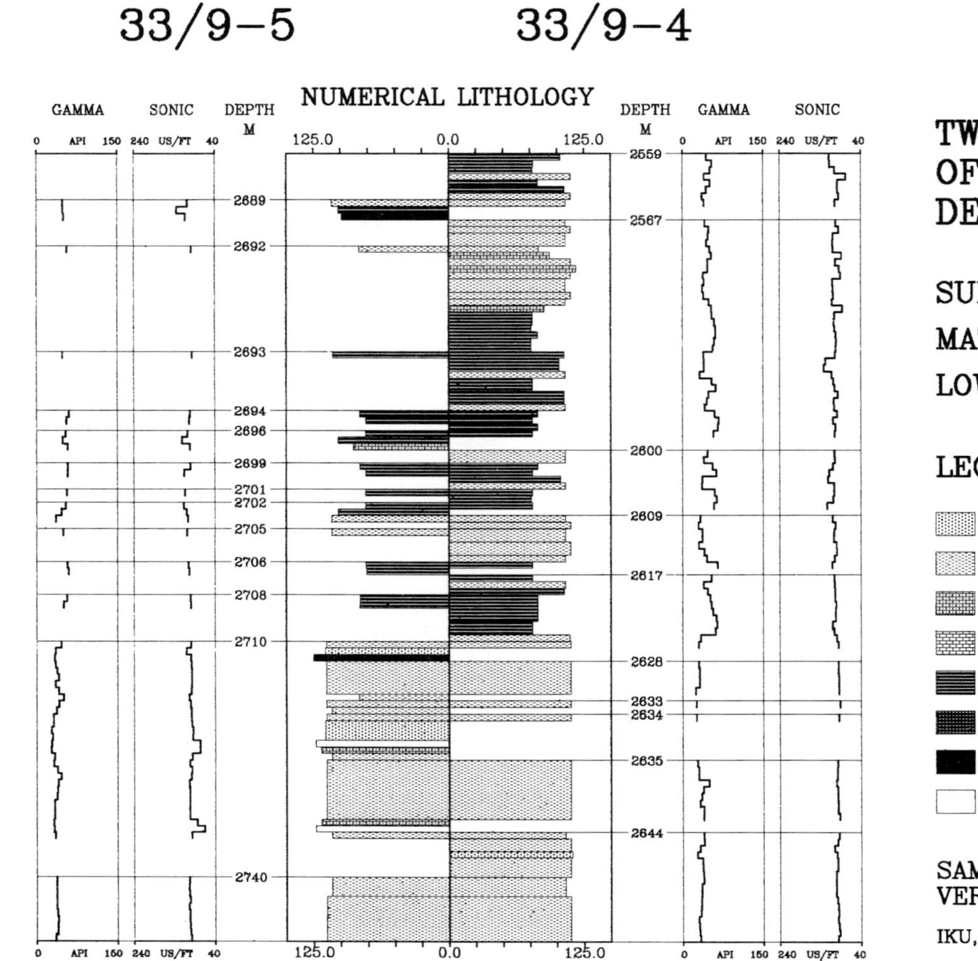

Fig. 3. Matching of the Etive and Ness formations for wells 33/9−5 and 33/9−4 at substitution level 0.

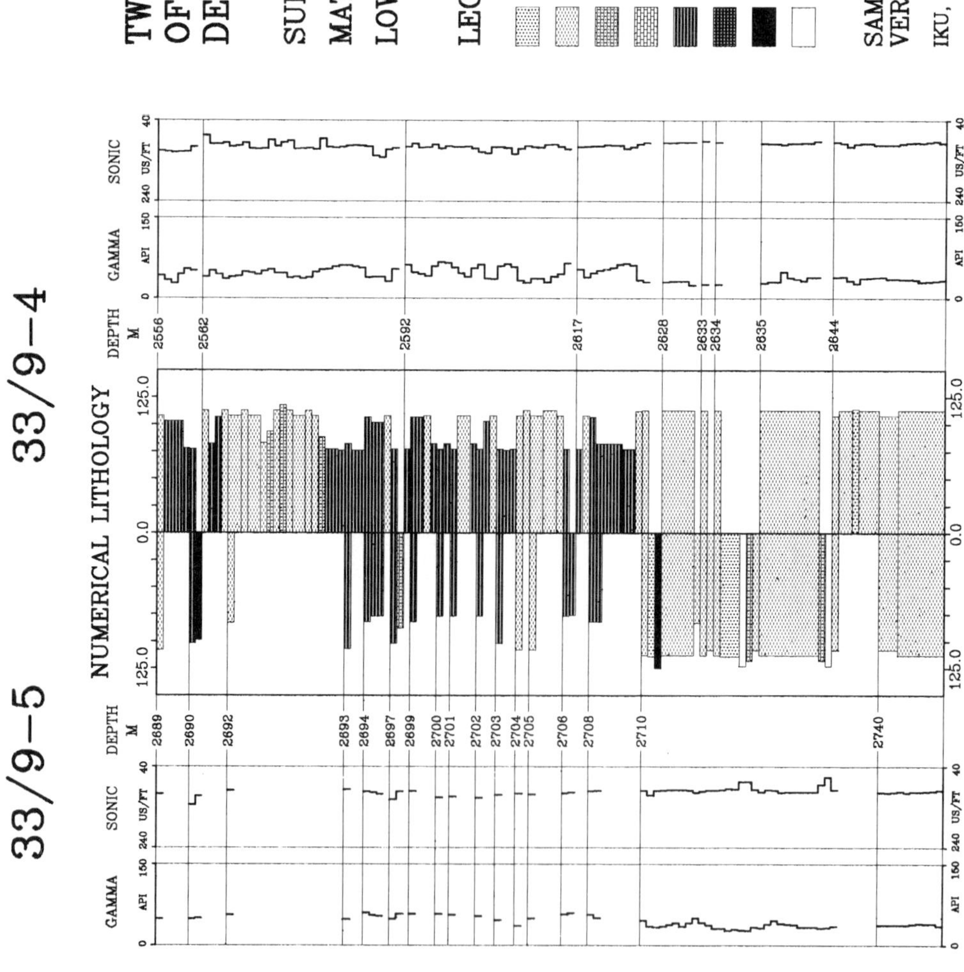

Fig. 4. Matching of the Etive and Ness formations for wells 33/9–5 and 33/9–4 at substitution level 1.

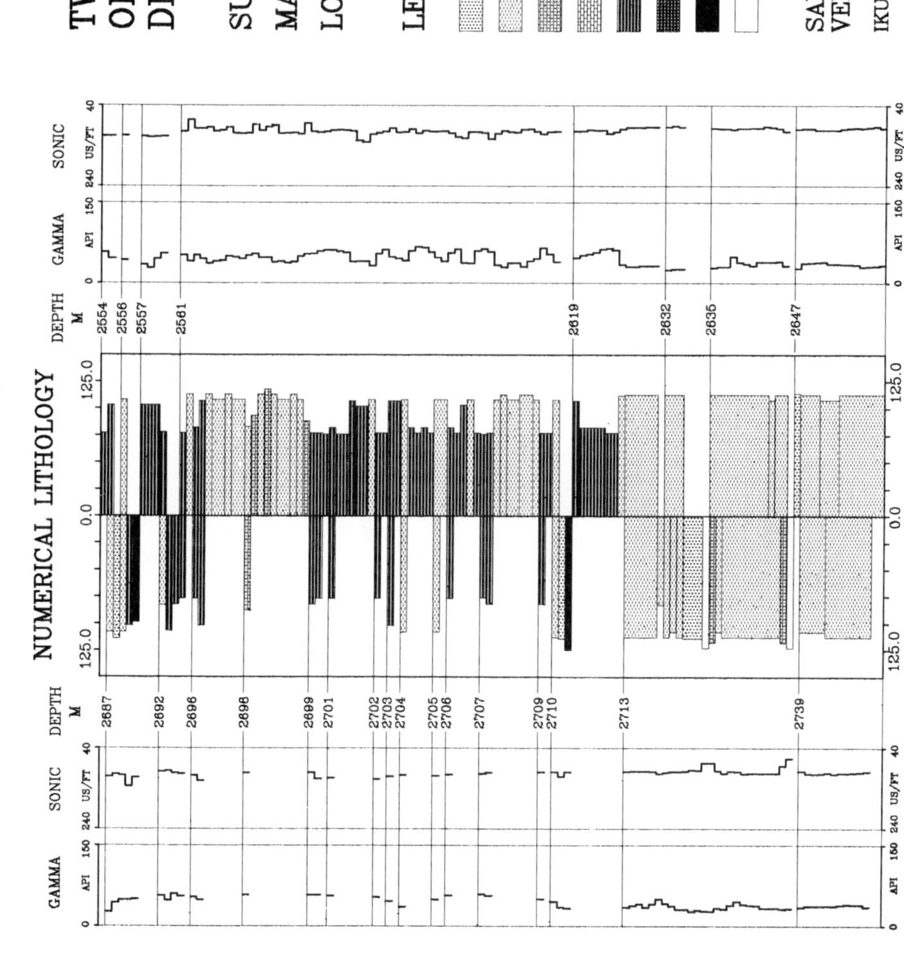

Fig. 5. Matching of the Etive and Ness formations for wells 33/9–5 and 33/9–4 at substitution level 2.

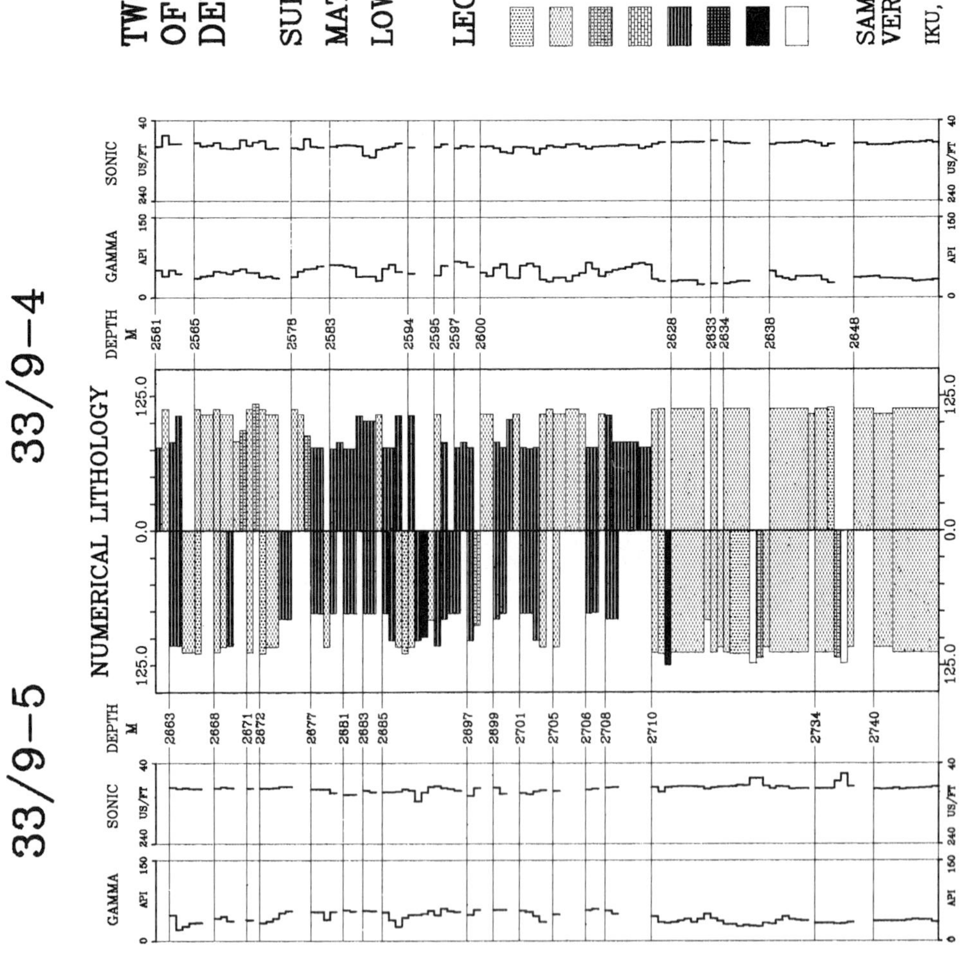

Fig. 6. Matching of the Etive and Ness formations for wells 33/9−5 and 33/9−4 at substitution level 3.

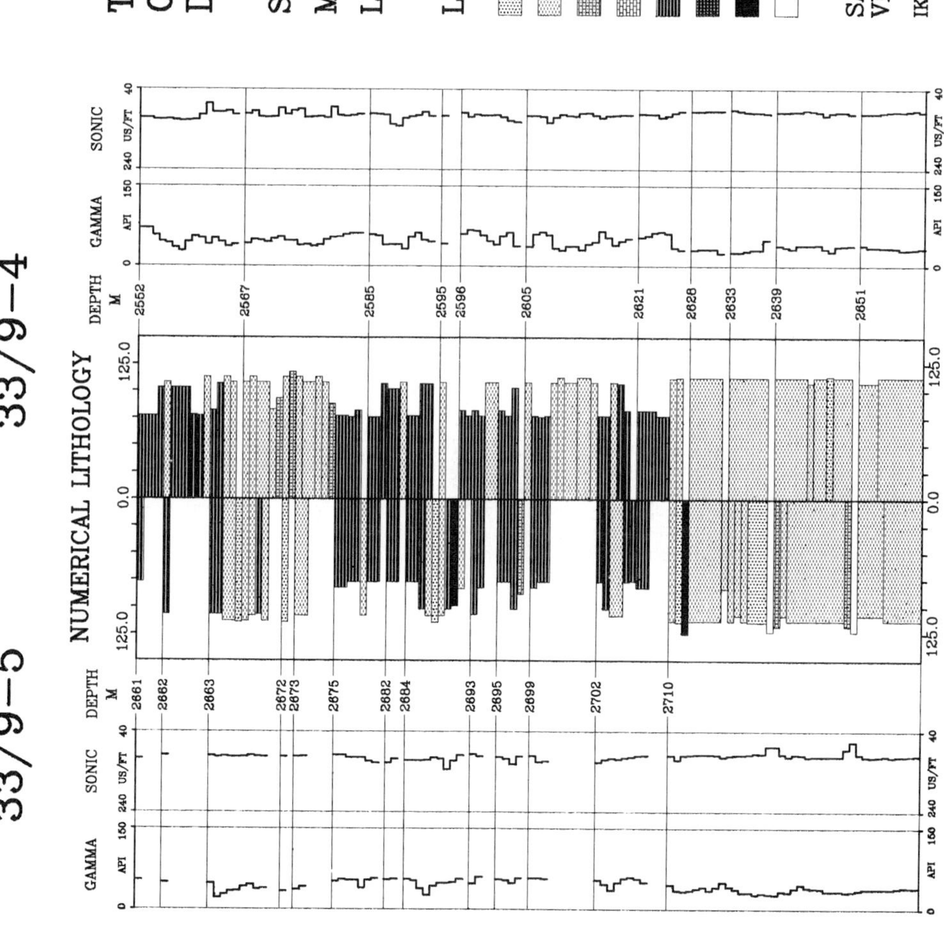

Fig. 7. Matching of the Etive and Ness formations for wells 33/9–5 and 33/9–4 at substitution level 4.

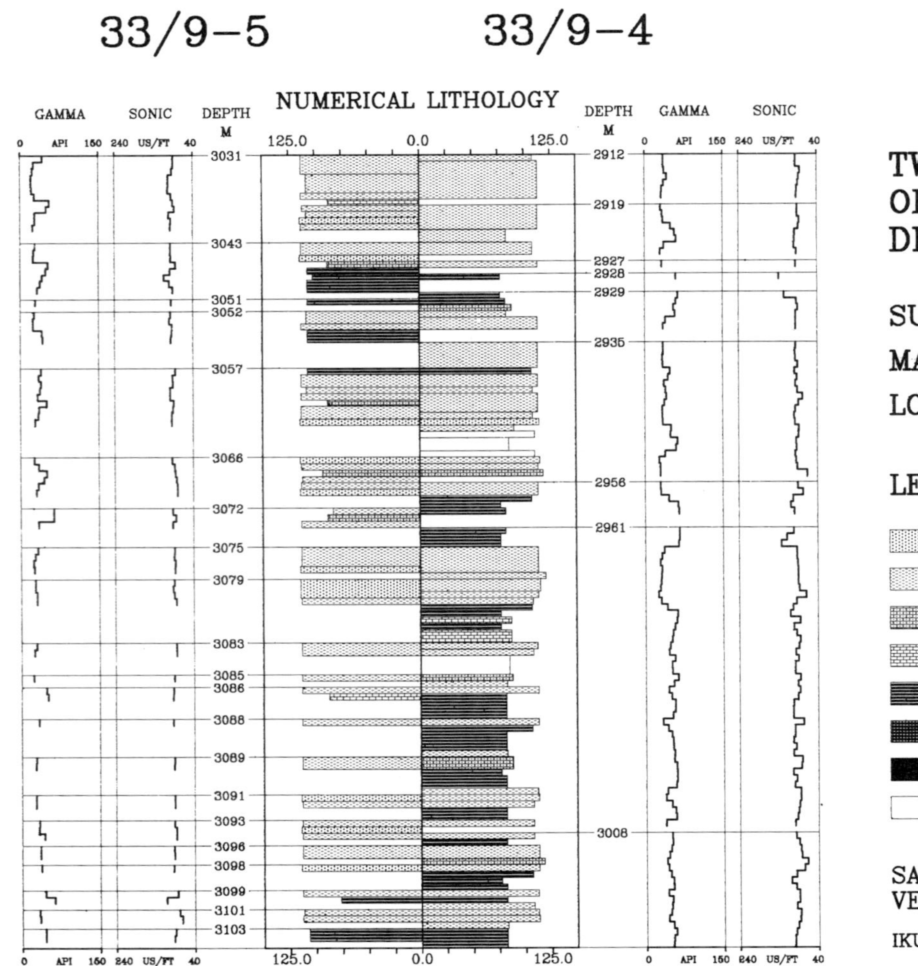

Fig. 8. Matching of the Statfjord formation for wells 33/9–5 and 33/9–4 at substitution level 4.

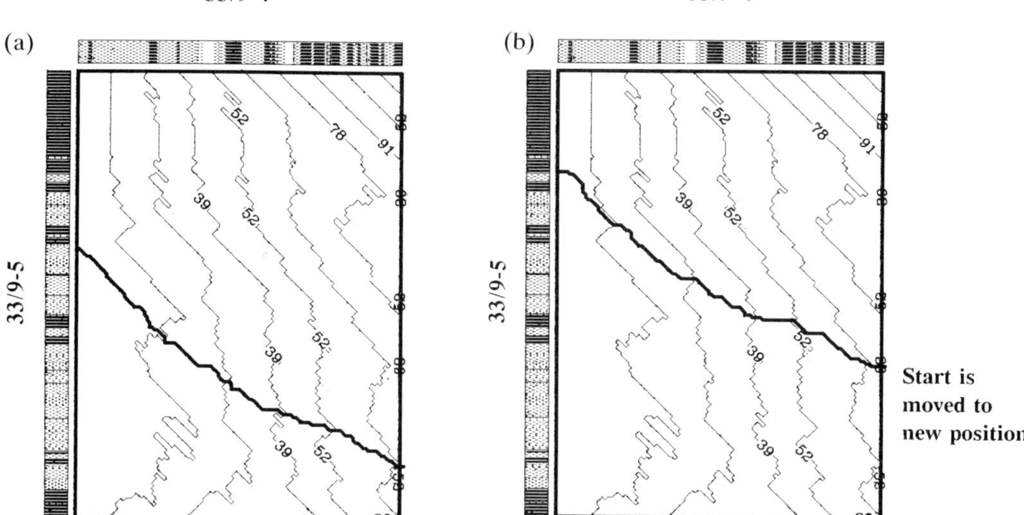

Fig. 9. (a) Original cost matrix with automatic selection of optimum alignment. (b) Cost matrix with alignment start-point selected by a geologist based on additional information. Contours are summed cost or distance values.

the M value is variation in V_{cl} producing oscillation around the $M = 0.7$ value).

Level 3 is concerned with variation in both carbonate content (dominated by the N value) and shalyness. This is useful for matching carbonate cemented equivalents of sands and silts.

Level 4 considers lateral variation in shalyness, quartz, and carbonate content, and autocyclic oscillation on all three axes.

Results

In Figs 3–8 the wells are plotted back-to-back to emphasize the location of stratigraphic gaps, and the location of matching lithologies.

In the example shown in Fig. 3 (wells 2.5 km apart) the top Etive (2710 mTVD in 33/9–5) is correctly matched in both wells, but only shales are matched in the overlying Ness Formation. Allowing gamma variation (Fig. 4) has little effect on the match, nor does allowing for M variation (Fig. 5). Level three substitution (N and gamma, Fig. 6), gives a more complete matching of the Ness sand shale sequence. A level 4 substitution (Fig. 7) produce matches that are more directly comparable with the published stratigraphic zonations.

In the example illustrated in Fig. 8, an automatic level 4 match between Statfjord Formation (alluvial) sequences from wells 33/9–5 and 33/9–4 has given a lowest cost of 64. However, this global optimum may not agree with prior biostratigraphic information, in which case a new match can be generated through the known point. The result of such an operation can be seen in Fig. 10. The original and revised cost matrices can be seen in Fig. 9a, b. Note that, in reality, all we are doing here is testing the hypothesis that, using a chosen set of petrophysical descriptors, two depth sections in different wells can be called equivalent according to a specified set of rules. If we change the rules, the results differ, but the algorithm forces us to state explicitly the rules of the game before it is played.

The change of match position with changing information should not cause concern. We are imposing external control on the system (i.e. biostratigraphic information) and should expect the system to respond. The resultant match is not the global optimum lithostratigraphic match, but should be a better stratigraphic correlation. This emphasizes the difference between matching and correlation.

In Figs 11, 12 & 13 the global optimum sequence matches using level 4 substitution for 300 m sections of all three wells can be seen. The stratigraphic units compared include the Brent Group, Dunlin Group and Statfjord Formation. Due to space considerations the major part of the Dunlin Group shales have been omitted in each diagram. These matches were completely automatic and yet have produced a matching of sequences that is comparable with the cuttings and log-based manual

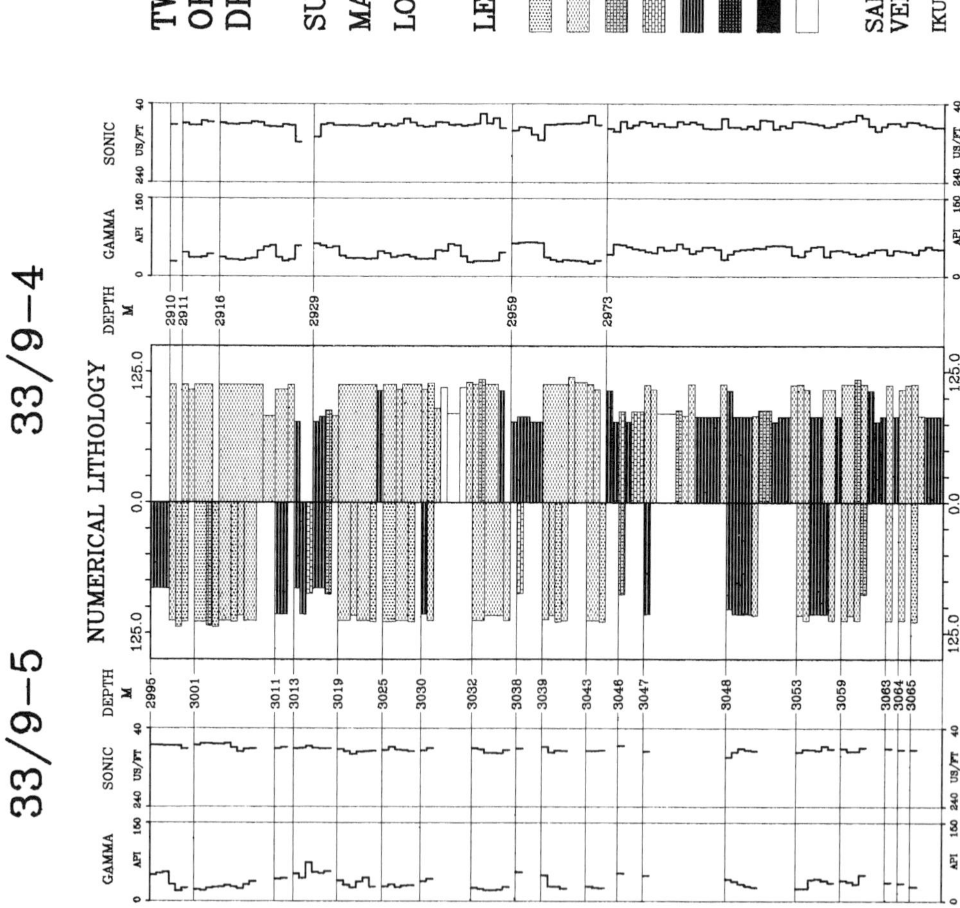

Fig. 10 Matching of the Statfjord formation for wells 33/9–5 and 33/9–4 at substitution level 4, with modified alignment as in Fig. 9b.

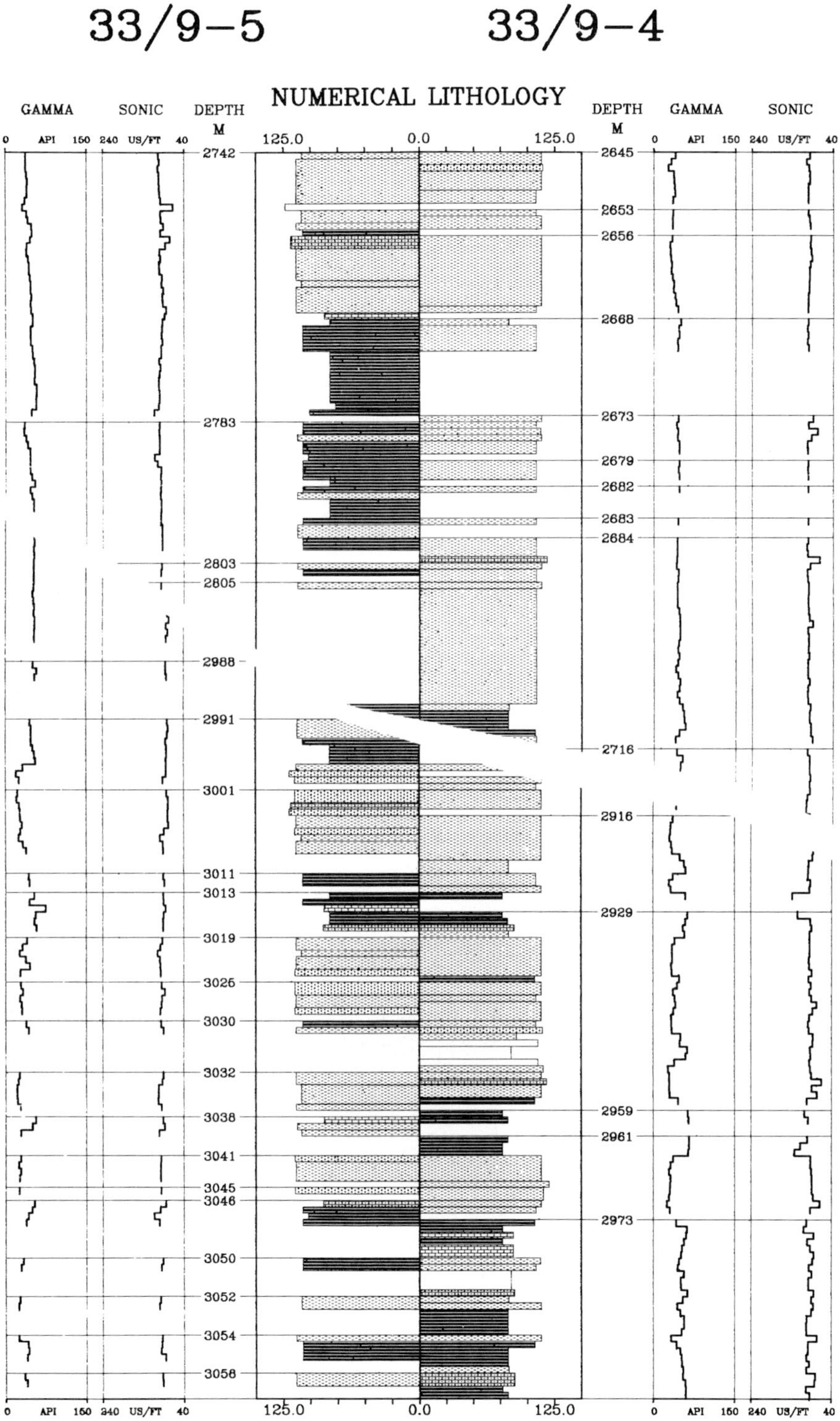

Fig. 11 Matching of Jurassic strata for wells 33/9−5 and 33/9−4 at substitution level 4. Legend as in previous figures. (Dunlin section has been truncated due to lack of space.)

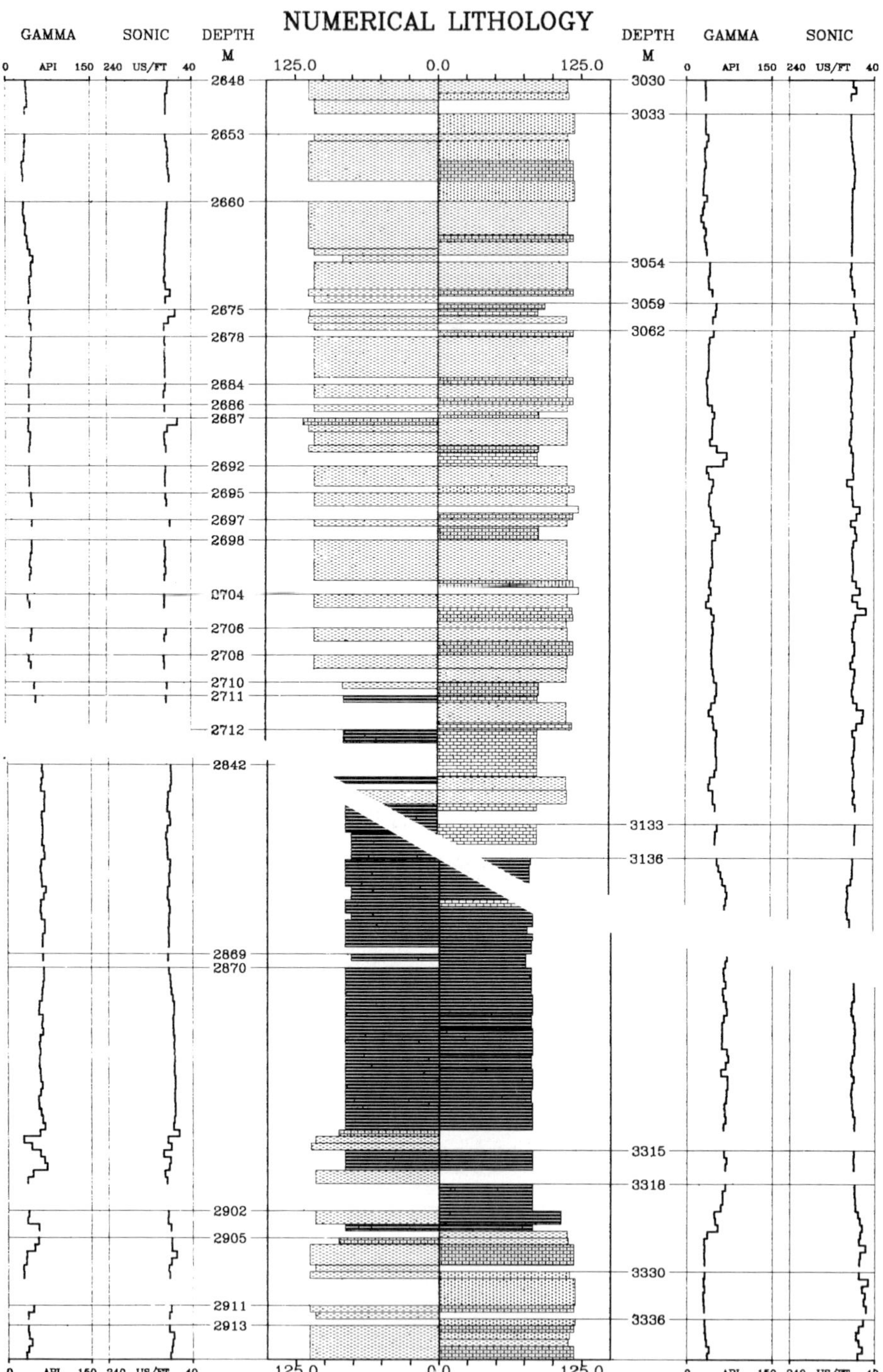

Fig. 12 Matching of Jurassic strata for wells 33/9–4 and 33/9–6 at substitution level 4. Legend as in previous figures. (Dunlin section has been truncated due to lack of space.)

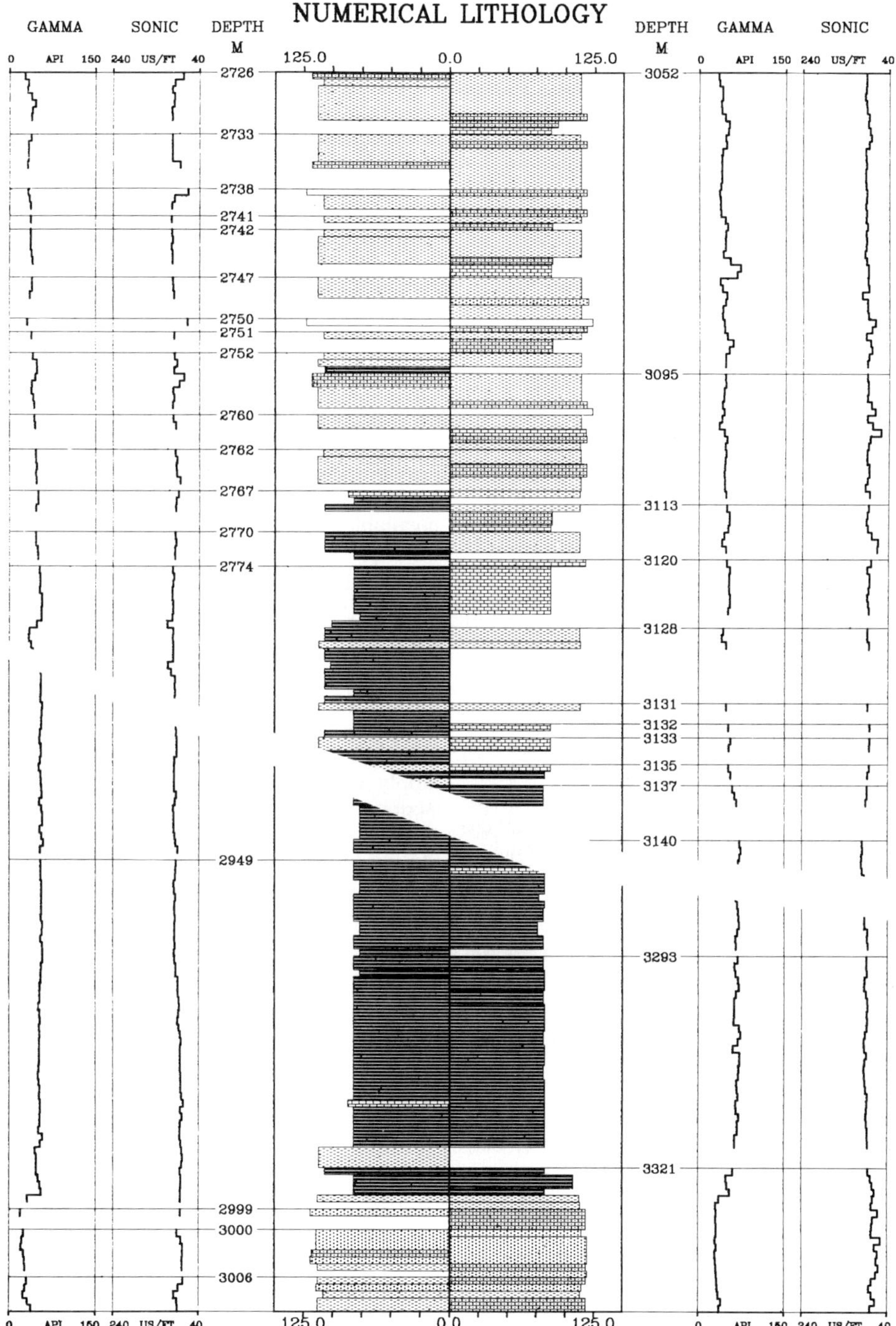

Fig. 13 Matching of Jurassic strata for wells 33/9–5 and 33/9–6 at substitution level 4. Legend as in previous figures. (Dunlin section has been truncated due to lack of space.)

correlations shown in Fig. 2. The technique has proposed gaps in the sequence that can be tested using biostratigraphic data. In particular it can be seen that, petrophysically, most of unit F and E in 33/9–5 is equivalent to the upper two thirds of unit F in 33/9–4, and is absent in 33/9–6. It is compensated for by the increase in thickness of the mid part of the overlying 33/9–5 unit G in both 33/9–4 and 33/9–6. On the other hand unit H (Etive) remains identifiable and of similar thickness in all three wells.

The general success of fully automated matching at substitution level 4 can be seen over these reservoir units. Matching of Well 33/9–6 with the other two wells has been possible at this substitution level despite significant changes in lithology. This well is from the Murchison field about twelve kilometers away from the other two Statfjord wells.

Comments

The effects of noise in the petrophysical data should be taken into consideration. The precision of wireline logs is rarely measured in quantitative fashion, and is not usually a part of any logging contract. The variation in gamma readings over two runs can be in excess of 10%, and the boundaries between lithostates defined by wireline log parameters should reflect this. The width of the class boundaries in *M–N–* gamma space and in spectrometer space (Collins & Doveton 1988) reflect this uncertainty. The possible effect of noise at a class boundary is ameliorated by including contiguity bias in the classification routine. This will modify class membership if a value lies close to a class boundary, *and* the near neighbours belong to a neighbouring class. The choice of a global matching path is influenced by the local structure of the correlation matrix, but since all possible matches are available to the pointer, it is possible to build in a 'look-ahead' function to look for more likely matches on different match paths. Given the stability of the type of lithostate that has been chosen here, noise is not considered the major problem.

The data used for the study are typical of data available in a reservoir correlation study. The matching technique presented here is suitable for proposing quantitative matching hypotheses that can be tested using other data. The matching process itself can incorporate information from other sources, and will be improved as more information becomes available.

The sections shown above illustrate the quantitative and precise matching hypotheses that can be generated by using a combination of numerical lithologies and the gene-typing class of matching algorithm. In the matching game (cf. Waterman & Raymond) the 'truth' can be elusive. The best we can often achieve is an objective test of matching hypotheses based on the most complete geological knowledge available. If that knowledge consists largely of petrophysical data then we are interested in matching (and ultimately correlating) both lithologies (i.e. hydraulic conductivity units) and sequence boundaries. Wu & Nyland (1987) have shown some of the possibilities concerning sequence matching using dynamic programming methods, while this paper has shown some of the possibilities of quantitative formation matching using logs. A third technique is to apply the gene-typing algorithms to bio-zone sequences derived from unitary association processing of biostratigraphic data. A combination of all three methods would then give the possibility of true quantitative stratigraphic correlation rather than matching.

The extensive support of this work by the Norwegian Continental Shelf and Petroleum Technology Institute is gratefully acknowledge.

References

Blackbourn, G. A. 1984. Diagenetic history and reservoir quality of a Brent sand sequence, *Clay Minerals*, **19**, 377–389.

Burke, J. A., Campbell, R. L. & Schmidt, A. W. 1969. The lithoporosity crossplot, *The Log Analyst*, Nov-Dec. 1969.

Collins, D. R. & Doveton, J. H. 1988. Application of colour information theory to the display of lithologic images from wire-line logs *Transactions of the 29th SPWLA Annual Logging Symposium*, San Antonio, Texas, June 5–8, 1988 Paper II. *Analyses*. San Antonio, Texas, June 5–8, 1988.

Evans, I. S. 1977. The selection of class intervals, *Transactions of the Institute of British Geographers*, New Series, **2**, 98–124.

Ghosh, B. K. 1984. STRETCH: a subroutine for stretching time series and its use in stratigraphic correlation. *Computers and Geosciences*, **10**, 137–147.

Gordon, A. D. & Reyment, R. A. 1979. Slotting of borehole sequences, *Journal of the International Association of Mathematical Geologists*, **11**, 309–327.

Griffiths, C. M. 1982a. A proposed geologically consistent segmentation and reassignment algorithm for petrophysical borehole logs. *In*: Cubitt,

J. M. & Reyment, R. A., (eds) *Quantitative Stratigraphic Correlation*. John Wiley London.
—— 1982b. *Pigeonholes and Petrography. Computer Applications in Geology III*, Geological Society of London Miscellaneous Papers, **15**.
—— 1983a. Strata recognition from a computer analysis of drilling variables, *PhD thesis*, University of Newcastle upon Tyne.
—— 1983b. An M-N-Gamma crossplot as the basis of a numerical lithostratigraphy with geotechnical implications. *Proceedings of International Symposium on Recent Advances in Quantitative Stratigraphic Correlation, Kharagpur, Dec. 12–13, 1983*.
—— & Bakke, S. 1988. Semi-automated well matching using gene-typing algorithms and a numerical lithology derived from wire-line logs. Paper GG, *Transactions of the 29th SPWLA Annual Logging Symposium*, San Antonio, Texas, June 5–8, 1988.
Hawkins, D. M. & Merriam, D. F. 1975. Segmentation of discrete sequences of geologic data, *Geological Society of America, Memoir* **142**, 311–315.
—— & Ten Krooden, J. A. 1978. A review of several methods of segmentation. *In*: Gill, D. & Merriam, D. F. (eds) *Geomathematical and Petrophysical Studies in Sedimentology*. Pergamon, Oxford.
Hedberg, H. D. (ed) 1976. *International Stratigraphic Guide: a guide to stratigraphic classification, terminology and procedure*. John Wiley, New York.
Howell, J. A. 1983. A Fortran 77 program for automatic stratigraphic correlation, *Computers and Geosciences*, **9**, 311–327.
Kulinkovich, A. Ye, Sokhranov, N. N. & Churinova, I. M. 1966. Utilization of digital computers to distinguish boundaries of beds and identify sandstone from electric log data. *International Geology Review*, **8**, 416–420.
Levenshtein, V. I. 1966. Binary codes capable of correcting deletions, insertions, and reversals, *Cybernetics and Control Theory*, **10**, 707–710.
Sankoff, D. & Kruskal, J. B. (eds.) 1983. *Time warps, string edits, and macromolecules: The theory and practice of sequence comparison*. Addison- Wesley, Reading, Ma.
Schwarzacher, W. 1980. Models for the study of stratigraphic correlation, *Journal of the International Association of Mathematical Geologists*, **12**, 213–234.
Serra, O. 1977. Methode rapide d'analyse faciologique par diagraphies differees. *Transactions of SPWLA 5th European Symposium, Paris,* Paper 9.
Shaw, B. R. & Cubitt, J. M. 1979. Stratigraphic correlation of well logs: an automated approach. *In*: Gill, D. & Merriam, D. F. (eds.) *Geomathematical and petrophysical studies in sedimentology*. Pergamon Oxford 127–148.
Smith, R. G. & Young, R. L. 1984. The design of the Dipmeter Advisor system. *Proceedings of the Association of Computational Mathematics National Conference* San Fransisco.
Smith, T. F. & Waterman, M. S. 1980. New stratigraphic correlation techniques, *Journal of Geology*, **88**, 451–457.
Spencer, A. M. *et al*. 1987. *Geology of the Norwegian oil and gas fields*. Graham & Trotman, London.
Startzman, R. A. & Kuo, T-B. 1987. A rule-based system for well log correlation, Society of Petroleum Engineers Formation Evaluation, Sept., 311–319.
Waterman, M. S. & Raymond, R. 1987. The match game: New stratigraphic correlation algorithms, *Journal of the International Association of Mathematical Geologists*, **19**, 109–127.
Wolfe, M & Pelissier-Combescure, J. 1982. FACIOLOG: Automatic electrofacies determination. *Transactions of SPWLA Annual Logging Symposium*, July 6–9, 1982, Paper FF.
Wu, X., & Nyland, E. 1987. Automated stratigraphic interpretation of well log data, *Geophysics*, **52**, 1665–1676.

Geophysical logging of water exploration boreholes in the Deccan Traps, Central India

D. K. BUCKLEY[1] & D. OLIVER[2]

[1]*British Geological Survey, Keyworth, Nottingham NG12 5GG, UK*

[2]*Sondex Geophysical Equipment, Finchampstead, Berks, RG11 4QW, UK*

Abstract: Drilling and geophysical logging of Deccan basalt during groundwater exploration in Central India demonstrated that the Deccan lavas comprise a series of simple lava flows, 15–20 m thick, having a porous upper cooling surface with thin clay development and a layered structure. Geophysical logging has proved capable of resolving the layering and produced consistent log responses. Gamma-ray logs were especially useful to identify the weathered flow surfaces but further revealed certain lavas have distinctly higher radioactivity. The higher radioactivity flows were found to be of regional significance. The resolution of the layering and positive identification of flows by geophysical logging permitted individual lavas to be recognized and traced over distances of at least 100 km with no apparent displacement. The logs provided a framework for a detailed stratigraphic sub-division of the lava sequence, and also a key to investigate lateral and vertical variation in aquifer properties.

A vast area of Central India is covered by basalt lava known as the Deccan Trap which was produced in late Cretaceous to Tertiary times by fissure eruption during the break-up of Gondwanaland. The present area of Deccan lava in India exceeds 500 000 km^2 but it is thought to have originally covered a much larger area probably in excess of 2.5 x 10^6 km^2 (Pascoe 1963). Typical lava is a low-potassium tholeiite with Fe- and Ti-enrichment. A marked feature of the lavas is their chemical, petrological and physical uniformity both laterally and vertically. The lava flows are extensive and some can be traced using traditional methods over distances exceeding 60 km (West 1959) but their marked similarity often makes correlation uncertain. The purpose of this paper is to show how geophysical logging has been able to assist geological and hydrogeological studies of the lavas and to emphasize how certain logs have a particular value in correlating the different lavas and establishing a stratigraphy of the sequence.

The lava sequence is thicker near the presumed source in western India. At least 1830 m of lava is proved in the Cambay region and in excess of 2 km is believed to exist offshore, north of Bombay in trough-faulted basins. Onshore the thickness is much less and over much of the main outcrop in Gujarat, Maharashtra, Andra Pradesh, Karnataka and Madhya Pradesh 300 to 500 m of lava is present. In the study area in Madhya Pradesh on the northern outcrop (Fig. 1) 310 m of lava and 18 individual flows were penetrated by exploration. Over most of the outcrop dykes and volcanic centres representing feeder fissures are not recognized. However gravity surveys have indicated rift structures up to 50 km wide beneath the lavas in the Cambay, Narmada, Koyna and Kuruvadi regions and it is speculated (Krishna Brahaman & Negi 1973) that these represent feeders.

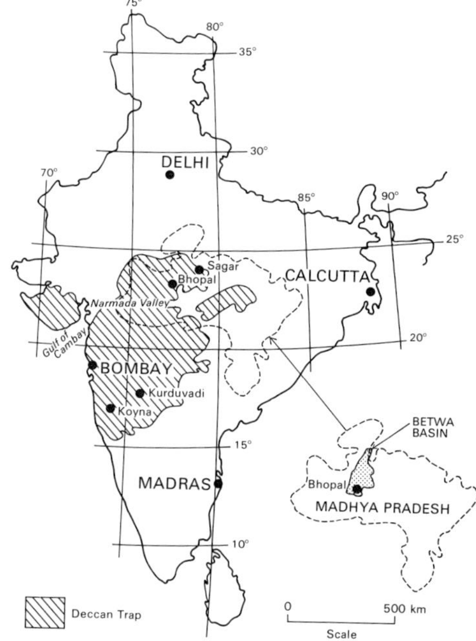

Fig. 1. The Deccan Traps.

From HURST, A., LOVELL, M. A. & MORTON, A. C. (eds), 1990, *Geological Applications of Wireline Logs* Geological Society Special Publication No. 48, pp. 153–161

The lavas have been the subject of several palaeomagnetic studies aimed at relating their magnetic record to extrusive history and documenting the northerly drift of the continent during their extrusion. The interpretation of the palaeomagnetic record and the number of reversal events is, however, in dispute. Several studies recognize only one magnetic reversal event in the onshore lavas, giving a simple division into an upper normal polarity group (lavas above 610 m) and a lower reversed polarity group (below 610 m) (McElhinny 1968, Wensink & Klootwijk 1971). This single reversal is an indication that the lavas were deposited rapidly within probably no more than 5 Ma or so between reversals (McElhinny 1968). A single reversal event is however disputed by Pal & Bhimasankaram (1971) who identified up to five reversals in the record.

The marked similarity and uniformity of the Deccan lavas has frustrated a conventional stratigraphic subdivision of the sequence so that only a broad division into Lower, Middle and Upper lavas, based on the presence of two well developed sedimentary layers, is possible (Wadia 1926, Krishnan 1956).

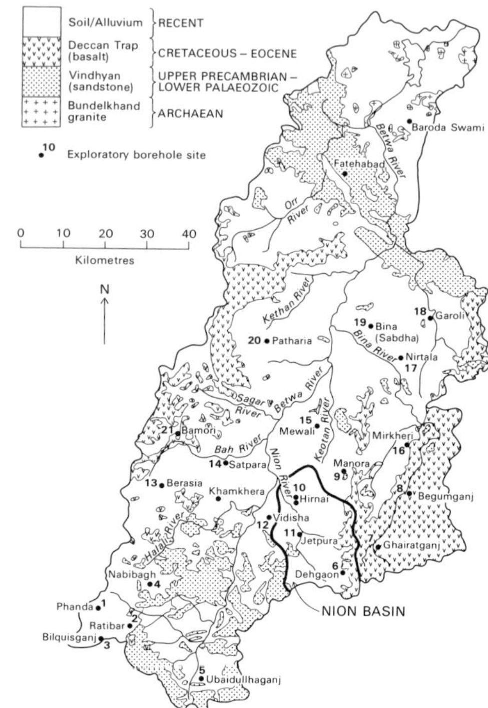

Fig. 2. Geological map of Betwa Basin showing exploratory borehole locations.

Groundwater exploration

Groundwater is abstracted from the lavas by many thousands of hand-dug wells penetrating the weathered layers close to the surface. Prior to 1976 no large scale examination of groundwater potential of the Deccan lava sequence had been undertaken but between 1976 and 1980 a collaborative study between the Central Groundwater Board of the Government of India (CGWB), the British Geological Survey (BGS) and the British Institute of Hydrology, known as the BETWA project systematically examined surface and groundwater resources, including deep aquifers, in the Betwa basin in the state of Madhya Pradesh. (Fig. 1). A main summary of project results was presented in the literature by Versey & Singh (1982).

The Betwa Project drilled more than 20 exploration holes to an average depth of 300 m over a large area (Fig. 2) using air-hammer methods. Using this technique it was possible to show that individual lava flows have a layered structure which could be recognized simply by the drilling penetration rate. The tops of each flow were soft and weathered and comprised green, grey or red clay and amygdaloidal material and usually 1.5 to 3.0 m thick. This graded down into an intermediate hardness amygdaloidal basalt (5 to 8 m thick) which in turn became harder and more compact at depth to merge into massive basalt (5 to 10 m thick). Individual flows were thus generally 15 m thick, but were occasionally as thick as 50 m. The abrupt contact between massive basalt and the clayey top of the underlying flow was easily recognized and was resolved clearly in all log responses. Occasionally a fourth layer of vesicular basalt might be developed at the base of the flow.

Geophysical log responses

Geophysical logs were run using portable logging equipment designed by the BGS for water well logging. The measurements included caliper, gamma-ray, point resistance, resistivity, fluid-conductivity, fluid-temperature, spinner and thermal flowmeter fluid velocity. Subsequently neutron logging and closed-circuit television inspection was carried out. Typical log responses obtained in Deccan basalt are shown in Fig. 3 where point-resistance, short normal resistivity and gamma ray logs of eight flows overlying pre-lava basement are compared with lithology and penetration rate.

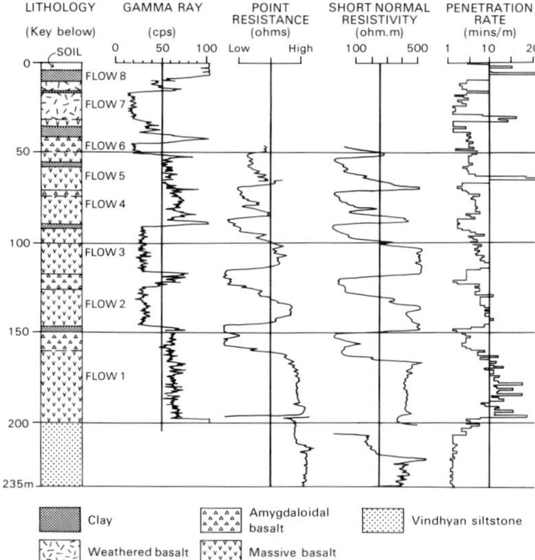

Fig. 3. Typical geophysical log responses in Deccan Trap (Jetpura borehole).

The lithologies and consistent log responses confirm a sequence of simple lava flow each having a layered structure with an upper cooling surface crust and soil formation.

Resistivity logs

The resistivity logs run were somewhat unconventional in that they were obtained using a flexible string of electrodes measuring resistance. The response was not fully linear tending to invert at resistivities in excess of 2000 ohm m; nevertheless over the normal range of aquifer resistivity (less than 500 ohm m) linearity was acceptable. Figure 3 shows that the resistivity log profile closely matches the flow structure with a conductive zone at the top of each flow (1 to 5 m thick) below which resistivity increases through the less vesicular and amygdaloidal portions as the lava becomes more compact and dense. The highest values record against the massive portions. Long normal resistivities of the flow tops are less than 20 ohm m, 15 to 50 ohm m within amygdaloidal layers and 500 to >4000 ohm m against massive basalt. True resistivity of the massive basalt could not be read with the equipment but will be considerably higher.

The point resistance log in Fig. 3 gives good vertical resolution and shows a general trend of increasing hardness and thickness of each flow with depth which is matched by the penetration rate log. The decrease in resistance and resistivity of the upper parts of each flow is due not only to weathering but to natural increased porosity and secondary mineral content associated with the greater frequency and size of pressure-release expansion vesicles near the surface of the flow. Neutron logs respond to these features particularly well.

Gamma-ray logs

Gamma ray responses are shown in Fig. 3. The response to a single flow comprises two parts: an upper flow top response showing high count rate reflecting the concentration of clay minerals (strictly potassium, uranium and thorium content) and a lower part of much reduced count against the massive unweathered basalt. The sharp change between the low count rate massive basalt and the high count rate of the weathered top of the underlying flow served to identify the boundaries between flows. Gamma-ray count rates of the weathered flow top layers were typically 70 to 90 counts per second but only 10 to 30 counts per second against the remainder of the flow. In places some flow surfaces were without clay development and Fig. 4 is an example from Ghairatganj borehole where gamma-ray logging (through drillpipe) failed to identify a junction at 32 m resolved by neutron and point resistance logging. Examination of the cuttings showed a layer of diatomite at the flow boundary.

The gamma-ray logs do not generally resolve detailed lithological variation within the amygdaloidal and massive sections of the flow but they do provide additional important information. Two distinct gamma count rates were observed against massive basalt. Most massive basalt was logged at less than 50 counts per second but certain flows gave significantly higher rates within the range 55 to 80 counts per second and these were termed *high gamma basalts*. In Fig. 3 flows 1, 4 & 5 are *high gamma basalts*. The higher count rate is not due to amygdaloidal material, secondary alteration or secondary minerals but to a higher primary potassium and thorium (and probably uranium) content. X-ray fluorescence and gamma spectrometry confirmed the *high gamma basalts* contained approximately four times the potassium (K_2O) content and 2 to 3 times the thorium content of the normal basalt (Table 1).

The *high-gamma* lavas were found to be present throughout the study area at intervals in the sequence and thus represented a feature of

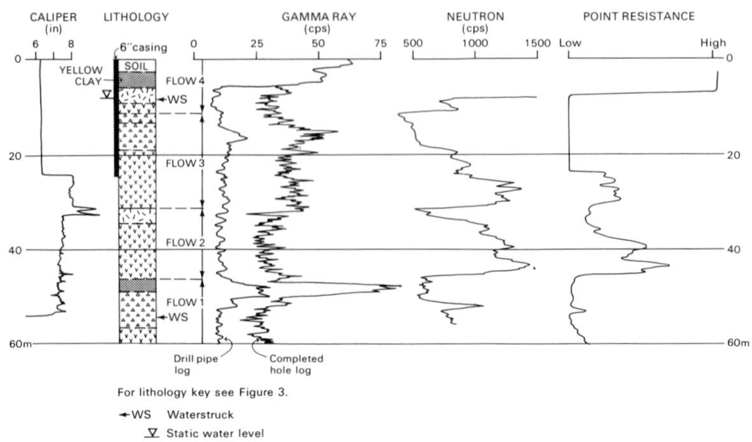

Fig. 4. Geophysical logs, Ghairatganj borehole.

Table 1. *Thorium and potassium content of some Deccan lavas*

Sample	location	depth (m)	basalt type	Th (ppm)	K_2O (%)
1	Dahgaon	50–55	normal	1.35	0.12
2	Dahgaon	111–119	normal	1.68	0.18
3	Dahgaon	173–188	high	3.81	0.64
4	Hirnai	55–62	high	3.45	0.57

Chemical analyses by BGS, samples supplied by BK Singh (CGWB).

regional significance related to a chemical difference between lava flows with time, rather than a local variation in magma composition.

Neutron logs

Neutron logs were run using a single detector tool and representative responses are shown in Figs 4, 5, & 6. The neutron response reflects the layered structure of the flows indicating higher porosity of the flow top, intermediate values of the amygdaloidal units and low porosity of the massive basalt. The neutron tool responds particularly well to vesicularity and presence of hydrated secondary minerals so that subtle variations in lithology, often difficult to describe by examination of the drill cuttings is revealed. In Fig. 4 the neutron log identifies a thin amygdaloidal layer at 40 to 42 m (within flow 2) and further such layers can be inferred near the top of flow 3. The caliper confirms these variations are not simply related to hole diameter changes.

Figure 5 is a composite of gamma ray, neutron and resistivity logs of five lava flows in the Hirnai borehole. Two *high-gamma* flows (4 & 5) at 30 to 50 m and 50 to 68 m are present but the neutron and resistivity logs show they differ in properties. Flow 5 is more dense, and has a lower porosity and higher resistivity than flow 4 which is more amygdaloidal as noted by its moderate neutron porosity and lower resistivity.

Figure 6 shows typical neutron responses in three holes 0.5 km apart showing that they can be used for correlation. The position of water strikes is also shown and it is noted that there are important differences from site to site over such distances. This proved to be a characteristic of the deeper aquifers. Figure 6 suggests that neutron logs could be used to examine lateral and vertical variation in vesicularity though a compensated tool would be preferred to minimize the effect of borehole diameter variation. The consistent log responses shown by the Deccan lava flows is by no means a feature of all basalt logs. In Fig. 7 gamma-ray, neutron and resistivity logs of Stormberg basalt (Jurassic), examined during groundwater exploration in Botswana by Cheney (1981), suggest that possibly four units are present. A clear weathered flow surface is identified at 52 m but elsewhere

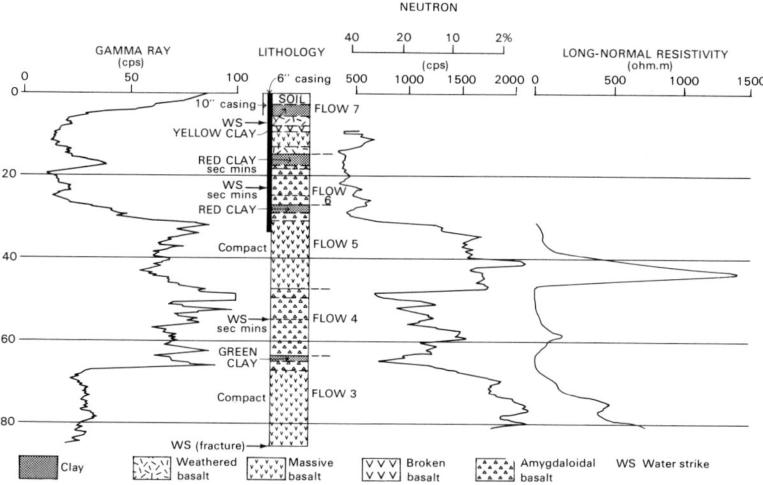

Fig. 5. Geophysical logs, Hirnai borehole.

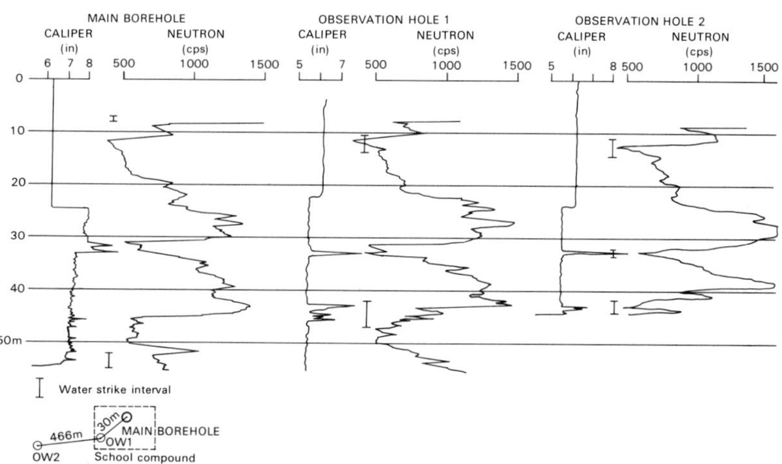

Fig. 6. Neutron and caliper logs, Ghairatganj boreholes.

changes in vesicularity inferred from the log responses reflect brief cooling intervals without soil development. Such lavas are relatively more complex though neutron and resistivity logs resolve the changes. This example illustrates the potential use of geophysical logs in a study of compound lava flows. The Stormberg basalts examined represent a late stage in the lava deposition. Some similar late-stage compound flows also occur in parts of Western India as described by Walker (1971).

Correlation of Deccan Trap lavas

A correlation across the Nion basin of the Project area using gamma ray logs is given in Figs 8 & 9. The high-gamma flows 1, 4 & 5 are the key to correlation and allow the lavas to be traced from Dahgaon to Gyaraspur over a distance of 50 km with little displacement.

Of local interest is that the gamma logging allows the Betwa flows to be linked to an established sequence of ten flows around Sagar described in the literature in detail by West (1958) and Alexander (1980). Geological considerations suggested that flows 1 & 2 at Sagar should correlate with Betwa flows 4 & 5. Published analyses of the Sagar flows 1 & 2 show elevated potassium content (Table 2) and this corresponds to the *high-gamma* response of flow 4 & 5 in the Betwa Basin. In addition the slightly higher potassium content of flow 1 is

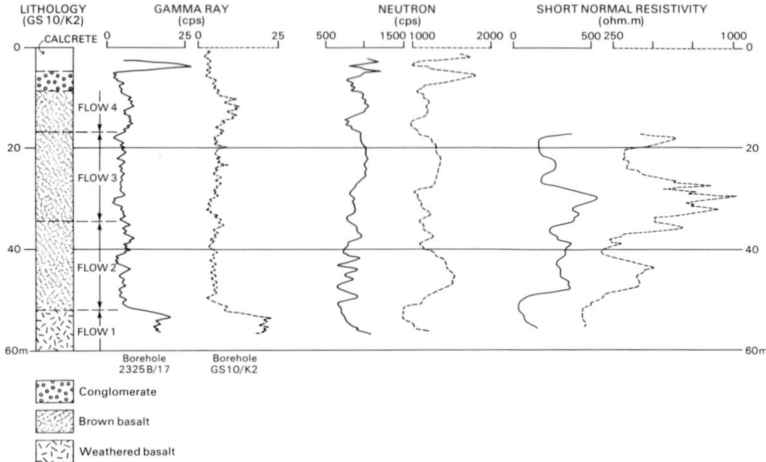

Fig. 7. Geophysical log responses, Stormberg basalt (Jurassic) Botswana, (after Cheney 1981).

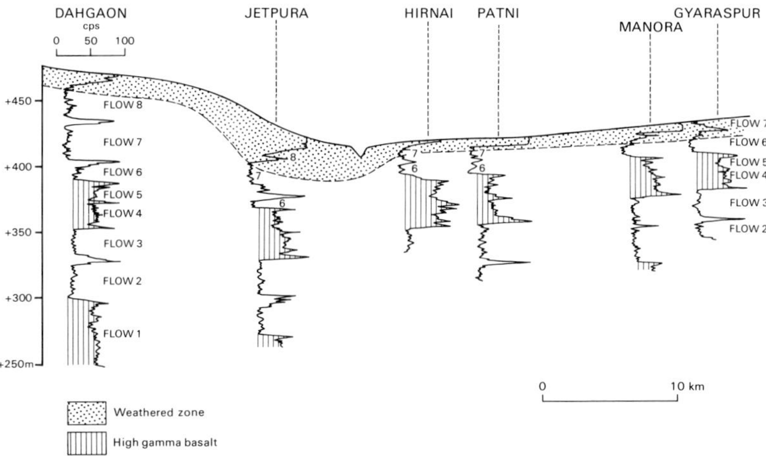

Fig. 8. Gamma-ray log correlation of basalt flows, Nion Basin.

Table 2 *Chemical analyses of Deccan basalt flows near Sagar (after Alexander & Paul 1977)*

Sagar Flows	1	2	3	4	5	6	7	8	9
SiO_2	49.27	48.80	47.99	46.92	47.66	48.68	47.10	48.70	49.22
TiO_2	2.62	2.33	2.31	2.39	2.27	2.65	2.62	2.63	3.48
Al_2O_3	14.50	13.71	13.56	14.33	14.23	14.32	14.45	13.71	12.07
Fe_2O_3	2.76	3.21	4.57	5.81	4.12	4.70	1.85	3.43	3.48
FeO	9.95	10.57	8.96	7.11	8.66	8.89	9.96	10.20	10.96
MnO	0.02	0.21	0.22	0.24	0.20	0.22	0.21	0.27	0.23
MgO	4.66	6.23	6.00	5.16	6.57	5.98	5.92	6.19	5.00
CaO	9.98	10.79	10.49	10.55	11.04	10.52	9.13	10.63	9.63
Na_2O	2.58	2.30	2.30	2.36	2.26	2.38	2.53	2.66	2.45
K_2O	0.57	0.33	0.22	0.21	0.16	0.23	0.26	0.30	0.61
P_2O_5	0.29	0.23	0.22	0.23	0.22	0.25	0.27	0.28	0.38
H_2O	2.09	1.23	2.67	3.98	2.43	0.52	4.59	1.01	1.18
Others	0.25	0.36	0.34	0.43	0.31	0.67	0.82	0.08	0.02

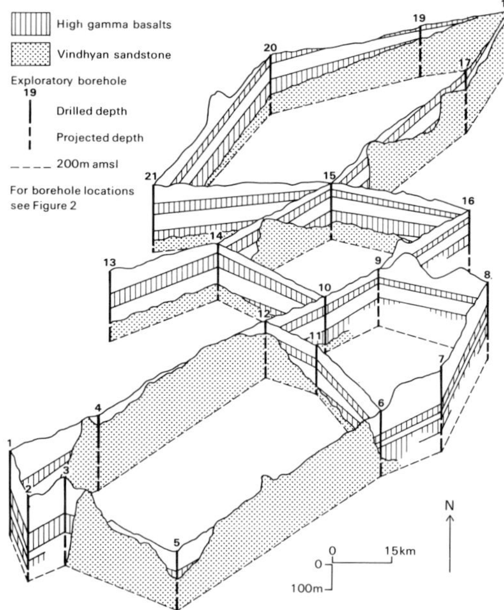

Fig. 9. Fence diagram of Deccan basalt, Betwa Basin (after Versey & Singh 1982).

reflected by the greater gamma response of Betwa flow 4 (Fig. 8). Sagar flow 9 is also a *high-gamma* flow though it is younger than those represented in the Nion Basin.

Chemical analyses of other Deccan basalts (Alexander 1980, Alexander & Paul 1977, Bose 1972, Ghose 1976, Kaneoka & Hamamura 1973) show that a variable potassium content is a feature of the Deccan lavas. In fact the Betwa and Sagar lavas are relatively potassium-deficient compared to later flows. A similar variable potassium content which would probably be manifest in the gamma-ray response is evident for other flood basalts e.g. the Ethiopian flood basalts reported by Mohr (1983).

Correlation of the lavas through geophysical logging demonstrated a regional dip which showed that within a basin the present-day weathered zone could cut across several lava flows as a diachronous unit. This proved to be important hydrogeologically because it introduced variations in the properties of the shallow aquifer.

Groundwater occurrence in the Deccan Lavas

Project studies identified two types of aquifer in the lava sequence:

(a) Water-bearing layers associated with present-day weathering representing a shallow aquifer.
(b) Water-bearing layers associated with the weathered surfaces of deeper flows and representing a deep aquifer.

In fact both groundwater occurrences relates to weathering. The shallow aquifer is associated with present-day weathering products and water production in the deep aquifer is obtained from the fossil weathered surfaces between lava flows.

Figure 10 is a typical example of water occurrence revealed by borehole flowmeter, fluid-temperature and fluid-conductivity logs in Dahgaon borehole. In this example the first water strike (WS) is from the weathered and jointed basalt at the base of the weathered zone of the first flow. The second strike is from a jointed and amygdaloidal unit at the base of the same flow and the third is from soft amygdaloidal layers of the underlying flow. Significantly the first and third water strikes represent an equivalent position in the weathering profile, that is a logical position towards the base of the weathered material where the rock is competent enough to support open joints, though not sufficiently weathered as to be clayey or contain clay-filled joints. Hydraulic continuity of the deeper aquifers was maintained over large areas but permeability was low and extremely variable so that productive horizons of the deep aquifer were of patchy occurrence. Details of variations in hydraulic properties and testing of the deep aquifers are given by Versey & Singh (1981) and by Herbert *et al.* (1981).

The fluid logs in Figure 10 were run whilst pumping and therefore refer to the upward movement of fluid. A reduction in both fluid temperature and electrical conductivity occurs at 43 m where water enters from a deep aquifer. This water moves up to the pump and is cooled by shallow aquifer water at 29–30 m from the jointed and amygdaloidal unit at the base of the flow, and by further addition at 17 m where water of similar temperature, but lower conductivity (and therefore different water) enters. The flowmeter log confirms these entries and shows approximately 50% of the discharge comes from the shallow aquifer at the prevailing pumping rate. The lower conductivity and cooler (and hence more rapidly circulating) shallow aquifer water is evident from the fluid logging.

Because the present day weathered zone and its shallow aquifer is a diachronous unit the components of the aquifer comprise different flows or parts of a flow at different locations in a basin, and this introduced a variation in aquifer

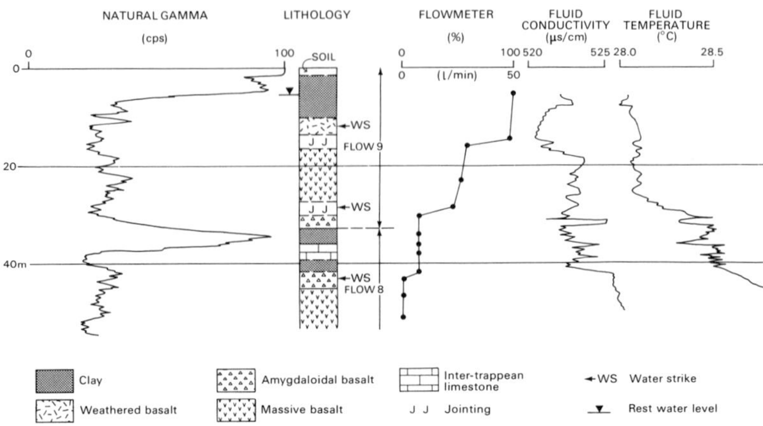

Fig. 10. Gamma-ray and fluid logs, Dahgaon borehole.

properties which was examined in some detail by the Project, (see for example Lawrence & Ansari (1980), Lawrence (1985) and Herbert *et al.* (1981)). For example the weathered zone at Jetpura (Fig. 8) is relatively thick (24 m) and the log correlation reveals that two flows, the lower part of flow 8 and the already weathered upper layers of flow 7 go to form the weathered zone. Here and elsewhere such thicker weathered sections locally provide more aquifer storage and higher aquifer transmissivity.

Through the geophysical log correlations particular amygdaloidal or productive flow surfaces at depth were targeted for drilling or could be traced to regions where they entered the weathered zone to provide optimum locations for hand dug wells. Features of lavas traditionally associated with high permeability such as scoriaceous layers, lava tubes, pipes, tunnels etc., were not found in the area studied nor are they perhaps likely to be encountered due to the implicit fluidity of the magma. The long-distance correlations that can be established also imply that vertical structures capable of displacing the units and thereby facilitating a groundwater circulation to depth, do not exist or are rare. As a consequence of the restricted groundwater circulation the permeability of the deep aquifer was often not adequately developed.

Summary and conclusions

Geophysical logging provides a useful means of examining the lava sequence of the Deccan Trap. Each flow generally has a three-layer structure which can be resolved by simple logs including electrical resistivity, gamma-ray and neutron. Neutron logs have proved particularly useful for examining vesicularity due to their particular response to hydrogen and hydrated secondary minerals.

Gamma-ray logging has proved useful for identifying individual lavas by recognition of their clayey flow top surfaces. Gamma-ray logs also reveal that certain lavas have relatively greater natural radioactivity. Such flows were termed *high-gamma basalts* and were found to be a regional feature related to a chemical difference between certain flows.

Over much of the Deccan the lavas are very uniform and monotonously similar making identification and correlation difficult. The presence and recognition of *high-gamma basalt* by geophysical logging has provided a key to a stratigraphic subdivision.

Such correlations show that individual lavas extend for distances exceeding 100 km with little or no displacement or variation in thickness. They must therefore have been very fluid and extruded very rapidly.

Water occurrence in the lavas was found to be closely related to weathering affects. Near the surface groundwater is associated with the present day weathered zone. At depth beyond the range of present weathering, groundwater is generally associated with the fossil weathering layers developed near the top of the deeper flows.

The geophysical logging and correlations were able to demonstrate that the present weathered zone is a diachronous unit which cuts across various flows or parts of a flow, and this provided a framework to examine variations in the aquifer properties and made it possible to predict optimum locations for borehole and hand dug well construction.

Logging of fluid temperature, fluid electrical conductivity and fluid velocity was able to locate producing horizons in the lava flow structure and to demonstrate that the deep aquifer units are neither continuous, nor have consistent permeability.

This paper was published with the permission of the Director, BGS (NERC).

References

ALEXANDER, P. O. 1980, Petrogenesis of low-potassium quartz normative Deccan tholeiites. *Journal of the Geological Society of India*, **21**, 261–272.

—— & PAUL, D. K. 1977. Geochemistry and strontium isotopic composition of basalts from the eastern Deccan volcanic province, India. *Mineralogical Magazine*. **41**, 165–172.

BOSE, M. K. 1972. Deccan Basalts. *Lithos*, **5**, 131–145.

CHENEY, C. S. 1981. *Hydrogeological investigations into the Stormberg basalts of the Lephepe/Dibete area.* Report GS10/13, Botswana Geological Survey.

GHOSE, N. C. 1976. Composition and origin of Deccan basalts. *Lithos*, **9**, 65–73.

HERBERT, R., BUCKLEY, D., ADAMS, B. & EDMUNDS, W. M. 1981. *The hydrogeology of the Deccan Traps of the Betwa Basin, India*, Indo-British Betwa Report 7, Betwa Report Series BGS

KANEOKA, I. & HARAMURA H. 1973. K/Ar ages of successive lava flows from the Deccan traps, India. *Earth and Planetary Science Letters*, **18**, 229–236.

KRISHNA BRAHMAN, N. & NEGI, J. G. 1973. Rift valleys beneath Deccan Traps. *Geophysical Research Bulletin*, **11**, 3

KRISHNAN, M. S. 1956. *Geology of India and Burma*, Madras.

LAWRENCE, A. H. 1985. An interpretation of dugwell performance using a digital model. *Groundwater*, **23**, 449–464.

—— & ANSARI, S. U. 1980. *The shallow aquifer of the Deccan basalts of the Betwa Basin.* Indo-British Betwa Report WD/OS/80/2, Betwa Report Series

MCELHINNY, M. W. 1968. Northward drift of India – examination of recent Palaeomagnetic results. *Nature*, **217**, 342–344.

MOHR, P. 1983. Ethiopian flood basalt province. *Nature*, **303**, 577–583.

PAL, P. C. & BHIMASANKARAM 1971. Volcanism, orogeny and drift of the Indian Place. *Bulletin of the National Geophysical Research Institute* **9**, 3+4 Hyderabad.

PASCOE, E. H. 1963. *Manual of the geology of India and Burma*.

VERSEY, H. R. & SINGH, B. K. 1981. *Deep groundwater in the Deccan basalts of the Betwa valley.* Indo-British Betwa Report 8. Betwa Report Series BGS.

—— & ——, 1982. Groundwater in Deccan basalts of the Betwa Basin, India. *Journal of Hydrology*, **58**, 276–306.

WADIA, D. N. 1926. *Geology of India*, Tata-McGraw-Hill, New Delhi.

WALKER, G. P. L. 1971. Compound and simple lava flows and flood basalts. *Bulletin Volcanologique*, **35**, 570–590.

WEST, W. D. 1958. The petrography and petrogenesis of 48 flows of Deccan Trap penetrated by borings in Western India. *Transactions of the National Institute of Science, India*. **4**, 1–56.

—— 1959. The source of the Deccan Trap flows. *Journal of the Geological Society of India*. **1**, 44–52.

WENSINK, H. & KLOOTWIJK C. T. 1971. Palaeomagnetism of the Deccan Traps in the Western Ghats near Poona India. *Tectonophysics*, **11**, 175–190.

Mineralogy and geochemistry

Geological applications of geochemical well logging

M. M. HERRON & S. L. HERRON
Schlumberger-Doll Research, Old Quarry Road, Ridgefield, Connecticut 06877–4108, U.S.A.

Abstract: Recent advances in geochemical logging and interpretation have made it possible to obtain in situ concentration logs for at least ten of the chemical elements present in sedimentary formations: Al, Si, Ca, Fe, S, Ti, K, Th, U, Gd and possibly Mg. Each of these elements is concentrated in the solid portion of the formation as opposed to the pore fluids, and together these elements provide an array of measurements with a large dynamic range and tremendous diagnostic strength for geological interpretation.

The combination of a few diagnostic elements, such as silicon, aluminum, and calcium, provides sufficient information for a rapid but accurate lithological description. In the case of siliciclastic reservoir rocks, it is possible to discriminate between sand and shales and to determine types of sandstones using the ratios of SiO_2/Al_2O_3 and Fe_2O_3/K_2O. Calcium is used in conjunction with these ratios to differentiate between non-calcareous, calcareous, and carbonate rocks. On a more sophisticated level, a set of chemical abundances can be incorporated into a sedimentary normative analysis to determine quantitatively both the framework and clay mineralogy of siliciclastic formations. Derived mineral assemblages can provide valuable information for the interpretation of depositional environments and diagenesis. In shales, elemental data can be used alone or in conjunction with derived mineralogy to derive total organic carbon and thereby begin to evaluate source rock potential. By using individual elemental concentration logs or any of the interpreted formation units it is possible to enhance the characterization of vertical sequences and the recognition of well-to-well correlations.

Geological interpretation frequently begins with the description of a rock in terms of its composition and texture. For rocks located in the subsurface, the lithological or compositional interpretations available from wireline logs are of limited accuracy because the input logs are more sensitive to porosity or fluid composition than to rock properties. As a result it is nearly impossible to interpret definitively changes in porosity from changes in lithology. Therefore, most geologists prefer core lithological description for its reliability.

Recent advances in geochemical logging and interpretation now make it possible to obtain in situ concentration logs for most of the important rock-forming elements in sedimentary formations. The concentration logs enable geologists to describe the rock composition in previously unobtainable detail; the potential for using this information for geological interpretation is enormous. Some techniques have been developed to use the data to describe the formation accurately in terms of its lithology, to classify siliciclastic reservoirs, and under favourable circumstances, to determine detailed mineralogy. From the log-derived mineralogy it is possible to derive or infer a number of other formation properties including matrix density and porosity, cation exchange capacity, intrinsic permeability, and grain size (Herron, 1987a,b). The next step of using these composition interpretations and derived properties for geological interpretation of depositional environment, facies, and geologic history has barely begun.

The Geochemical Logging Tool (GLT*) string uses three types of nuclear measurements in combination with a geochemical formation model to provide elemental concentration logs of ten elements: aluminum, silicon, calcium, iron, sulphur, titanium, gadolinium, potassium, thorium, and uranium (Hertzog *et al.* 1987). Aluminum concentrations are measured by delayed neutron activation analysis using a californium-252 source of neutrons. Potassium, thorium, and uranium concentrations are determined from the natural gamma-ray activity spectrum. Relative concentrations of the remaining elements are determined from the prompt capture gamma-ray spectrum measured after a burst of 14 MeV neutrons. These relative concentrations are then converted to absolute weight per cent using a geochemical model which assumes that elemental oxides sum to unity. Although magnesium is not measured by these techniques, Mg concentrations can be inferred from a comparison between measured and derived photoelectric factor. Details of these techniques and comparisons with core chemistry are

* Mark of Schlumberger.

From Hurst, A., Lovell, M. A. & Morton, A. C. (eds), 1990, *Geological Applications of Wireline Logs* Geological Society Special Publication No. 48, pp. 165–175

available (Hertzog *et al.* 1987). Carbon concentrations can also be derived from carbon–oxygen ratios measured by inelastic gamma-ray spectroscopy (Roscoe & Grau, 1985; S. Herron, 1986).

Geochemical lithology

One of the most straightforward applications of elemental logs is continuous lithological identification. While many other wireline logs are used conventionally for estimating lithology, none offer the accuracy, dynamic range, porosity independence, and consequently, diagnostic strength provided by elemental concentrations. As an example, consider two rock-forming minerals: quartz, a major component of sandstones; and calcite, a major component of carbonates. Although geophysical log measurements such as density, neutron porosity, or compressional transit time can be used to discriminate between formations containing large quantities of quartz versus calcite, they are actually more sensitive to changes in porosity. Consequently as porosity varies and the formation becomes more compositionally complex, it becomes increasingly difficult to describe lithological variations accurately from geophysical logs. In contrast, most elemental concentrations are primarily sensitive to variations in the matrix composition, and they retain their diagnostic quality even for complex lithologies with varying porosities.

The three elements Si, Al, and Ca can be used to create a rapid screening of concentrations logs for major lithological categories. Table 1 shows average concentrations for Si, Al, and Ca in sandstone, shale, and carbonates as well as a general range of values for each of the major lithologies. In general, sandstones have high Si, carbonates have high Ca, and shales contain more of almost every other element including Al. The ranges of concentrations for Si, Al, and Ca were used to rapidly identify lithological variations in the Conoco 33–1 test well in Oklahoma. Logs of these three elemental concentrations are presented in Fig. 1, with concentrations measured on core samples included for comparison. The lithological column resulting from the lithology screening values in Table 1 is also presented. It is clear that not only are the sand, shale, and carbonate lithologies easily found, but it is also easy to identify the shaly sand sections, such as that at 760 m, and the shaly carbonate above the carbonate at 840 m. Geophysical data from these complex intervals cannot be reliably interpreted in terms of lithology. The geochemical data provide a fundamentally sound basis for lithological classification that does not require the extensive assumptions necessary for similar types of interpretation from conventional geophysical logs.

Table 1. *Average elemental concentrations for sandstone, shale and carbonate lithologies (Turekian & Wedepohl, 1961) and ranges for lithological screening*

Element	Sandstone	Shale	Carbonate
Si (%)	36.8 (30–47)	27.3 (20–30)	2.4 (<5)
Al	2.5 (0–4)	8.0 (>4)	0.42 (0–2)
Ca	3.9 (<4)	2.2 (<4)	30.2 (>15)

Sandstone classification

The next step in interpretation of geochemical well log data is the inclusion of additional elemental concentrations for further classification of the lithologies into categories which have depositional or diagenetic significance. For example, Herron (1988) presents a means of mapping chemical concentrations of siliciclastic

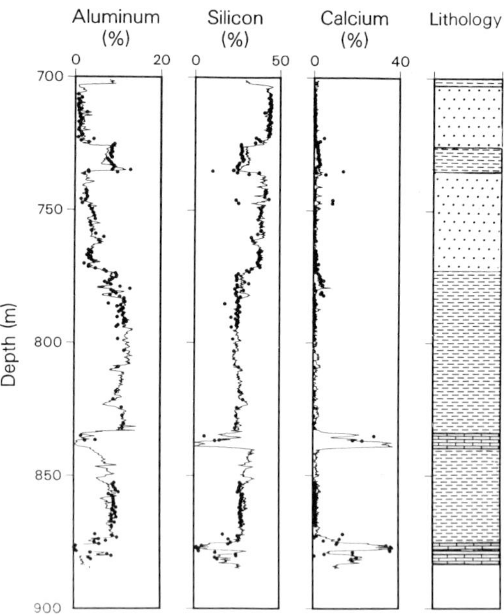

Fig. 1. GLT concentration logs of Al, Si, and Ca from the Conoco 33–1 test well. Core chemistry values are shown as filled circles for comparison. The far right column depicts the major sandstone, shale, and carbonate lithologies determined from the lithological screening values listed in Table 1.

sediments to basic sandstone classification. The SandClass* system (Fig. 2) uses the SiO_2/Al_2O_3 ratio to separate quartz arenites from shales with other sandstones having intermediate values. The Fe_2O_3/K_2O ratio separates lithic sands from feldspathic sands. The Ca concentration is used to differentiate non-calcareous from calcareous sandstones and shales and to separate siliciclastic from carbonate rocks.

The classification system, developed from core analyses, has been applied to logs obtained from a well in Kern County, California where the sediments are composed primarily of Plio-Pleistocene arkosic sands and shales deposited in an alluvial fan environment. The elemental concentration logs are presented in Fig. 3 along with core data. The concentration logs show very good agreement with the core values, except in two thin, calcareous intervals at about 600 and 675 m depths. Such discrepancies demonstrate the consequences of different measurement scales for the two methods between the 2 cm thick core plugs and the *c.* 60 cm vertical averaging of the GLT data. The sandstone classifications derived from the geochemical log data (Fig. 4a) are presented in approximate order of decreasing reservoir quality from highest quality quartz arenite at the left grading through sublitharenite, subarkose, litharenite, and arkose to wacke and shale. Non-calcareous samples are plotted on the main class divisions; calcareous samples are displaced one half division to the right. The geochemical logs indicate that the formation is composed primarily of arkosic sands and shales. Core petrographical analyses, presented in Fig. 4a as solid dots, confirm that the sand units are exclusively arkosic.

Successful identification of these and other arkosic, or 'granite wash' sands from wireline logs has frequently been difficult because the large K-feldspar content produces a high, shale-like, gamma-ray signature (Fig. 4b) such that many sand units are difficult to distinguish from shales using the gamma-ray log alone. Inclusion of other log data will usually improve the reliability of lithological estimations. However, the geochemical log data alone permit an unambiguous distinction of sands from shales and they also identify correctly the sand type as arkosic arenite. This sensitivity to sandstone composition permits evaluation far beyond simple lithology and provides an opportunity for enhanced sandstone evaluation and mapping throughout fields and basins.

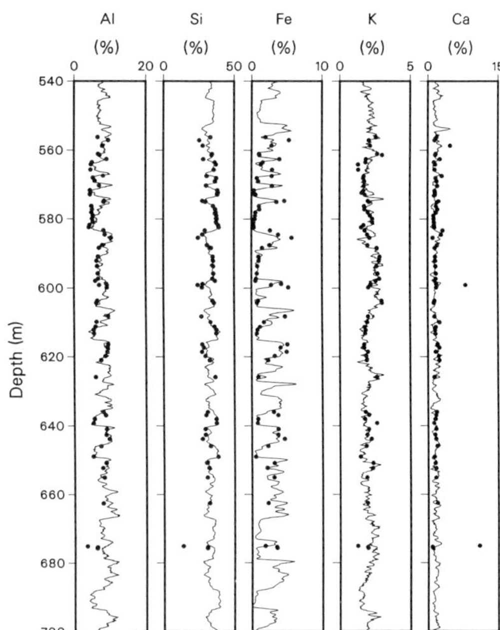

Fig. 3. Elemental concentration logs and core chemistry values (filled circles) from Kern County, California well. Core data are from neutron activation and X-ray fluorescence analyses.

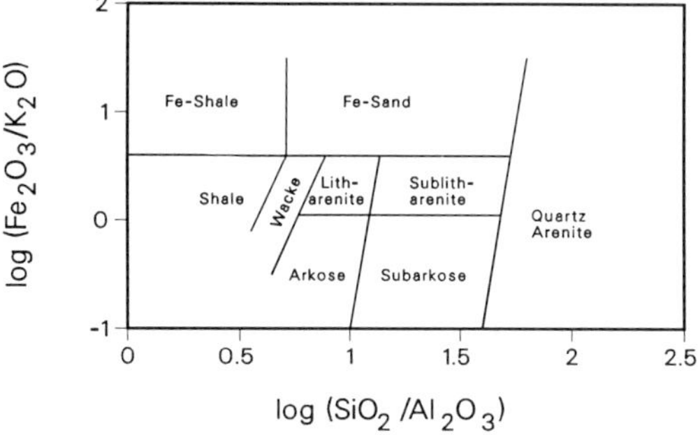

Fig. 2. SandClass System for relating chemical concentrations in clastic environments to basic sandstone classifications (Herron 1988).

Chemical mineralogy and diagenesis

A more sophisticated utilization of geochemical well log data is the transformation of elemental data into abundances of chemical mineral equivalents, or 'chem-minerals.' Chem-minerals, like normative minerals, are alternate expressions of chemical data in a form which is easier and more meaningful to use. The chem-mineralogy may or may not reflect the true formation mineralogy, but it is a useful distillation of elemental data into a more useful and familiar

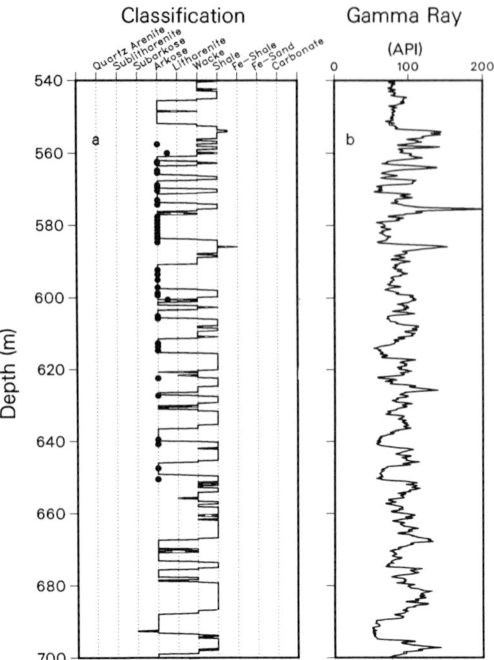

Fig. 4. A) SandClass classifications derived from the geochemical logs presented in Figure 3. Petrographic classifications analyzed from core plugs are shown as filled circles. B) Gamma ray log for the same Kern well. Note that many sand units seen from the Sand Class presentation would not be identified as sands in a conventional gamma ray interpretation.

form which may include framework minerals, clay minerals, and even trace minerals on a quantitative basis. There are two basic types of element-to-mineral transforms: the locally calibrated model and the normative mineral analysis.

Local element-to-mineral transforms can be constructed from core data and used to describe the mineralogy of a given well (e.g. Flaum & Pirie 1981; Peveraro & Russell 1984; M. Herron 1986). For example, a four-mineral geochemical model was presented for a Venezuelan well where the concentrations of quartz, feldspar, illite and kaolinite were derived from three simultaneous equations using Al, Fe, and K concentrations (M. Herron 1986). Such transforms are of great value in the local area for which they have been calibrated, allowing vital petrophysical parameters such as average matrix density to be reliably determined from the calculated mineralogy. In the larger sense, however, such models usually involve only a few minerals and the mineralogy estimated in one study is not directly comparable to that estimated in another study because of differences in the models' mineral suites, elements, dimensionality, and constraints, among other factors.

Normative transformation of elements to minerals provides a means of calculating the mineralogy when a locally calibrated model is unavailable. This can be critical in understanding trends in mineralogy that are difficult to see from the raw element logs. Normative chem-mineral abundances will agree with the abundances of the actual minerals present if two conditions are met (M. Herron 1986). Firstly, the chemical composition of each mineral to be considered must be fairly constant and different from the composition of other minerals. Obviously, two minerals of identical composition cannot be distinguished. The limits of 'constant' composition are determined by the degree of difference from other mineral compositions as well as by the required degree of accuracy. Secondly, all the minerals residing in the formation and containing elements used in the model must be present in the model.

Since over 2000 minerals have been recognized, converting elemental concentrations to abundances of meaningful chem-minerals seems like an impossible task. However, a vast majority of the sediments contain only ten minerals: quartz, four clay minerals, three feldspar minerals, and two carbonate minerals, and this greatly simplifies the problem. Table 2 provides approximate ranges of mineral concentrations generally observed in the common sedimentary rock types (Blatt *et al.* 1972; Pettijohn 1975). Given this limited suite of minerals and the array of rock-forming elements measured by the GLT logs, the problem of estimating in situ mineralogy from geochemical well logging becomes more manageable. Indeed, as long as the two conditions are not seriously violated, chem-mineral abundances will be close to mineral abundances determined from X-ray diffraction (XRD) or other mineralogical analysis. In any event, chem-mineral abundances will be readily comparable between wells, fields, and basins, and will be available with a vertical detail unimaginable from conventional core analyses.

Table 2. *Average mineral composition of sandstones, shales, and carbonates*

Mineral	Sandstone	Shale	Carbonate
Quartz	65–70	20–30	0–5
Feldspar	10–25	5–20	0
Clay	5–10	50–60	0–5
Carbonate	0–5	0–10	95

A ten-mineral sedimentary normative mineral analysis

Here we present a general element-to-mineral transform that can serve the function of a normative mineral analysis for many siliciclastic sedimentary environments. The model uses the elements Al, Si, Ca, Fe, S, K, and Ti, in conjunction with epithermal (or corrected thermal) neutron porosity and bulk density, to determine the following chem-minerals: quartz, feldspar, calcite, kaolinite, illite, smectite, rutile, pyrite, siderite and residual. In addition, if a photoelectric factor log indicates the presence of magnesium, dolomite can also be determined. This model is presented and demonstrated using geochemical logs from a well in Utah.

Currently, the model involves four steps, three pre-processing steps followed by the solution of a series of simultaneous equations.

(i) Calculation of excess sulphur. In this model, the only mineral containing sulphur is pyrite, FeS_2, and that requires a certain amount of associated iron. Any sulphur in excess of the measured iron multiplied by the S/Fe ratio of pyrite (1.15) is subtracted from the measured sulfur in this pre-processing step. In this well, there was no excess sulphur.

(ii) Calculation of excess iron. In this model, the most iron-rich mineral after pyrite is illite. All the iron that can be accommodated by S is considered pyrite iron. Non-pyrite iron (Fe−0.89S) is then compared to the concentrations of other elements (typically Al, K and Ti) and the elemental ratios of poorly ordered illite (Table 3) to see how much iron might be present in illite. Any excess iron is subtracted in this step and modelled as siderite. For example, one sample had concentrations of 3.9% Al, 2.8% Fe, 1.2% K and no sulphur. The excess Fe from the Al comparison is 2.8−(8/12) (3.9) or 0.2%, where 8/12 is the Fe/Al ratio of illite from Table 3. Similarly, the Fe and K data yield an excess iron of 2.8−(8/4) (1.2) or 0.4%. The final excess iron is the maximum of these computations, or 0.4%. The siderite determined from the chemical concentrations using this procedure is then the excess Fe divided by the Fe content of siderite, 48.2%. The siderite estimated from the excess Fe computation for the Utah well compares well to the siderite measured by XRD (Fig. 5).

(iii) Calculation of water in the minerals, WMIN. In order to make quantitative determination of the amounts and types of clay minerals present in the formation, it is desirable to have an estimate of the water

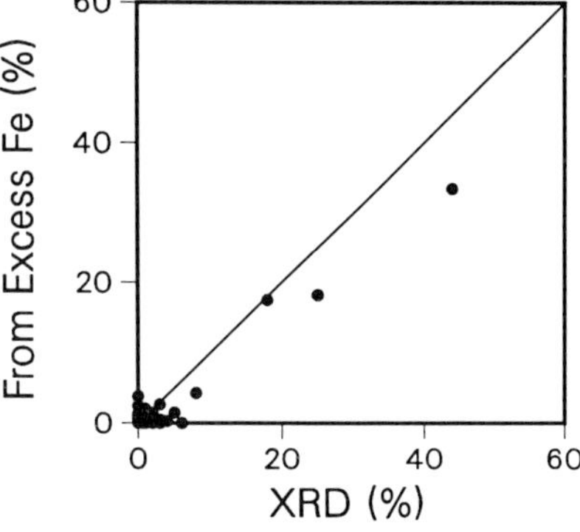

Fig. 5. Siderite concentrations in the Utah well determined from X-ray diffraction and from the calculation of excess iron strictly from core chemistry data. The close agreement provides support for the excess iron calculation.

Table 3. *Elemental concentration matrix for nine chem-minerals*

Mineral	Al (%)	Si (%)	Fe (%)	K (%	Ti (%)	S (%)	Ca (%)	XSFe (%)	WMIN (%)
Feldspar	10	30	0	10	0	0	1	0	0
Quartz	0	46.7	0	0	0	0	0	0	0
Calcite	0	0	0	0	0	0	40	0	0
Kaolinite*	19	22	0.8	0	0.9	0	0	0	14
Illite*	12	24	8	4	0.8	0	0	0	8
Smectite	8.5	21.1	1	0.5	0.2	0	0.2	0	32
Pyrite	0	0	47	0	0	53	0	0	0
Rutile	0	0	0	0	60	0	0	0	0
Siderite	0	0	0	0	0	0	0	48	0

* refers to the poorly ordered mineral phase (after M. Herron 1986).

content of the minerals corresponding to core measurements of H_2O^+. This is estimated from wireline log measurements from the difference between the epithermal neutron porosity, which is a measurement of the total hydrogen expressed as H_2O, and density porosity, which is an approximation of the hydrogen located in the pore space expressed as H_2O. For this calculation, the epithermal neutron porosity is computed on a sandstone matrix and density porosity is determined from the measured bulk density using an assumed matrix density of 2.65 g cm^{-3}. The difference, converted to weight fraction of dry rock, is called WMIN, the water content of the minerals. The difference is dependent in a complex way on the absolute value of the matrix density and for formations with matrix densities close to 2.65 g cm^{-3}, the WMIN parameter equals the sum of H_2O^+ from associated hydroxyls, hydration water and interlayer water in swelling clays (Fig. 6b).

(iv) Finally, the chem-mineral abundances are determined by solving nine (or ten if Mg and dolomite are included) simultaneous equations relating elemental concentration logs to the chem-mineral compositions (Table 3). If the sum of chem-mineral abundances is less than unity, an unidentified residual tenth (eleventh) mineral is calculated. For the Utah well, the residual was always less than 2%.

Example from Utah

For the Utah well, elemental concentration logs required for the model are compared with chemical concentrations measured by neutron activation, X-ray fluorescence, or wet chemistry on over eighty core samples (Fig. 6). The comparisons have good agreement although the log Si is overestimated slightly at some depths. The log data were then processed through the ten-mineral model and the resulting chem-mineral abundances are compared with core mineralogy provided by a major oil company research laboratory in Fig. 7. The degree of agreement is again good, despite the slight overestimation of quartz in the log data. No interpretation of conventional geophysical wireline data can compare with the richness and accuracy of the GLT interpretation shown in Fig. 7.

The derived mineralogy has been used to provide several petrophysical properties including a matrix density which, when combined

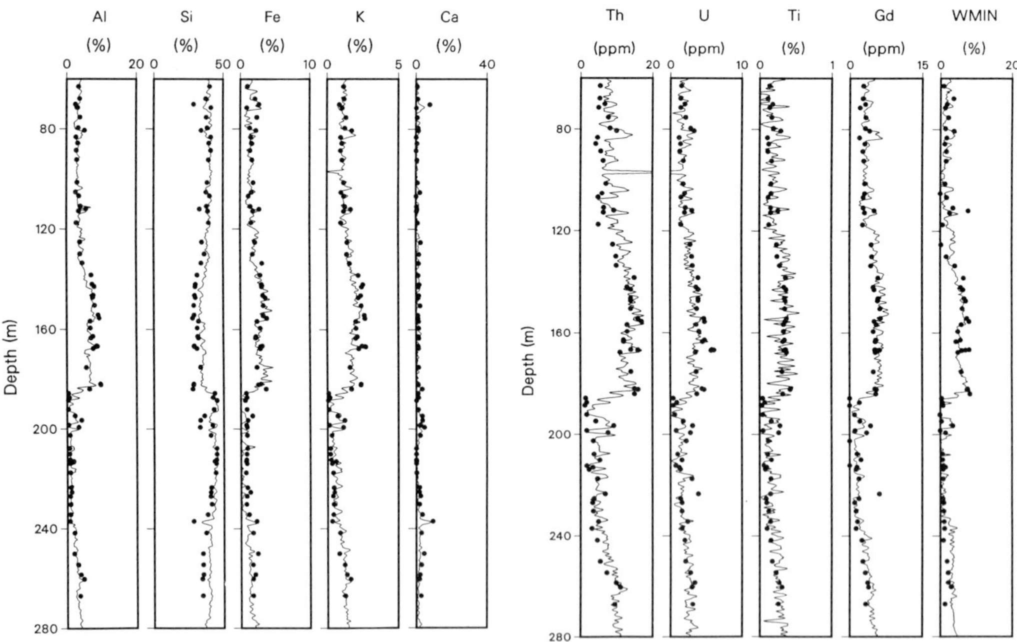

Fig. 6. GLT concentration logs for the Utah well. Core plug chemistry data are shown as filled circles. WMIN is derived from the epithermal neutron porosity and bulk density logs and is compared to core H_2O^+ measurements.

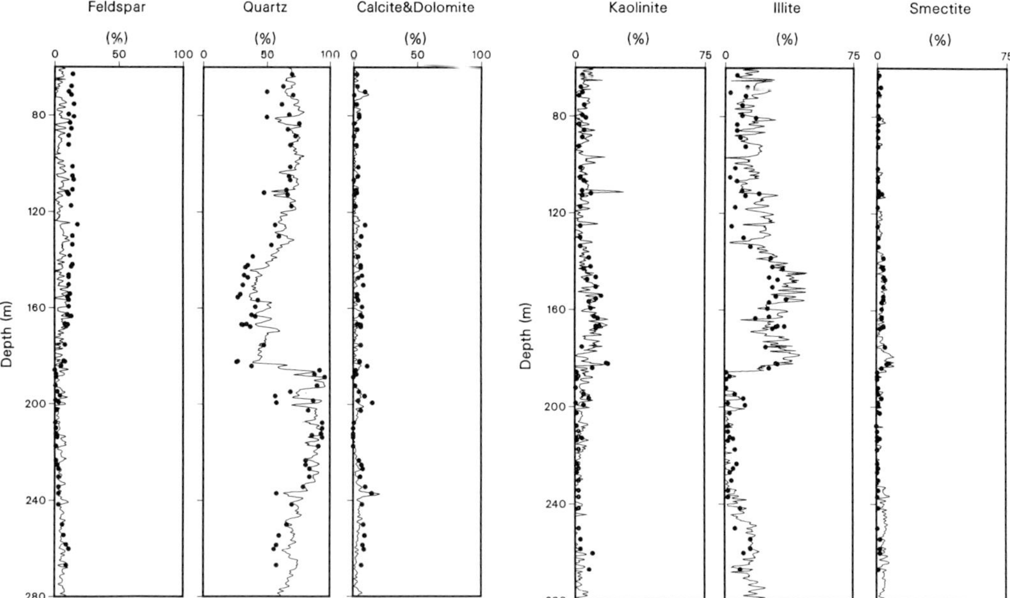

Fig. 7. Chem-mineral abundances derived from the GLT concentration logs in Fig. 6 and the chem-mineral composition matrix presented in Table 3. Core XRD mineral abundances are shown as filled circles.

with the bulk density log, yields an accurate log of total porosity (Herron 1987a). The chem-mineralogy and porosity (Fig. 8) are a distillation of many different data sources into a unified, normative presentation of the formation composition. At a glance, it is possible to see the major lithological units, the types of transitions from one unit to another, the detailed reservoir mineralogy including types and amounts of the three clay species, and the high degree of mineralogical maturity of the sediments as indicated by quartz contents exceeding 95%, low smectite/illite ratios, and low feldspar abundances. Using the model, it is also possible to infer porosity reduction due to compaction in the more clay-rich sediments and to identify the diagenetic overprint from the calcite and siderite cements.

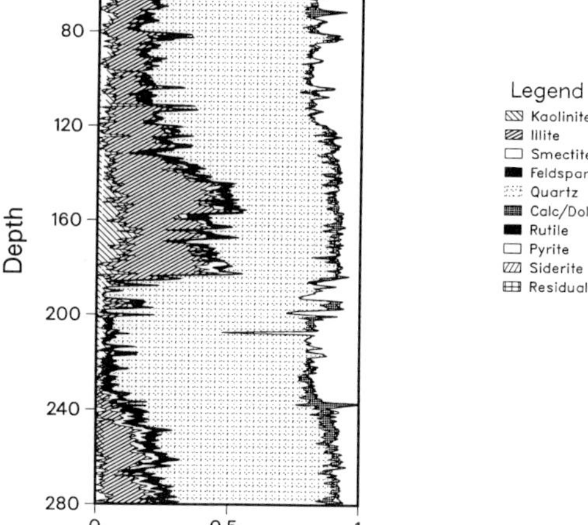

Fig. 8. Composite mineralogy and porosity for the Utah well. Bed at 207 m depth is a coal. Combined carbonate/siderite streak at 237 m has been identified as being primarily ankerite.

Applicability of the ten-mineral model

The ten-mineral model has been applied to wells in a variety of locations around the world representing widely varying sedimentological environments, ages, and degrees of diagenesis. Included in the suite of examples are a Venezuelan well containing mineralogically mature deltaic sandstones of Miocene age (M. Herron 1986), a California well of Plio-Pleistocene alluvial fan deposits (Herron & Grau 1987), and several wells from the North Sea and Gulf of Mexico. It has not been necessary to alter the matrix of chem-mineral compositions (Table 3) and the chem-mineral abundances have usually closely matched mineralogy determined by such techniques as XRD and petrographic analysis.

When the two assumptions required for accurate element-to-mineral conversion are not met, the chem-mineralogy is perturbed in predictable ways. First, some minerals vary in composition but are treated as though their composition is fixed. For example, feldspar exists as K-feldspar or as a solid solution between Na- and Ca-plagioclase. Since Na is not yet available from geochemical logs, a fixed mixture of potassium feldspar, albite, and anorthite is assumed (Table 3). Where chem-mineralogy and core mineralogy have differed, the problem has generally been that the second assumption is violated; that is, the formation contains minerals which contribute to the suite of elements but are not part of the model. In this case, the minerals not contained in the model will be calculated as combinations of one or more of the minerals in the model. For example, muscovite in the formation will appear as one-third kaolinite and two-thirds feldspar. Similarly, iron-rich sedimentary chlorite will appear as a combination of kaolinite, illite, and siderite; and ankerite will be a combination of dolomite and siderite. In such cases, the discrepancies occur not because the chem-mineral concentrations in Table 3 are incorrect, but rather because the model is incomplete. Note, for example, that well ordered illite, common in many North Sea environments, has a different chemical composition from the poorly ordered illite in Table 3. Appropriate regional models can be effectively constructed using the ten-mineral model as a starting point and replacing minerals as desired.

To date, the ten-mineral model has predicted formation mineralogy successfully in a variety of geologic settings, and it has provided a more detailed composition picture of the formation than could ever be routinely obtained from cores or conventional wireline logs. Although it is limited by the number of inputs to a specified suite of minerals, it will identify correctly the majority of the minerals present in many siliciclastic formations. Even when some minerals are misidentified, the total amounts of framework minerals, clay minerals, and carbonate minerals remain fairly accurate. The incorrect mineral identifications are both predictable and understandable. Furthermore, the compositional essence of the formation (e.g., Fe-rich sandstone) is never lost, even when some minerals are misidentified.

Source rock evaluation

Geochemical logs can also be used to determine quantitatively the total organic carbon content (TOC) of a formation, and thus they provide an important new tool for source rock evaluation and basin analysis (S. Herron 1986). The technique for determining TOC takes the measured carbon/oxygen ratio of the formation and multiplies it by an estimated oxygen content to obtain the total amount of carbon, organic plus inorganic, in the formation. The inorganic carbon is then estimated using the calcium and/or magnesium concentration logs or geochemically-derived mineralogy, and it is subtracted from the carbon to obtain total organic carbon.

The key to this technique is the estimation of formation oxygen, accomplished previously by modelling the formation as two components: a solid mineral matrix and a pore space filled with water. The model works well for relatively low TOC values, but for very high TOC contents (>5 wt%), it tends to overestimate the formation carbon. The reason for this is that the organic matter, assumed to be kerogen, has a density of about 1 g cm^{-3}, comparable to water, and consequently when there is a large quantity of kerogen in the formation the calculated density porosity is also large. Since the porosity is assumed to be water-filled, the formation oxygen is overestimated, and the calculated TOC is too high.

In a more realistic model, the formation is composed of three components: a solid mineral matrix, a water-filled pore space, and kerogen. The density porosity now represents a volume containing organic matter and fluid. Since the volume of kerogen is not known, TOC is estimated using the two-component system described above and that value is converted to weight of organic matter using a rough conversion factor of 1.25 (Tissot & Welte 1978), and then converted to volume. For the carbon computation, it is necessary to assign an oxygen content to the kerogen. This value varies from about 2 to 20 wt% depending on kerogen type and maturity (see Tissot & Welte 1978); for this application a value of 6 wt% was selected.

The final step in obtaining TOC is to correct for the presence of inorganic carbon which resides primarily in carbonate minerals. If the mineralogy has been derived, the carbon contributions of individual carbonate species can be summed and subtracted from the total carbon. If mineralogy is not available, the Ca or Ca and Mg logs can be used to estimate inorganic carbon. If both Ca and Mg are present in the formation, the Mg can be attributed to dolomite, the appropriate amount of Ca can be apportioned, and the remaining Ca can be attributed to calcite. Alternatively, if there is no magnesium, the inorganic carbon can be estimated by simply assuming that all the calcium belongs to calcite and that there is no other source of inorganic carbon.

The technique has been applied to a section of the Conoco 33–1 borehole which is described as a black, slightly calcareous shale. Figure 9 shows the two and three component model results for total carbon. Where the values are low, around 1 wt%, the two models produce nearly identical results. In contrast, at the high levels there is a difference of up to 1 wt%. The correction for inorganic carbon is made to the three component model using calcium (Fig. 10). Core measurements of TOC, provided by Conoco's Research Division, show very good agreement with log-derived TOC for both high and low values. The major advantages of this technique over other wireline approaches include the ability to evaluate low levels of organic carbon and the fact that it does not require calibration with core.

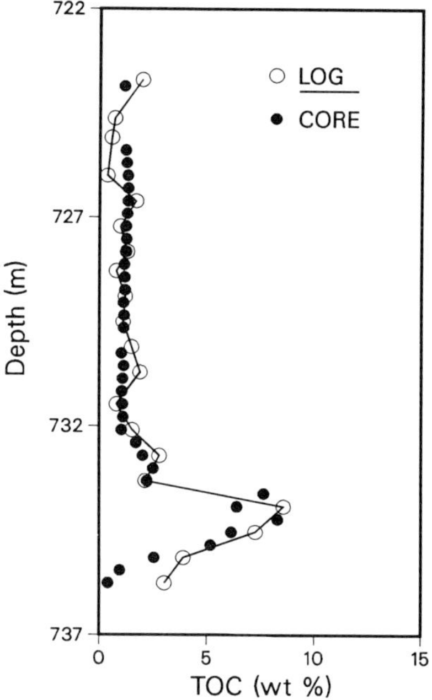

Fig. 10. Total organic carbon (TOC) log from Conoco 33–1 well using the three-component model. Core data (filled circles) are shown for comparison.

Inter-well correlation

One of the most common geological applications of well log data is inter-well correlation. Here, wireline measurements are used to map the horizontal extent of sedimentary beds as seen in the vertical sequences provided by logs. This task is sometimes quite difficult because formations can undergo many types of spatial sedimentological variations which affect porosity, fluid content, and composition. Consequently, a very efficient way to map the actual rock units is to use the geochemical logs which respond to the compositional changes in the rock, not in the fluid. This is accomplished by using individual concentration logs or any of the geochemical interpretations described above.

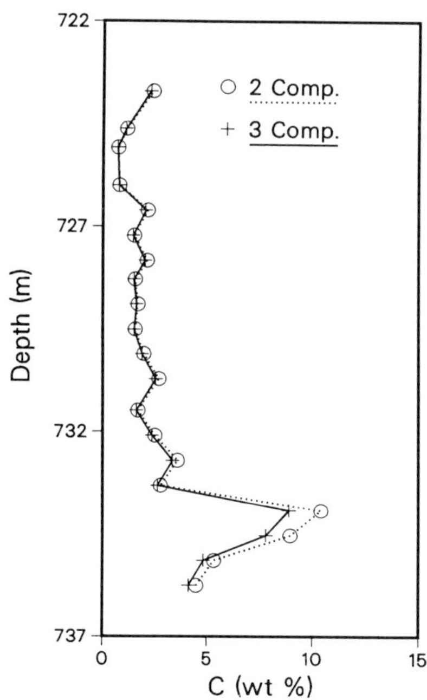

Fig. 9. Total carbon logs in the Conoco 33–1 well derived from five minute stationary measurements using the two-component and three-component models.

The potential of geochemical log data in interwell correlation is demonstrated in Figs 11 & 12 for two Californian wells with a separation of about 1000 m. The wells penetrate Plio-Pleistocene alluvial fan sequences composed of alternating feldspar-rich sands and shales. From the gamma-ray curves (Fig. 11), it is difficult to see any patterns of correlation that might suggest levels of continuity of deposition. In these wells, the high potassium content of the feldspar tends to obscure the boundaries between sands and shales on the gamma-ray logs. On the other hand, some of the elemental concentration logs and other geochemical interpretation logs may be useful for correlation between the two wells. The iron concentration curves

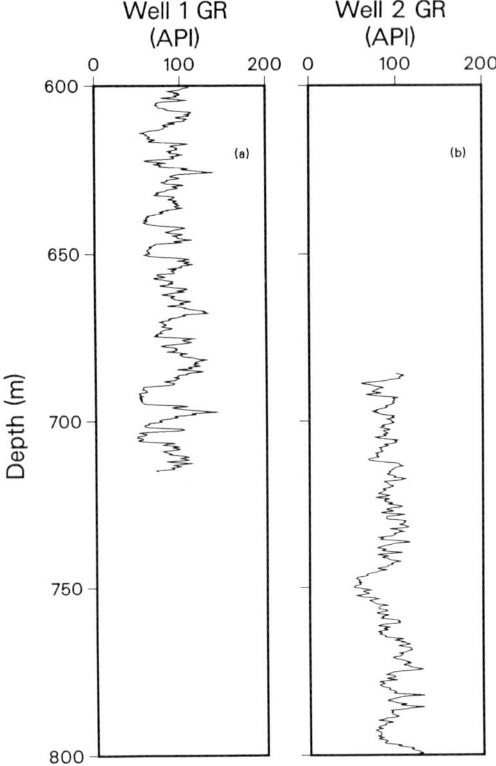

Fig. 11. Gamma ray logs from two California wells separated by less than 1000 m. Well-to-well correlation using these data is quite difficult due to high gamma ray signals in the sands.

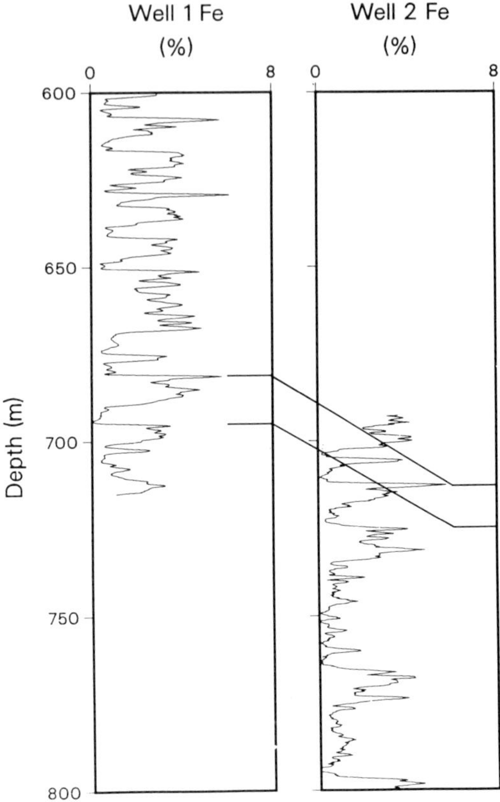

Fig. 12. Iron concentration logs from the source wells shown in Fig. 11. A possible interwell correlation is shown.

(Fig. 12) are typical and indicate a potential correlation as shown. It is clear that inter-well correlations using chemical signatures will be less prone to error than when based on geophysical data alone.

Conclusions

The sensitivity of geochemical well log data to subtle changes in formation composition provides a new opportunity to describe sedimentary strata in a detail and volume never before available. The data can be utilized in the form of elemental concentrations to discern major lithologies, bed boundaries, and lithological transitions. For a more detailed analysis, a combination of elements may be used either to construct new geochemical classification schemes or, as is the case with the SandClass system, to link geochemical data to existing petrographic classifications. It is easy to visualize the development of similar classification schemes for other lithological groups such as carbonates, evaporites, volcaniclastics, and shales. The concentration logs can also be used for determining organic carbon which represents an important step in organic facies characterization.

The elemental data can also be transformed into new variables of chem-mineral abundances. For most siliciclastic sediments, the ten-mineral model presented here is successful for the interpretation of the information mineralogy in a strictly normative sense; frequently it also serves to accurately describe the true formation mineralogy. The application of a general normative analysis provides an objective basis for comparing rock compositions on a local, regional, or global basis. Regional models have also been developed while geochemical research continues into expanded general models relating sedimentary minerals and elements.

The enhanced formation evaluation made

possible by geochemical logs, will provide a new foundation for geological interpretation. The logs provide a continuous and unbiased sampling of the vertical sedimentary column. The detailed formation composition information they provide makes them ideal for further geological investigations in such areas as depositional environment, facies, diagenesis, and reservoir quality, extent and continuity.

References

Blatt, H., Middleton, G., & Murray, R. 1980. *Origin of sedimentary rocks* Prentice-Hall, New Jersey.

Flaum, C. & Pirie, G. 1981, Determination of lithology from induced gamma-ray spectroscopy, *Transactions of 24th SPWLA Annual Meeting*, Mexico City, Paper H.

Herron, M. M. 1986. Mineralogy from geochemical well logging, *Clays and Clay Minerals*, **34**, 204–213.

—— 1987a. Future applications of elemental concentrations from geophysical logging, *Nuclear Geophysics*, **1**, 197–211.

—— 1987b. Estimating the intrinsic permeability of clastic sediments from geochemical data, *Transactions of 28th SPWLA Annual Meeting*, London, paper HH.

—— 1988. Geochemical classification of terrigenous sediments using log or core data, *Journal of Sedimentary Petrology*, **58**, 820–829.

—— & Grau, J. A. 1987. Clay and framework mineralogy, cation exchange capacity, matrix density, and porosity from geochemical well logging in Kern County, California. *American Association of Petroleum Geologists Annual Meeting, June 7–10, Los Angeles.*

Herron, S. L. 1986. A total organic carbon log for source rock evaluation, *Transactions of 27th SPWLA Annual Meeting*, Houston, paper HH, revised in 1987 *Log Analyst,* **28**, 520–527.

Hertzog, R., Colson, L., Seeman, B. O'Brien, M., Scott, H, McKeon, D., Wraight, P., Grau, J. Ellis, D., Schweitzer, J. & Herron, M. 1987. Geochemical logging with spectrometry tools: *Society of Petroleum Engineers Annual Technical Conference and Exhibition, Dallas, Paper* **16792**.

Pettijohn, F. J. 1975. *Sedimentary Rocks*, 3rd edition, Harper & Row, New York.

Peveraro, R. C. A. & Russell, K. J. 1984. Interpretation of wireline log and core data from a mid-Jurassic sand/shale sequence, *Clay Minerals*, **19**, 483–505.

Roscoe, B. A. & Grau, J. A. 1985. Response of the carbon/oxygen measurement for an inelastic gamma ray spectroscopy tool, *1985 Society of Petroleum Engineers Annual Technical Conference and Exhibition*, Las Vegas, paper 14460.

Tissot, B. & Welte, D. H. 1978. *Petroleum formation and occurrence,* Springer, New York.

Turekian, K. K. & Wedepohl, K. H. 1961. Distribution of the elements in some major units of the earth's crust, *Bulletin of the Geological Society of America*, **72**, 175–182.

Geochemical well logs: calibration and lithostratigraphy in basaltic, granitic and metamorphic rocks

R. N. ANDERSON, R. E. DOVE* & E. PRATSON
Borehole Research Group, Lamont-Doherty Geological Observatory, Columbia University, Palisades, New York 10964, USA

Abstract: Elemental analyses of rock compositions have been measured *in situ* with a geochemical logging tool (GLT) in drillholes through igneous and metamorphic terrains. Comparisons of dry weight percent oxide analyses from laboratory measurements on core samples and the log-derived results demonstrate that accuracies are fair in Ocean Drilling Program Hole 504B basalts, good in Cajon Pass, California granites, and very good in Moodus, Connecticut metamorphic rocks and Palisades, New York diabases. Improved accuracies of GLT measurements are attainable through the use of a boron sleeve to isolate the borehole fluid from wellbore effects, and in smaller diameter, land-based boreholes. Precisions within each well are very good in all cases. That is, only a linear calibration shift was required to bring core and laboratory analyses into general agreement in all cases. Variations in mineralogy, derived from inversion models of the log-derived elemental abundances at each well, produce a lithostratigraphic interpretation that is useful for determining the structure of the oceanic crust (Hole 504B), the fractionation sequence of basaltic intrusions in a new rift basin (Palisades Sill), the deformation history near the San Andreas Fault (Cajon Pass), and the variations in degree of metamorphism across a shallow decollement surface caused by collision tectonics (Moodus).

The Geochemical Logging Tool (GLT†) was developed for use in sedimentary basins by the oil industry and is now beginning to be used in scientific investigations in igneous and metamorphic rocks. The GLT records the concentrations of major and minor elements continuously at 0.15 m depth increments throughout a well, which allows the calculation of elemental abundances. Should geochemical logging provide relatively precise elemental analyses, the method will become one of the most important geological tools for the determination of not only lithology, but stratigraphy, structure, rock-characterization, and degree of alteration in hard-rock wellbores. The elemental abundances derived from these geochemical logs will add greatly to the geological interpretation capabilities within a drillhole because they are fast and complete. In contrast with the time required for laboratory analysis of the cored borehole sections, the log results are available as continuous records of relative elemental dry weight percent oxides within days. One can easily envisage a future drilling plan for scientific boreholes which begins with the most economical and rapid drilling of the hole itself, followed by geochemical and geophysical logging to select locations for sidewall coring, fluid sampling and downhole experimentation.

It is the purpose of this paper to test the accuracy and precision of GLT measurements in wellbores into igneous and metamorphic rock. We present below a comparison between geochemical logging results and laboratory chemical analyses on core samples from (1) the basalts of the oceanic crust at Hole 504B, Costa Rica Rift, Eastern Equatorial Pacific Ocean, (2) diabases of the Palisades Sill, New York, (3) granitic basement rocks in the Cajon Pass Scientific Drillhole, California, and (4) metamorphic rocks of the Moodus Deep Drillhole, Connecticut (Fig. 1). We find that the log-derived elemental abundances vary in quality from fair (in ocean drilling holes), to very good (at Cajon Pass), to very good (at Moodus and the Palisades Sill) in terms of the accuracy of their determinations compared to core analyses. Further, the geochemical logging measurements are excellent discriminators of chemical changes in all these wellbores. In order to demonstrate the scientific utility of such geochemical logging, we then use the GLT-derived chemical analyses to determine the lithostratigraphy of these wells.

* On sabatical leave from Schlumberger Offshore Services, New Orleans, La., USA.

† Mark of Schlumberger.

From HURST, A., LOVELL, M. A. & MORTON, A. C. (eds), 1990, *Geological Applications of Wireline Logs* Geological Society Special Publication No. 48, pp. 177–194

Wells used in this Study

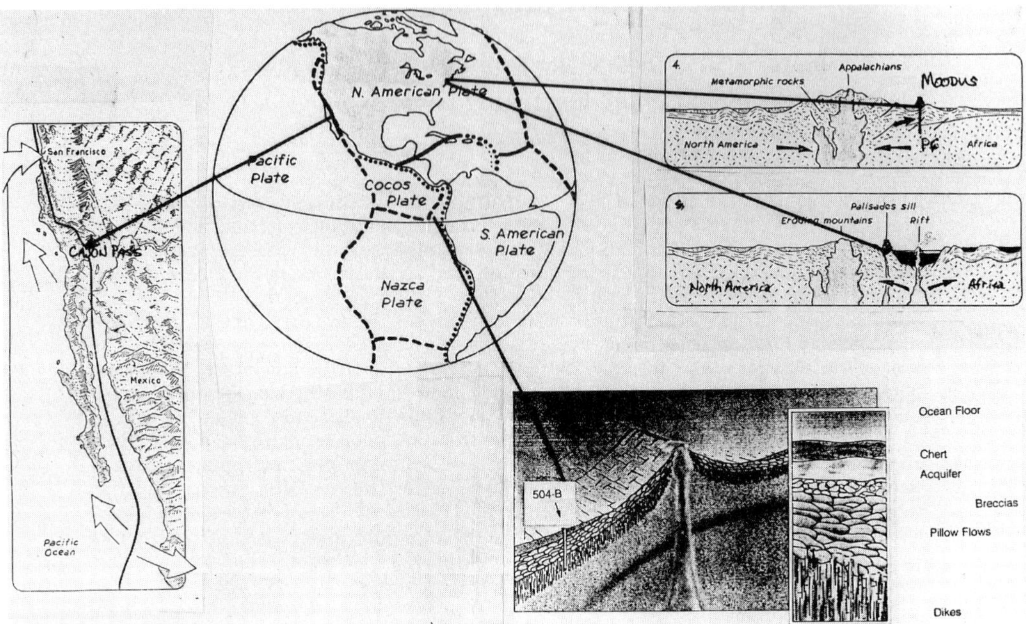

Fig. 1. Schematic representations of the tectonic settings of each of the logging sites reported in this paper: Cajon Pass, California, Moodus, Connecticut, Palisades, New York, and Hole 504B in the Eastern Equatorial Pacific Ocean south of the Costa Rica Rift on the Nazca Plate.

Geochemical log analysis techniques

The Geochemical Logging Tool (GLT) (Fig. 2) carries four measurement devices into the borehole. A NaI crystal, natural gamma-ray spectral analyser detects the energy levels of naturally emitted gamma rays from the formation, allowing the determination of weight percentage of K present in the rock, and the total content of Th and U in parts per million. A Cf neutron source, NaI photoaccumulator combination activates Al atoms in the formation and records the emitted gamma rays. The weight percent of Al is measured. A minitron bursts neutrons for the activation of Si, Ca, Fe, S, and Ti as well as the trace abundances of the heavy rare earth element (HREE) Gd. H and Cl contents of the wellbore fluid and formation are also measured. The combination of these analyses with the measurement of the photoelectric capture cross section and bulk density from a Cs source of gamma rays allows for the further estimation of the total content of Mg plus Na present in the rock (Fig. 2).

The variations in dry weight percent of the major elemental oxides are estimated by assuming that approximately 50% of all rock is oxygen by weight (Herzog *et al.* 1987). The gamma-ray spectroscopy measurements made in the well detect a fraction of the total spectrum for each activated element which is linearly proportional to the concentration of that element within the volume of the measurement. Thus, the relative yields from all the elements that are measured are renormalized to calculate elemental oxides which are, in turn, forced to sum to 100% at each sampling depth-interval. Since the elements measured by either capture activation or natural spectroscopy make up almost all of the significant oxides comprising most rock, the calculations should be reliable in any geological formation (Herzog *et al.* 1987). We test this supposition below in basaltic, granitic and metamorphic rocks.

Schlumberger has demonstrated that the GLT measures elemental yields that can be accurately converted to dry weight percent oxides in sediments (Herzog *et al.* 1987), but the accuracy in basalts has not yet been verified. For example, the distribution of O is different in igneous rocks from that in sediments (e.g., in carbonate instead of silicate lithologies). We have modified

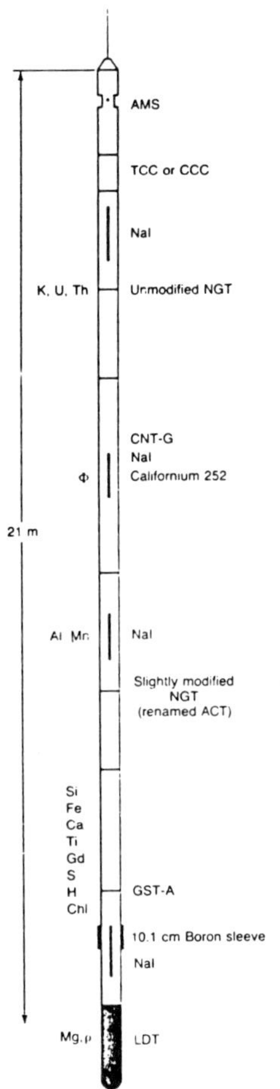

Fig. 2. Configuration of the Geochemical Logging Tool (GLT) used in this study. At the top are the AMS (auxiliary measurement sonde), which determines temperature and cable tension, and the TCC or CCC (both of which telemeter digital information to the surface much as a complex telephone modum). K, U, and Th are measured by the NGT (natural gamma-ray spectrometer). The CNT-G (compensated neutron tool) measures porosity and activates Al with a Cf 252 source. The ACT (Aluminum clay tool) follows to measure Al concentration. The GST (gamma ray spectrometer tool) bursts the formation with neutrons from a minitron, then passively collects emitted gamma rays for the determination of the concentrations of Si, Fe, Ca, Ti, Gd, S, H, and Cl. A LDT (lithodensity tool) then either follows, as in the Ocean Drilling Program, or is lowered separately to measure density and photoelectric capture cross-section using a Cs source. Mg + Na are calculated as residual abundance from the combination of all the measurements above.

the O distribution coefficients of Herzog *et al.* (1987) from their CO_3-based system to a SiO_2-based system. The elemental oxide calculations are consistent with an igneous oxide model and should produce results that stand up well in comparison to chemical analyses of core samples.

Basalts of Hole 504B, eastern Equatorial Pacific

Hole 504B on the south flank of the Costa Rica Rift (Fig. 1), is the deepest penetration yet into the oceanic crust. As such, it is the first reference section of ocean lithosphere for comparison with geophysical and geochemical models based upon remote sensing and ophiolite models from land. However, the core recovery in Hole 504B was less than 20% of the drilled section. Consequently great emphasis has been placed upon geophysical, as well as geochemical logging, to locate the recovered basalts within their proper geological framework during Deep Sea Drilling Project (DSDP) Legs 69, 70, 83, and 92 and Ocean Drilling Program (ODP) Leg 111.

Two major difficulties exist with the data-quality of GLT measurements in ODP boreholes. First, the drilling technique employed aboard the JOIDES RESOLUTION Drillship requires the use of a small diameter logging tool (3 5/8 in) in a large diameter drillhole (9 7/8 in bit). The GLT must pass through a 4-inch inside-diameter drillpipe to reach the hole. Thus, there is a large borehole-fluid component in the GLT measurement spectrum. This noise is further compounded by the lack of a boron sleeve on ODP logging sondes because of the small diameter requirement for the tool. The boron sleeve is used to further lower the borehole-fluid signal in the GLT spectrum. Thus, calibration of GLT results with core-sample analyses is of even more importance in ODP wells than with measurements made in land-based wells. This situation has been corrected with the deployment of a boron sleeve beginning on ODP Leg 126.

Structure of Hole 504B

The geophysical logging suites run at Hole 504B revealed three distinct layers within the upper km of basaltic basement (Fig. 1): an upper 150 m section of pillow basalts with very low seismic velocity and resistivity (Layer 2A), a middle layer of pillows with increasing velocity and slightly higher resistivity (Layer 2B), and a deep zone of intrusive dikes of high velocity and resistivity (Layer 2C) (Anderson *et al.* 1985a). The upper unit in this young crust is highly altered, as indicated by the contrast in porosities determined from neutron and density logs (Anderson *et al.* 1985a).

At the pillow-dike transition between layers 2B and 2C, a sulphide mineralization, stockwork-zone was cored during Leg 83. Hot, upwelling, mineralized hydrothermal fluids mixed with cold sea water at this depth in the crust to deposit this stockwork (Honnorez *et al.* 1985). The mixing and deposition probably occurred at the ridge axis as part of an axial hot spring system (Alt *et al.* 1985), active long before sealing of the crust by chert deposition in the sediments which occurred at approximately 1 Ma in this 5.9 Ma oceanic crust.

Geochemical log-derived elemental abundances in Hole 504B

Major element, dry weight percent oxide abundances were calculated from geochemical logging results at Hole 504B (Fig. 3). These log-derived, elemental abundances were calculated with no calibration or other reference to core analyses.

The compositions of major elements were then compared with core analyses (Fig. 3), remembering that the 20% of the section recovered during coring was largely from the

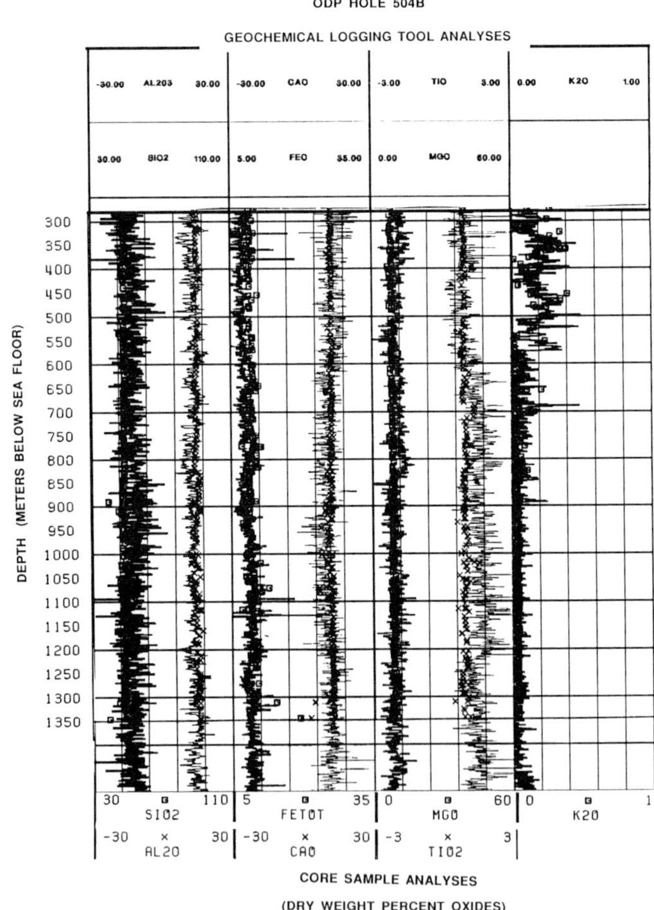

Fig. 3. Dry weight percent oxide concentrations of Si, Al, Ca, Fe, Ti, Mg+Na, and K from geochemical logs (Fig. 2). Symbols are core analyses from Hole 504B (analyses from Initial Reports, DSDP, Legs 69, 83, and ODP Leg 111). Large positive spikes in Fe are from mineralization zones encountered in the well.

freshest basalts in Hole 504B (Alt *et al.* 1986). Whereas K agrees well between the two analysis techniques, Fe, Ca, and Al contents from the logs are lower than those from the core analyses, and Si, Ti and Mg values from the logs are high, especially toward the top of the well (Fig. 3). The Si and Mg contents are high because coring systematically undersampled the alteration zones in the hole which are high in Mg (Anderson *et al.* 1989). Such zones are also relatively low in Fe and Ca (Alt *et al.* 1985). Much of the difference in Al is caused by porosity changes in the well (Anderson *et al.* 1989). Only the accuracy of Ti appears to be in question, but then only in the interval of the well from 600 to 1200 m below the sea floor. Precision of the Ti measurements from the GLT appear to be much better than accuracy in this interval, as can be seen from log results compared to core analyses in the interval from 600 to 750 m (Fig. 3).

The upward increase in Mg content in the hole, shown by the log-derived Mg curve, is the result of the measurement of a country rock that has absorbed increasing amounts of Mg from sea water during hydrothermal alteration (Anderson *et al.* 1989). Ca content should, and does decrease correspondingly because hydrothermal alteration has occurred extensively in the hole. Consequently, the log-derived analyses show a wider variation uphole than do the core analyses. This again argues for preferential sampling of the freshest basalts by the coring program at Hole 504B, instead of some bias in the precision of the log measurements. The increased Ti content versus that of the core analyses is also likely to be real because of a concommittant increase in S associated with pyrite mineralization (see mineralogy model in Fig. 4). Sphene is also noted to increase in this interval from thin-section investigations (Honnorez *et al.* 1983).

Mineralogy model of Hole 504B

Once we have established the accuracy and precision of the geochemical measurements from the GLT, the elemental concentrations can be inverted using a correlation matrix to determine the volumes of ideal mineral compositions that might be present in the well (Herzog *et al.* 1987). The average elemental compositions of selected minerals are input into the model. In the Hole 504B case, plagioclase (An_{70}), clinopyroxene, smectite, chlorite, calcite and pyrite minerals were used (Anderson *et al.* 1989). The mineral compositions were then input as end members into an inversion matrix with the weight percent oxide compositions of Si, Al, Fe, Ca, K, Ti, Na+Mg and S from the geochemical logs (Fig. 4).

The geochemical variations measured by the GLT in Hole 504B allow us to model the mineralogy of the drilled section. The elemental abundances indicate the presence of two major mineralogical components, plagioclase and clinopyroxene, which form the bulk of the formation. This is the fresh basalt of the oceanic crust itself. A second component appears as large spikes of K-rich zeolites in Layer 2A (this composition has been included within the smectites in the mineralogy model), Mg-rich clays such as smectites in the pillows of Layers 2A and 2B, and Fe- and Mg-rich chlorites in the dikes of Layer 2C. These alteration components are present extensively in Layer 2A, but are confined to breccia and fracture zones along individual eruption-event boundaries in Layer 2B, and appear as spikes along contacts between largely unaltered basaltic dikes of Layer 2C (Fig. 4).

Lithostratigraphy at Hole 504B

The combination of geochemical and geophysical logs and the mineralogical model allows the detailed lithostratigraphy of Hole 504B to be examined. An understanding of the hydrogeology of the upper oceanic crust requires that the fracture system bounding the fresh basalts and the degree of infilling of these fractures with alteration products also be quantified.

The extensive amount of smectite (and K-rich zeolites) in Layers 2A and 2B appears to be the consequence of hydrothermal alteration (higher Mg and K and lower Ca abundances in the pillows than in the dikes). The degree of fracture infilling and basaltic alteration decreases into the dikes, correlating with an increase in temperature of alteration as shown by the appearance of high-grade chlorites in Layer 2C. This change in mineral abundances may explain the permeability and porosity variations which show the dikes to be much tighter, lower porosity rocks than the pillow units. Permeability measurements from flow tests below packers show, however, that the permeability, though high in the upper pillows, remains constant and low within the lower pillow and dike sections even though the apparent porosity is high in the lower pillows (Anderson *et al.* 1985b; Becker *et al.* 1988). There is much void space in the Layer 2A pillow sequence in spite of the extensive low temperature precipitation of clay and zeolite minerals. Fractures are present in the lower pillows, but they are filled with smectites, and so, like the Layer 2C dikes, are of relatively low permeability. The resistivity of these smectites is low relative to fresh basalt (closer to that of pore water in fact), and, thus, the apparent porosity is caused by the presence of hydroxyl-bearing alteration minerals within plugged fractures in layer 2B (Pezard & Anderson 1989).

Cyclicity in aluminum content in Hole 504B

The above analysis has demonstrated that the precision of the Hole 504B GLT results is sufficient to derive significant lithostratigraphic insights into geological processes in the oceanic crust. For an additional example, consider one of the puzzles revealed by the geophysical logging in Hole 504B: the detection of cyclicity in the wellbore (Fig. 5). These cycles affect all the logs, including velocity, resistivity, and the reflectivity and degree of fracturing imaged by the ultrasonic borehole televiewer (Newmark *et al.* 1985). The variations in wavelength of this cyclicity within the wellbore can be seen qualitatively by plotting the thickness of constant-composition Al layers recorded by the GLT (Fig. 5). A Fourier analysis of Al and porosity logs provides a more quantitative

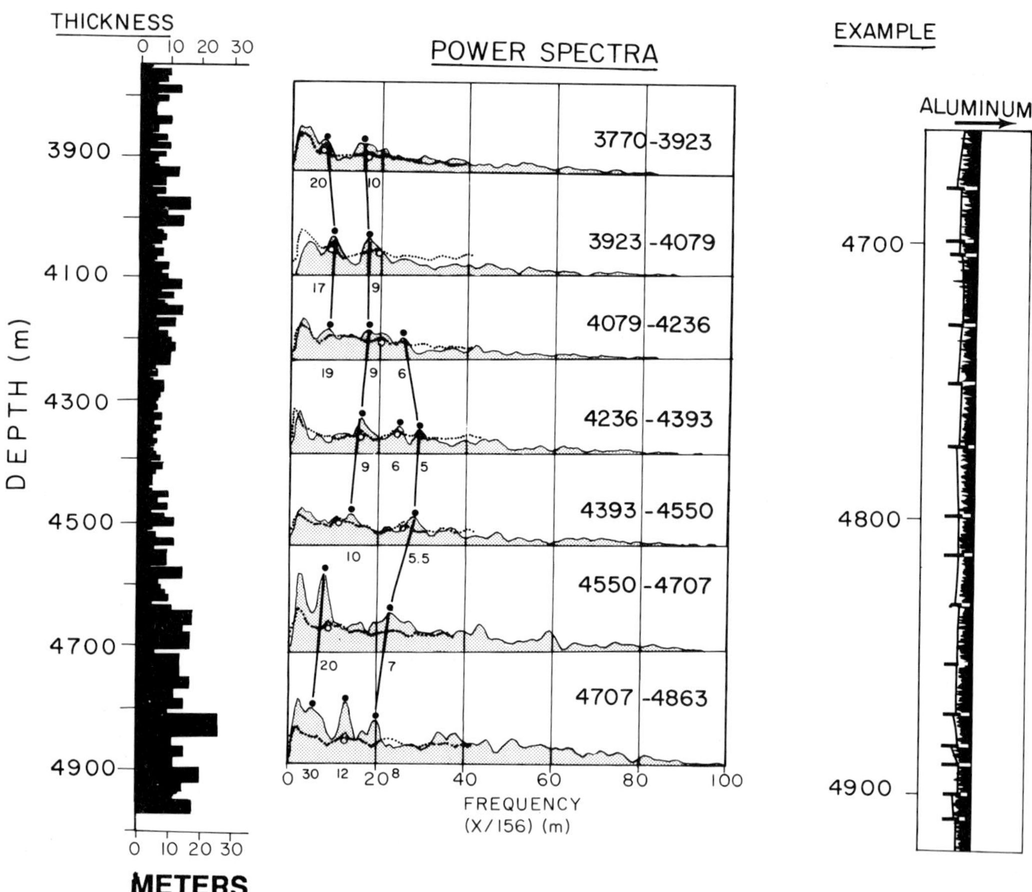

Fig. 5. Cyclicity in Al logs from Hole 504B. Left, variation in thickness of major units determined by abrupt changes in Al content, as in example at right. Centre, Fourier transform of Al curve (shaded spectrum) and total porosity curve (dashed spectrum) over the depth intervals indicated. Black dots identify prominent wavelengths of coherent power in the Al log, and open circles are peaks in the porosity spectra. Numbers indicate the wavelengths of correlatable energy peaks.

examination of the variation in the wavelength of cyclicity with depth.

This structural cyclicity was found to have predominant periods of 9–10 m and 17–20 m thicknesses in Layer 2A (panel 1, Fig. 5), decreasing to 5–9 m thicknesses in the lower pillows of Layer 2B, then increasing to 20–30 m thicknesses in the dikes of Layer 2C (panels 5–7, Fig. 5). The wavelengths of the dike units are considerably larger than those between pillow units in Layer 2A, and particularly in Layer 2B (Fig. 5). This is thought to be due to the vertical drillhole penetrating through horizontally stratified units in the pillows of Layers 2A and 2B, in contrast to the nearly-vertical dikes of Layer 2C. This long-period cyclicity in Layer 2C can be converted to an average width of 3.5 to 5.2 m for individual dikes, it we assume a dip of 80°.

The observed variation in Al-content can be attributed to the plagioclase variations observed in core samples (Adamson 1985). High Al-content, plagioclase–olivine phyric and plagioclase–olivine-clinopyroxene phyric basaltic units alternate with low Al-content, aphyric basalts. Although changes in plagioclase composition that account for the primary Al-content variations are locally abrupt, the total volume of aphyric versus phyric basalts erupted in Hole 504B is about equal (Adamson 1985). Also, the geochemical logs show that zeolite-, smectite-, and chlorite-rich fracture zones are found along aphyric and phyric basaltic-unit contacts.

A reasonable scenario for the lithostratigraphic evolution of the Hole 504B oceanic crust that combines the observations of cyclicity and hydrogeology contained within the geochemical log and mineralogy model results is as follows. Aphyric-phyric eruption events alternated during original intrusion and extrusion at the Costa Rica Rift ridge axis. Subsequently, interaction with sea water began as the crust cooled, spread and aged away from the ridge crest. High-temperature alteration appears to have been present throughout the dikes whereas low-temperature alteration trends are present within the pillow section (the zeolite to smectite to chlorite transition).

The geochemical log measurements identify relative changes in major element chemistry rather well even when the absolute accuracies of the analyses are compromised by the very large borehole and lack of a boron sleeve. Below, we consider the usefulness of higher-quality GLT data obtainable in wells on land.

Diabases of the Palisades Sill, New York

The Palisades sill, along the lower Hudson River in New York, consists of more than 350 m of diabase (coarse-grained, basaltic composition igneous rock) that was intruded into the Triassic sandstones and shales of the Newark Basin at approximately 200 Ma during the initial rifting of North America away from Africa (Fig. 1). The sill is a classic example of closed-system, basaltic magma fractionation, and has been extensively studied (e.g., Walker 1940, Shirley 1987).

The Palisades Sill is most famous for the existence of a sharply-defined olivine layer located approximately 10 m above the lower contact of the sill (Walker 1940). This olivine-rich zone is thought to be the result of crystal fractionation within the magma of the newly-forming sill. The heavy olivine crystals sank to the bottom of the sill as lighter plagioclase crystals floated toward the top. An upward increase in Al-content of the diabases of the sill is the result of increased abundance of plagioclase toward the top. Abrupt increases in Th-content upward within the sill were used by Shirley (1987) to establish the existence of at least two distinct magma intrusion events that formed the sill.

The Palisades sill outcrops along the western cliffs of the Hudson River along the northern New Jersey and southern New York boarder (Fig. 6). The Lamont #2 test well drilled through the lower 250 m of the sill. Geochemical logs were recorded in the well so that the log-derived geochemical analyses could be compared to laboratory analyses of samples taken from outcrops along the cliff face. Of particular interest were whether the geochemical logs could record (a) the olivine layer, (b) the increased plagioclase content toward the top of the sill, and (c) the existence of multiple intrusion events within the sill.

Geochemical log-derived elemental abundances in the Palisades Sill well

Elemental analyses from the GLT locate clearly both the olivine layer and the boundaries between at least three intrusion events in the lower section of the sill (Fig. 7). Furthermore, the comparison of geochemical log-derived elemental abundances to the laboratory study of Shirley (1987) shows both excellent precision and accuracy (Fig. 7).

The accuracy of the correlation between

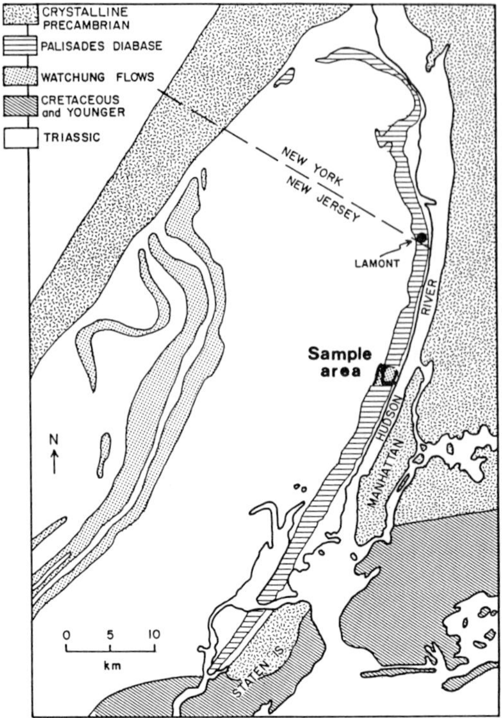

Fig. 6. Location map of Lamont #2 well (arrow) along the Palisades Sill (shading) in southeastern New York (figure 1). Outcrop area where samples of outcrop were taken is under the western end of the George Washington Bridge (from Shirley 1987).

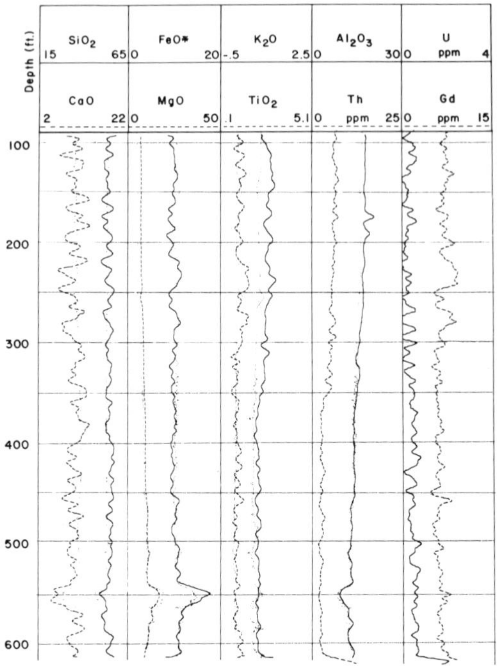

Fig. 7. Dry weight percent oxide analyses of major elements and elemental abundances of trace elements from geochemical logs and sample analyses (the latter from Shirley 1987). Prominent olivine-rich layer at 8–20 m above base depth is evident from logs, as is another intrusion event at 70 m above the base of the sill, particularly from the Al and Th logs.

sample-derived and log-derived chemical analyses is considerably better in the Palisades Sill Hole than for Hole 504B, even though the basalts are of very similar compositions. The main reasons for the better results were the lack of alteration in the Palisades Sill, the use of a boron sleeve and the smaller hole diameter in the Palisades Sill well (6 inches).

Lithostratigraphy of the Palisades Sill well

As noted above, the olivine-rich layer was observed on all major elemental curves, with Fe and Mg increasing and Si, Ca, and Al decreasing from 8 to 20 m above the base of the sill (Fig. 7). The increased plagioclase content upward in the hole was also recorded by a corresponding Al-abundance increase, although the Al-values recorded by the GLT are even higher than those from core analyses from 90 to 160 m above the base of the sill (Fig. 7). In addition, the Th log displays an abrupt increase at the boundary between two intrusion events 70 m above the base of the sill. The Al-abundance curve also records this same boundary. The analyses from samples of the outcrop 20 miles to the south placed this intrusive contact at 90 m above the base of the sill.

The continuous nature of the geochemical log analyses provides useful information even when abundant samples can be taken from outcrop. For example, there appears to be considerable stratification within the olivine layer because the peaks in Mg- and Fe-abundances occur at different depths within the olivine-rich rock (the Mg-peak is 2 m higher than the Fe-peak in Fig. 7). These data leave no doubt of the usefulness of geochemical logging in basalts.

Granites of Cajon Pass, California

The Cajon Pass scientific borehole 3.5 km north of the San Andreas Fault in California (Figs 1 &

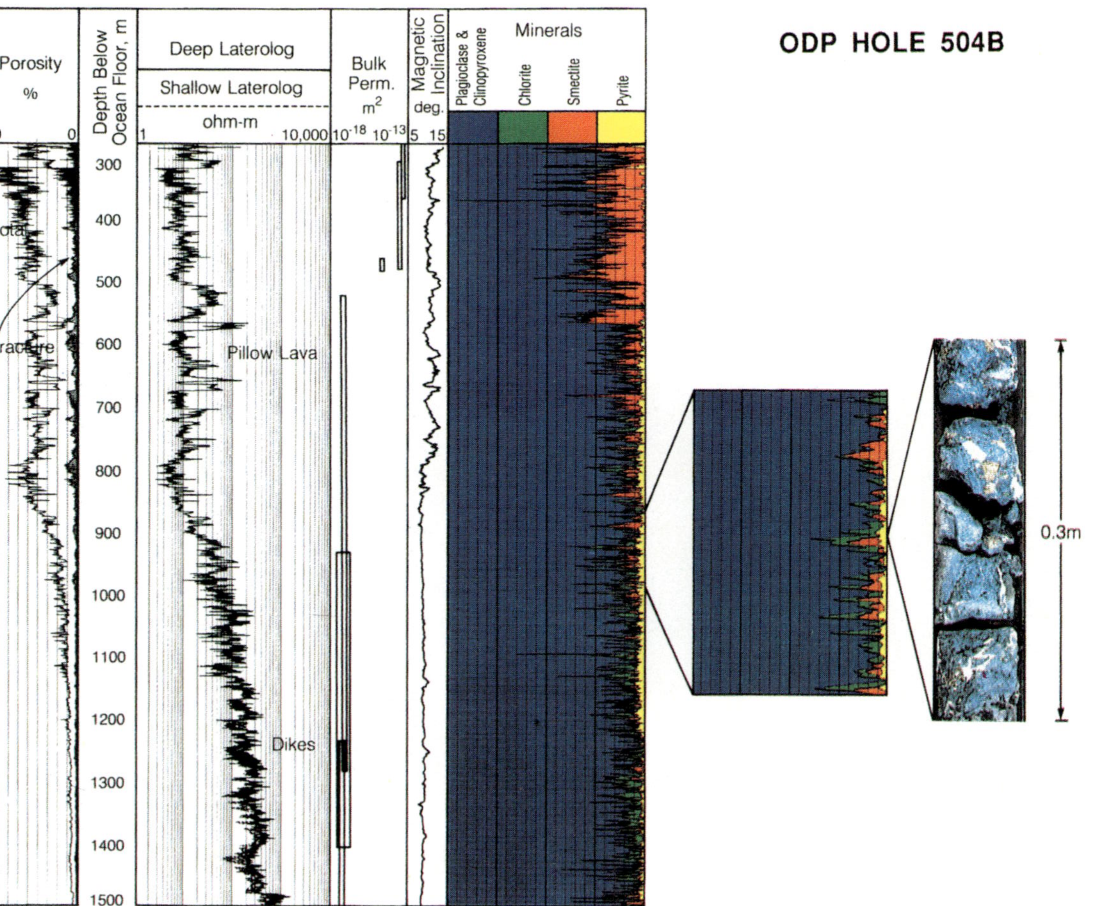

Fig. 4. Geophysical and geochemical logs from Hole 504B. At left, total porosity and fracture porosity logs derived from dual laterolog (left-centre) (Pezard & Anderson 1989). Centre, permeability measurements made in the Hole 504B by inflating packers against the wellbore and pumping pressure pulses (from Anderson *et al*. 1985b and Becker *et al*. 1988). Right-centre, magnetic inclination from three-axis borehole magnetometer. Right, mineralogy model derived from geochemical logs, with blow-up of stockwork mineralization zone at 4390 m. Far right, photo of core from stockwork located within the mineralogy model by the abundance of pyrite from the log and core (finger-sized crystals evident in the photo). Figure from Anderson *et al*. (1988d).

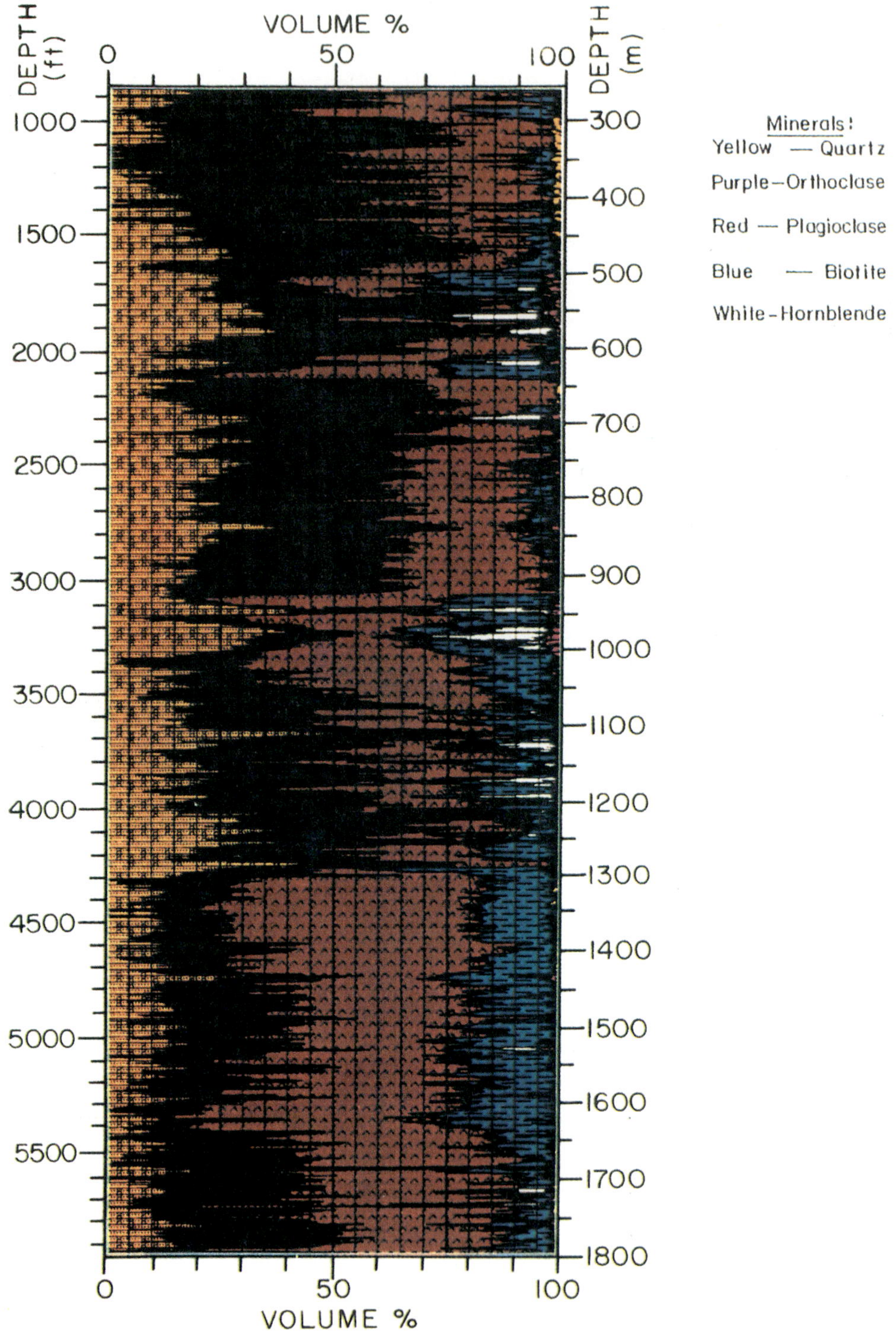

Fig. 11. Mineralogy model from matrix inversion of GLT elemental compositions of Fig. 9 in the Cajon Pass Scientific Drillhole (from Anderson *et al*. 1988b).

8) was drilled to measure the heat flow and state-of-stress at seismogenic depths in order to explain the 'heat flow-stress paradox'. The shear stresses of the San Andreas Fault are thought to be high yet the high heat flow from friction that should accompany such high stresses has never been measured in shallow wells (Zoback *et al.* 1980). Either the heat from the high stresses is being carried away before it reaches the surface (e.g., by groundwater convection), or the stresses on the fault are low (Lachenbruch & Sass 1980). The geophysical experiments planned for the Cajon Pass drillhole depend critically upon an understanding of the structure, stratigraphy and lithology of the country rock. Conventional geophysical logs do not display great contrast between granitic units which all have porosities of $< 2\%$ and densities of $>$ 2.7 gm cm^{-3} (Pezard *et al.* 1988). Geochemical logging, therefore, becomes a critical tool for determining the lithostratigraphy (Anderson *et al.* 1988a).

The combination of analyses from samples taken during a partial coring program and continuous geochemical logging allows the cross-correlation of elemental variations at several depth intervals of the Cajon Pass well. Below, we discuss the accuracy and precision of GLT measurements over core intervals for which preliminary laboratory analyses from Cajon Pass are completed (Anderson *et al.* 1988b).

Geochemical log-derived elemental analyses in the Cajon Pass drillhole

The variations in major element chemistry determined from the elemental yields measured by geochemical logging are shown in Fig. 9 for the Cajon Pass well. Major lithological discontinuities are readily identified from the elemental variations. In particular, there is a significant decrease in SiO_2 content at 1298 m from around 70 to 60 wt%. A corresponding increase in Fe, Ca and particularly Al_2O_3 is evident below this discontinuity, with Al increasing from around 25% to more than 30%. The arkosic sandstone sediments above the basement contact at 500 m, which are the weathering product of rocks with similar compositions to the Cajon Pass granitic basement, are also low in SiO_2. Mg+Na content is higher in the sediments than elsewhere in the hole, however, reaching 20% in clay-rich stringers.

Log-core analysis comparison in the Cajon Pass well

Laboratory core analyses are often difficult to locate in depth relative to log positions because of incomplete core recovery. If a 10 m interval is cored but only 2 m of rock is recovered, as

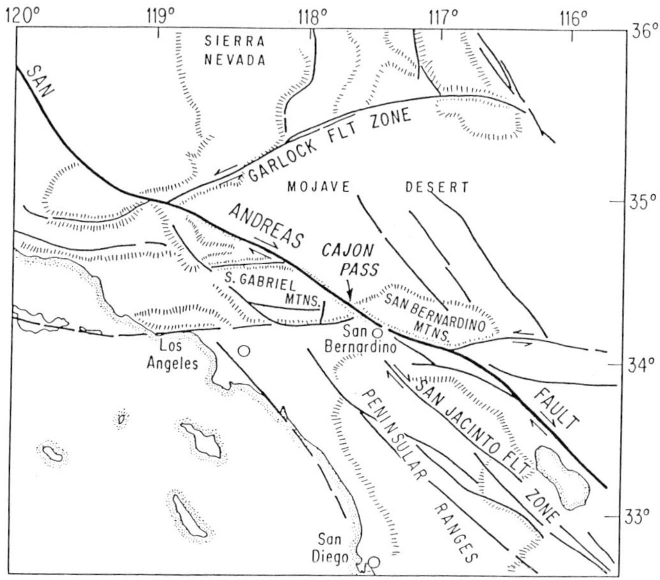

Fig. 8. Location and structure surrounding the Cajon Pass drillhole along the Squaw Peak Fault just north of the San Andreas Fault (Fig. 1) (from Anderson *et al.* 1988a).

frequently occurred in Hole 504B, the rock could have come from anywhere within that 10 m interval. Also, the sampling interval is different. The core is from a specific cylinder of rock whereas the geochemical log measurement is from the irradition of a volume of up to approximately 1 m^3 surrounding the tool.

As part of an extensive geochemical analysis program on the core from the Cajon Pass well (Silver *et al.* 1987, James & Silver 1987), Anderson *et al.* (1988b) sought to minimize these cross-correlation difficulties. Great care was taken to insure maximum recovery (always > 90%), and samples were selected for chemical analysis to ensure that the sample depth interval closely corresponds with the log measurement interval. Vertical wedges were spliced from top to bottom of each core sample. These wedges were then powdered, thoroughly, mixed and sampled several times. The elemental analyses from core are thus averages for each interval of core. The log analyses were then averaged over the same interval, and the data compared (Fig. 10).

The geochemical logs reproduce the core analyses with variable degrees of accuracy (Fig. 10). K, U, Si and Fe, provide the most linear of the cross-correlations. Ti appears to be the most poorly determined major element, but even this analysis is accurate to within 30%, except in a bad hole interval where the GLT results reach values 50% too high. Bad hole is a major cause of error in any logging data-set. Removal of geochemical logging data from depths that are clearly out-of-gauge, as determined by the four-arm caliper measurements, increases the accuracy of the Al and Ca comparisons, as well as that of Ti (Fig. 10). Th is reasonably accurate at concentrations greater than 5 ppm (accuracies of 20% in bad hole, and 10–15% in good hole), but a diorite core with only 0.32 ppm Th) is recorded as having 2.5 ppm by the GLT. Log measurements in low-abundance rock are obviously less accurate.

The abundance of Na+Mg from the geochemical log is calculated from a residual based upon the difference between the observed and a theoretical reconstruct of the photoelectric effect. As a true residual, the curve labelled MgO (Fig. 9) is in fact a representation of all elements unaccounted for in the GLT measurement analysis. Na and Mg add to fully 95% of these elements, so the log determinations should, and do, agree reasonably well with the core analyses for these elements (accuracies of 10–15%, except for one core, Fig. 10). These preliminary data verify that the geochemical logging results are reliable in granitic terrains.

Mineralogy inversion in the Cajon Pass well

A mineralogy model was constructed by inverting a matrix consisting of the elemental analyses from the geochemical logs and ideal mineral compositions of quartz, orthoclase, plagioclase (An_{45}), biotite, hornblende, pyrite, sphene, and calcite (Fig. 11) (Anderson *et al.* 1988b).

Preliminary modal analyses were done on representative thin sections from the core (Anderson *et al.* 1988b), and the cross-correlation of mineral modes observed in core, versus the predicted mineral volumes from the mineralogy model are shown for the upper half of the well in Fig. 12. The log-derived mineralogies display somewhat lower quartz contents (perhaps 5% low) and slightly higher orthoclase and plagioclase contents than the modes determined from thin sections.

The reconstruction of elemental compositions from the best-fit mineralogy model provides a good test of the reliability and self-consistency of the mineral inversion. Because the matrix inversion seeks the most statistically significant fit of elements to minerals, the exact reconstruct of that model back into predicted compositions of elements can show where elemental concentrations from the log were not well fitted by the ideal mineral compositions of the model. Whereas most elemental reconstructions were indistinguishable from the input log-derived values (Fig. 9), reconstructed Ti actually agrees with core analyses better than the input log data. This indicates that our input mineral composition was close to correct, but the input data for Na+Mg were not consistent with the other elemental compositions.

Lithostratigraphy of the Cajon Pass well

The major lithostratigraphic units in the upper half of the well are readily identifiable from variations in the mineralogy model (Fig. 11). The granodioritic rocks below the discontinuity at 1298 m have up to 50% plagioclase and 10–20% quartz. The granites above are low in biotite, but the granodiorites have high mafic contents. The biotite content decreases significantly uphole with the granite at 700–924 m having less than 5% mafic minerals. Up to 45%

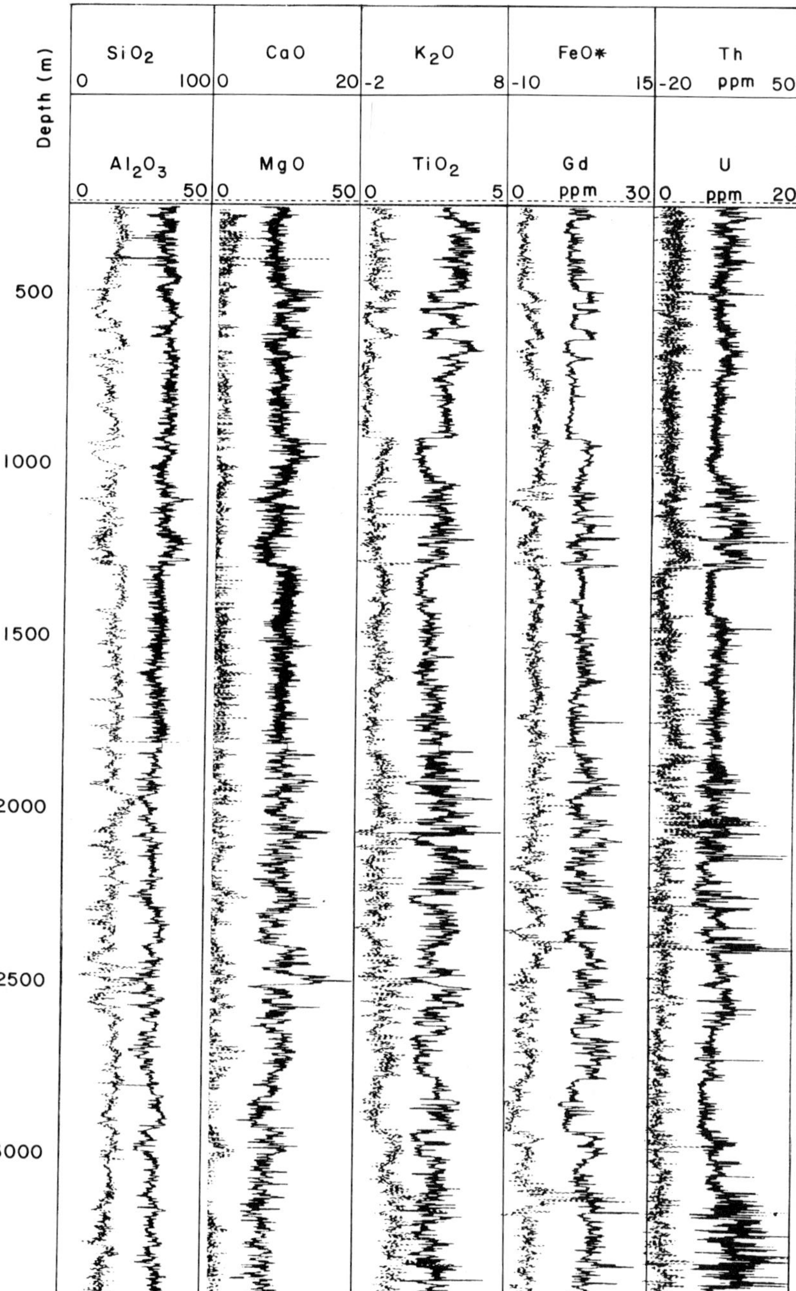

Fig. 9. Dry weight percent oxide abundances from the GLT measurements of the Cajon Pass drillhole.

of this unit is orthoclase. The sediments contain even more orthoclase. The sediments contain even more orthoclase and less quartz than this granitic unit. A dioritic core at 1800 m is shown by the mineralogy log to be an isolated dike within the lower granitic unit.

The mineralogy model determined from the GLT analyses in the Cajon Pass well proves to

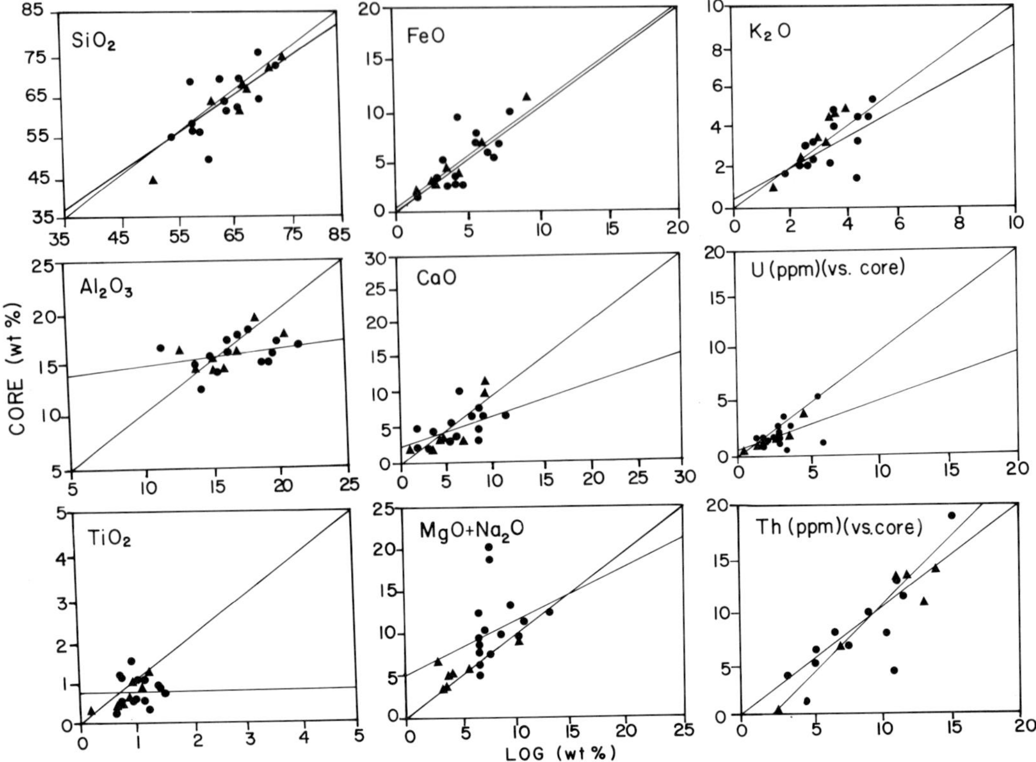

Fig. 10. Laboratory versus log-derived oxide abundances from core intervals in the Cajon Pass drillhole (from Anderson *et al.* 1988b).

be extremely useful for several of the major studies in progress at Cajon Pass. Thermal conductivity determined from the log-derived mineralogy contributes to a continuous description of the heat flow variation with depth when combined with thermal gradient measurements (Williams *et al.* 1988). Lithological units determined from elemental variations can be quantified in terms of stratigraphy of the well (Anderson *et al.* 1988a). Structural interpretations of granitic basement can be constrained by the lithological boundary-contrasts recorded by the mineralogy model (Pezard *et al.* 1988).

Metamorphic rocks from Moodus, Connecticut

The movement of great volumes of rock over large distances along shallow thrust faults is an integral part of the formation of continental terrains during collision events. Yet, the mechanics of thrusting along such decollement surfaces has remained one of the unsolved problems in our understanding of compressional tectonics. The occurrence of such displacements is widespread (from the Nappes of the Alps, to the east coast of the United States, to present-day displacement along the forelands of India), but the mode by which coherent blocks of rock slide long distances over very modest slopes has remained controversial. Hubbard & Rubey (1959) have proposed the most widely held explanation for such movements: very high pore pressures along such faults decrease frictional resistance to the point that very small differential stresses can result in large displacements.

While no direct measurements have been made which clearly demonstrate the existence in the past of such high fluid pressures along a decollement fault, the widespread metamorphism which accompanies such terrains requires large fluid fluxes to have been present along such faults. Thus, the timing of faulting versus metamorphism is crucial to our understanding of compressional tectonics on the continents.

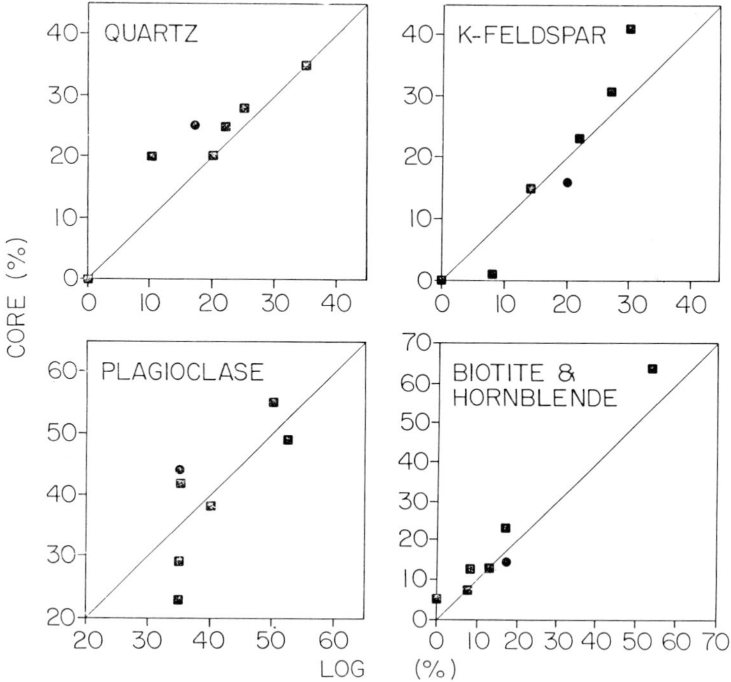

Fig. 12. Modal analyses of mineralogy model (Fig. 11) compared to thin section modal-analyses from core (from Anderson *et al.* 1988b).

Consequently, deep crustal drilling has aimed to penetrate at depth these low-angle thrust faults in a variety of geological environments around the world. It is clear that we must measure the geophysical, and particularly, the geochemical state of such fault zones *in situ* if we are to understand the mechanics of such movements. The West German 'KTB' project and the American 'ADCOH' program are two ultra-deep continental drilling plans which are designed to drill through old decollements at 10–15 km depths. For the present GLT calibration study, we obtained geochemical logging results from a shallower penetration of such a decollement surface under eastern Connecticut at the Moodus deep drillhole (Fig. 1).

The Moodus #1 well first penetrated 770 m of Merrimack Group Paleozoic metamorphic rocks, then encountered the Honey Hill Fault from 770 m to 820 m depth, and bottomed in Avalon Group Precambrian metamorphic rocks (Fig. 13). The Honey Hill Fault is the southern extension of the Lake Char thrust fault which represents the major decollement surface in eastern Connecticut (Fig. 1). The fault zone has shown evidence for repeated activation in the past (Wintsch & Ambers 1988), and may be the locus of present day, low-level, seismogenic activity.

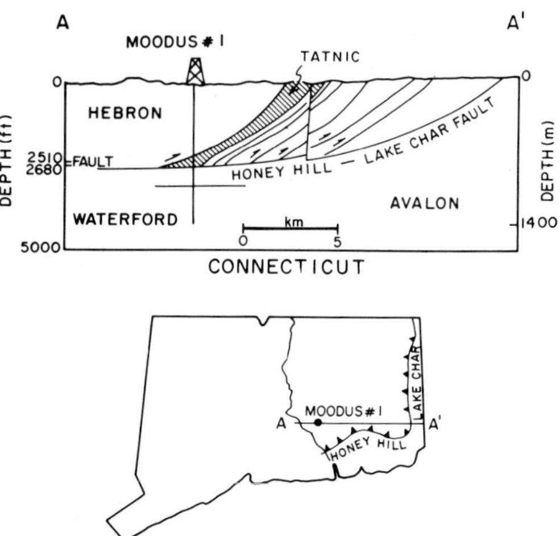

Fig. 13. Location map and structural cross-section surrounding the Moodus. Connecticut drillhole (Fig. 1).

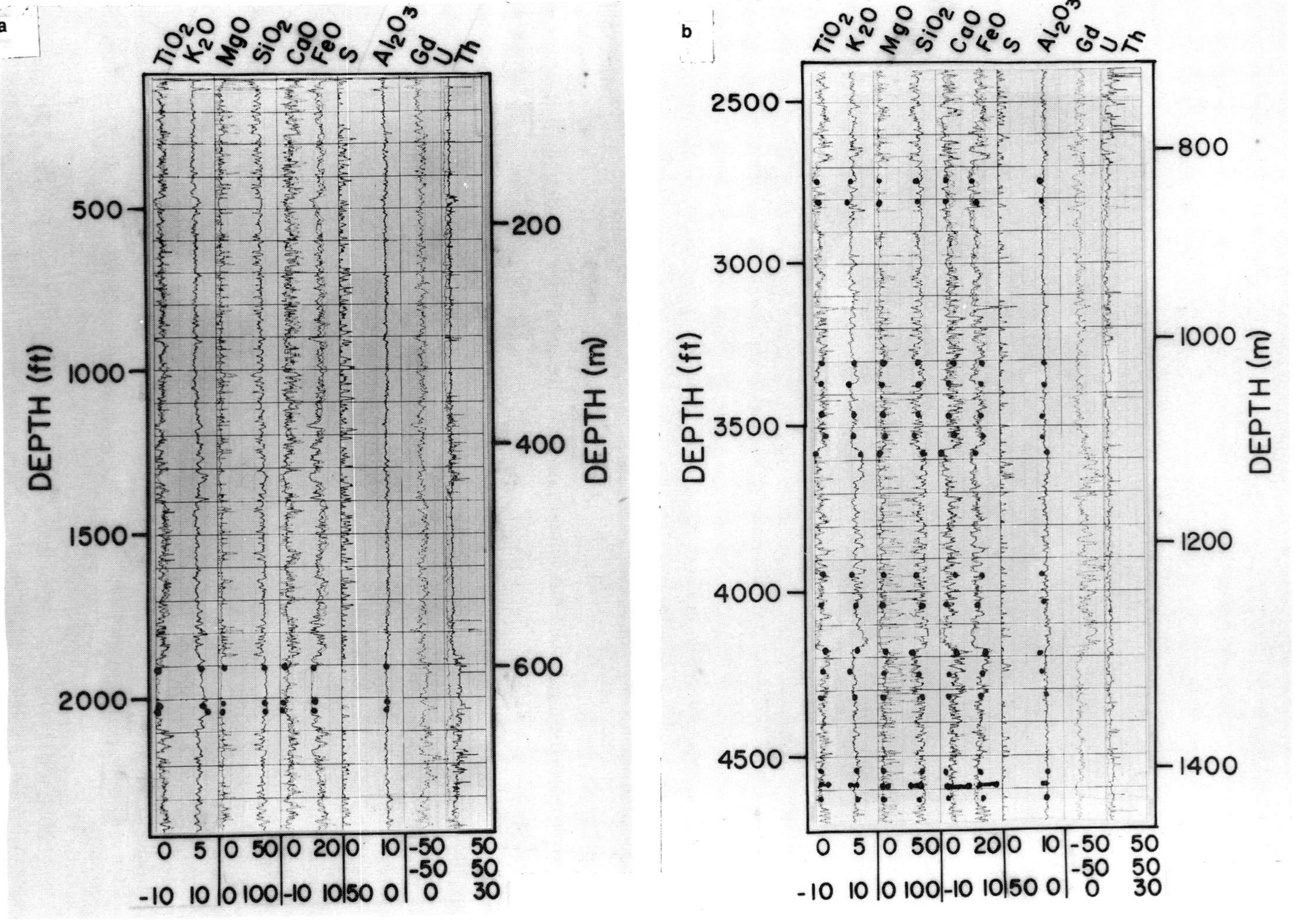

Fig. 14. Dry weight percent oxide and trace element abundances from geochemical logs and reconstructed predictions of elemental abundances from mineralogy model from Moodus #1 well. Solid circles are core and cutting analyses from Wintsch & Ambers (1988).

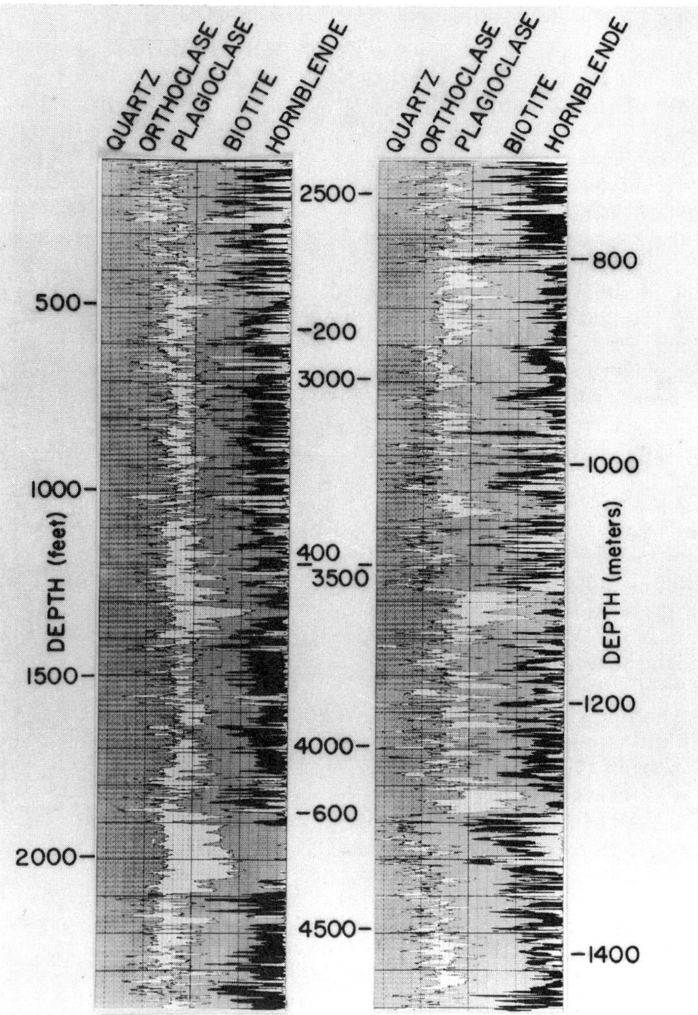

Fig. 15. Mineralogy model from log-derived major element analyses (Fig. 14) in the Moodus #1 well.

Geochemical log-derived elemental analyses from the Moodus well

The accuracy of the geochemical logging results in the metamorphic rocks of the Moodus #1 well were verified by comparing log and XRF analyses made on cuttings at 18 intervals within the well (Fig. 14) (analyses from Wintsch & Ambers 1988). The GLT log-derived elemental abundances are in excellent agreement with the laboratory analyses.

In order to establish a mineralogy-based lithostratigraphy of the Moodus #1 well, the elemental concentrations from the geochemical logs were inverted using a correlation matrix to determine the volumes of ideal mineral compositions which might be present in the well (Herzog *et al.* 1987). Anderson *et al.* (1988c) identified average dry weight percent oxide chemical compositions of quartz, orthoclase, plagioclase (An_{45}), biotite, hornblende, and pyrite. The mineral compositions were then input as end-members into an inversion matrix with the geochemical logs. The resulting best-fit mineralogy of the well is shown in Fig. 15. Reconstruction from the mineralogy determinations back to ideal elemental compositions demonstrates that the model fits the observational data very well (compare dashed and solid curves of Fig. 14).

Mineralogy model of the Moodus well

It is possible to distinguish the geological units from the mineralogy model of the Moodus well. The mineral variations (Fig. 15) show that the upper 590 m of the well consists of a quartz-biotite rich rock with the quartz-content increasing with the depth at the expense of plagioclase. This mineralogy fits with that observed in cuttings, core and outcrop for the Hebron Formation. The quartz–orthoclase–plagioclase rock from 590 to 650 m is the Canterbury granitic gneiss. The remaining Paleozoic rock unit above the Honey Hill Fault at 770 m is the quartz–biotite–hornblende Tatnic Hill Formation.

The Honey Hill Fault zone is composed of biotite-rich mylonites bounded at 780 m and 810 m by amphibolites. Beneath the fault zone, granitic gneisses grade downward into more biotite-rich gneisses of the Waterford Formation to 1030 m. Within the Waterford Formation is a zone from 1120 to 1270 m which appears to consist of interlayered and quartz-rich rocks. This interlayered unit is bounded top (1030 m to 1120 m) and bottom (1270 to 1300 m) by pegmatitic dikes (Fig. 15). The Waterford interval from 1300 m to 1480 m is an amphibole-rich gneiss grading into a more quartzitic rock. Amphibolite is found from 1480 m to the bottom of the wellbore.

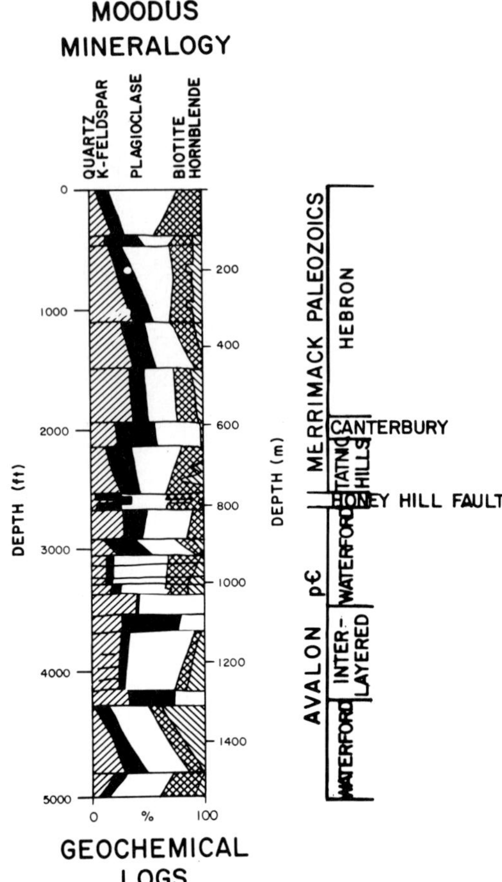

Fig. 16. Smoothed and simplified lithostratigraphy determined from the mineralogy model (Fig. 15) in the Moodus #1 well.

Lithostratigraphy of the Moodus well

The lithostratigraphic classification (Fig. 16) determined from the mineralogy model (Fig. 15) is further enhanced when the variations of U, Th and Gd are considered (Fig. 14). These trace elements clarify the lithostratigraphic interpretation from the geochemical logs. For example the Waterford Gneiss contains much lower abundances of U, Th, and Gd below the interlayered interval compared to that above (Fig. 14). Of particular interest is the interlayered zone itself (from 1030 to 1120 m), which has very high Gd values even though both end-members from which the interlayering presumably was derived (the two Waterford units above and below the interlayered unit) have low HREE values. This observation, coupled with the location of pegmatitic dikes at the boundaries of the interlayered zone, suggest an intrusive origin for the interlayering rather than a structural origin. This is because variations in HREE content within rocks of similar major elemental compositions must come from magmatic fractionation rather than from hydrothermal overprinting which might have accompanied shearing, because hydrothermal fluids are not believed to be able to mobilize Gd or other HREE.

The Honey Hill Fault contains high Th-content mylonites bounded by low Th-content amphibolites. Extremely high Th-content continues upward from the Honey Hill Fault into the Tatnic Hills Formation. The granites of the Canterbury Gneiss (From 590 to 650 m) have high U- as well as high Th-values (Fig. 14). These high U contents continue to the surface in the Hebron Formation, but the Th- and HREE-values are low in the Hebron compared to the Tatnic Hills or Canterbury Formations. The so-called 'silver bullet gneiss' from 340 to

420 m contains up to 20 ppm uranium, and may represent a highly enriched, late stage granitic intrusive. Th is only mobile within hydrothermal fluids at extremely high temperatures (> 400°C). Thus the identification of high Th content within the major low-angle thrust fault between Precambrian and Palaeozoic rocks indicates thrusting occurred in the presence of extremely high-temperature fluids.

Conclusions

Though we have examined geochemical logging results from a wide variety of igneous and metamorphic terrains, the resulting elemental abundances and mineralogy models were of sufficient accuracy and precision to yield geological insights into each environment. The precision of the measurements is good, whereas the accuracy between GLT analyses and laboratory analyses from core samples and cuttings varies from good (504B basalts), to very good (Cajon Pass granites), to excellent (Palisades Sill basalts and Moodus metamorphic rocks). In all cases, the lithostratigraphy of each well was sufficiently constrained by geochemical logging results. Future calibration of the GLT in igneous and metamorphic rocks promises even better results.

We wish to thank the officers and crew of the JOIDES RESOLUTION, ODP science management at Texas A&M, and Keir Becker and Akeho Sakai, co-chief scientists of ODP Leg 111 for support during logging of the Hole 504B. The Ocean Drilling Program, administered by the Joint Oceanographic Institutions, Inc., funded the Hole 504B logging program. The Empire State Electrical Energy Research Corporation of New York and Woodward-Clyde Consultants are thanked for their support of the Palisades Sill and Moodus Connecticut logging effort. We also wish to thank Mark Zoback, Principal Investigator at Cajon Pass and Lee Silver for support of the Cajon Pass logging program, which was funded by the National Science Foundation through Deep Observation of the Earth's Continental Crust, Inc.

References

ADAMSON, A. C. 1985. Basement lithostratigraphy, Deep Sea Drilling Project Hole 504B. *In*: ANDERSON, R. N., HONNOREZ, J., BECKER, K., *et al.*, Initial Reports of the Deep Sea Drilling Project **83**, 121–127.

ALT, J. C., LAVERNE, C. & MUEHLENBACHS, K. 1985. Alteration of the upper oceanic crust: Mineralogy and processes in Deep Sea Drilling Project Hole 504B, Leg 83. *In*: ANDERSON, R. N., HONNOREZ, J., BECKER, K., *et al.* Initial Reports of the Deep Sea Drilling Project, **83**, 217–247.

ANDERSON, R. N., ALT, J. C. & MALPAS, J. 1989. Integrated chemical fluxes from geochemical logs in Hole 504B, eastern equatorial Pacific, *Journal of Geophysical Research*.

—— BROGLIA, C., PEZARD, P. & WILLIAMS, C. 1988a. Lithostratigraphy determined from discriminant analysis of geochemical well logs from the Cajon Pass Scientific Drillhole, California, *Geophysical Research Letters*, **15**, 957–960.

—— DOVE, R., SILVER, L., JAMES, E. & CHAPPELL, B. 1988b. Elemental and mineralogical analyses using geochemical logs from the Cajon Pass Scientific Drillhole, California, and their preliminary comparison with core analyses, *Geophysical Research Letters*, **15**, 969–973.

——, —— & WILLIAMS, C. 1988c. Geochemical and geophysical logs from the Moodus Deep Drillhole, and the mass flux associated with low-angle thrust faults, *Transactions, American Geophysical Union*, **69**, 495.

—— O'MALLEY, H., NEWMARK, R. L. 1985a. Use of geophysical logs for quantitative determination of fracturing, alteration, and lithostratigraphy in the upper oceanic crust, Deep Sea Drilling Project, Holes 504B and 556. *In*: ANDERSON, R. N., HONNOREZ, J., BECKER, K., *et al.*, Initial Reports of the Deep Sea Drilling Project, **83**, 443–478.

——, —— HICKMAN, S. H. & NEWMARK, R. L., 1985b. Permeability versus depth in the upper oceanic crust: *In*: situ measurements, in Deep Sea Drilling Project Hole 504B, eastern equatorial Pacific. *In*: ANDERSON, R. N., HONNOREZ, J., BECKER, K., *et al.* Initial Reports of the Deep Sea Drilling Project, **83**, 429–442.

BECKER, K., SAKAI, *et al.* 1988. Initial Reports, Ocean Drilling Program (Part A) 111.

HERZOG, R., COLSON, L., SEEMAN, B. O'BRIEN, M., SCOTT, H., MCKEON, D., WRAIGHT, P., GRAU, J., SCHWEITZER, J. & HERRON, M. 1987. Geochemical logging with spectrometry tools. Society of Petroleum Engineers, Paper 16792.

HONNOREZ, J., ALT, J. C., HONNOREZ,-GUERSTEIN, B. -M., LAVERNE, C., MUEHLENBACHS, K., RUIZ, J., & SALTZMAN, E. 1985. Stockwork-like sulfide mineralization in young oceanic crust. Deep Sea Drilling Project Hole 504B. *In*: ANDERSON, R. N., HONNOREZ, J., BECKER, K., *et al.*, Initial Reports of the Deep Sea Drilling Project, **83**, 263–278.

HUBBARD, M. K. & J. RUBEY, 1959 Role of fluid. pressure in the mechanics of overthrust faulting. *Bulletin of the Geological Society of America* **70**, 115–166.

JAMES, E. W. & SILVER, L. T. 1987. Geologic column of the Cajon Pass deep drillhole *Transactions, American Geophysical Union*, **68**, 1489.

LACHENBRUCH, A. H., & SASS, J. 1980. Heat flow and energetics of the San Andreas Fault *Journal of Geophysical Research*, **85**, 6185–6223.

NEWMARK, R. L., ANDERSON, R. N., MOOS, D., & ZOBACK, M. D. 1985. Sonic and ultrasonic logging of Hole 504B and its implications for the structure, porosity, and stress regime of the upper 1 km of the oceanic crust. *In*: ANDERSON, R. N., HONNOREZ, J., BECKER, K., *et al.*, Initial Reports of the Deep Sea Drilling Project, **83**, 479–510.

PEZARD, P. A. & ANDERSON, R. N. 1989. Morphology and alteration of the upper oceanic crust from in situ electrical resistivity experiments in DSDP Hole 504B, *Journal of Geophysical Research*.

——, —— HOWARD, J. M. & LUTHI, S. M. 1988 Fracture distribution and basement structure from measurements of electrical resistivity in the basement of the Cajon Pass Scientific Drillhole, California, *Geophysical Research Letters*, **15**, 1021–1025.

SHIRLEY, D. 1987. Petrology of the Palisades Sill *Journal of Petrology*, **86**, 489–502.

SILVER, L. T., JAMES, E. W. & CHAPPELL, B. W. 1987, Regional geologic setting of the Cajon Pass deep drillhole, Transactions, *American Geophysical Union*, **68**, 1489.

WALKER, F. 1940. Differentiation of the Palisades Diabase, New Jersey *Bulletin of the Geological Society of America*, **51**, 1059–1106.

WILLIAMS, C. F., ANDERSON, R. N., BROGLIA, C., LACHENBRUCH, A. H. & SASS, J. H. 1988. In Situ investigation of thermal conductivity, heat production, and past hydrothermal circulation in the Cajon Pass Scientific Drillhole, California. *Geophysical Research Letters,* **15**, 985–988.

WINTSCH, R. P. & AMBERS, C. P. 1988. Regional correlation of subsurface lithologies and structures, Moodus, Connecticut. *Transactions, American Geophysical Union*, **69**, 495.

ZOBACK, M. D., TSUKAHARA, H. & HICKMAN, S. 1980. Stress measurements at depth in the vicinity of the San Andreas Fault: Implications for the magnitude of shear stress at depth. *Journal of Geophysical Research*, **85**, 6157–6173.

Preliminary geochemical results from DSDP/ODP Hole 504B: a comparison of core and log data

T. S. BREWER[1], M. A. LOVELL[1], P. K. HARVEY[1], R. PELLING[1], B. P. ATKIN[1] & A. ADAMSON[2]

[1]*Geology Department, University of Nottingham, Nottingham NG7 2RD, UK*

[2]*Ocean Drilling Program, Texas A&M University, College Station, TX 77840, USA*

Abstract: DSDP/ODP Hole 504B, located in the Eastern Equatorial Pacific in 5.9 Ma old crust, penetrates both basaltic pillows and dykes down to a depth of 1562.1 m below the sea floor. Core recovery is poor, averaging 23% but falling to a 12% average in the deepest sections. Geochemical logs give a continuous measure of elemental (Si, Ca, Fe, S, Ti, Gd, H, Cl, K, Th, U) abundances. Correlation between laboratory (XRF) geochemical data and log data is difficult because of the lack of accurate depth information for the cores. Pattern recognition techniques are of little help for locating sample depths in the uniform basaltic lithologies. Non-hierarchical clustering techniques prove the existence of relationships between core descriptions and gross geochemical variations; major litho-stratigraphic zones are readily identified from the log-derived data.

The Deep Sea Drilling Project (DSDP) attempts to sample directly the ocean crust are well documented and are now being expanded by the Ocean Drilling Program (ODP) which uses a larger and better equipped drillship, JOIDES Resolution (SEDCO/BP 471). Sampling of sediments of the seismically defined layer 1 has become routine; however, effective sampling of the crystalline basement of layer 2 is more problematic. Penetration through a complete oceanic sequence to reach layer 3 has not yet been achieved.

The most successful sampling to date of a complete vertical sequence of the supposed oceanic layer cake model, as characterized from seismic reflection data and the ophiolite sequences preserved on land, is at Hole 504B in the Eastern Equatorial Pacific (Fig. 1). Located some 200 km south of the Costa Rica Rift in 5.9 Ma old crust the hole has been drilled by three successive DSDP Legs (69 and 70 in 1979/80, and 83 in 1981/82) and by one ODP Leg (111 in 1986). Consequently, it now penetrates 274.5 m of sediments and 1287.8 m of oceanic basement (Fig. 2).

The specific location of the hole was originally chosen as being representative of an area in the ocean crust where the crustal temperatures agree closely with those predicted for conductive plate heat transfer (Anderson & Hobart 1976). It was suggested from the observed heat flow pattern that Site 504 may have rebounded recently to a conductive geothermal state after undergoing hydrothermal cooling in the past few million years. Consequently, Site 504 is well suited for studies of the effect of sealing the crust by sedimentation on a ridge-flank hydrothermal system, and the unit is young enough to have preserved the alteration record from the ridge-axis circulation. However, Largseth *et al.* (1988), from recent studies of heat flow and pore fluid movement, suggest that hydrothermal convection does still occur in the basement on a lateral scale of 3.5–4.0 km and may penetrate deeply into the crust.

Core recovery and laboratory analyses

Poor core recovery is one of the biggest problems encountered by both DSDP and ODP when attempting to sample the ocean floor. In Hole 504B the average recovery for the basement section is about 20%, though it rises locally to 70% and falls to 12% in the lowermost 200 m of the hole. Coring is carried out using rotary techniques with roller-cone bits using seawater as the drilling fluid. During the most recent drilling at 504B (ODP Leg 111), two steel rotary bits disintegrated after only 16–17 rotating hours each, with all four roller cones being left in the hole and requiring removal. In all, Hole 504B claimed parts of three coring assemblies during Leg 111 (Becker, Sakai *et al.* 1988). The problems encountered in both drilling and coring are no doubt the result of a combination of factors. Becker, Sakai *et al.* (1988) suggested four:

From Hurst, A., Lovell, M. A. & Morton, A. C. (eds), 1990, *Geological Applications of Wireline Logs* Geological Society Special Publication No. 48, pp. 195–202

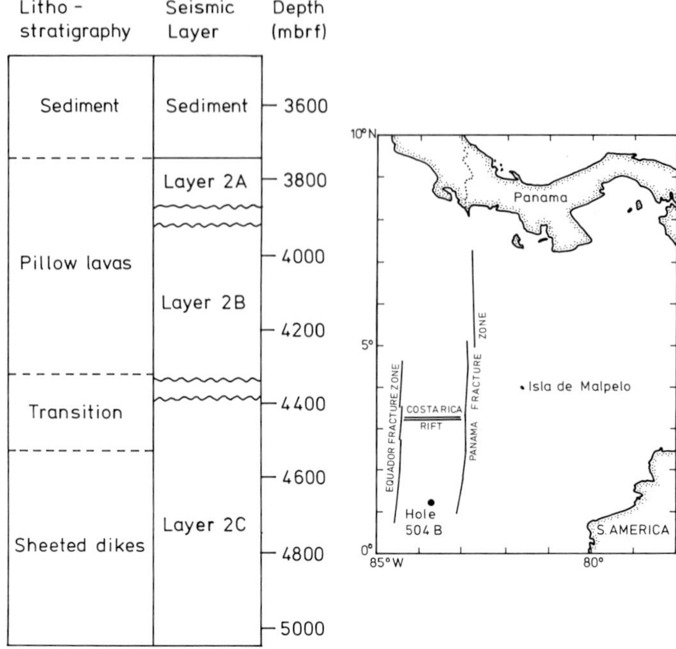

Fig. 1. The location of Hole 504B, together with the lithostratigraphy and seismic stratigraphy. Depth is in metres below rig floor.

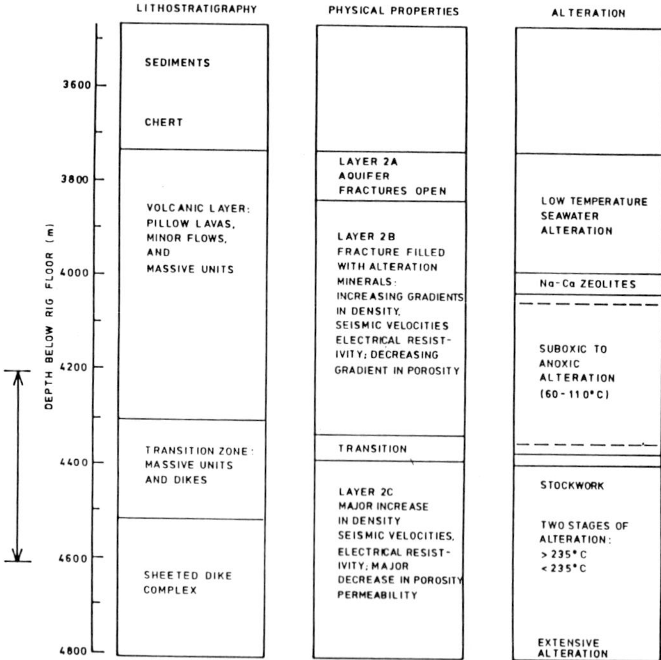

Fig. 2. Generalized lithostratigraphy and physical properties of Hole 504B (after Emmermann 1985 and Adamson 1985). The arrowed section represents the area of study.

1. An inability to flush cuttings completely from the very deep hole.
2. The steady accumulation of steel junk in the hole.
3. Spalling of wall rocks into the hole and around the coring assemblies.
4. The dense, crystalline nature of the dykes cored deeper than 1500 mbsf.

Additionally, the following may all contribute to poor core recovery: the heave of the ship, which, although compensated for, may still be significant, the lack of a suitably dense drilling fluid, and the absence of a riser system.

It is doubtful to what degree the recovered samples are representative of the in situ material. Given that the rocks exhibit a history of alteration in this geological setting, it is probable that the unaltered material will be sampled and altered material destroyed in the drilling/coring process. Thus, the sampled material constitutes a biased representation of the total in situ section.

Of equal importance in terms of the detailed lithostratigraphy is the problem of locating the recovered samples within each cored interval. For example, a typical ODP coring interval is 9.5 m, the nominal capacity of a core barrel. Given 20% recovery, then the total length of recovered material will be 1.9 m. By convention, the top of the recovered material is positioned at the top of the cored interval; this is unlikely to be realistic. The remaining core must then be interpreted in terms of the overall vertical succession. Two simple options, neither of which are particularly satisfactory, are:

(a) to attribute all the recovered material to the top of the hole, and then have substantial gaps between successive cored intervals;
(b) to expand the recovered material proportionately to fill the cored interval.

An alternative to both of these, and a possible means of evaluating the geochemistry of the hole throughout the vertical succession, is to use downhole measurements.

Downhole geochemistry

The quantitative measurement of geochemistry downhole relies on the application of nuclear logging techniques (Herzog *et al.* 1977, Anderson 1988, Lovell & Anderson 1988, Anderson *et al.* 1990). K, Th and U are measured using a natural gamma-ray spectroscope, Si, Ca, Fe, S, Ti, Gd, H, and Cl are measured by the gamma spectroscopy tool,

Table 1. *Comparison of the geochemistry for the different lithologies encountered in Hole 504B*

	Pillow Lavas		Thin Lava Flows	
	Range	Average	Range	Average
SiO_2	46.00–51.30	49.30	47.67–51.60	49.11
Al_{O3}	13.60–17.10	15.18	13.60–16.40	15.42
TiO_2	0.56– 1.58	0.98	0.81– 1.08	0.93
FeO	6.57–10.98	9.19	8.26–10.07	9.04
MgO	6.40–12.70	8.20	7.86– 9.15	8.61
CaO	2.20–13.61	12.31	11.60–13.00	12.45
Na_2O	0.02– 2.41	1.94	1.42– 2.45	2.01
K_2O	0.00– 0.10	0.02	0.00– 0.08	0.02
MnO	0.10– 7.98	0.30	0.16– 8.30	0.57
P_2O_5	0.04– 1.34	0.10	0.06– 0.12	0.08
	Massive Units		**Dykes**	
	Range	Average	Range	Average
SiO_2	47.55–51.70	49.37	45.64–50.70	48.99
Al_2O_3	14.00–16.50	15.16	12.90–16.60	15.24
TiO_2	0.77– 1.12	0.94	0.81– 1.28	0.95
FeO	8.34–10.46	9.17	1.65–10.88	8.89
MgO	7.72–10.23	8.56	7.69– 9.25	8.38
CaO	11.28–13.37	12.56	10.80–13.30	12.41
Na_2O	1.31– 2.48	1.97	1.08– 2.78	1.99
K_2O	0.00– 0.14	0.02	0.00– 0.04	0.01
MnO	0.14– 0.31	0.19	0.15– 0.30	0.19
P_2O_5	0.05– 0.12	0.08	0.05– 0.11	0.08

Data obtained by XRF (Sharaskin *et al.* 1983, Tual *et al.* 1985).

whilst Al is measured by the Aluminium activation tool. Additionally, measurement of the photoelectric capture cross-section allows for the determination of Mg and Na (Lithodensity Tool).

Geochemical results

The volcanic and magmatic components of the oceanic lithosphere can be divided into volcanics, dykes, gabbros and ultramafic complexes. Viewed collectively, the geochemistry of the volcanics, dykes and flows often fall within a very restricted range and the compositions of each frequently overlap, which makes separate geochemical classification extremely difficult (Table 1).

This study is concerned with a 400 m interval of Hole 504B from 4200 to 4600 m (brf). This interval has a particularly interesting and variable lithostratigraphy as observed in the core descriptions (Fig. 2). The hole first penetrates the pillows and flows of layer 2A before going through the transition zone into the dykes of layer 2C (Figs 2 & 3). Within this section the core recovery is poor and over the entire section it averages 23% (Fig. 3).

In order to compare the geochemical measurements derived by the logging tool and X-ray fluorescence (XRF) the elemental concentrations are plotted against depth (Fig. 4). If one assume the XRF measurements to be accurate, then one can assess the similarity of the log response for an individual element. The results of this analysis are presented in Table 2. In general, all of the log-derived oxides show a greater range than the XRF-derived data and the oxides showing the greatest variation are Fe_2O_3 and MgO. The log-derived TiO_2 is higher

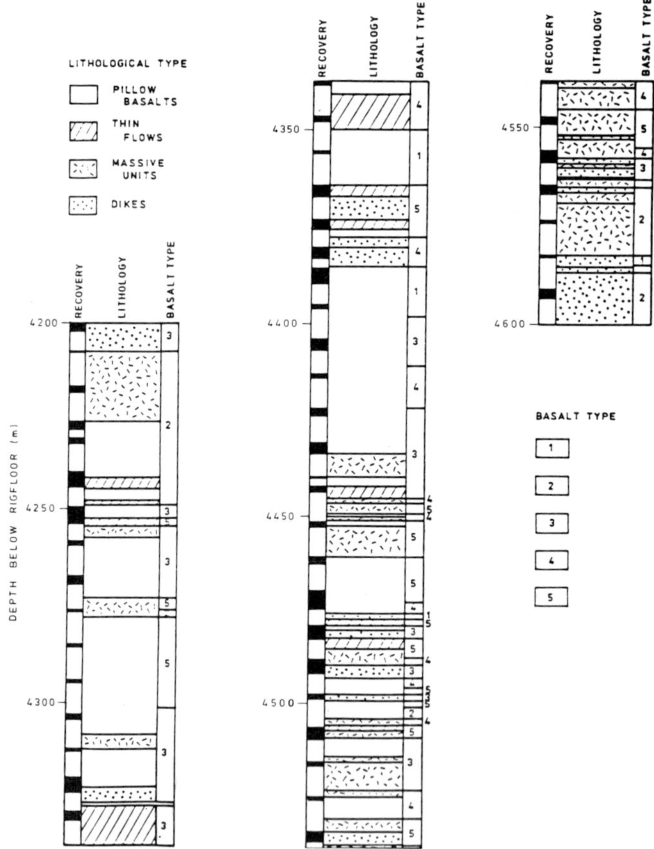

Fig. 3. Detailed lithostratigraphy of studied section. Note the poor core recovery. Basalt types 1: plagioclase-olivine phyric with essential chrome spinel, 2: plagioclase-olivine-clinopyroxene (augite) phyric, 3: plagioclase-olivine-clinopyroxene (emerald green) phyric, 4: olivine-plagioclase phyric, 5: aphyric (from Adamson 1985).

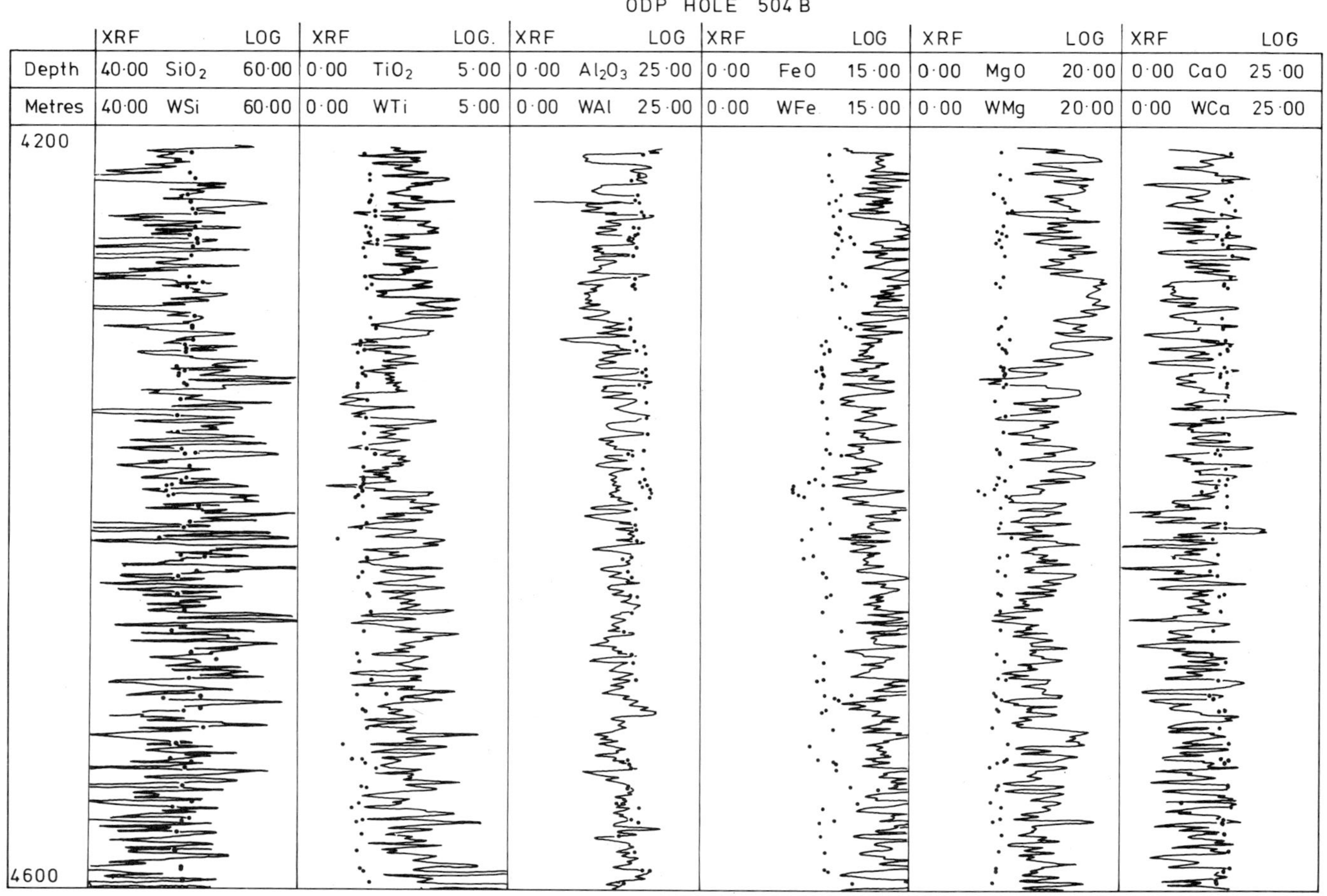

Fig. 4. Elemental concentrations plotted against depth for log responses and recovered material (XRF) within the studied section.

than the TiO_2 (XRF) which is somewhat alarming considering its low concentrations.

Comparison of the two data sets shows that there is no correlation between the two (Fig. 4, Table 2). The underlying assumption in this type of interpretation is that the XRF analyses are accurate in terms of both the composition and their supposed positions within the hole. If the depth matching is correct then the use of the log-derived data for more detailed geochemical calculations (i.e. normative mineralogy) cannot be validated.

Accepting the limitations of the data, we have attempted a preliminary geochemical interpretation in order to indicate the types of conclusions that may be reached. In order to classify the data into groups we have taken a non-hierarchical statistical approach based upon six elements obtained from the logging suite (Si, Ti, Al, Fe, Mg, Ca). Using this approach six groups are identified (Fig. 5). The non-hierarchical cluster analysis used a K-means method and is classified by assuming minimum euclidian distances between samples. The principles of the method are given by several authors including Lebart *et al.* (1984) and Spath (1980); Shepherd *et al.* (1987) have applied the technique in a geochemical context. The initial cluster seed points were chosen by the method of Ball & Hall (1967). Models with two to sixteen groups were run. A marked reduction of the within-group sum of squares at six groups was chosen for presentation of the log data. These six groups are not necessarily a function of the igneous geochemistry but may also reflect variations caused by low-temperature alteration. The encouraging feature of the classification is the basic division of the section into three main units, which correlates with the supposed positions of the pillow lavas, transition zone and sheeted dykes (Fig. 5). Within the major units the smaller variations may reflect small scale lithological variations (i.e. dykes, flows) although such features cannot as yet be resolved because of the lack of calibration of the data.

A slightly different approach to the interpretation is to consider the inter-element correlations of the two data sets (Fig. 6). In Fig. 6 the inter-element correlations for both the XRF and log-derived data show certain similarities. Firstly the general grouping of Ca and Al could be related to the plagioclase component of these basic magmas, whereas the grouping of Ti, Fe and Mg probably relates to the ferromagnesian component of the magmas.

Although the above interpretations are preliminary there is clearly great potential for using the logging suite in the understanding of the oceanic crust. Only further research into the calibration of the log measurements will realize their full potential and allow their application to the understanding of the detailed geochemical problems associated with the oceanic crust.

Conclusions

The nuclear logging suite has the potential to revolutionize the study of the oceanic litho-

Table 2. *Comparative statistics for the log derived and XRF core measurement data*

	SiO_2	Al_2O_3	TiO_2	FeO	MgO	CaO
XRF (n=206)						
Average	49.24	15.22	0.96	9.12	8.55	12.40
Standard Deviation	1.03	0.78	0.13	0.94	0.59	0.99
Minimum	45.64	12.90	0.56	1.65	6.40	2.20
Maximum	51.70	17.10	1.58	10.98	12.70	13.61
Log (n=2998)						
Average	48.08	13.12	1.42	12.84	13.18	9.93
Standard Deviation	5.04	2.08	0.38	1.75	2.32	2.87
Minimum	25.64	6.67	0.36	9.84	7.25	3.15
Maximum	59.21	17.64	2.71	24.35	18.27	16.37
Difference	+1.16	+2.10	−0.46	−3.72	−4.64	+2.46
r	−0.12	0.04	0.17[1]	0.27[2]	0.04	−0.03

[1] r significantly different from zero at =0.05.

[2] r significantly different from zero at =0.01.

Correlation coefficients for supposed equivalent position samples from Hole 504B. XRF data from Sharaskin *et al.* (1983) and Tual *et al.* (1985).

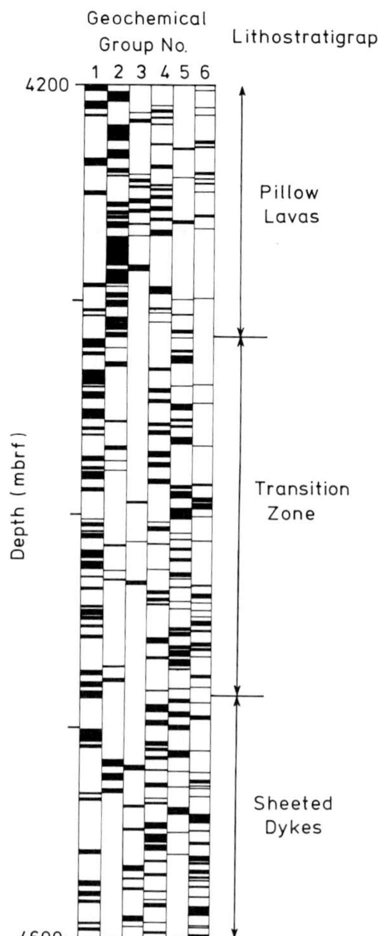

Fig. 5. Non-hierarchical cluster analysis results for the 400 m section based upon six elements. Note the subdivision of the interval into three major sections (X, Y, Z). The sections correlate very well with the lithological subdivisions of Fig. 2.

sphere and downhole geochemistry. At present the data are poorly calibrated and cannot be used to solve detailed geochemical problems. The most important features of the log data are that they produce results which are consistent with other observations and that they can be used to make coarse-scale classifications of the oceanic basement. The use of the logs for the calculation of normative mineralogies should be

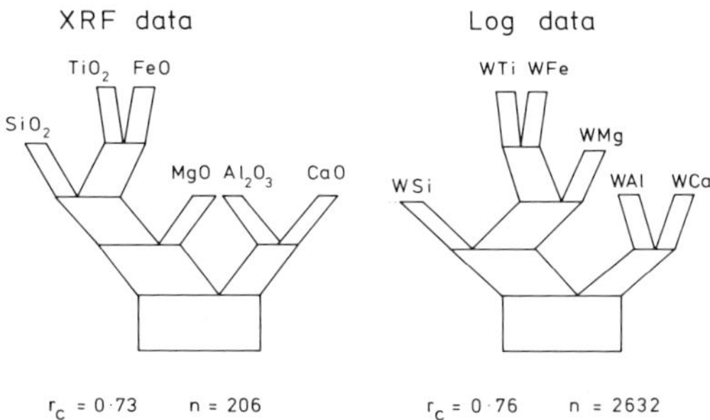

Fig. 6. A Kleiner–Hartigan tree for the XRF and log response data. The correlation structure of the two data sets show a striking similarity.

discouraged since the norms may represent features such as alteration, primary igneous variations or the error associated with the log responses. Further research will allow the development of satisfactorily calibrated data which can then be used to study detailed petrogenetic problems and also to locate accurately cored sections within holes with poor recovery.

This paper was written in early 1988 as the result of an initial appraisal of the downhole nuclear measurements made on Leg 111 of the ODP. The discrepancies highlighted between some of the elemental abundances determined from laboratory and in situ measurements may be explained by a variety of mechanisms as described in the text. From the substantial offsets evident in Fig. 4, it is suggested that there may be fundamental problems with the initial processing of the downhole nuclear data when producing elemental abundances. This has since been shown to be true and the data have since been reprocessed to compensate for this effect which may be due to the largely unknown environment (at least in nuclear terms) of basaltic ocean crust. The reprocessed data are presented in Anderson *et al.* (1990).

The authors wish to thank Roger Anderson, Cristina Broglia, and Beth Pratson at Lamont Doherty Geological Observatory, and Roy Dove, Jeff Schweizer, and Jim Grau of Schlumberger-Doll Research for helpful discussions. In Nottingham Dave Jones, Josie Wilkinson and Jean Pearson greatly assisted in the production of this paper, which also benefitted greatly from advice provided by an anonymous reviewer. The research forms part of a programme of work funded by the Natural Environment Research Council under the Ocean Drilling Program Special Topic.

References

ADAMSON, A. C. 1985. Basement Lithostratigraphy, Deep Sea Drilling Project Hole 504B. *Initial Reports of the Deep Sea Drilling Project*, **83**, 121–127.

ANDERSON, R. N. 1988. Nuclear logs. *Proceedings of the Ocean Drilling Program, Initial Report* (Part A), Leg 111.

——, DOVE, R. E. & PRATSON, E. 1990. Geochemical well logs: calibration and lithostratigraphy in basaltic, granitic and metamorphic rocks. In: HURST A., LOVELL M. A. & MORTON A. C. (eds) *Geological Applications of Wireline Logs*. Geological Society, London, Special Publication **48** 177–194.

—— & HOBART, M. A. 1976. The relation between heat flow, sediment thickness and age in the Eastern Pacific. *Journal of Geophysical Research*, **81**, 1968–1989.

BALL, G. H. & HALL, D. J. 1967. A clustering technique for summarizing multivariate data. *Behavioral Science*, **12**, 153–155.

BECKER, K, SAKAI, H. *et al.* 1988. *Proceedings of the Ocean Drilling Program, Initial Report* (Part A), Leg 111.

EMMERMANN, R. 1985. Basement geochemistry, Hole 504B. *Initial Reports of the Deep Sea Drilling Project*, **83**, 183–199.

HERZOG, R., COLSON, L., SEEMAN, B. O'BRIEN, M., SCOTT, H., MCKEON, D., WRAIGHT, P., GRAU, J., SCHWEITZER, J. & HERRON, M. 1987. Geochemical logging with Spectrometry Tools. *Society of Petroleum Engineers*, SPE 16972.

LANGSETH, M. G., MOTTL, M. J., HOBART, M. A. & FISHER, A. 1988. The distribution of geothermal and geochemical gradients near site 501/504: implications for hydrothermal circulation in the ocean crust. *Proceedings of the Ocean Drilling Program, Initial Report* (Part A), Leg 111.

LEBART, L., MORINEAU, A. & WARWICK, K. M. 1984. *Multivariate descriptive statistical analysis*. John Wiley and Sons. New York.

LOVELL, M. A. & ANDERSON, R. N. 1988. When downhole logging turns to geochemistry. *Geology Today*, **4**, 164–166.

SHARASKIN, A. Ya., MIGDISOV, ROSTCHINA, I. A. & MIKLISHANSKY, A. Z. 1983. Major and trace element chemistry of Hole 504B, basalts and their alteration products (Costa Rica Rift, Deep Sea Drilling Project Leg 70). *Initial Reports of the Deep Sea Drilling Project*, **69**, 775–790.

SHEPHERD, A., HARVEY, P. K. & LEAKE, R. C. 1987. The geochemistry of residual soils as an aid to geological mapping; a statistical approach. *Journal of Geochemical Exploration*, **29**, 317–331.

SPATH, H. 1980. *Cluster analysis algorithms for data reduction and classification of objects*. Ellis Horwood, Chichester.

TUAL, E., JAHN, B-M, BOUGAULT, H. & JORON, J-L. 1985. Geochemistry of basalts from Hole 504B, Leg 83, Costa Rica Rift. *Initial Report of the Deep Sea Drilling Project*, **83**, 201–216.

Feasibility of in situ elemental analysis in mudrock evaluation

T. J. PRIMMER, S. A. KERR & K. J. MYERS
B. P. Research, Sunbury-on-Thames, TW16 7LN, UK

Abstract: Geochemical logging now offers the possibility of performing quantitative in situ elemental analysis. Laboratory measurements of element abundances on selected sections of mudrocks were made to simulate the potential use of data available from geochemical logging. The analyses centre on mudrocks from the Magnus Field (North Sea) and Dorset Kimmeridge Clay Formation sequences (Southern England). Subsequent multivariate statistical processing of these analyses showed that pre-defined units within the sections studied can be distinguished geochemically without recourse to mineralogical description. This is demonstrated firstly by cross plotting elemental abundances to pick out significant sympathetic trends and thereafter by clustering and factor analysis techniques which utilize a more extensive dataset. This study shows that the direct application of statistical tests to raw element abundances determined downhole is an appropriate methodology by which conventionally impractical topics, such as the subsurface discrimination of mudrock facies, may be tackled. This conclusion will, no doubt, be tested further as geochemical logging develops over the next few years.

Recent developments in wireline nuclear logging tool technology now offer the possibility of performing quantitative in situ elemental analysis (Hertzog *et al.* 1987). Although applications of geochemical logging data to the study of sedimentary rocks (M. Herron 1988), hydrocarbon source rocks (S. Herron 1987) and igneous rocks (Grau *et al.* 1987) have been described, the interpretation of the data is still in its infancy. For the characterisation of sedimentary rocks of importance to the oil industry, element abundances recorded downhole are transformed into equivalent normative mineral abundances. The derived mineralogy is further manipulated into parameters considered to represent clay mineral volume, cation exchange capacity, grain size and even permeability, the aim being to enhance reservoir description (M. Herron 1987). This methodology requires basic assumptions to be made (without any prior knowledge) on which minerals to include in the normative scheme, their chemical compositions and their individual contribution to any subsequent physical property which is to be determined (e.g. cation exchange capacity). This places specific formation (or even single well) constraints on the final solution.

For applications such as reservoir zonation or facies recognition and correlation, elemental data can be manipulated directly without recourse to a derived mineralogy. Thus, for a given interval (in different wells) or a given sequence (in the same well), an interpretation of the data could be made in terms of the geological processes that give rise to the various element abundance patterns observed. This technique would be universally applicable if the elemental data recorded are accurate and precise. Subsequent interpretation based on the processes that give rise to any observed variation in element abundances (e.g. depositional or diagenetic, physical or chemical) need be the only subjective aspect of the study.

This paper aims to demonstrate the applicability of this style of approach to the study of mudrocks. Although this is not a new concept, most geochemical studies aimed at determining the provenance or the depositional history of mudrocks have concentrated on either their organic geochemistry or trace element geochemistry (Potter *et al.* 1980). Variations in major element geochemistry should also be considered as they are nearly always controlled by aspects of sediment provenance and depositional history (Pettijohn 1975). A few studies have successfully developed models incorporating major elements into geochemical classifications aimed at determining variations in mudrock facies (e.g. Bjørlykke 1974). In this context, the suite of ten major and trace elements that can be currently measured downhole with reasonable precision (namely Si, Ca, Fe, S, Ti, Gd, Al, K, U and Th) (Hertzog *et al.* 1987) may well provide a suitable combination of element abundances that can be used for obtaining a convenient and rapid assessment of mudrock geochemistry for subsequent facies discrimination.

Good quality elemental data, measured downhole on mudrocks, has not yet been made available. For this reason, a study on outcrop and core material was made in order to simulate

From HURST, A., LOVELL, M. A. & MORTON, A. C. (eds), 1990, *Geological Applications of Wireline Logs* Geological Society Special Publication No. 48, pp. 203–210

the potential use of data obtained from geochemical logging techniques. Element abundances were measured on material from sections of the Kimmeridge Clay Formation (KCF), a laterally persistent mudrock unit in NW Europe. Two vertical profiles from geographically distinct areas (Fig.1) were analysed for a comprehensive range of major and trace elements but with particular attention being paid to those elements currently available from geochemical logging. In order to distinguish trends in element abundances data were initially appraised simply in terms of the range of different elements. Any qualitative correlations observed were then treated more quantitatively in the form of two or three element cross plots. In addition, more complex statistical analyses (using multivariate techniques) were also used to see if there were any recognizable trends in those elements which showed the greatest spread in abundance throughout the whole dataset.

Materials and methods

Samples of KCF were taken from core retrieved from a well in the Magnus field (Block 211/12a) and from outcrop on the Dorset coast (Fig. 1). The Magnus material consisted of 31 samples taken from three distinct mudstone units (i.e. above, below and within the Magnus Sandstone) buried to depths greater than 2.8 km. Each unit consists of uniform dark grey organic-rick mudrocks although some samples towards the base of the lowest unit are silty. Material from the Dorset coast was selected to cover approximately the same age range (late Kimmeridgian–early/middle Volgian) as that identified for the Magnus samples. Twenty-eight samples were collected in total from two localities; at Kimmeridge Bay (the *autissiodorensis* zone, Lower Kimmeridge Clay) and at Freshwater Steps (the *pectinatus* zone, Upper Kimmeridge Clay). The Kimmeridge Clay in Dorset consists of an alternating succession of organic rich and organic poor mudrocks with occasional limestones and diagenetic dolomites. The Upper Kimmeridge Clay mudrocks are appreciably more calcareous than those from the Lower Kimmeridge Clay (Myers & Wignall 1988).

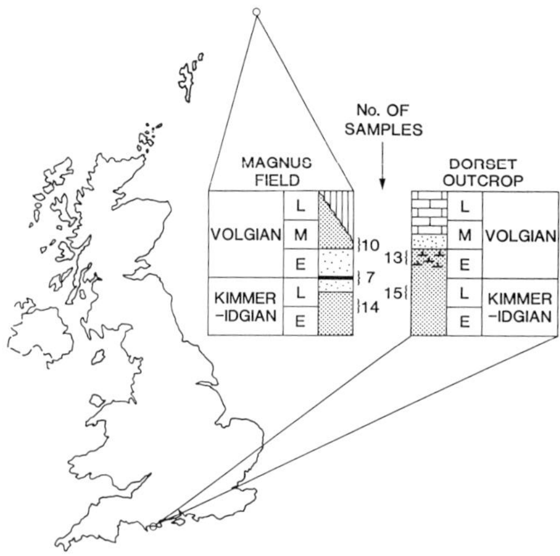

Fig. 1 Location and stratigraphic setting of the Kimmeridge Clay samples used in the study.

Material was sampled in quantities of approximately 200–300 g at between 0.5 and 2 m intervals in each section (Table 1), cleaned of any weathered material, crushed and powdered. Major element analyses were performed using standard XRF techniques on fused glass beads of rock and trace elements were determined by instrumental neutron activation analysis, XRF and prompt neutron capture gamma ray analysis on powdered samples. Full major element analyses (reported on a traditional equivalent oxide basis) summed consistently to between 99.5 and 100.5%. Analysis of the USGS standard reference material BHVO-1 (basalt) served to monitor the accuracy of the measurements. As geochemical logs conventionally report their raw data in elemental form, analyses are presented similarly here (i.e element wt% for majors and ppm for traces) in preference to the equivalent oxide form.

Element abundances and their statistical manipulation

A preliminary assessment of the range in element abundances is shown by boxplots for three example distributions (Fig. 2a). This form of statistical representation clearly demonstrates the 'loggable' variables which show the greatest range of values between the sample sets studied. For example, the abundance of Ca is negatively correlated to Si. It seems that a strong dilution of silicate-hosted elements occurs (particularly in the Upper Kimmeridge Clay of Dorset) is a direct result of the presence of Ca-bearing minerals. Subsequent X-ray diffraction proves this to be the case; those samples with high Ca abundances contain appreciable quantities of calcite. Other non-silicate-hosted variables (e.g. U) have no such behaviour; neither does the Al/(Al+Si) ratio. This illustrates how large local variations in carbonate content can mask the underlying trends in all

Table 1. *Renormalized Ca-free Kimmeridge clay formation mudrock analyses*

Magnus (Block 211/12A) samples											
	Si %	Al %	Ti %	K %	Fe %	S %	Gd ppm	Th ppm	U ppm	Th/K	Al/(Al+Si)
MAG–1	22.78	9.96	0.46	2.05	3.84	8.37	7.33	12.22	29.73	5.97	0.304
MAG–2	22.08	9.35	0.41	1.80	5.27	14.14	4.85	12.12	27.58	6.72	0.298
MAG–3	16.73	6.99	0.30	1.42	13.41	37.35	5.35	9.39	27.76	6.63	0.295
MAG–4	21.19	8.36	0.35	1.82	7.29	20.46	5.17	10.13	24.72	5.58	0.283
MAG–5	23.40	9.29	0.39	1.99	3.62	8.62	4.49	12.25	24.19	6.15	0.284
MAG–6	22.66	9.53	0.40	2.04	3.58	8.93	5.29	12.21	24.72	6.00	0.296
MAG–7	23.71	9.38	0.39	2.02	3.88	9.00	6.03	12.26	25.74	6.07	0.283
MAG–8	19.13	7.40	0.31	1.60	10.35	31.54	4.06	10.04	20.08	6.28	0.279
MAG–9	20.64	8.14	0.35	1.75	7.62	23.35	4.89	11.20	24.54	6.40	0.283
MAG–10	20.88	7.43	0.32	1.66	7.40	23.93	6.62	10.18	24.85	6.15	0.263
Magnus sandstone member											
MAG–11	19.85	9.94	0.43	2.44	8.29	21.91	5.23	15.08	18.59	6.17	0.334
MAG–12	21.83	9.42	0.42	2.17	5.65	14.87	4.75	14.16	27.82	6.51	0.301
MAG–13	19.85	8.82	0.36	2.22	8.36	24.12	6.13	13.06	18.29	5.89	0.308
MAG–14	20.41	8.08	0.34	2.23	9.27	25.51	3.72	12.05	15.47	5.39	0.284
MAG–15	20.65	8.19	0.34	2.18	9.27	25.43	4.72	12.06	15.48	5.54	0.284
MAG–16	19.56	8.13	0.34	2.22	10.81	27.32	4.22	12.05	15.57	5.44	0.294
MAG–17	27.73	7.61	0.34	2.42	4.41	9.69	4.76	11.15	9.52	4.60	0.215
Magnus sandstone member											
MAG–18	26.91	7.68	0.36	2.11	4.51	9.95	6.90	13.19	9.94	6.27	0.222
MAG–19	24.44	8.60	0.42	1.85	4.43	10.10	5.05	14.14	12.63	7.63	0.260
MAG–20	26.86	6.68	0.4	1.39	3.61	7.33	5.66	10.11	9.00	7.26	0.199
MAG–21	25.50	8.90	0.44	2.06	3.82	7.82	5.27	12.16	8.31	5.90	0.259
MAG–22	26.21	8.61	0.41	1.95	3.67	6.97	5.80	13.22	7.93	6.78	0.247
MAG–23	25.75	8.71	0.42	1.94	3.62	7.52	6.57	13.13	7.07	6.78	0.253
MAG–24	26.02	9.04	0.45	2.03	3.56	3.82	7.28	13.14	5.67	6.47	0.258
MAG–25	25.46	8.74	0.55	2.21	4.35	7.54	6.18	14.19	8.41	6.41	0.256
MAG–26	27.19	8.87	0.43	1.98	3.31	3.31	8.64	12.19	6.10	6.15	0.246
MAG–27	22.37	8.71	0.36	1.51	6.64	17.37	6.23	12.05	18.07	7.99	0.280
MAG–28	21.35	8.77	0.38	1.63	7.09	18.39	7.74	14.07	19.80	8.61	0.291
MAG–29	12.79	6.66	0.28	1.35	8.80	28.79	4.33	9.46	15.80	6.99	0.252
MAG–30	21.09	7.34	0.32	1.53	10.26	25.55	4.73	11.06	12.57	7.24	0.258
MAG–31	26.03	7.27	0.28	1.63	5.27	14.44	5.05	9.80	11.11	5.99	0.218
Kimmeridge Bay (*autissiodorensis* zone) samples											
KB–1	25.21	10.99	0.54	2.68	4.53	2.04	4.93	12.91	4.34	4.82	0.304
KB–3	25.04	10.09	0.57	2.77	4.92	2.45	5.58	12.64	4.95	4.56	0.287
KB–5	26.22	10.22	0.57	2.81	3.22	1.86	3.56	12.21	3.26	1.34	0.281
KB–7	25.29	11.07	0.57	2.73	3.30	1.82	3.35	13.21	4.06	4.83	0.304
KB–9	21.07	9.46	0.48	2.40	5.02	4.20	6.95	11.39	5.01	4.74	0.310
KB–11	23.45	9.62	0.50	2.29	4.29	3.07	3.59	10.89	4.14	4.76	0.291
KB–13	26.40	9.63	0.58	2.75	3.82	2.54	6.10	13.92	5.03	5.05	0.267
KB–15	17.71	7.84	0.40	2.01	4.29	13.37	5.18	9.03	9.87	4.50	0.307
KB–17	28.59	8.08	0.58	2.39	3.18	2.25	8.01	11.75	4.27	4.91	0.220
KB–19	23.58	9.64	0.49	2.67	5.78	2.02	6.03	12.07	3.54	4.52	0.290
KB–21	28.40	9.08	0.58	2.70	3.49	2.00	4.97	11.88	3.56	4.40	0.242
KB–23	24.79	10.01	0.54	2.62	3.35	2.41	3.89	11.56	3.78	4.42	0.289
KB–25	28.20	9.18	0.61	2.47	2.84	1.69	5.24	11.30	3.59	4.57	0.246
KB–27	20.26	8.71	0.44	2.07	7.26	5.10	5.03	10.74	5.17	5.20	0.301
KB–29	26.08	9.96	0.59	2.68	2.89	1.87	5.45	12.34	3.39	4.60	0.276
Freshwater steps (*pectinatus* zone) samples											
FS–1	15.82	6.42	0.35	1.72	3.53	19.92	2.45	8.52	7.24	4.97	0.289
FS–2	19.70	8.30	0.35	2.00	5.46	7.81	3.01	9.64	7.83	4.82	0.296
FS–3	25.81	9.51	0.49	2.72	4.76	2.45	4.72	12.08	5.75	4.45	0.269
FS–4	21.50	8.58	0.45	2.21	3.97	9.12	5.23	10.07	6.54	4.55	0.285
FS–5	23.55	9.36	0.34	1.77	8.57	11.06	5.34	10.02	7.68	5.65	0.284
FS–6	22.81	9.40	0.36	2.25	6.97	8.38	2.87	10.71	8.62	4.75	0.292
FS–7	9.61	3.46	0.15	0.33	6.82	34.75	2.14	5.00	16.43	15.33	0.265
FS–8	11.53	5.39	0.26	1.16	3.32	25.93	6.76	6.48	9.93	5.61	0.319
FS–9	23.31	10.37	0.47	3.21	5.60	5.37	4.40	11.36	4.40	3.54	0.309
FS–10	10.54	4.87	0.24	0.94	4.88	31.65	2.82	5.80	7.36	6.19	0.316
FS–11	21.47	8.59	0.32	1.53	7.29	10.22	3.54	9.90	15.56	6.49	0.286
FS–12	21.35	9.29	0.34	1.31	8.15	9.36	4.85	10.60	15.14	8.11	0.303
FS–13	22.77	9.67	0.39	1.93	6.43	7.48	1.53	10.42	9.50	5.39	0.298

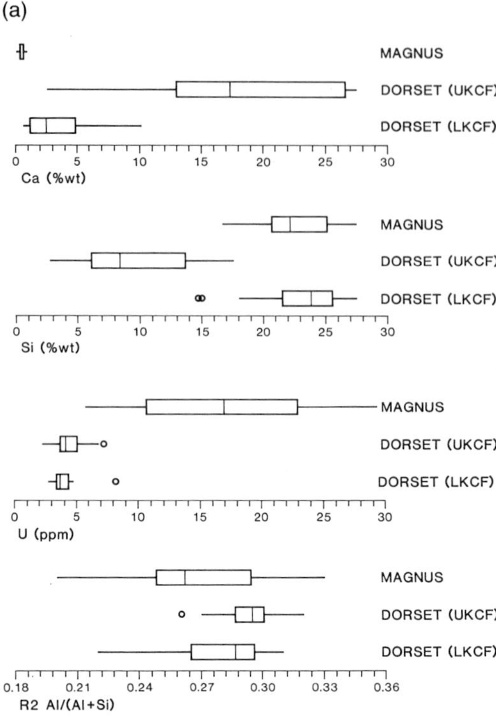

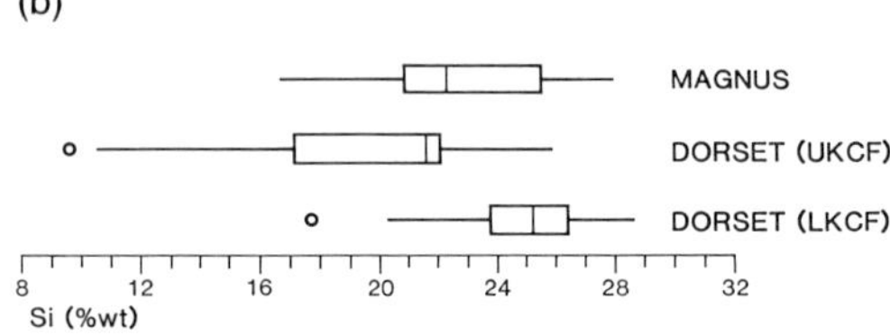

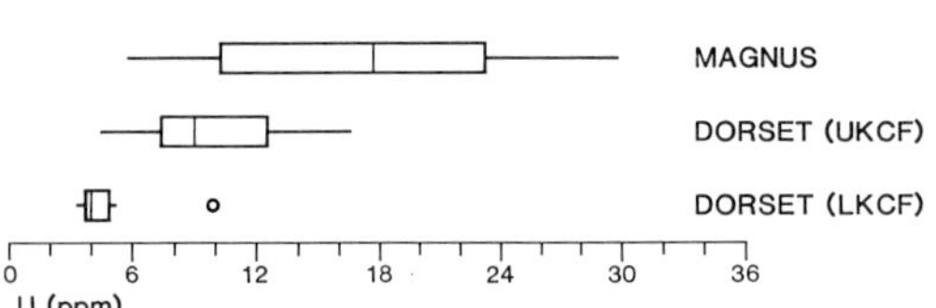

Fig. 2. (a) Boxplot representation of raw elemental abundances of selected elements from the three sub-groups constituting the dataset (boxes enclose 25% to 75% of the data values; thin line within box, median; whiskers, effective range; circles, single point outliers). (b) Boxplot representation of renormalised Ca-free elemental abundances of Si and U from the three sub-groups illustrated in Fig. 2a (symbols as in Fig. 2a).

the other elements. A renormalized Ca-free data set eliminates the effects of calcite dilution and enables better comparison of material from separate geographical locations. Evidently the occurrence of other carbonate minerals (e.g. ankerite or dolomite) would require alternative renormalization methods to overcome carbonate dilution effects. Table 1 lists the renormalized data for eleven 'loggable' variables comprising nine elements and two additional ratios (Al/(Al+Si) and Th/K). Visual assessment of the renormalized data using similar boxplots shows that of the eleven variables in Table 1, Si, Al, Ti, Fe, S, U and Th/K account for the majority of the variance observed. Examples of renormalized data are shown in Fig. 2b.

Progressive cross plotting of one variable against another can expose some important trends and differences between the sample sets studied. Figure 3 shows one particular example where trends in the abundance of Fe and S may be interpreted in terms of different plausible geological processes. Three trends can be identified. Data points in the central linear trend are mainly from the Magnus samples and illustrate a specific (but non-stoichiometric) relationship between Fe and S. A similar relationship was noted by Rainey (1987) who suggested that the excess S (i.e. that over the amount required to form the abundant pyrite identified by X-ray diffraction) was present in the native, elemental form. Rainey suggests that S, in addition to that being produced by the microbial activity of sulphate reducing bacteria (Berner 1971), was produced by the oxidation of H_2S in anoxic bottom waters at the interface with oxygenated water. It appears that the rate of Fe supply, although high, was not only insufficient to convert all of the elemental S to pyrite but was also controlled at such a rate that the amount of excess S remaining was proportional to the amount of pyrite precipitated (such that a linear Fe–S relationship was maintained). There is no apparent explanation as to why this should be so.

The samples from Dorset show a different picture. In the older, Lower Kimmeridge Clay, Fe is present in amounts greater than that stoichiometrically held in pyrite. Either there was insufficient sulphate reduction taking place as a result of too little organic matter being available, or that the rate of Fe^{2+} supply from the reduction of Fe^{3+} oxides and hydroxides by bacterial action or the halmyrolysis of Fe-bearing clay minerals (e.g. chlorite, phengitic micas) was too slow (Berner 1971). However, by the time of deposition of the Upper

Kimmeridge Clay, intervals appear where the supply of S greatly outstripped the supply of Fe, both in terms of that required for stoichiometric pyrite formation and that seen in the systematic trend observed in the Magnus samples. It is also interesting to note that in contrast to the Magnus samples, these particular samples appeared to contain more organic matter (confirmed by the highest losses on ignition values recorded during analysis of the dataset) implying that the excess S may be held as organic S instead of native S.

In terms of bottom water conditions, rate of deposition and rate of early burial, the Magnus mudrock section has experienced a very specific, yet consistent depositional and early burial history, whereas the Dorset section records a marked fluctuation in local conditions. This is confirmed in Fig. 3 by considering a third component, uranium. In the Dorset samples U abundance increases with increasing Fe and S content. This pattern is not seen in Magnus, where a high U abundance occurs independently of Fe and S content. One hypothesis may be to relate the euxinic nature of the Magnus depositional basin to the specific type of organic matter present (i.e. terrigenous or marine) and a critical rate of sedimentation and early burial such that rates were perhaps neither too fast (insufficient time for sulphide precipitation) or too slow (for excessive degradation of organic matter and loss of U to occur) (Curtis 1980; Rainey 1987; Myers 1987). An exhaustive discussion on the topic of specific Fe, S and U geochemistry with respect to the Kimmeridge Clay Formation is outside the scope of this paper. However, what is important is that geochemical logging information can indicate differences in Fe, S, and U geochemistry which are then are open to interpretation in terms of the environment of deposition of these mudrocks.

Simultaneous examination of more variables requires the use of a multivariate statistical technique. With an array of n samples which has m logged variables, it is possible to define some measure of resemblance between each of the samples (see Davis 1986) for a full discussion on this class of statistical methods). In this study, the Euclidian distance in m-dimensional space was represented in two dimensions using non-linear mapping (NLM) (Howarth 1973). The principle of the technique is illustrated in the inset to Fig. 4 which also shows the results of using NLM on the seven loggable variables recognised as accounting for the majority of the spread in the Ca-free dataset. Samples should cluster together according to their degree of geochemical similarity; consequently the relationships identified in Fig. 3 reappear as trends in the NLM analysis. Although more variables are involved, the similarity in the nature of the clustering may indicate that the actual processes invoked to explain data distribution in Fig. 3 may similarly control the other four (silicate hosted) variables included in the multivariate analysis. Corresponding cross-plots of NLM coordinates and specific variables could be used to establish the physical significance of the clustering seen in Fig. 4.

Inclusion of the additional variables exposes further sub-grouping within the Magnus sample set. This was investigated using a varimax

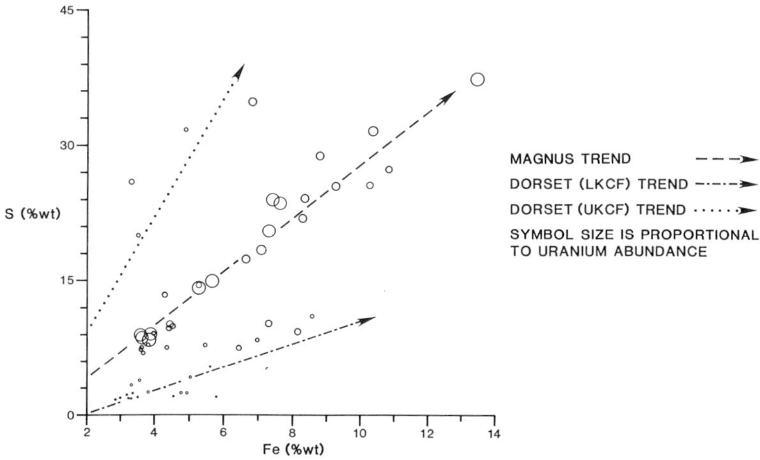

Fig. 3. Cross plot of abundance of Fe with respect to S illustrating three distinct trends and their relationship to U abundance.

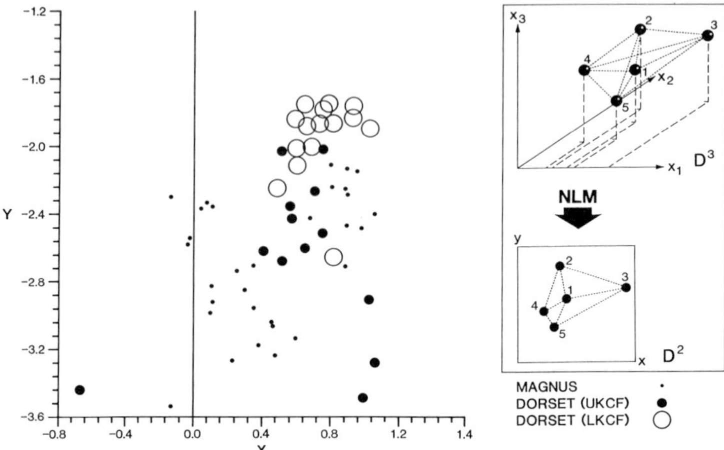

Fig. 4. Multivariate analysis of whole dataset reduced to two dimensions (X & Y) by a non-linear mapping (NLM) procedure (Howarth 1973). Inset shows the principal of the operation in reducing three variables to two for simple graphical representation.

rotated factor analysis technique (Davis 1986). This technique allows correlated variables to be linearly transformed into a new set of independent variables, or factors. The factors are derived in decreasing order of importance such that the first factor accounts for the most variation in the original data. If the original variables are highly correlated, the effect of factor analysis is to describe variations in the original data using fewer parameters. In applying this technique to the Magnus sample set, all loggable variables were used (but again on a Ca-free renormalized basis) leaving 11 variables with which to construct the correlation matrix. Four factors (which accounted for over 90% of the total variance of the Magnus subset) were extracted from the correlation matrix (Table 2). Factor scores were then calculated with respect to each factor for each sample. These identify the sample in terms of a linear combination of the original variables weighted by their contributions (i.e. loadings) to a given factor. Fig. 5 shows a biplot of the first and second factor scores (plotted as crosses) together with the factor loadings (the broken lines — labelled for each variable). As might be expected, there appears to be a clear delineation in the factor loadings between mainly silicate hosted and specific non-silicate hosted (e.g. Fe, S and U)

Table 2. *Varimax rotated factor loadings for the first four factors calculated from the Magnus sample set*

	Factor 1	2	3	4
Si	0.842	0.046	0.047	−0.518
Al	0.372	0.768	0.083	0.471
Ti	0.570	0.692	−0.077	−0.008
K	0.103	0.738	0.643	0.058
Fe	−0.955	−0.185	−0.016	−0.098
S	−0.933	−0.275	−0.005	0.186
Gd	0.556	0.218	−0.441	−0.042
Th	0.178	0.944	−0.120	−0.083
U	−0.206	−0.106	0.028	0.943
Th/K	0.015	0.019	−0.969	−0.043
Al/(Al+Si)	−0.412	0.532	0.029	0.720

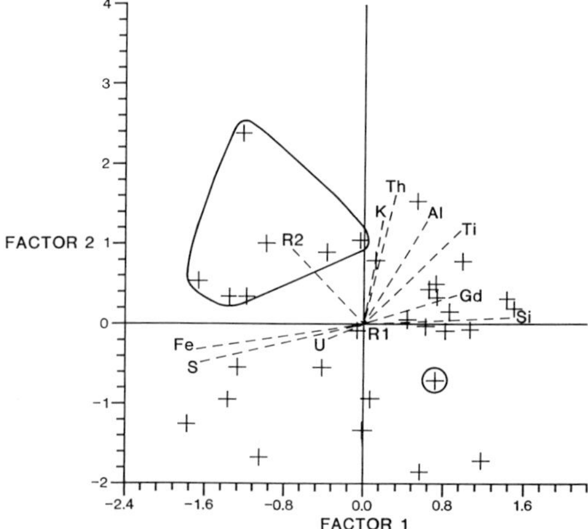

Fig. 5. Plot of factor scores (crosses) and varimax rotated factor loadings (dashed lines) for the Magnus subset of samples. Samples from the mudrock within the Magnus Sandstone are enclosed by solid lines (N. B. R1 Th/K ratio; R2, Al/(Al+Si) ratio).

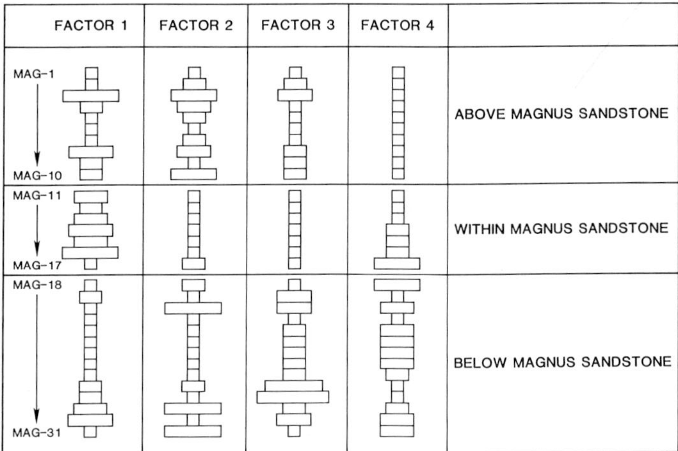

Fig. 6. Factor scores for the first four factors extracted from the Magnus subset of samples arranged in order of increasing sample depth (top to bottom). Column width is proportional to factor score.

variables. Similarly, all the factor scores associated with the mudrock within the Magnus Sandstone appear (with one outlier) to cluster in distinctive fashion in that quadrant of the plot around the single silicate-hosted factor loading (Al/(Al+Si)ratio), R2.

Herron (1986) employed a similar technique to relate four independent factors (which accounted for 86% of the total variance in his dataset) to four specific minerals. He then used this association to establish a minimal-input model whereby the abundances of three of the minerals (K-feldspar, illite and kaolinite) could be calculated from the abundances of just three elements (K, Fe and Al respectively), each of which were uniquely related to one of the four initially extracted principal factors.

No attempt was made in the present study to quantitatively relate factors or factor loadings directly to mineralogy through a minimal-input model. Semi-quantitative evaluation of X-ray diffraction data did provide tentative proof that the middle mudstone unit may contain less kaolinite and more mixed layer illite-smectite than either of the mudstone units above or below the Magnus Sandstone. Apart from this, subsequent X-ray diffraction work could not provide any other clear mineralogical reasons for the clustering seen in Fig. 5.

Examination of all four factor scores (Fig. 6) and factor loadings (Table 2) embellishes the observations noted above. This schematic plot shows that the principal factor alone neatly delineates the mudrock within the Magnus Sandstone (this time on the basis of a high positive loading due to Si and negative loadings due to Fe and S). Similarly, factors three (with a negative loading due to Th/K) and four (with a high positive loading due to U) highlight the mudrock below the Magnus Sandstone.

This last example shows that substantial additional structure has been uncovered by exposing the sub-sample of data from a single well to further statistical scrutiny. The approach can be envisaged as an iterative process. Obtaining more detailed statistically derived information over smaller intervals or units increases the accumulated data for incorporation into the emergent geological framework. The advantage of this rather more flexible approach is that it can be undertaken without the constraints imposed by a minimal-input model which forces the elemental analysis to be described in terms of a few normatively derived minerals.

Conclusions

This paper outlines an approach geologists could consider for the interpretation of raw downhole geochemical data. The ten elements currently measureable downhole yield sufficient variation for statistical manipulation to characterize different mudrock units. Scope clearly exists for applying geochemical logging technology to hitherto unfashionable or impractical topics, such as the sub-surface discrimination of mudrock facies. Multiwell studies could do much to enhance zonation and correlation basinwide and with improved technology, additional log-derived trace elements could further refine statistical manipulation. There are additional practical advantages in this approach over traditional studies on core or cuttings. For instance, there is no chance of bias through irregular or unrepresentative sampling and a much larger volume of rock is investigated per data point by this means. Nevertheless, it is essential that the appropriate environmental corrections are applied to ensure the accuracy of the measurements. In conclusion, geochemical logging represents a significant advance in the application of wireline logs to geology.

We wish to thank the British Petroleum Company plc. and Partners in the Magnus Field for permission to publish this paper. We particularly wish to thank Dr R. J. Howarth (BP) for his help and guidance on the various statistical applications used and Dr J. M. Huggett (BP) for supplying the XRD analyses.

References

BERNER, R. A. 1971. *Principles of Chemical Sedimentology*. McGraw-Hill, New York.

BJØRLYKKE, K. 1974. Geochemical and mineralogical influence of Ordovician island arcs on epicontinental clastic sedimentation. A study of Lower Paleozoic sedimentation in the Oslo region, Norway. *Sedimentology*, **21**, 251–272.

CURTIS, C. D. 1980. Diagenetic alteration in black shales. *Journal of the Geological Society, London*, **137**, 189–194.

DAVIS, J. C. 1986. *Statistics and Data Analysis in Geology* (2nd Edition). Wiley, New York.

GRAU, J. A., ELLIS, D. V., HERRON, M. M. & SCHWEITZER, J. S. 1987. Geochemical logging in the Siljan Ring. *In*: *3rd International Symposium on Observation of the Continental Crust through Drilling*. Mora & Orsa, Sweden.

HERRON, M. M. 1986. Mineralogy from geochemical well logging. *Clays and Clay Minerals*, **34**, 204–213.

—— 1987. Future applications of elemental concentrations from geophysical logging. *Nuclear Geophysics*, **1**, 197–211.

—— 1988. Geochemical classification of terrigenous sands and shales from core or log data. *Journal of Sedimentary Petrology*, **58**, 820–829.

HERRON, S. L. 1987. A total organic carbon log for source rock evaluation. *Log Analyst*, **28**, 520–527.

HERTZOG, R., COLSON, L., SEEMAN, B., O'BRIEN, M., SCOTT, H., MCKEON, D., WRAIGHT, P., GRAU, J., ELLIS, D., SCHWEITZER, J. & HERRON, M. 1987. Geochemical logging with spectrometry tools. *SPE Paper 16792*.

HOWARTH, R. J. 1973. Preliminary assessment of a Nonlinear Mapping Algorithm in a geological context. *Mathematical Geology*, **5**, 39–57.

MYERS, K. J. 1987. Onshore-outcrop gamma-ray spectrometry as a tool in sedimentological studies. *Ph.D. thesis*, University of London.

—— & WIGNALL, P. B. 1987. Understanding Jurassic organic-rich mudrocks- new concepts using gamma-ray spectrometry and palaeoecology: examples from the Kimmeridge Clay of Dorset and Jet Rock of Yorkshire. *In*: LEGGETT J. K. & ZUFFA G. G. (eds) *Marine Clastic Sedimentology: Concepts and Case Studies* Graham & Trotman, London, 172–189.

PETTIJOHN, F. J. 1975 *Sedimentary Rocks* (3rd Edition). Harper & Row, New York.

POTTER, P. E., MAYNARD, J. B. & PRYOR, W. A. 1980. *Sedimentology of Shale: Study Guide and Reference Source*. Springer, New York.

RAINEY, S. C. R. 1987. Sedimentology, diagenesis and geochemistry of the Magnus Sandstone Member, Northern North Sea. *Ph.D. thesis*, University of Edinburgh.

Natural gamma-ray spectrometry in hydrocarbon-bearing sandstones from the Norwegian Continental Shelf

ANDREW HURST
Statoil, Postboks 300 Forus, 4001 Stavanger, Norway

Abstract: Concentrations of thorium (Th), uranium (U) and potassium (K) are measured using natural gamma-ray spectrometry (NGS) logs. The precision of NGS measurements in sandstones varies: K measurement is in general moderately good; Th measurement varies with the service company involved; U measurement is unreliable. Concentrations of Th, U and K in standard tables of minerals used commonly in log evaluation are not representative of the actual elemental concentrations found in sandstone minerals. Th/K cross-plots are shown to be unsuitable for the identification of clay minerals in sandstones. Illite cement cannot under normal geological circumstances be identified from the K content of the rock. Authigenic kaolinite has no particular affinity for concentration of Th. Th in sandstones is concentrated predominantly in detrital heavy minerals with up to 20% of the total Th present in silt-sized and finer authigenic REE-phosphates and Th salts. Variation in the K content of micas from some Jurassic sandstones demonstrates the need for using formation-specific chemical data for the minerals of interest (micas, feldspars, clay minerals) as input parameters in normative calculations. Appropriate statistical control of the validity of Th, U and K contents for specific minerals is required.

Natural gamma-ray spectrometry (NGS) logs measure the concentrations of the three most common naturally occurring radioactive elements, potassium, thorium and uranium. NGS logs have been a regular feature of wireline log suites for over ten years and many publications (e.g. Hassan *et al.* 1976; Marett *et al.* 1976; Fertl 1979; Suau & Spurlin 1982) have proclaimed the value of NGS logs for the identification of minerals. Generally low natural radioactivity in sedimentary rocks, combined with measurement in an often hostile borehole environment, gives an inherent uncertainty to the measurements. In the geological literature however, there are examples of optimism regarding the application of NGS data to sedimentary rock descriptions (Gautier 1983; Peveraro & Russell 1984; Herron 1986; Myers & Wignall 1987). None of these take a critical standpoint regarding the validity of standard radioactive elemental data for commonly occurring sedimentary minerals.

Accuracy and precision of NGS logs, and the comparability of data from NGS logs run by different service companies, has been subject to criticism (Wiley & Zittel 1982; Dion & Allen 1985; Heinze 1985). Accuracy of NGS measurements can be evaluated both by comparison of log data with laboratory gamma-ray spectrometers and by comparison with laboratory geochemical analyses. In both cases, laboratory measurements may vary by a factor of two or more compared with the log data and, considerable variation occurs in the NGS data from different service companies. Accuracy is, however, difficult to evaluate between core and log data because of the large differences in sample size. The precision of NGS log measurements is also criticised (Wiley & Zittel 1982; Dion & Allen 1985) because the repeatability of logs in open-hole conditions is apparently very poor over short ($\leqslant$1.5 m) intervals. Analytical precision is controlled ultimately by the random nature of the radioactive decay process and hence counting statistics. Measured values of radioactivity are thus a function of the radioactivity of the formation and the logging speed. Poor repeatability is attributable to low count rates associated with low concentrations of Th, U and K.

Identification of the clay mineralogy of sandstones is in particular considered tenable from NGS data (Fertl 1983; Herron 1986). This is important in the petroleum industry because of the known relationships between clay content, type and texture and water saturation and permeability. Quantification of the clay mineralogy of sandstone reservoirs is difficult using traditional geological methods (Hurst & Archer 1986; Hurst 1987) and derivation of mineralogy from log data may be at least as reliable (Flølo 1988; Herron & Herron 1990). However, the use of geochemical data from logs for deriving mineralogy is strongly dependent on the use of suitable chemical compositions for the minerals defined in the normative calculations. Several reviews of mineral parameters are found in the petrophysical literature (Edmundson & Raymer

From Hurst, A., Lovell, M. A. & Morton, A. C. (eds), 1990, *Geological Applications of Wireline Logs* Geological Society Special Publication No. 48, pp. 211–222

1979; Fertl 1979; Ellis *et al.* 1988) which have to date provided the input for normative mineral calculations.

The validity of some of the standard mineral compositional data are evaluated here and their relevance to calculation of sandstone mineralogy from NGS data assessed. Comparison of NGS data from two service companies is presented including an evaluation of the precision of the measurements.

Methods and data

Before embarking on an analysis of NGS data, it is important to note that measurement of photoelectric absorption cross section (P_e) of the formation is usually impossible in Norwegian shelf wells because of the high barite content of drilling muds which may dominate the P_e measurement. Therefore, the promising combination of NGS and P_e data for mineral identification (Suau & Spurlin 1982; Peveraro & Russell 1984) is usually impossible.

NGS logs are interpreted to measure the natural gamma radiation from the radioisotopes K^{40}, Th^{232} and U^{238}. For example, Th^{232} and U^{238} are alpha emitters, their daughter series being the source of the gamma emission spectra (Tittman 1986). The emission spectra of Th, U and K give discrete lines (Fig. 1a). In practise, however, line broadening and valley filling (Fig. 1b) are caused both by the poor resolution of scintillation detectors (Tittman 1986) and by the effects of multiple Compton scattering of the gamma rays prior to reaching the detector (Bertozzi *et al.* 1981).

Conversion of emission spectra into elemental concentrations is made by dividing the spectra with a series of pulse-height windows (Fig. 1b). A relationship exists between the partial gamma-ray flux of a given element at the detector and its mass fraction in the formation. Logging tools may thus be calibrated in artificial formations with known elemental concentrations of Th, U and K.

Two major sources of error occur when using NGS logs in an offshore drilling environment. These are contamination due to the use of potassium-bearing additives in drilling mud (often potassium chloride) and adsorption of gamma rays in barite introduced to increase the mud weight. Correction factors can be applied for potassium contamination. High logging speed is a contributary error factor when gathering data which have low count rates.

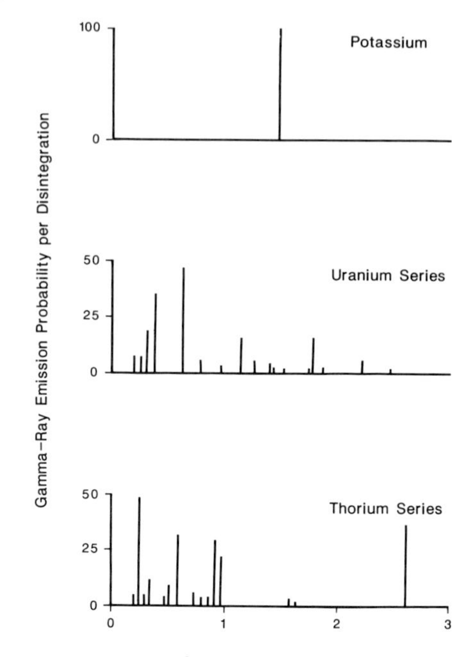

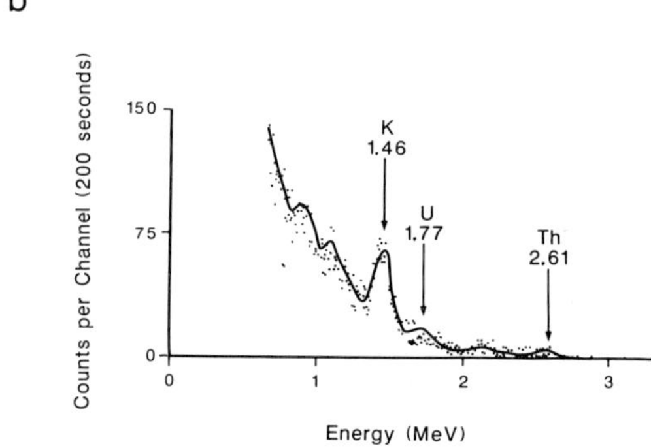

Fig. 1. (a) Characteristic emission spectra of potassium, uranium and thorium (after Tittman 1986). (b) A typical natural gamma-ray pulse height spectrum with the major radioactive sources marked (after Serra *et al.* 1980).

Accuracy and precision

Comparison between NGS elemental contents and chemical (e.g. neutron activation) analyses of Th, U and K give a favourable impression of the accuracy of NGS logs under North Sea conditions, despite the reservations expressed by Wiley & Zittel (1982) and Dion & Allen

(1985). The problems of comparing geological analyses, which are essentially point data, with running averages of data from logs which measure a large volume of a formation should not, however, be underplayed (Humphreys & Lott 1990).

The precision (repeatability) of NGS logs is assessed by comparing the main run with the repeat run section (Fig. 2). Several problems exist when using repeat run data. Most importantly, depth correlation between the two runs is problematic. It is reasonable to assume that even in the best possible situation some degree of error is present in a depth correlation between two runs. Tidal effects, wireline stretching and general movement on the wireline are the main sources of error.

An inherent problem when evaluating the precision of NGS measurements is that emission of natural radioactivity is random in nature (a Poisson distribution). Thus, measurements of radioactivity made on the same formation over different time periods will inevitably be different. In practice, the speed for NGS logging is determined by the recommended speed for running formation density logs (460 m h^{-1}). Variations in precision are thus caused by the random nature of radioactivity, the quality of the logging tool, logging speed (approximately constant in these examples) and in particular, the barite content of the drilling mud. No attempt to calibrate for mud composition is made.

If one assumes that the depth correlation is reliable (some minor error is invariably present) precision may be estimated by calculation of correlation coefficients for cross-plots of main run and repeat run data. Unfortunately, correlation coefficients are not well suited to giving an expression of precision in data sets with marked deviations from an ideal one-to-one relationship (Fig. 2a, U). Precision is here estimated by calculating a relative error (RELOR) $= \overline{A}/\overline{RS}$, the average distance of all data points from a 1:1 curve ($\overline{A}$) divided by the average value of all points from the repeat section ($\overline{RS}$). RELOR does not give an absolute value of precision (0 to 1) but the relative precision, and the potential usefulness of the different data can be assessed (Table 3). Clear differences are observed between the precision of Th, U and K data from the service companies "A" and "B". Thus, subsequent treatment of the data should not assume that elemental analyses from "A" and "B" from similar lithologies are directly comparable.

Applications and interpretation

Th and K concentrations and the Th/K ratio are the most important NGS data when characterizing sandstones. Th/K cross-plots may be used for the identification of clay mineralogy (Hassan & Hossin 1975; Hassan *et al.* 1976) and concentrations of both Th and K are useful as input parameters in calculations of bulk rock mineralogy (Everett *et al.* 1983; Herron 1986). Application of Th and K concentrations in the determination of mineralogy is dependent on the ability to partition the elements into minerals of known chemistry.

The geological data on which log-based interpretation of Th and K (and U) are based come largely from Murray & Adams (1958), Rodgers & Richardson (1964), Rodgers *et al.* (1969/1970) and Hassan & Hossin (1975). Additional data, generally regarding the Th, U and K content of minerals, are taken from Deer *et al.* (1966) and Clark *et al.* (1966). All chemical analyses of single minerals are from coarse-grained samples, i.e. pegmatites, kaolin pits, mineral veins, geodes; none are from single minerals separated from sandstones. Implicit is the assumption that sandstone minerals have compositions similar to the standard data (Table 1). Publications which attempt any critical assessment of the assumption are lacking, as is the application of analytical data from minerals separated from sandstones for the calibration of NGS data.

Potassium in sandstones

Table 1 lists the commonly occurring potassic minerals in North Sea reservoir sandstones. It is reasonable to expect that the total potassium content (K_T) of a sandstone is closely related to the abundance of some or all of those minerals. Illitic cements in reservoir sandstones have a destructive effect on permeability. Small volumes of authigenic illite (<2 wt%) can, because of their highly microporous character, virtually seal a rock with around 20% porosity (Pallatt *et al.* 1984). To be of any value for illite detection and quantification, a potassium log must be able to detect concentrations of <2% illite in the presence of any other potassic minerals.

North Sea reservoir sandstones may typically contain a few to 20% potassium feldspar, and anything from ≤1% to 20% mica. If one supposes a potassium feldspar content of 10% and an illite content of 2%, what contribution to $K_{T,sandstone}$ does each mineral make? In the absence of chemical analyses of K-feldspars and

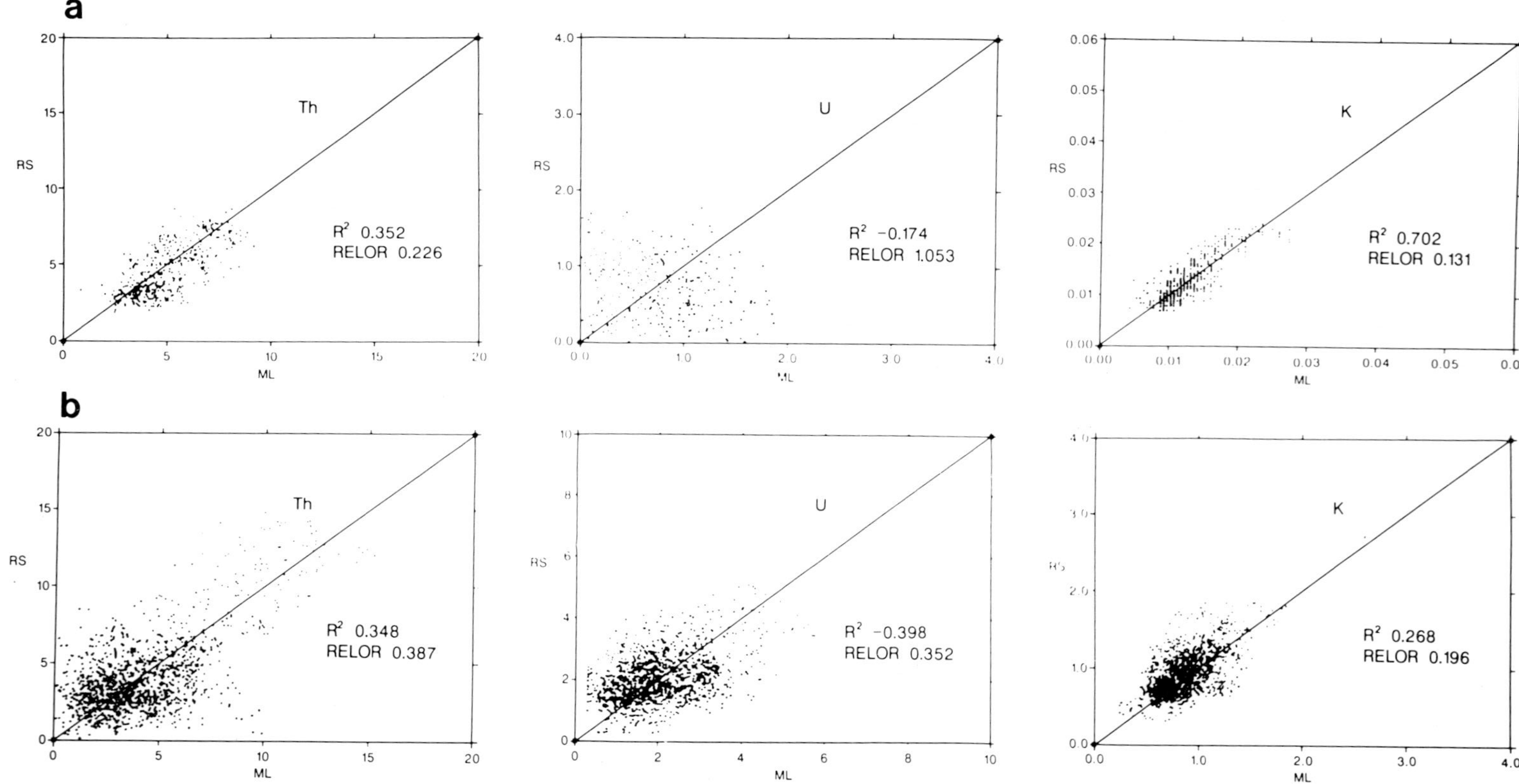

Fig. 2. Cross-plots of main log (ML) and repeat run sections (RS) for Th, U and K for service companies 'A' (a) and 'B' (b) operating on the Norwegian Continental Shelf. R^2 = root mean squared correlation coefficient. RELOR = relative error (see text for explanation) used for grading precision. For a perfect match of ML and RS, RELOR = 0 (see Table 2).

illites from a specific North Sea reservoir independent data for K-feldspars (Cocco *et al.* 1969) and illites (Nadeau & Bain 1986) are used as follows: K-feldspar, (a) K_T in 146 K-feldspars from rocks other than pegmatites gives a $\bar{x}$ = 10.68, (b) K_T in 254 K-feldspars from pegmatites gives a $\bar{x}$ = 12.13, or when (a) and (b) are combined a mean K_T for K-feldspar of $\bar{x}$ = 11.4% with σ = 1.99%; 12 illites from sedimentary rocks gives a mean K_T, $\bar{x}$ = 5.2 with σ = 1.51. The mean K_T from K-feldspar analyses is thus around 3% lower than the value quoted by Ellis *et al.* (1988). 10% K-feldspar contributes approximately 1.2% K (σ = 0.2%) and 2% illite contributes approximately 0.1% K to $K_{T,sandstone}$. $K_{T,illite}$ is in fact less than the standard deviation of the K_T of K-feldspar (Fig. 3). Thus, use of K_T for identifying illite in the presence of K-feldspar is impractical, unless illite is approximately an order of magnitude more abundant than K-feldspar.

A major weakness of the published K_T data (Edmundson & Raymer 1979; Fertl 1979; Serra 1984; Ellis *et al.* 1988) is that their statistical and the geological validity is not investigated. Neither the number of analyses on which the compositional ranges are based, nor mean values and standard deviations are given; thus the data have no proven statistical validity. Comparison of K_T from muscovites and biotites from sandstones (Fig. 4) with the standard K_T values (Table 1) reveals that the standard values are generally higher than the actual values. When the data set for both micas are broken down into data sets from specific formations (Table 5) it is clear that even for sandstones of similar age there can be considerable variation in the K_T of micas. The mica data (Fig. 4) are a very restricted data set which should not be regarded as globally applicable. However, they represent many more analyses than the database of Table 1.

Table 1. *Th, U and K concentrations from some typical sandstone minerals*

Mineral	Th(ppm)	U(ppm)	K(%)
illite	10–25	1–5	3.5–8.3
kaolinite	6–47	1–12	0–0.6
chlorite	3–5		0–0.3
smectite	6–44	1–21	0–1.5
muscovite	0–25	2–8	7.8–9.8
biotite	0.5–50	1–40	6.2–10
K-feldspar	3–12	0.2–3	10.5–16
plagioclase			

Data from Clark *et al.* (1966). Deer *et al.* (1966) and Ellis *et al.* (1988).

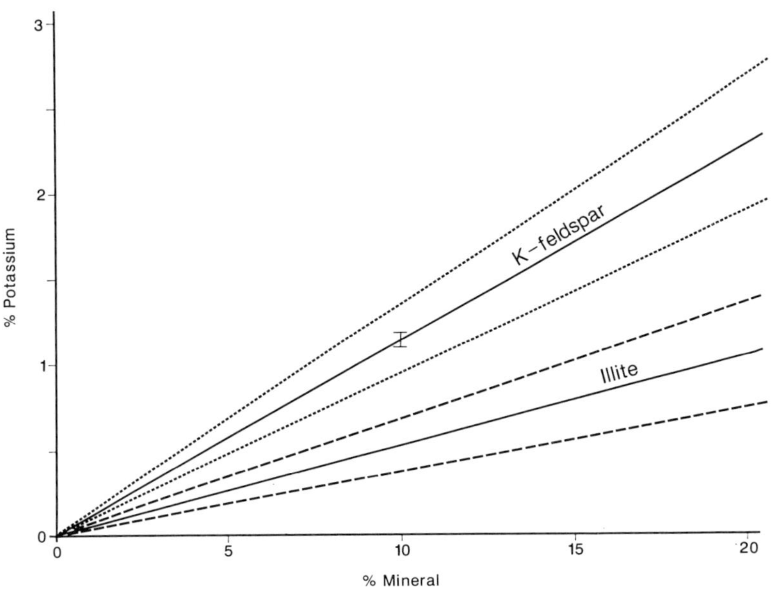

Fig. 3. Potassium contribution of K-feldspar and illite. Solid lines represent the mean K_T of the minerals, broken lines are single standard deviations from the mean. The scale-bar at 10% K on the mean K-feldspar line represents the likely range of % K from a 2% volume of illite. K-feldspar data are from Cocco *et al.* (1969). Illite data are from Nadeau & Bain (1986).

a

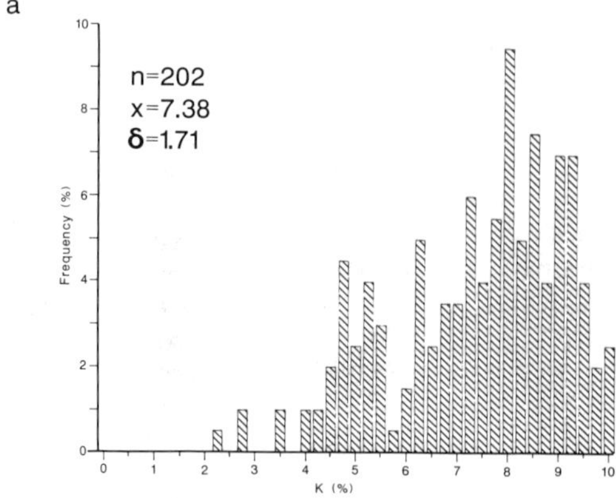

b

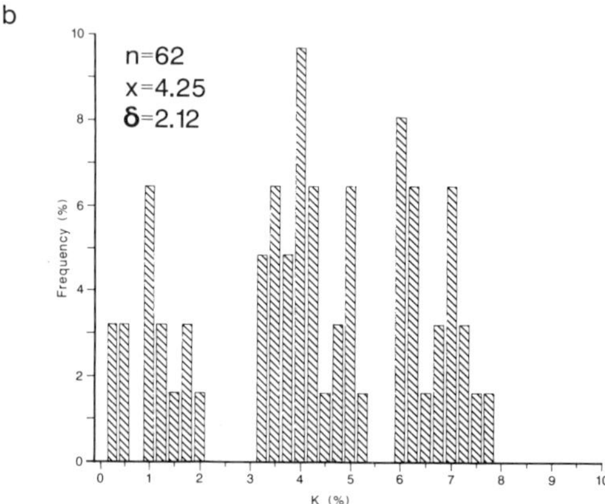

Fig. 4. Potassium contents of (a) muscovites and (b) biotites from some Jurassic sandstones (Hurst 1980). n = number of analyses, x =mean, σ = standard deviation. Data are from energy-dispersive microprobe analyses.

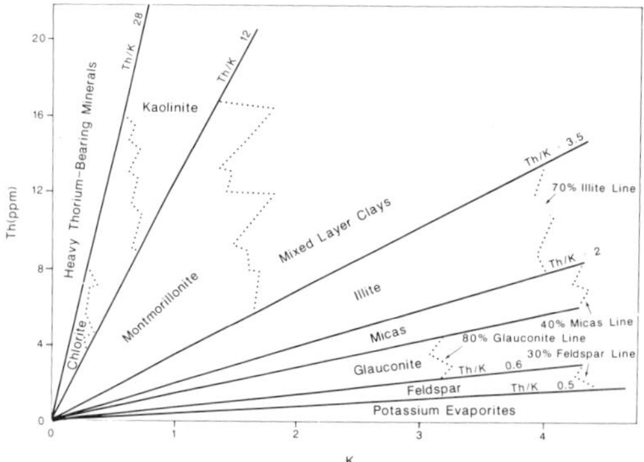

Fig. 5. Thorium (Th) versus potassium (K) plot for identification and quantification of clay mineralogy (after Quirein *et al.* 1982).

Thorium: potassium (Th/K)

Application of Th/K cross plot to mineral identification (Fig. 5) is an attractive method for identification of clay minerals. Quirein *et al.* (1982) construct their Th/K crossplot from standard mineral data (Edmundson & Raymer 1979), a data set which has been updated subsequently (Table 1). When the new data are plotted (Fig. 6) several complications are apparent, despite only plotting data for kaolinite, illite, muscovite and K-feldspar. Firstly, the compositional field for muscovite covers approximately half the compositional field of illite. If the data for biotite are included the mica compositional field spreads over the entire illite field and half of the K-feldspar field. Secondly, the kaolinite compositional field is much broader and has a boundary along the Th axis (K = 0%). A similar trend would be shown by chlorite but with a correspondingly lower Th content. As commented by Eslinger & Pevear (1988), the potassium contents of both kaolinite and chlorite are negligible. Therefore, the standard data (Table 1) and the cross-plots (Figs 5 & 6) are misleading.

Probably the greatest single weakness of Th/K cross-plots is that they are the product of bland generalizations about mineral chemistry. A better understanding of the validity of the cross-plots can be made by examining the database from which the plots are made. Without exception the chemical data for each mineral are sparse. Furthermore, in very few cases are Th, U and K measured from the same mineral sample. If compositional ranges (Table 1) are to be useful for mineral determination their statistical value should also be given, e.g. number of analyses, mean composition, standard deviation (Fig. 7). The identification of K in kaolinite and chlorite is caused both by the presence of micron-scale impurities, e.g. mica, and by the limitations (sample size, analytical accuracy) of the techniques used for analysis.

A second weakness of Th/K cross-plots is that the behaviour of Th and K in mineral chemistry are fundamentally different. With respect to clay minerals, Th is present as an adsorbed ion whereas K is included as part of the mineral structure. Some minerals have K_T and Th_T which are determined by their primary composition in a source area, with possible additions or subtractions caused during diagenesis. For example, diagenetic dissolution of feldspar and apatite will tend to change their K and Th contents respectively. Minerals which contain structural K and Th tend to have compositions which fall within given compositional ranges.

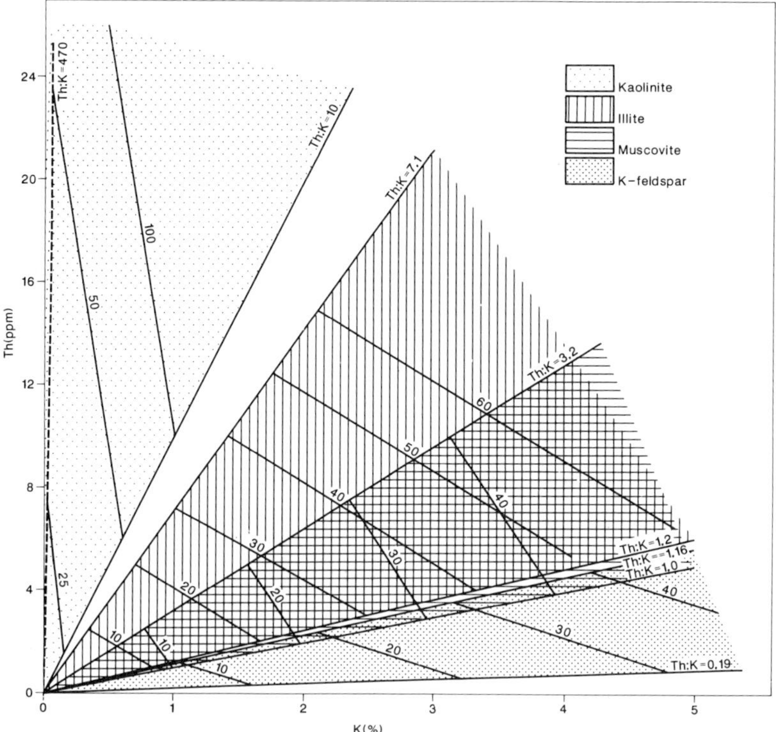

Fig. 6. Th versus K plot based on the updated data from Table 1. Note the considerable areas of overlap between potassic minerals and the broad kaolinite field. Kaolinite normally has a negligible K-content.

The Th_T of minerals which do not contain Th in structural sites, i.e. clays, feldspars, micas, varies dependent on the availability of dissolved Th in pore waters and the adsorption capacity of mineral surfaces, processes related closely to diagenesis and, thus, likely to vary in different sedimentary facies. The low solubility of Th in formation waters favours crystallisation of authigenic Th-salts near minerals releasing Th during diagenetic dissolution (e.g. apatite dissolution, Morton (1986)) rather than transport and eventual adsorption onto clay mineral surfaces. Detrital clay clasts are commonly not monomineralic. For example, detrital kaolinite may contain inclusions of resistate heavy minerals such as monazite or zircon (Myers & Bristow 1989) which give the 'clay' clasts high natural radioactivities.

Experiments to test the Th-adsorption capacity of sandstone minerals, performed on sandstones, prove that kaolinite has a very low adsorption capacity (similar to quartz) relative to micaceous minerals (Statoil, unpublished data). Further analytical work on samples from North Sea sandstones has proved that detrital heavy minerals contain approximately two orders of magnitude greater Th than any other minerals present. 10–20% of Th is found in the < 10 μm fraction (fine silt) located in authigenic REE-phosphate minerals or other Th salts. Kaolinites in which Th is detected by chemical analysis probably contain inclusions of silt or finer sized authigenic Th-rich minerals.

There are no sound reasons for expecting Th/K from NGS logs to be useful for identifying and quantifying the clay mineralogy of sandstones, despite the apparent success of the method (Fertl 1979; Quirein *et al.* 1982; Fertl 1983). Justification of a Th/K cross-plot with mineral compositional fields which delineate clay mineralogy (Fig. 5) is impossible (Table 2). A high degree of uncertainty should be applied to all compositional fields in Fig. 5 between the lines Th/K = 28 and Th/K = 0.5.

Discussion

Measurement

Attainment of high levels of precision for elemental analyses from NGS logs should be

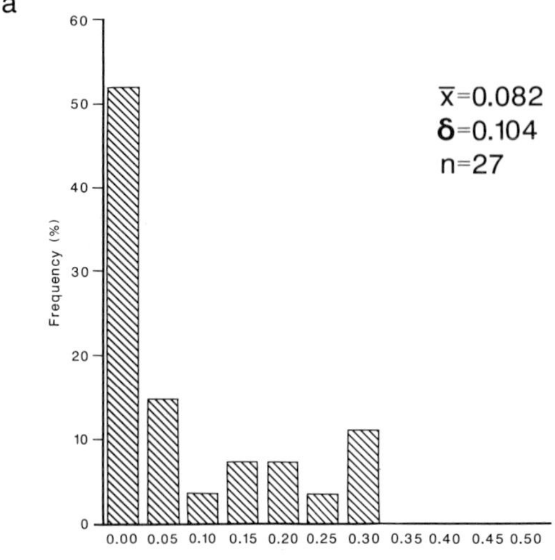

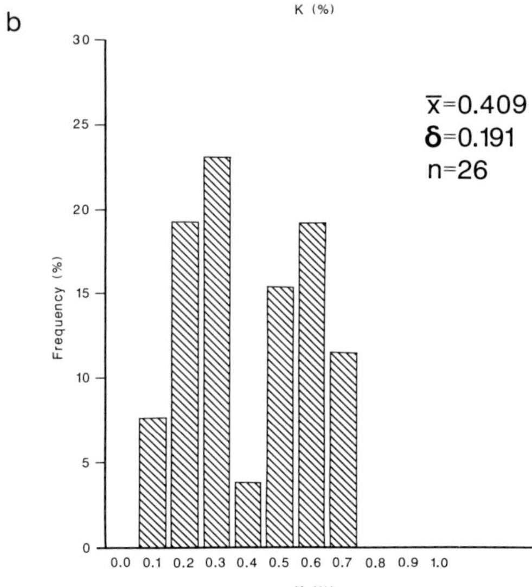

Fig. 7. Potassium contents of authigenic kaolinite (a) and chlorite (b) from Jurassic sandstones. Data are from energy-dispersive microprobe analyses. $\bar{x}$ = mean, σ = standard deviation, *n* = number of analyses.

Table 2. *Reasons for the lack of justification of Th/K cross plots for identification and quantification of the clay mineralogy of sandstones.*

- kaolinite and chlorite contain negligible K, which when present in detectable quantities has no relation to the mineral chemistry of either kaolinite or chlorite.
- Th tends to form discrete authigenic minerals of silt-size rather than adsorb onto clay mineral surfaces;
- the chemical behaviour of Th and K in minerals are unrelated;
- the existing data base for Th:K is inadequate and unrepresentative for sandstone characterisation;
- no statistical control (e.g. mean compositional values) of the Th- and K- contents of minerals is available.

Table 3. *Analytical precision as defined by RELOR (relative error)*

>0–0.150	very good
0.150–0.200	good
0.200–0.250	moderate
0.250–0.300	poor
>0.300	unreliable

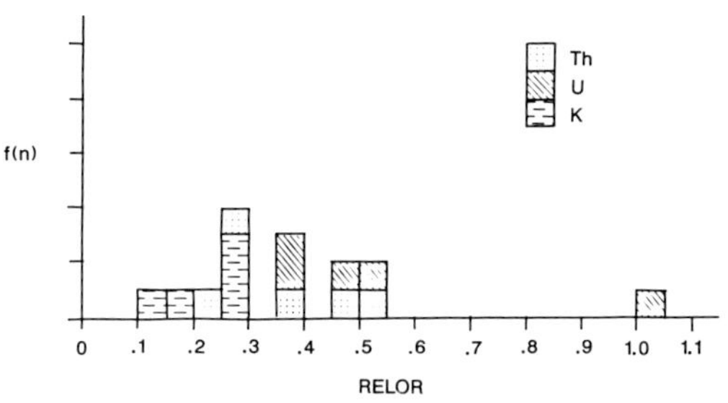

Fig. 8. Level of precision (RELOR) obtained for Th, U and K from NGS measurements. Data from Table 3. *n* = number of data = 15.

given highest priority. Absolute accuracy is less important so long as calibration of precise measurements is possible. However, satisfactory levels of precision are not readily attainable. It is reasonable to hope that two of the three elements should, for any one NGS log, have comparably good precision, with RELOR values of 0.250 or better (Table 3). Only 20% of the data in this study have moderate or better precision (Fig. 8). K analyses are in general better than Th and U, the latter of which is consistently unreliable. Most surprising (and worrying) are the low levels of precision for Th, an elemental concentration used widely in mineral identification (e.g. Quirein *et al.* 1982; Fertl 1983; Herron 1986). Significant to note, however, is that Th analyses by company 'A' have a far higher degree of precision than those of company 'B' (Table 4). For characterization of sandstone mineralogy Th and K contents are the most useful information to obtain. On the basis of the limited data analysed, company 'A' provide Th and K contents with an acceptable degree of precision (with some reservations about the precision of Th). Company 'B' how-

Table 4. *Ranges of precision (RELOR) from five repeat sections*

Well	Company	Th	U	K
1	'A'	0.298	0.491	0.275
2	'A'	0.226	1.053	0.131
3	'B'	0.492	0.541	0.262
4	'B'	0.387	0.352	0.196
5	'B'	0.541	0.352	0.297

ever, only approach a similar degree of precision for K. Use of measurements which have RELOR values greater than 0.250 is not recommended for use in mineralogical computations.

Having established the precision of the Th, U and K measurements, they then must be calibrated against actual contents of Th, U and K in the rock, i.e. accuracy determination (Willey & Zittel 1982; Dion & Allen 1985; Heinze 1985). It is possible that company 'B', despite producing analyses of low precision, actually attain better levels of accuracy than company 'A'. For the purposes of reservoir characterization, however, accuracy is less important than precision so long as the exact composition of the formation is not a major goal. As NGS data are normally going to be used in computations where other variables are present (e.g. Quirein *et al.* 1982; Herron 1986; Flølo 1988), it is most important that all variables present have similar levels of precision, or that variables are given a weighting determined by their relative precision.

Mineral chemistry

The inadequacy and probable unsuitability of standard compositional (Th, U, K) data (Table 1) for characterizing sandstone mineralogy are apparent from this study. Furthermore, the traditional application of compositional ratios (e.g. Th/K) for (clay) mineral identification seem to be unfounded. NGS logs may however, give elemental concentrations with high degrees of precision which should be very useful for determining mineralogy.

Probably the most useful method of applying NGS data is in combination with geochemical data from GST (gamma spectral tool) logs (Herron 1986; 1987; Fløló 1988). When log data are calibrated against core data quantitative mineralogy can be obtained by using multivariate statistical analysis (Herron 1986) or stochastic analysis (Quirein *et al.* 1986). The sandstones studied by Herron have rather simple mineralogy, and it remains questionable to what extent it is possible to define compositional end-members for the mineralogy under investigation.

There are at least two approaches to the problem of defining compositional end-members. It may be assumed that the minerals present have compositions which approximate to ideal compositional formulae, e.g. K-feldspar should contain approximately 14% K, muscovite should contain approximately 9.8% K etc. Alternatively, it may be assumed that the major minerals present do not have ideal compositions, that compositional ranges are present and that unique mineral compositional solutions will exist for any specific sandstone.

In the light of the success of initial studies which assume ideal mineral compositions (Herron 1986; Herron & Herron 1989) it is disarming to note that all the available analytical data conflict with the assumption. For example as quoted earlier, the global average K-feldspar contains 11.6% K and not 14% K. If ideal end-member compositions are forsaken and actual compositional data are used a change in methodology is required which allows incorporation of components with variable compositions. The methodology should include provision of frequency diagrams for mineral chemistry which either are 'global' (all available analyses) field, area, formation or even reservoir zone specific.

Examination of the data for micas from Jurassic sandstones from a restricted geographic area (Table 5) reveal a considerable range of K_T. Grouping data into a 'global', or even a semi-regional, database is unreasonable as the mean K_T values are significantly different in each stratigraphic level. Indeed, the two sets of

Table 5. *Potassium content (%)* (K_I *of micas from Jurassic sandstones, Inner Moray Firth (from Hurst 1980)*

Age	Mineral	$\overline{X}$	σ	*n*
Hettangian	muscovite	6.78	2.00	26
Sinemurian	muscovite	7.63	1.04	48
Sinemurian a	biotite	3.99	2.22	49
Sinemurian b	biotite	5.25	1.37	13
Pliensbachian	muscovite	6.79	2.09	58
Toarcian	muscovite	6.96	0.76	22
Callovian	muscovite	8.32	1.59	16
Kimmeridgian	muscovite	8.40	1.29	32

a and b are data from opposing sides of the same basin.
$\overline{X}$ = arithmetic mean, σ = standard deviation. *n* = number of analyses.

biotite data come from different sides of the sedimentary basin (approximately 70 km apart), are of the same age, are from similar sedimentary facies but have quite different mean K_T. This illustrates the importance of accounting for possible local geological variations before assuming the chemistry of a given mineral. Clearly, in the absence of previous experience in an area when evaluating a well, 'global' values may be the only possible chemical input in an evaluation. However, if 'local' values from areas thought to be geologically similar to the area of interest are used in addition to global values, ranges of possible mineralogies could be derived and the significance of varying the chemistry of specific minerals can be evaluated.

Identification and even quantification of illite in sandstones typical of the Norwegian Continental Shelf using NGS data seems improbable. The presence of significant K-feldspar will overshadow the K_T of any illite present (Fig. 3 and Cowan & Myers 1988) irrespective of possible changes in the chemistry of feldspars and micas.

Thorium and kaolinite

A positive relationship has previously been supposed between Th and kaolinite (Fertl 1979; Quirein *et al.* 1982; Herron 1986). As shown (Fig. 6), there is little reason to believe that Th/K cross-plots are useful for evaluating the occurrence of kaolinite in sandstones. The prevalence of detrital heavy minerals as sources of Th in North Sea sandstones may preclude the application of Th_T in clay mineral evaluation. However, the association between authigenic kaolinite and fine-grained authigenic Th-rich minerals may be of some help. As both kaolinite and the Th-rich salts are authigenic, their cooccurrence depends on diagenesis. As diagenetic reactions in sandstones are strongly dependent on sedimentary facies and burial history there is considerable potential for the authigenesis of kaolinite and Th-rich salts not to coincide. At present too little is known about the parageneses of the minerals.

Despite the apparent usefulness of Th/K crossplots, there is a lack of mineral chemical data to substantiate them. Similarly, the relationship between kaolinite and Th content (Th_T) is not well-documented, the evidence referred to here being interpreted to favour that authigenic kaolinite and Th_T are independent variables. As for K, the measurement of Th from NGS logs may often be of sufficiently high precision to be of great use in sandstone characterization (Table 4).

That kaolinite of detrital origin may generally contain more Th, probably as mineral inclusions, than kaolinite of authigenic origin (alluded to in Schlumberger 1982), is confirmed by Myers & Bristow (1989). Detrital kaolinite quite probably becomes less Th-rich with progressive recycling during weathering, erosion and deposition (Dombrowski & Murray 1984). The chemical differences may often coincide with textural differences and petrophysical characteristics in sandstones; detrital kaolinite tends to be structural clay, authigenic kaolinite tends to be dispersed clay (structural and dispersed used as defined by Frost & Fertl 1981).

Conclusions

Acceptably good measurements, in terms of analytical precision, are obtainable for potassium (K) and, in the case of one service company, for thorium (Th) under Norwegian offshore conditions. The levels of precision for both Th and K are very variable dependent on the service company involved and the borehole environment. Uranium measurement has a very poor level of precision. Direct correlation between NGS data from different service companies is unwise.

Illite cannot be identified from the K content of sandstones unless the K-feldspar (and other common K-bearing minerals) content is insignificantly low. Standard values for the K-content of K-feldspar and micas from sandstones are too high when compared with analytical data. Considerable ranges of K contents are found in micas from sandstones even within restricted areas and from similar stratigraphic levels.

There is no theoretical, and negligible analytical, basis for using Th/K cross-plots for mineral identification. Existing standard data for the Th and K contents of minerals produce confusing compositional fields when plotted. Insufficient data are available on the location and diagenesis of Th in sandstones. This restricts the value of Th-content for mineral evaluation even when measured with high degrees of precision.

There need be no direct relationship between the kaolinite and Th contents of sandstones. On the Norwegian shelf, approximately 80% of Th is contained in detrital heavy minerals while up to 20% is in the silt fraction. Authigenic REE-phosphates and other Th-salts are identified as locations of Th in sandstones. As both kaolinite and Th-rich minerals are authigenic they can only be assumed to have common parageneses under certain diagenetic conditions. Thus, Th-salts need not be associated with authigenic kaolinite although they often are.

Lars Helge Flølo is thanked for helpful discussions and criticism of a draft manuscript. Erling Siring wrote the RELOR program. Refereeing by Keith Myers and Bernard Humphreys was corrective and enlightening. Den Norske Stats Oljeselskap (Statoil) are acknowledged for supporting publication of the paper.

References

BERTOZZI, W., ELLIS, D. V. & WAHL, J. S. 1981. The physical foundation of formation lithology logging with gamma rays. *Geophysics*, **46**, 1439–1455.

CLARK, S. P., PETERMAN, Z. E. & HEIER, K. S. 1966. Abundances of uranium, thorium and potassium. *Section 24, Handbook of Physical Constants, Geological Society of America Memoir*, **97**, 521–541.

COCCO, G., FANFANI, L., ZANAZZI, P. F., HEIER, K. S. & BILLINGS, G. K. 1969. Potassium. *In*: WEDEPOHL, K. H. (ed.), *Handbook of Geochemistry*, Section 19 E-1 to E-19. Springer, Berlin.

COWAN, D. R. & MYERS, K. J. 1988. Discussion-Surface gamma ray logs: a correlation tool for frontier areas. *Bulletin of the American Association of Petroleum Geologists*, **72**, 634–663.

DEER, W. A., HOWIE, R. A. & ZUSSMAN, J. 1966. *An introduction to the rock-forming minerals*. Longman, London.

DION, E. P. & ALLEN, L. S. 1985. The accuracy of commercial natural gamma-ray spectral logs. 55th Annual SEG International Meeting, Technical Program, Abstracts and Biographies, Paper RS4, 283–285.

DOMBROWSKI, T. & MURRAY, H. H. 1984. Thorium-a key element in differentiating Cretaceous and Tertiary kaolins in Georgia and South Carolina. Proceedings of the 27th International Geological Congress **15**, 305–317.

EDMUNDSON, H. & RAYMER, L. L. 1979. Radioactive logging parameters for common minerals. *Transactions of the SPWLA 22nd Annual Logging Symposium*, June 3–6, Paper O.

ELLIS, D., HOWARD, J., FLAUM, C., MCKEON, D., SCOTT, H., SERRA, O. & SIMMONS, G. 1988. Mineral logging parameters: nuclear and acoustic. *The Technical Review*, **36**, 38–52.

ESLINGER, E. & PEVEAR, D. 1988. Clay minerals for petroleum geologists and engineers. *SEPM Short Course Notes No. 22*.

EVERETT, R., HERRON, M. M. & PIRIE, G. 1983. Log responses and core evaluation case study technique field and laboratory procedures. *Transactions of the SPWLA 24th Annual Logging Symposium*, June 27–30, Paper OO.

FERTL, W. H. 1979. Gamma ray spectral data assists in complex formation evaluation. *The Log Analyst*, 3–38.

—— 1983. Gamma ray spectral logging: a new evaluation frontier, Part 1, Principles. World Oil (March 1983), 79–82, 86, 90–91.

FLØLO, L. H. 1988. Improved reservoir description by use of induced gamma spectroscopy. *Transactions of the SPWLA 11th European Logging Symposium*, Sept. 14–16, Paper EE.

FROST, E. & FERTL, W. H. 1981. Integrated core and log analysis concepts in shaly clastic reservoirs. *The Log Analyst* (March-April), 3–16.

GAUTIER, D. L. 1983. Marine shales. *In*: Rice, D. D. & Gautier, D. L., Patterns of sedimentation diagenesis, and hydrocarbon accumulation in Cretaceous rocks of the Rocky Mountains, Chapter 8, *SEPM Short Course No. 11*.

HASSAN, M. & HOSSIN, A. 1975. Contribution a l'etude des comportements du thorium et du potassium dans les roches sedimentaires. *Comptes Rendue Academie Science, Paris* **280**, 533–535.

—— & COMBAZ, A. 1976. Fundamentals of the differential gamma log-intrepretation technique. *Transactions of the SPWLA 17th Annual Logging Symposium*, June 9–12, 1976, Paper H.

HEINZE, D. M. 1985. Clay identification and amount measured by laboratory techniques compared to well log responses: application to tight gas sands and shales. Sandia Report, Sand 85–0312, Sandia National Laboratories, Albuquerque.

HERRON, M. M. 1986. Mineralogy from geochemical well logging. *Clays and Clay Minerals*, **34**, 204–213.

—— 1987. Future applications of elemental concentrations from geophysical logging. *Nuclear Geophysies*, **3**, 197–211.

—— & HERRON, S. L. 1990. Geological applications of geochemical well logging. *In*: HURST, A., LOVELL, M. A. & MORTON, A. C. (eds) *Geological Applications of Wireline Logs*, Geological Society, London, Special Publication, **48**, 165–175.

HUMPHREYS, B. & LOTT, G. K. 1990. An investigation into nuclear log responses of North Sea Jurassic sandstones using mineralogical analysis. *In*: HURST, A., LOVELL, M. A. & MORTON, A. C. (eds) *Geological Applications of Wireline Logs*, Geological Society, London, Special Publication **48**, 223–240.

HURST, A. 1980. *The diagenesis of Jurassic rocks of the Moray Firth, NE Scotland*. PhD. thesis, University of Reading.

—— 1987. Problems of reservoir characterisation in

some North Sea sandstone reservoirs solved by the application of microscale geological data. *In*: KLEPPE, J., BERG, E. W., BULLER, A. T., HJELMELAND, O. & TORSAETER, O. (eds). *North Sea Oil and Gas Reservoirs*, pp. 153–167, Graham & Trotman, London.

—— & ARCHER, J. S. 1986. Some applications of clay mineralogy to reservoir description. *Clay Minerals*, **21**, 811–826.

MARETT, G., CHEVALIER, P., SOUHAITE, P. & SUAU, J. 1976. Shaly sand evaluation using gamma-ray spectrometry applied to the North Sea Jurassic. *Transactions of the SPWLA 17th Annual Logging Symposium*, June 9–12, Paper DD.

MORTON, A. C. 1986. Dissolution of apatite in North Sea Jurassic sandstones: implications for the generation of secondary porosity. *Clay Minerals*, **21**, 711–733.

MURRAY, E. G. & ADAMS, J. A. S. 1958. Thorium, uranium and potassium in some sandstones. *Geochimical et Cosmochimica Acta*, **13**, 260–269.

MYERS, K. J. & BRISTOW, C. 1989. Detailed sedimentology and gamma-ray log characteristics of a Namurian deltaic succession. Part II, gamma-ray logging. *In*: WHATELY, K. & PICKERING, K. T. (eds), *Deltas: Sites and Traps for Fossil Fuels*. Geological Society, London, Special Publication, **41**, 81–88.

—— & WIGNALL, P. B. 1987. Understanding Jurassic organic-rich mudrocks-new concepts using gamma-ray spectrometry and palaeoecology: examples from the Kimmeridge Clay of Dorset and the Jet Rock of Yorkshire. *In*: LEGGETT, J. K. & ZUFFA, G. G. (eds.), *Marine Clastic Sedimentology*. Graham & Trotman, London 172–189.

NADEAU, P. H. & BAIN, D. C. 1986. Composition of some smectites and diagenetic illitic clays and implications for their origin. *Clays and Clay Minerals*, **34**, 455–464.

PALLATT, N., WILSON, M. J. & MCHARDY, W. J. 1984. The relationship between permeability and the morphology of diagenetic illite in reservoir rocks. *Journal of Petroleum Technology*, **46**, 2225–2227.

PEVERARO, R. C. A. & RUSSELL, K. J. 1984. Interpretation of wireline log and core data from a mid-Jurassic sand/shale sequence. *Clay Minerals*, **19**, 483–505.

QUIREIN, J., GARDNER, J. S. & WATSON, J. T. 1982. Combined natural gamma ray spectral/lithodensity measurements applied to complex lithologies. SPE 11143, 57th Annual Fall Technical Conference and Exhibition of SPE of AIME, New Orleans, Sept. 26–29.

——, KIMMINAU, S., LAVINGE, J., SINGER, J. & WENDEL, F. 1986. A coherent framework for developing and applying multiple formation evaluation models. *Transactions of the SPWLA 27th Annual Logging Symposium*, June 9–13.

ROGERS, J. J. W., CONDIE, K. C. & MAHAN, S. 1969/70. Significance of thorium, uranium and potassium in some early Precambrian greywackes from Wyoming and Minnesota. *Chemical Geology*, **5**, 207–213.

—— & RICHARDSON, K. A. 1964. Thorium and uranium contents of some sandstones. *Geochimical et Cosmochimica Acta*, **28**, 2005–2011.

SCHLUMBERGER 1982. *Natural gamma-ray spectroscopy. Essentials of N. G. S. interpretation.* Schlumberger C. E. A. Marketing Publication.

SERRA, O. 1984. Fundamentals of well-log interpretation. 1. The aquisition of logging data. *Developments in Petroleum Science 15A*. Elsevier.,

—— BALDWIN, J. & QUIREIN, J. 1980. Theory, interpretation and practical applications of natural gamma-ray spectroscopy. *Transactions of the SPWLA 21st Annual Logging Symposium*, July 8–11, Paper Q.

SUAU, J. & SPURLIN, J. 1982. Interpretation of micaceous sandstones in the North Sea. *Transactions of the SPWLA 23rd Annual Logging Symposium*, July 6–9, Paper G.

TITTMAN, J. 1986. *Geophysical well logging*. Academic Press.

WILEY, R. & ZITTEL, R. J. 1982. Natural gamma ray data of various sandstones. *Transactions of the SPWLA 23rd Annual Logging Symposium*, July 6–9, Paper K.

An investigation into nuclear log responses of North Sea Jurassic sandstones using mineralogical analysis

B. HUMPHREYS AND G. K. LOTT[1]

Stratigraphy and Sedimentology Research Group, British Geological Survey, Keyworth, Nottingham, NG12 5GG, UK

[1]Present address: Lemigas, P.O. Box 89/Jkt., Cipulir–Kebayoran Lama, Jakarta 10002, Indonesia

Abstract: The results of laboratory mineralogical analyses have been compared with downhole log responses to determine the possibility of detecting individual minerals using the gamma-ray, natural gamma-ray spectroscopy, formation density and litho-density logs. Cored sequences from the Murchison, Thistle, Kittiwake, Gannet and Tartan Fields, North Sea, are used as case examples. K-feldspars, micas, detrital clays (illite, and possibly kaolinite) and phosphatic cements are shown to produce a measurable response on the gamma-ray and natural gamma-ray logs. Authigenic kaolinite did not impart radioactivity to the sandstones in the examples studied. The photoelectric absorption index (*Pe*) was found to be particularly sensitive to Fe-bearing minerals and baryte. A number of problems constrain full characterization of the mineralogy of a sandstone: specific detection of quartz and plagioclase feldspars is not possible by nuclear logs, distinction between minerals is problematical when a log responds to several minerals simultaneously, and calibration for minerals with a variable composition such as ferroan dolomite and mixed-layer illite-smectites is difficult.

Wireline logging is a standard but crucial operation in oil exploration. A large variety of logging tools have been developed in the past 60 years that measure properties of the rocks and/or their pore fluids. The results from these tools are used in the oil industry to identify potential reservoir rocks, to determine their porosity, structure (dip of beds, fractures etc) and the nature of the fluids present, and to correlate between boreholes. Log signatures are also commonly used for facies interpretation and sedimentological modelling. Log analysts now use complex computer modelling and computer-generated cross-plots of spectral gamma-ray data plotted against the photoelectric absorption index (*Pe*) to determine the presence of individual minerals (e.g. Suau & Spurlin 1982; Quirein *et al.* 1982). Mineral determinations are becoming increasingly important for reservoir engineers who need to anticipate the reaction of the formation to fluids during drilling, production and reservoir stimulation operations (Lever & Dawe 1987; Lievaart & Davies 1987). In view of the marginal economic nature of many smaller North Sea oilfields, there are considerable cost and time advantages of using wireline logs for mineralogical determinations instead of extracting cores. However, log results can be ambiguous without core data for calibration purposes.

This paper evaluates the use of some nuclear logs for the purpose of mineral identification, using case examples of sandstone reservoirs of both Middle and Upper Jurassic age from the UK Central and Viking Grabens, North Sea. Particular emphasis is placed on two categories of lithology-sensitive nuclear logs; those measuring natural radioactivity, the gamma-ray (GR) log and the Schlumberger natural gamma-ray tool (NGT), and those measuring rock density, the formation density (RHOB) log and the PEF log from the Schlumberger litho-density tool (LDT).

Previous detailed core calibration studies are few in number and have had mixed results. Peveraro & Russell (1984) found that detection of individual minerals from log cross-plots was often only possible when a mineral was a major component of the rock. Some recent studies show significant disagreement between laboratory-derived and core-derived data (e.g. Heinze 1985). A major problem of all core to log comparative studies is that data of contrasting physical and statistical character are usually compared. Whereas mineralogical data are traditionally measured on a microscopic scale using point samples from cores such as petrophysical plugs, wireline log data are running averages, in which the vertical resolution varies from tool to tool, made over approximately 0.25 m by a tool travelling at an approximately constant speed up a borehole (Hurst & Archer 1986; Rider 1986).

From HURST, A., LOVELL, M. A. & MORTON, A. C. (eds), 1990, *Geological Applications of Wireline Logs* Geological Society Special Publication No. 48, pp. 223–240

Methods of mineralogical analysis

To attempt to provide a meaningful mineralogical data set for comparison with the moving average data of log responses, detailed petrographical analyses have been undertaken on a continuous thin slice of rock cut from cores. This continuous slice is divided into one foot (0.33 m long) intervals, and each interval is treated as a bulk sample for grain size and X-ray diffraction (XRD) analysis. This approach eliminates the effects of small scale (i.e. thin-section scale) lithological variations. Each sample is gently disaggregated using a mortar and pestle (many of the sandstones are weakly cemented, and grain fracturing is generally not a problem), followed by ultrasonic treatment in distilled water, which ensures that all the clay particles are separated from the framework grains. The disaggregated sediment is then separated into size fractions using wet sieving and centrifuge techniques. In selected core studies, the amount of clay size (<4 μm) material was determined by pipette analysis (Folk 1974).

Obtaining the weight per cent values of the <32 μm fraction (clay to medium silt size fraction) is one of the prime objectives of our studies. The <32 μm size fraction is considered to represent the non-framework minerals present in sandstones, including detrital clays, silt-sized quartz and feldspar grains and a varied suite of cements. It is also close to the arbitrarily defined limit of formation fines of 37 μm given by Muecke (1979). When the <32 μm fraction consists largely of detrital clay minerals, it could be used directly by petrophysicists as an upper estimate of the V_{clay} parameter to evaluate water and hydrocarbon saturation in a reservoir.

The mineralogy of the <32 μm fraction has been determined by XRD. In Brent Group sandstones from the Thistle and Murchison oilfields, where the <32 μm fraction consists dominantly of authigenic kaolinite, siderite cement and quartz silt (with minor feldspar), an attempt has been made to quantify mineral proportions using internal X-ray standards. The standards used were selected from Well 211/18A-A33; a siderite nodule, hand-picked quartz grains (the presence of extraneous feldspar was determined by chemical staining), and a large separate of authigenic kaolinite, extracted by pipette analysis during grain size analyses (Folk 1974). The purity of the clay sample, over 95% kaolinite, was verified by XRD and scanning electron microscope (SEM) analyses. By mixing the three essentially pure mineral components in different proportions, calibration curves have been constructed from which the percentage of a mineral present in a sample can be estimated from measurements of major peak areas on X-ray diffractograms. With uniform sample preparation, the results are considered accurate in a relative sense, and are probably within 5–10% in absolute terms. The data obtained from analysis of the core slice are augmented by closely-spaced spot analyses using thin sections and SEM observations.

Depth matching of mineralogical and log data

It is very difficult to assess the effects of specific minerals on log responses without accurate matching of core depths to log depths. Core depths (drillers depths) rarely correspond to log depths; although core losses may occur at any level, they are usually assumed by the driller or on site geologist to have taken place at the bottom of each core. Depth matching of cores to logs can be achieved by locating distinct features seen in slabbed cores, such as discrete shale beds and carbonate-cemented horizons, to marked responses on the gamma and sonic logs respectively (e.g. Knox *et al.* 1981; Hurst & Archer 1986). Alternatively, laboratory gamma-ray spectrometry can be used to duplicate a down-hole log. Perfect depth-matching is not likely to be achieved by these methods, mainly because of shoulder effects on log responses caused by statistical error in the counting procedures (Rider 1986, p. 61) and stretching of the wireline, but also because of contamination by drilling mud which obscures primary mineralogy. Nevertheless, a good working match is usually possible.

The mineralogical analyses undertaken by the authors also allow core to log depth matching even where core recovery is incomplete and marker horizons are absent (it is usually the least resistant beds such as shales or coals that are not sampled when recovery is poor), or when only a portion of the core is available for study. By plotting a profile of laboratory-derived V_{clay} values against the downhole GR log, the depth of the V_{clay} profile can be adjusted to obtain the best fit. In the process, zones of core loss and wireline stretching may be identified. The method is most successful where the values of V_{clay} vary significantly and the <32 μm fraction is composed largely of naturally radioactive detrital clays. If the <32 μm fraction consists largely of authigenic kaolinite with a low natural radioactivity, there may be a considerable discrepancy between the core data and the GR response. This approach, though more laborious than laboratory GR logging, identifies the minerals actually inducing the log responses and is thus more useful in circumstances where log interpretation is of paramount importance.

An example of core depth to log depth correlation using mineralogical data for sandstones of Middle Jurassic age illustrates the potential of our method (Fig. 1). In this example, a downhole profile of the <4 μm clay separation before depths were adjusted showed poor correlation with the gamma log (Fig. 1A). An optimal core depth to log depth match was achieved by applying a depth correction of up to +13 ft to core 1, and of up to +5 ft to core 2 (Fig. 1B). Some stretching of the sample intervals, 7% and 3% for cores 1 & 2 respectively, was necessary to compensate for poor recovery. The clay content profile matches closely the gamma-ray curve in core 1, but less closely in core 2, the result of a change in clay mineralogy in the vertical section. XRD analyses prove that detrital illite is a significant component of the clay assemblage in core 1, whereas authigenic kaolinite is dominant in core 2 (see also Fig. 3).

Detection of specific minerals: some case examples

Clay minerals

The GR response is traditionally used to assess the degree of 'shaliness' of a sandstone with cursory regard to the mineralogy of the clays present. Some authors relate the GR curve to changes in grain size without a sound mineralogical basis for doing so. To test the sensitivity of the GR response to the amount and mineralogy of non-framework detrital clays, mineralogical and grain size analyses were undertaken on a continuously cored interval of muddy, highly bioturbated sandstones from the Upper Jurassic Fulmar Formation (Fig. 2). Combined XRD and thin section analyses show that the clay mineral assemblage consists of kaolinite, mostly authigenic but some detrital, with variable but usually subordinate amounts of detrital illite. A profile of fluctuating values of V_{clay} and grain size shows good correlation with the GR

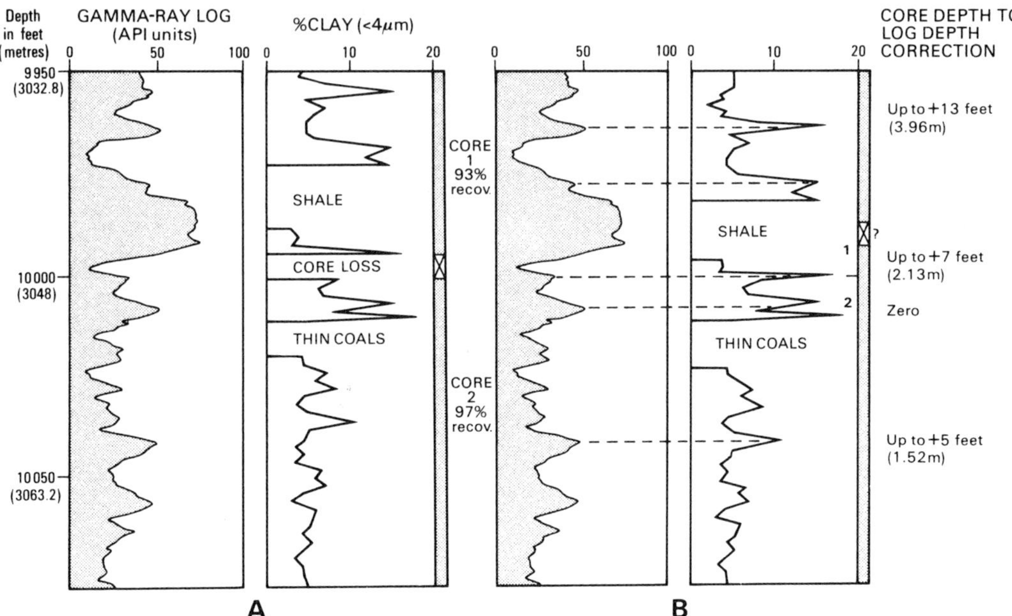

Fig. 1. Matching of profiles of core data to downhole log responses. (a) Profile of the clay content of sandstones, plotted using drillers depths, shown against the downhole GR log. Note the poor correlation between the two profiles. Core loss (X) was indicated by the drillers at the bottom of core 1. (b) Core data from (a) adjusted to log depths by matching high clay contents to high GR responses. Different core depth corrections at different levels indicate that core loss must have occurred in the shale bed, and not at the bottom of core 1.

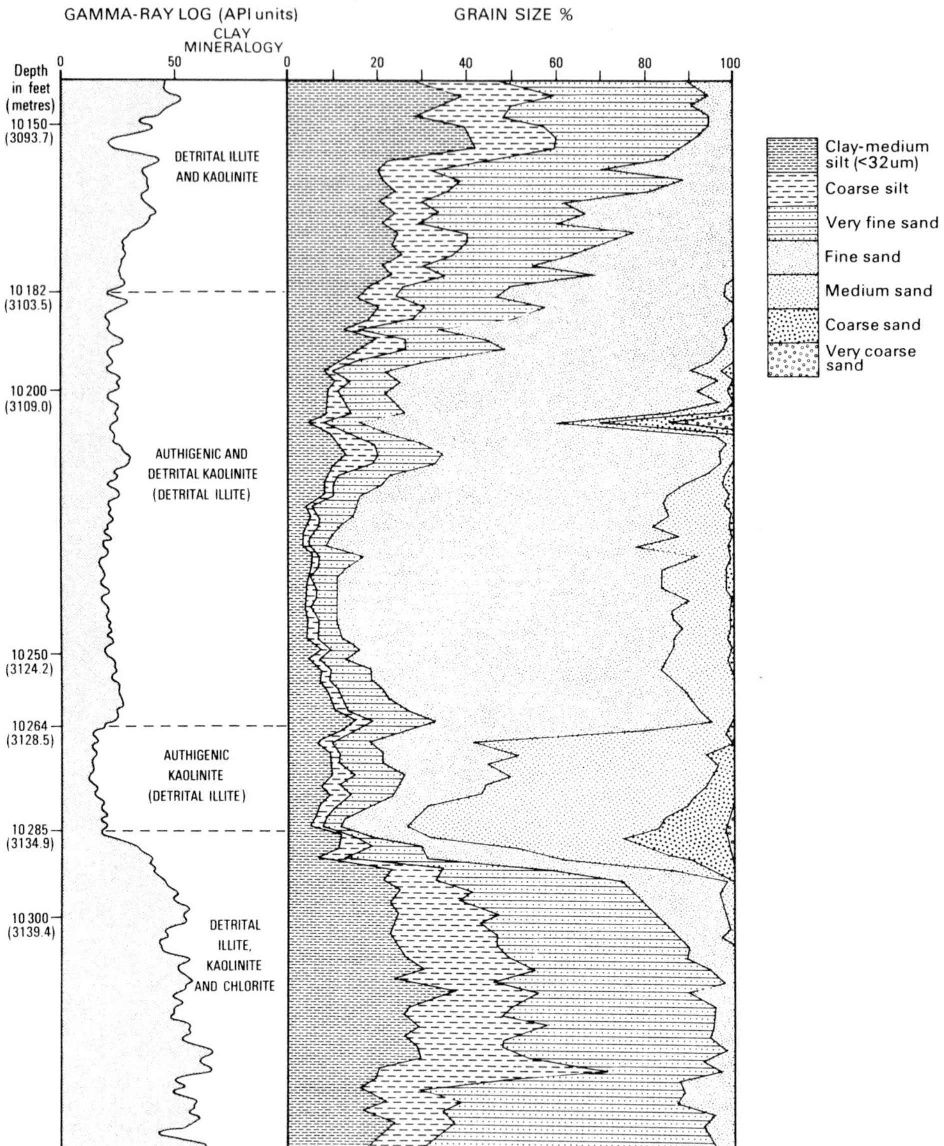

Fig. 2. Grain size profile through a sequence of highly bioturbated, Upper Jurassic Fulmar Formation sandstones, plotted against the downhole GR log. Clay mineral assemblages are indicated with minor phases in brackets. Although the GR log responds to the clay content and mineralogy of the sandstones, it is also apparent that there is a correlation with grain size in this example.

log (Fig. 2). Between log depths 10264–10285 ft (3128.5–3134.9 m) the clay and fine silt content is low (5–9%) and the clay assemblage consists predominantly of authigenic kaolinite (80%+) with small amounts of illite which, together with *c.*10% K-feldspar account for the radioactivity at this level. Below 10285 ft. (3134.9 m) the sandstones contain more than 20% detrital clay and fine silt with an increased proportion of detrital illite. While the GR log responds to the change in clay content, it also reflects a change in grain size from medium-grained sandstones to very fine-grained sandstones. Although the measured clay content is sometimes lower above 10264 ft (3128.5 m) than in the interval 10264–10285 ft (3128.5–3134.9 m), the sandstones contain thin clay laminae comprising detrital kaolinite as well as illite. The clay assemblage contains little more detrital illite than below 10264 ft (3128.5 m), and the higher background radioactivity may be a response to detrital kaolinite. Note also that some quite small increases in clay content of 4 to 5% at 10213 ft (3112.9 m) and *c*.10260 ft (3127.2 m) induce a measurable response on the GR log. The contribution of radioactive K-feldspars, which occur throughout the interval in variable amounts (but usually forming between 10 and 20% of the total mineralogy), further complicates the overall pattern. The lowest K-feldspar proportions occur in the fine- to medium-grained sandstones, between *c*.10190 and *c*.10290 ft (3105.9–3136.4 m), which accentuates the relationship between the GR log and grain size. The K-feldspar content probably accounts for much of the background radioactivity in these sandstones (see also Fig. 5). Nevertheless, the overall fluctuations of the GR curve are a response to changing proportions of detrital illite and possibly also detrital kaolinite. Authigenic kaolinite seems to have little effect on the formation radioactivity. Other case examples can be used to illustrate the variable radioactivity of the different clay minerals.

Kaolinite. Authigenic kaolinite is one of the commonest and most abundant cements in sandstones of the Middle Jurassic Brent Group. A profile of kaolinite proportions determined by XRD is rarely mirrored by the downhole GR log response in the Etive Formation of the Murchison oilfield (Fig. 3). The kaolinite, which, on purely mineralogical grounds, should have a negligible inherent radioactivity (Eslinger & Pevear 1988), has no significant effect on the GR log in this example.

Illite, mixed-layer illite-smectites. Potassium (K) occurs within the structure of illite and illite-bearing mixed-layer clays and these clays, whether detrital or authigenic, will always contribute to the GR response. Referring again to the Brent sequence shown in Fig. 3, illite of both detrital and authigenic origin is more common than kaolinite in the Ness Formation, and a good match is obtained between clay volume and log response. The relative effects of authigenic and detrital illite are, however, impossible to separate.

The possibility of quantifying the natural radioactivity attributable to detrital illitic clay has been investigated with reference to a Fulmar Formation sequence from the Gannet West Field, Well 21/30–3. The Upper Jurassic sandstones typically contain between 15% and 40% non-framework minerals (<32 μm), a large proportion of which consists of clay minerals, either detrital illite or kaolinite (both detrital and authigenic), together with traces of detrital chlorite. Although the sequence includes levels rich in glauconite grains and mica, intervals were chosen where the only significant K-bearing minerals were illite and K-feldspar. The detrital kaolinite, judging by Fig. 2, could be slightly radioactive, but this effect is impossible to quantify. Thus, to further simplify matters, the radioactivity of kaolinite was assumed to be negligible. Point counting of thin sections was used to determine proportions of K-feldspar grains (selectively stained yellow by treatment with a sodium cobaltinitrite solution). By multiplying the amount of K present in each mineral by the quantity of the mineral present, and knowing the amount of K detected by the NGT log, it is possible to check predicted proportions of illite and K-feldspar with those actually measured. The %K in K-feldspar was determined by electron microprobe analyses of Fulmar Formation feldspars, whereas a value of the average %K in illite was taken from Rider (1986). It was found that the log values of K were lower than would be expected given, in particular, the amount of K-feldspar present. It was nevertheless possible to predict the amount of illite by using a constant correction factor to suppress the effect of feldspar (Fig. 4). A correlation coefficient of 0.73 was obtained between measured and calculated values of illite in Fig. 4. An additional factor to be considered is that the K content of the illite may be lower than average due to interstratification of up to *c*.40% non-radioactive smectite layers. The 40% value is an approximation based on the interpretaton of glycerolated XRD traces. If this is the case, then K-feldspar has a very much greater effect on the GR log than illite in the Fulmar Formation.

Chlorite. Chlorite is not particularly abundant in the reservoirs studied by the authors. Chlorites have a negligible radioactivity (Cowan & Myers 1988), but Fe-chlorites should be easily detected by the photoelectric absorption index, *Pe*, when present in sufficient quantities. However, due to the compositional variability (particularly the Fe content) of authigenic and

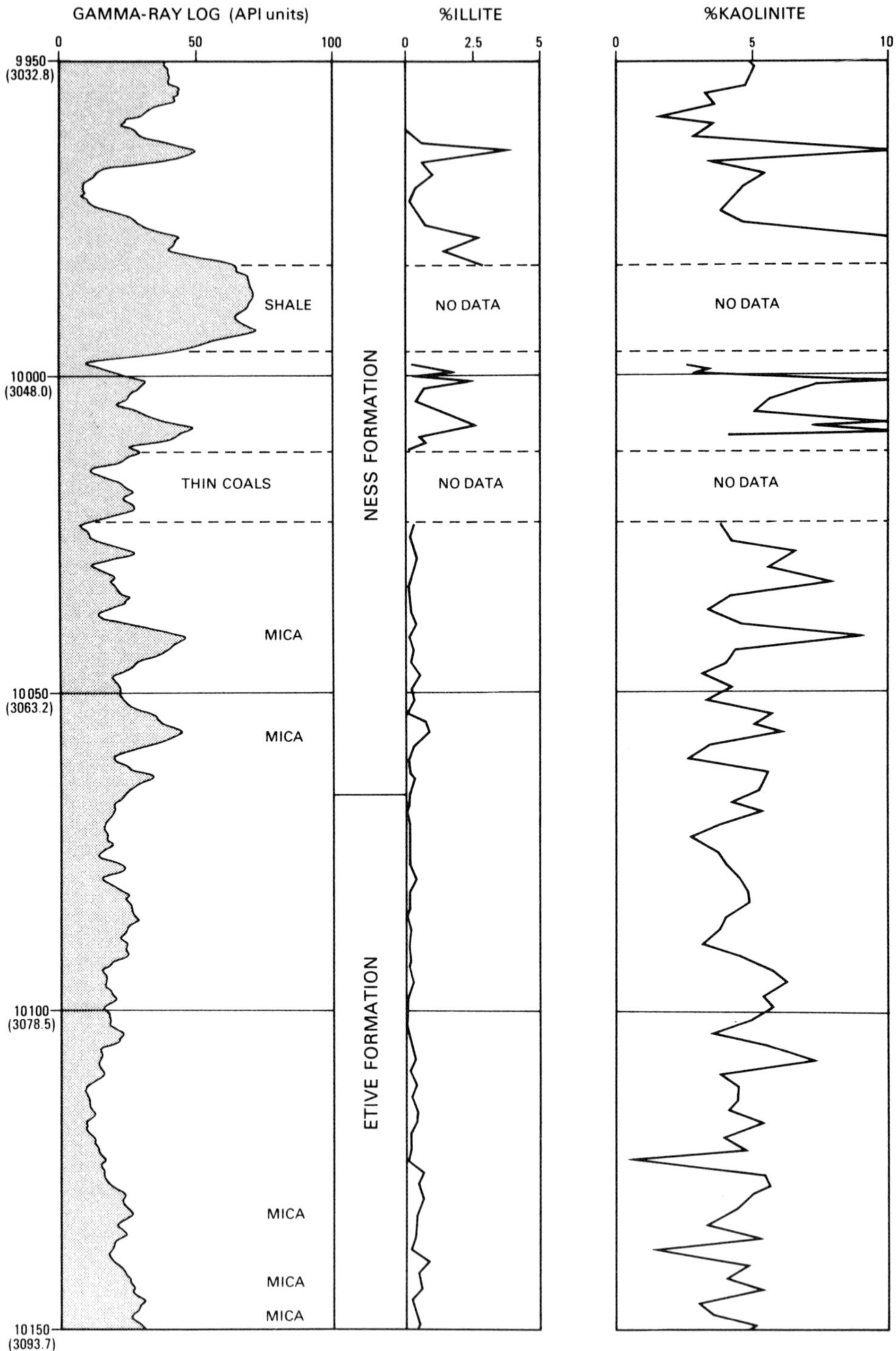

Fig. 3. Effect of clay minerals on the GR log in the Middle Jurassic Brent Group. Detrital illite and kaolinite are dominant in the Ness Formation and a close match is seen between profiles of clay content and the GR log. Authigenic kaolinite is the dominant clay type in the Etive Formation where variations in radioactivity are due to differing amounts of K-feldspar, mica, and, as shown, very small quantities of detrital illite, but not kaolinite cement.

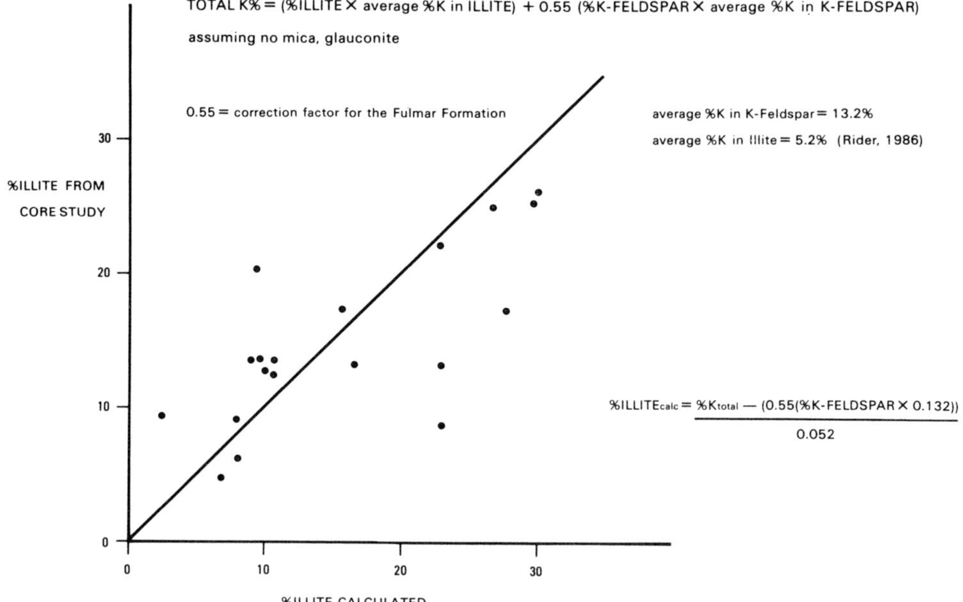

Fig. 4. Proposed method of calculating % illite in a sandstone reservoir using spectral GR log values of potassium and average values of % K-feldspar from point counts of thin sections.

detrital chlorites (Curtis *et al.* 1985), it is anticipated that detection of Mg-chlorite may sometimes be difficult.

Muscovite and biotite micas

The occurrence of micaceous sandstones in the Middle Jurassic Brent Group is well known (Morton & Humphreys 1983; Peveraro & Russell 1984). The effect of mica on the GR and total K logs is shown in Fig. 5. In the mica-poor Etive Formation, the clay content is almost wholly authigenic kaolinite and thus K-feldspar is almost entirely responsible for the radioactivity. In the Rannoch Formation, a marked increase in the GR (up to +25 API) and K responses (up to +1.5%) relates to the occurrence of up to 12% muscovite and up to 5% biotite, together with a slight increase in the average K-feldspar content. Muscovite has a greater effect than biotite on this increase in radioactivity, not just because of its greater abundance, but probably because of its higher K content (Hurst 1990).

Ferroan carbonate and baryte

If dense mineral cements such as baryte and various Fe-rich minerals are present in significant quantities, they are detected by the formation density (RHOB) log. Examples of the response of the density log to baryte and ferroan dolomite are shown in Figs 6 & 7, although it must be emphasised that these minerals are present in unusually large amounts. Poikilotopic baryte forms up to 21% of whole rock (according to thin section point counts) in the example shown, but detection of baryte is possible where the cement exceeds 5% of the total mineralogy. Ferroan dolomite and minor ankerite form up to 32% of the total mineral content (based on thin section point counts and chemical determinations) of a carbonate cemented sandstone in the Upper Jurassic Piper Formation of the Tartan oilfield. An average composition for the dolomite of $(Fe_{0.23}Mg_{0.76}Mn_{0.01})_{0.97}Ca_{1.03}(CO_3)_2$ has been determined from 27 micro-probe analyses on different dolomite rhombs, but the Fe content is highly variable, from 3.9 to 22.3 mol% $FeCO_3$. Fig. 7 shows that the RHOB log clearly responds to the presence of this dolomite but its variable iron content makes mineral quantification problematic. Because thin section data over this interval are limited, the amount of dolomite is presented in terms of XRD peak heights, a crude but nevertheless relative indicator of the abundance of dolomite. Absolute proportions of dolomite (up to 32% of total mineralogy) would not necessarily give a better

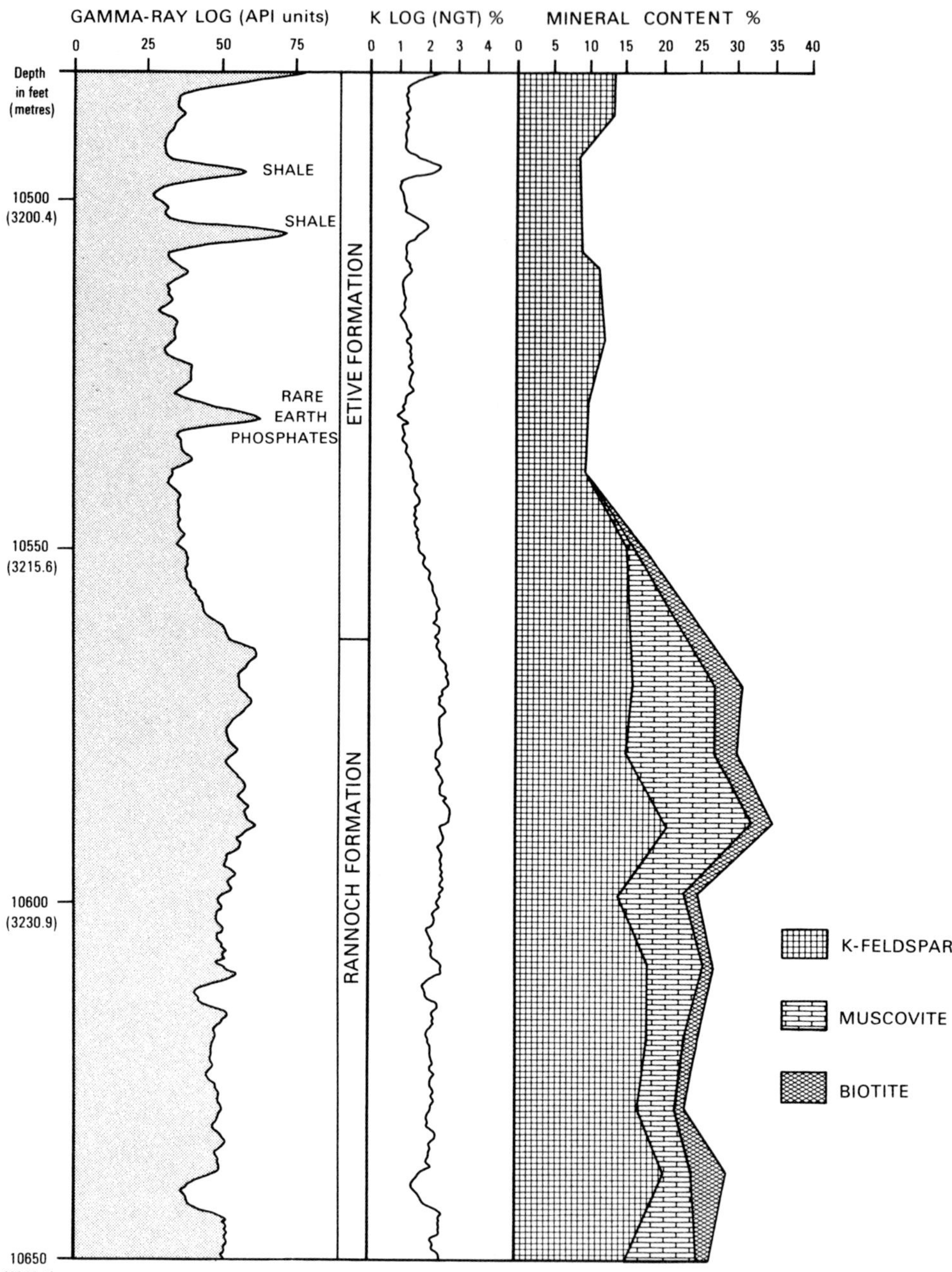

Fig. 5. Effect of micas and K-feldspar on the GR and K logs. Mineral proportions were determined by thin section point counts. The radioactivity of the kaolinite-cemented Etive Formation is almost wholly caused by K-feldspars, except for two thin shales and an occurrence of rare earth phosphate cements as shown. K-feldspar, muscovite and biotite are the main contributors to the radioactivity of the Rannoch Formation. Small amounts of detrital and authigenic illite also impart some radioactivity to the Rannoch Formation.

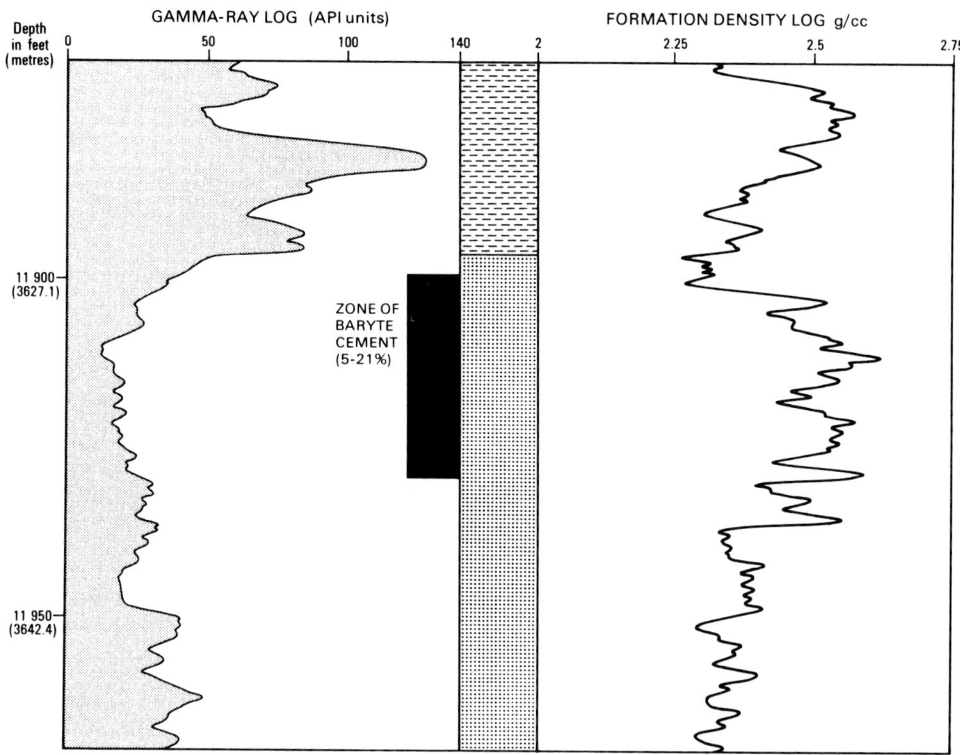

Fig. 6. An example of how the presence of up to 21% baryte cement is detected by an increase of up to 0.3 g/cc on the formation density log.

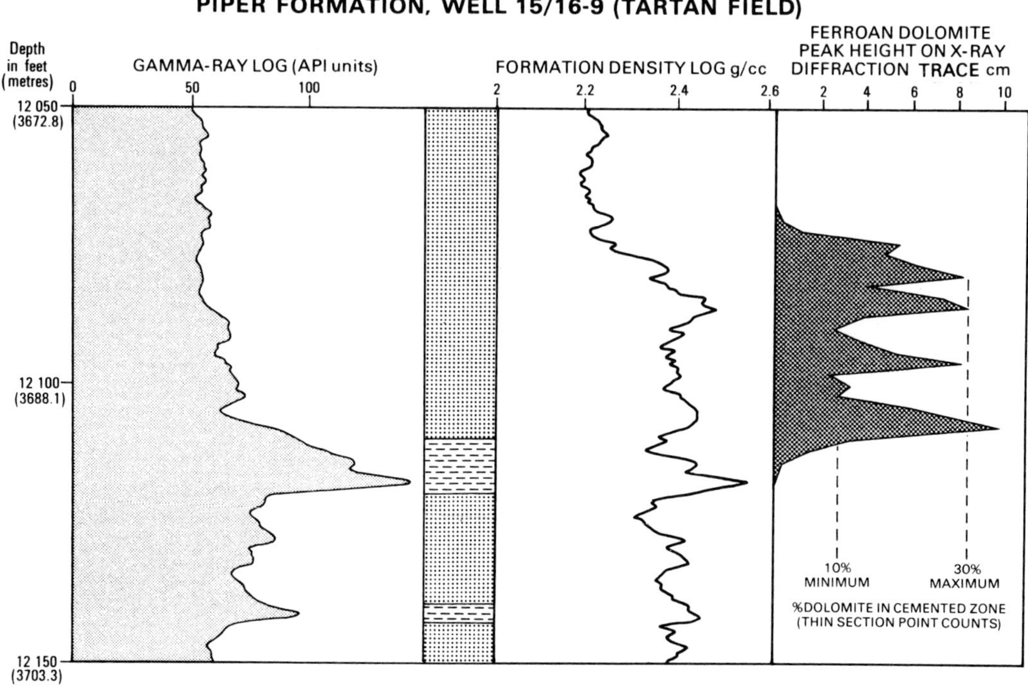

Fig. 7. Ferroan dolomite cement can also increase the response of the formation density log if present in sufficient quantities. In this particular example, the variable iron content of the dolomite makes accurate calibration difficult.

correlation with the RHOB log, because it is the iron content of the dolomite that is critical, and not the absolute amount of dolomite. Given the variations in chemistry of dolomite from just one horizon, an average composition for a dolomite is thus a meaningless parameter for log calibration.

The *Pe* (photoelectric absorption index) curve, a much more sensitive indicator of Fe-rich minerals, is measured simultaneously with apparent bulk density by the LDT log. Figure 8 shows the occurrence of siderite in the Ness and Rannoch Formations of a Brent sequence from the Thistle oilfield. It is particularly noticeable that the *Pe* curve responds dramatically to the presence of thin beds of sideritic mudstone, only 10–15 cm thick, which border shale beds in the Ness Formation. However, detection of siderite cement is of greater significance in sandstone intervals for the purpose of determining reservoir quality. The common occurrence of siderite in the Rannoch Formation is reflected by a clear increase in the background *Pe* response. However, muscovite, biotite, pyrite, ferroan calcite cements and rare-earth phosphatic cements are also known to contribute to the *Pe* response in this sequence. Responses attributable to the latter two minerals are indicated on Fig. 8. Pyrite cements are common in North Sea Jurassic sandstones, but generally only in small amounts, typically 1–2% by volume, and only rarely near 10%. Although no case examples are presented here, allowances should be made for the effect on the *Pe* curve of pyrite.

In order to assess the relative effect of siderite and mica on the *Pe* curve, two peaked responses in Ness Formation sandstones containing both minerals were investigated. A profile of mica content tends to mirror the profile of total siderite content (Fig. 9). Mica, especially biotite, has a close textural association with siderite in this sequence. Siderite has often replaced micas during diagenesis (Morton & Humphreys 1983; Peveraro & Russell 1984). The ratio of muscovite to biotite varies between laminae, but an average ratio for all micaceous sandstone in this cored Jurassic sequence is 2.46 : 1 (mean of 22 point counts from thin sections). Some idea of the relative effect of siderite and mica on these peaks can be obtained by consideration of their relative values of Uma, the effective photoelectric absorption cross section per unit volume (Fig. 9). Knowing the amount of siderite and mica, and assuming muscovite is 2.5 times more abundant than biotite, then through a simple calculation of ratios of Uma times volumes it can be determined that siderite is responsible for more than 80% of the two peaked reflections. As little as 3% siderite may thus induce a *Pe* response of 1 barn/electron.

Feldspars

Apart from quartz, feldspars are the most common components of North Sea sandstones, in our experience forming between 5% and 40% of the total rock volume. Some of the lowest feldspar contents are recorded from coarse-grained Middle Jurassic Brent Group sandstones, which have undergone extensive feldspar leaching during early diagenesis (Morton & Humphreys 1983), where the present assemblage is dominated by K-feldspars (orthoclase and microcline) with occasional untwinned albites. The highest feldspar to quartz ratios are found in fine-grained, muddy, commonly arkosic Upper Jurassic sandstones (e.g. Fulmar Formation, Johnson *et al.* 1986) where the feldspar assemblage is more variable, and includes varieties of plagioclase as well as K-feldspars.

Orthoclase, microcline, sanidine and perthitic feldspars all contain ^{40}K and thus contribute significantly to the natural gamma radiation of the formation (Figs 4 & 5). The actual %K in K-feldspars can be determined by electron microprobe analyses. The average %K for K-feldspars from Middle Jurassic sandstones in the Murchison Field is 12.73 (range 12.0 to 13.45 from 43 analyses). The average value for Upper Jurassic Fulmar Formation sandstones from the Gannet Field is 13.20 (range 11.43 to 14.28 from 41 analyses). These values do not discriminate between the varieties of K-feldspar, principally orthoclase and microcline, which are both abundant in these sandstones. The %K in perthites is obviously lower. However, plagioclase feldspars are also present, often as untwinned albites, which are sometimes difficult to distinguish from quartz without the aid of chemical staining or use of an electron microprobe. In many of the Upper Jurassic sandstones studied, plagioclase feldspars exceed potash feldspars in abundance.

The difficulties of characterising a rock in terms of its feldspar content are amply illustrated with reference to Fig. 10. Firstly, the closeness of plagioclase and quartz on a variety of petrophysical cross-plots makes them very difficult to distinguish using logs (Fig. 10d). Secondly, the true amount of plagioclase is often incorrectly recorded by geologists (untwinned feldspars can be mistaken for quartz in thin section) which only compounds the problems of calibrating mineral data with logs. The

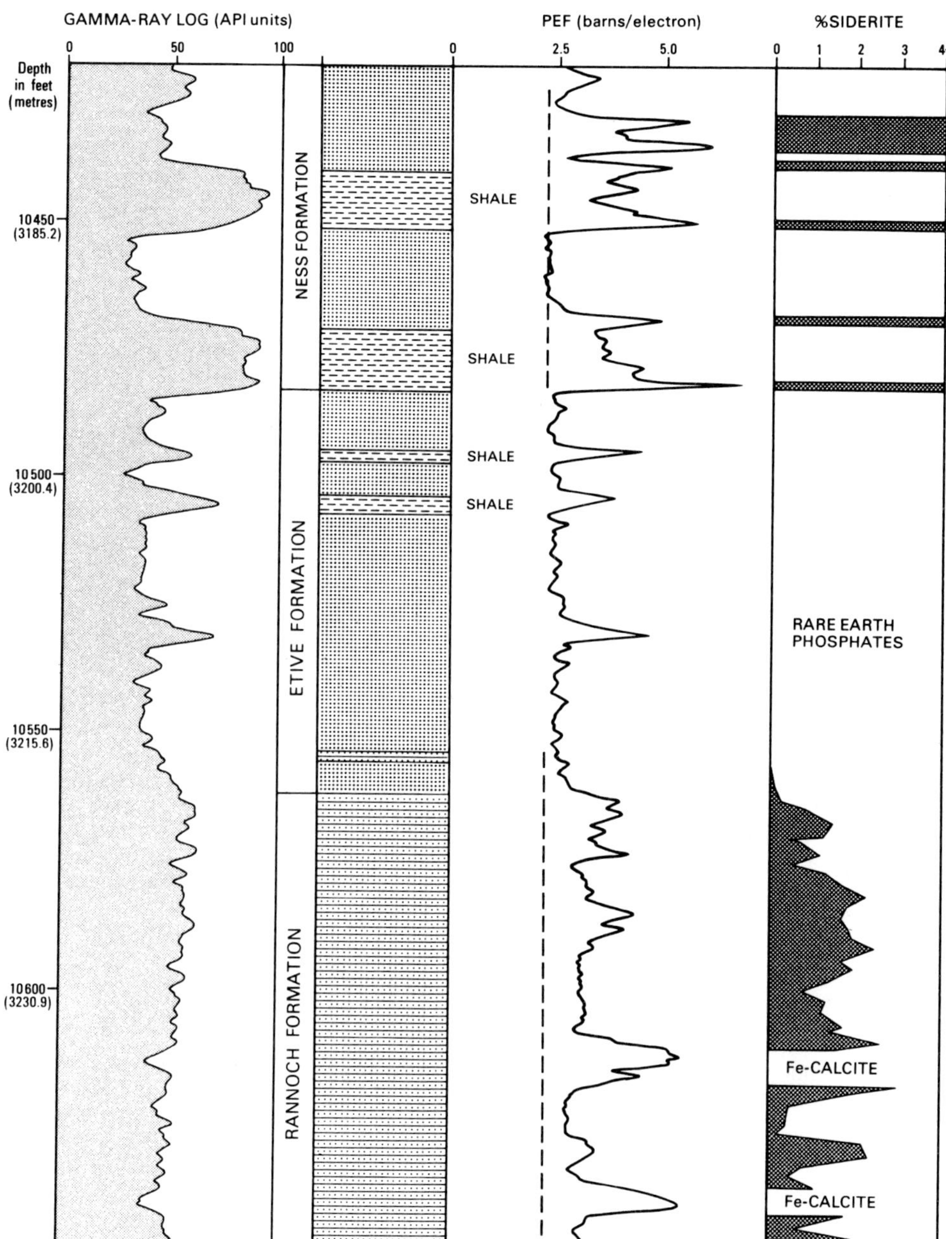

Fig. 8. Detection of siderite, ferroan calcite and rare-earth phosphates using the Pe curve. Mica also contributes to the Pe response in the Rannoch Formation and at some levels in the Ness Formation (see also Fig. 9).

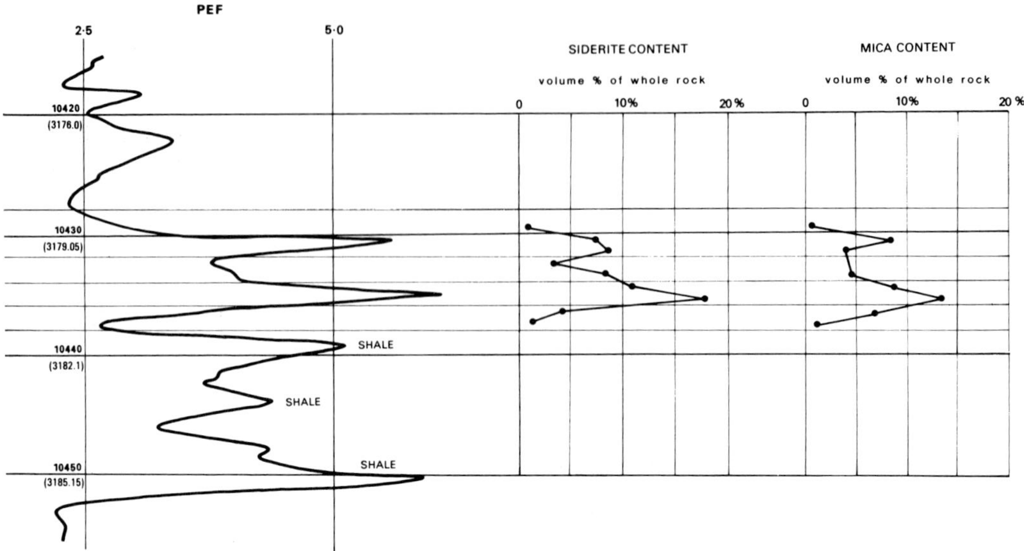

Fig. 9. Comparison of Pe curve (PEF) with mica and siderite contents. The response of the Pe curve is largely (80%+) due to siderite because of its larger effective photoelectric absorption cross-section per unit volume, Uma. Uma values for siderite (56.0), biotite (21.03) and muscovite (7.90) were presented by Suau & Spurlin (1982).

presence of K-feldspar in a typical Upper Jurassic sandstone feldspar assemblage (Fig. 10a) can be readily obtained from the K log of the NGT (Fig. 4), but albite would not be detected. When albite is the dominant feldspar (Fig. 10b), the GR response is almost entirely attributable to clay minerals and micas and thus is unsuitable for identifying the feldspar content. In sandstones with a very complex feldspar mineralogy (Fig. 10c) the problems of feldspar detection are compounded. In the example shown, detrital clays tend to mask the radioactive effects of the feldspars.

Quartz grains and quartz cement

While RHOB and CNL logs can indicate a quartz-rich rock, quartz is not specifically detected by any logs presently in use, but rather is deduced to be present after other minerals have been detected, i.e. by elimination. A pure quartz rock would be indicated by an insignificant radioactivity and a constant density of 2.65 g/cm^3. This is never encountered in practice because, for example, feldspars are always present. Nevertheless, quartz may form *c.*50 to 90% of a North Sea sandstone. A hypothetical pure quartz rock may contain just rounded detrital grains, or varying quantities of quartz cement or, as often is the case in Upper Jurassic sandstones, sponge spicule debris. The only indirect clues to this variation in quartz components may be variations in the porosity values obtained from neutron porosity logs.

Discussion

The previous case examples illustrate both the potential and also the limitations of using the NGT and LDT logs for mineralogical determination. In our experience, examples where a single mineral is sufficiently abundant for its log effects to be isolated are rare; in practice, highly complex mineral assemblages are normal. It is nevertheless important for reservoir characterisation to try to identify all minerals present and to recognize their potential effect or log readings (Nyberg *et al.* 1978; Quirein *et al.* 1982). In particular, successful identification and quantification of clay mineralogy is a major goal of geological analysis of logs. The conductive properties of rocks are affected by clay minerals (Eslinger & Pevear 1988) which, if not corrected for, may lead to inaccurate estimations of the hydrocarbon saturation. Clay

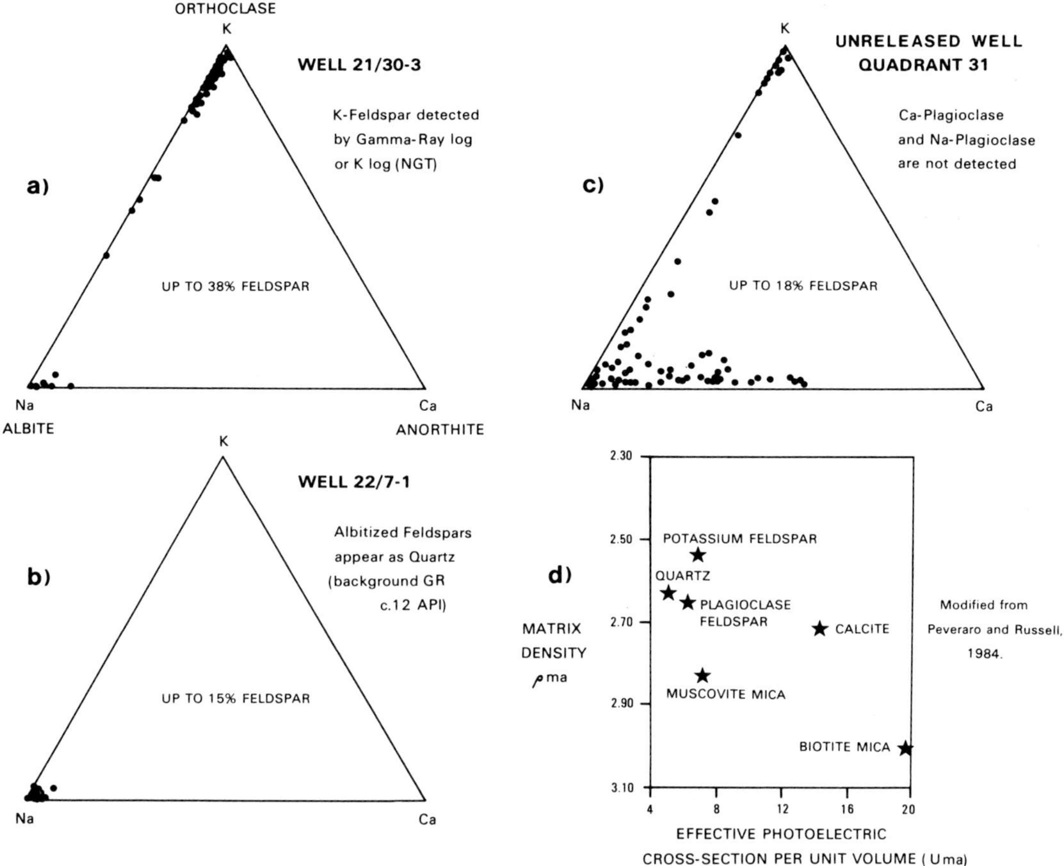

Fig. 10. Electron microprobe analyses of feldspar compositions presented on K–Na–Ca triangular plots showing examples of the variety of feldspar assemblages in North Sea Upper Jurassic sandstones (a–c) Plagioclase feldspar would be confused with quartz on most log cross plots (d).

cements also have a well documented deleterious effect on reservoir flow properties; fibrous illite cements can severely impede permeability (Stalder 1973; Pallatt *et al.* 1984) and reservoir stimulation can induce the migration of kaolinite plates which can block pore throat passages (Muecke 1979; Lever & Dawe 1987).

Our core studies undertaken over a number of years have shown that a clear distinction needs to be made between clays of detrital and clays of authigenic origin, particularly with respect to kaolinite. Detrital clay minerals have invariably proved to be radioactive and affect the GR response. In the case of detrital kaolinite, which on purely mineralogical grounds should have a negligible inherent radioactivity, the observed radioactivity, albeit less significant than that attributable to detrital illite, may result from the adsorption of thorium (Th) or the presence of microscopic inclusions of radioactive heavy minerals. Authigenic kaolinite, however, cannot be reliably identified from the GR log alone. Our attempts to measure the radioactivity of large separates of authigenic kaolinite in the laboratory have failed to detect more than a background level of radioactivity. Utilization of microresistivity logs could help with detection of kaolinite if the pore-filling clay affects the electrical conductivity of the formation (Hurst 1987), but again core analyses would be required for quantitative calibration. Our observations show that the traditional assumption of log analysts, that clay properties in reservoir sandstones and adjacent shales are identical, is clearly erroneous. Indeed, such an assumption is clearly unrealistic when clays of different morphologies are present in sandstones (Ruhovets & Fertl 1982, p. 68).

Our observation also have important implications for V_{clay} determinations using the total GR response, or the Th values from NGT log following the general formula

$$V_{clay} = \frac{\text{Log reading} - \text{minimum reading}}{\text{maximum reading} - \text{minimum reading}}$$

where the maximum reading = 100% shale for a given cored interval and where the minimum reading = lowest log value for a sandstone in the chosen interval.

The formula assumes a linear relationship between log response and V_{clay}. Due to the presence of radioactive K-feldspar and micas, there is usually a significant discrepancy between log-derived values of V_{clay} and mineralogical data (Fig. 11). Although the effect of detrital mica on the GR log is recognized (Suau & Spurlin 1982), the presence and effect of K-feldspars is still often underestimated, even though K-feldspars are major mineral components of sandstones (Cowan & Myers 1988). Because of the negligible radioactivity of authigenic kaolinite, log-derived values of V_{clay} could underestimate V_{clay} in kaolinite-cemented sandstones where the feldspar content is depleted, but overstimate V_{clay} in arkosic or micaceous sandstones (Fig. 11a). Log-derived V_{clay} values will give an upper limit of the actual amount of clay present in muddy sandstones where the radioactivity of K-feldspars and mica supplements the radioactivity of detrital clays (Fig. 10b).

According to Fertl (1979, p. 19) it should be possible to derive quantitative estimates of

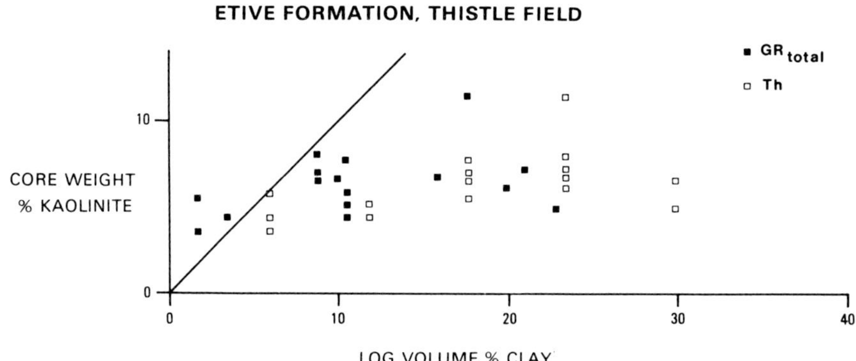

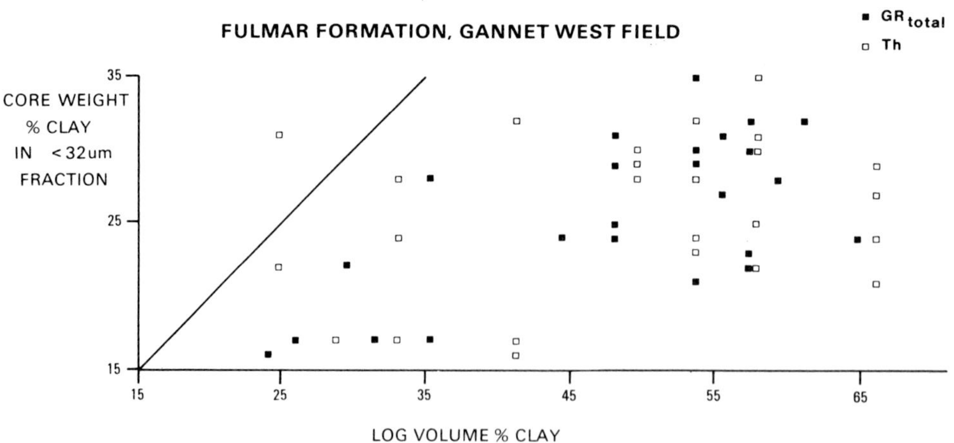

Fig. 11. Log estimates of clay volume, using either total GR or thorium values, are persistently high compared to laboratory measured values. The discrepancy is greater in shaly Upper Jurassic arkoses or sub-arkoses of the Fulmar Formation (b) than in the Middle Jurassic sub-arkoses of the Brent Group (a).

V_{clay} in sandstones using the Th curve from the NGT Log. The authors have found that estimates of V_{clay} using Th values were, if anything, more inaccurate than those derived from total GR values in the North Sea samples (Fig. 10). The unsuitability of using the Th-log for V_{clay} determination is not surprising because knowledge of the distribution of Th in sedimentary rocks and specific minerals is sparse (Rider 1986), the Th content of sandstones is often low, and the precision of downhole Th measurements is often poor (Hurst 1990). The Th-log is, however, useful for detecting rare earth phosphatic cements, which occur locally in some Middle Jurassic Brent sandstones (Fig. 7).

Without core calibration, V_{clay} determinations from GR and NGT logs must therefore always be suspect. In practice, this problem is reduced by combining the results from a variety of logs (e.g. microresistivity, neutron porosity and GR logs) to evaluate this parameter (Frost & Fertl 1981; Hurst 1987; Quirein *et al.* 1981). Use of cation-exchange measurements on petrophysical core plugs to determine clay content is another alternative (Thomas 1976).

Detecting the presence of dense mineral cements is of considerable importance both for specific porosity determination and also for mineral calibration of porosity estimates from the density log (Throckmorton & Al-Shaieb 1986). These minerals can also have an adverse effect on reservoir quality. In extreme cases, dolomite can form laterally extensive cemented horizons (e.g. Irwin 1980) which form barriers to fluid flow in a reservoir. The zone of dolomite shown in Fig. 7, which can be correlated between wells, is also inferred to be laterally extensive. Due to their hydrodynamic properties, flakes of biotite and Fe-chlorite, often in close association with muscovite, occur aligned along bedding planes and can impede vertical permeability.

Although the RHOB log responds to a high concentration of Fe-bearing minerals and barite, the *Pe* curve has been found to be a more useful and sensitive indicator of the presence of minerals with a density greater than 2.65 g cm^{-3}. For example, as little as 2 or 3% siderite can be detected in some of the northern oilfields of the Brent Province such as Murchison and Thistle (Fig. 8). Although the quantity of 'heavy' cements can be assessed with considerable accuracy, there are still problems in discriminating between siderite, pyrite, Fe-chlorite, biotite, ferroan dolomite and ferroan calcite using just the *Pe* response. This is important for enhanced recovery considerations, because the various minerals might react differently to, for example, acid flushing treatments (Lievaart & Davies 1987). Tightly cemented carbonate horizons would be indicated on the sonic log. Measurement of Fe by the GLT log and, perhaps in the future, Mg, will aid in the detection of pyrite and may help to distinguish between the other Fe-bearing minerals.

The abundance and state of preservation of feldspars in a sandstone reservoir is not only of importance for log interpretation, but also of importance for reservoir quality considerations. Dissolution of feldspar grains during burial diagenesis is probably the main origin of secondary porosity in the North Sea oilfields. Plagioclases are generally been more susceptible to dissolution than potash feldspar, although feldspar assemblages consisting of pure plagioclase (albite) do occur in the North Sea area. If logs indicate low K-feldspar contents, zones of secondary porosity might be suspected. Authigenic K-feldspar cements are rare in Jurassic North Sea sandstones and are unlikely to make a significant contribution to the GR response.

Mineral determinations from nuclear logs can be usefully corroborated by geochemical data. For example, the new GLT log improves the detection of feldspars in that SiO_2/Al_2O_3 ratios, used in conjunction with the GR response, would indicate an arkose or sub-arkose composition. However, Na is not yet recorded by the GLT log and current GLT interpretations assume a fixed mixture of K-feldspar, albite and anorthite (Herron & Herron 1990).

Fig. 12 is an attempt, based on the authors' experience, to summarise the relative ease of detection of different minerals, using only the density and gamma-ray suite of logs available to the authors at the present time. Where possible, the detrital and authigenic phases of the same mineral are separated. Detection of certain mineral phases of course depends on the amounts present, and the detection thresholds presented are minimum values at which the mineral could be detected if its log effects were not overwhelmed by other minerals.

Conclusions

In this brief survey it has been demonstrated that there is considerable scope for further utilization of detailed mineralogical studies for the calibration of logs. A detailed core calibration during the early phase of exploration is essential. However, there are clearly still many problems with deriving geological data from wireline logs, and these problems would be

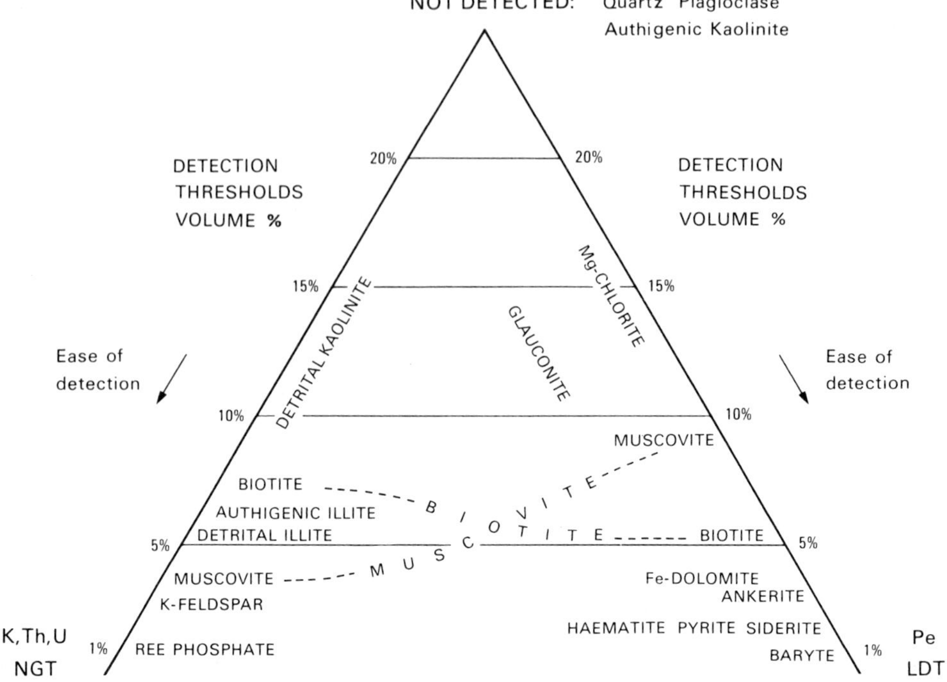

Fig. 12. Summary diagram showing the relative ease of detection of different minerals. Detrital and authigenic forms of the same mineral are distinguished where it is merited.

compounded when interpretations are attempted in geologically unfamiliar territories. Difficulties arise in the case of complex mineral assemblages, because minerals differ in their ease of detection, and because most of the common mineral components of reservoir sandstones can occur both as detrital grains and as cement and which obviously exert differing effects on reservoir properties. A number of specific problems with mineral interpretations from logs have been identified in the course of our study.

(1) It is often difficult to unravel the effects of individual minerals when a log is responding to several minerals simultaneously. For example, siderite, muscovite, biotite, ferroan calcite and pyrite all make significant contributions to the Pe curve in micaceous sandstones from the Brent Group.

(2) It is particularly difficult to calibrate nuclear logs for minerals with a variable composition such as dolomite or chlorite. Detrital illites are often interstratified with non-radioactive or poorly radioactive smectite. An elemental rather than mineralogical approach might be more appropriate in such circumstances.

(3) Authigenic and detrital forms of the same clay mineral may have different radioactivities, largely the result of adsorbed radioactive material.

(4) The log effect of some minerals may be overwhelmed by more abundant or more easily detected mineral phases. For example, despite its potentially disastrous effect on permeability, fibrous illite cement is unlikely to be detected by logs; the K content of fibrous illite would be overshadowed by the effect of K in the ubiquitous and more abundant K-feldspars.

(5) The non-radioactive feldspars in sandstones cannot be identified by nuclear logs.

Some of these problems can be alleviated by utilizing all available log data including geochemical data from the GLT log in conjunction with nuclear log responses. Finally, it should be emphasised that reservoir quality is not merely determined by the amounts and nature of the cements present, grain size and porosity that can be interpreted from core-calibrated logs. Petrological and petrophysical studies of core material will still be required to determine factors such as the distribution and morphology of cements, and the shape and degree of connection of pores, and lateral and vertical permeability.

We are indebted to the Petroleum Engineering Division of the UK Department of Energy for permission to publish the results of this work which is part of the ongoing hydrocarbons programme. We are grateful for the technical support given by staff at the former BGS office in Leeds, and our present office in Keyworth. Special thanks are due to Judith Renwick, Debbie Ettershank, Rosalind Price and Claire Hallsworth for undertaking many of the routine analyses over the years that have proved crucial for this study, and to Tony Myers for drafting many of the figures. Andy Morton is thanked for drawing our attention to the rare earth phosphate anomaly shown in Fig. 9. Andrew Hurst, Helen Steward, Andy Morton, Robert Knox & Chris Evans are all thanked for helpful reviews and advice on various drafts of this paper. This paper is published with the approval of the Director of the British Geological Survey, NERC.

References

COWAN, D. R. & MYERS, K. J. 1988. Surface gamma ray logs: a correlation tool for frontier areas: discussion. *American Association of Petroleum Geologists Bulletin*, **72**, 634–636.

CURTIS, C. D., HUGHES, C. R., WHITEMAN, J. A. & WHITTLE, C. K. 1985. Compositional variation within some sedimentary chlorites and some comments on their origin. *Mineralogical Magazine*, **49**, 375–386.

ESLINGER, E. & PEVEAR, D. 1988. Clay minerals for petroleum geologists and engineers. *SEPM Short Course Notes* **22**.

FERTL, W. H. 1979. Gamma ray spectral data assists in complex formation evaluation. *The Log Analyst*, **20**, 3–37.

FOLK, R. L. 1974. *Petrology of Sedimentary Rocks*. Hemphill Publishing Company, Austin, Texas.

FROST, Jr., E. & FERTL, W. H. 1981. Integrated core and log analysis concepts in shaly clastic reservoirs. *The Log Analyst*, **22**, 3–16.

HEINZE, D. M. 1985. Clay identification and amount measured by laboratory techniques compared to well log responses: application to tight gas sands and shales. *Sandia report SAND 85–0312*. Sandia National Laboratories, Albuquerque, New Mexico.

HERRON, M. M. & HERRON, S. L. 1990. Geological applications of geochemical well logging. *In*: HURST A., LOVELL, M. A. & MORTON, A. C. (eds). *Geological applications of wireline logs*. Geological Society, London, Special Publication, **48**, 165–175.

HURST, A. 1987. Mineralogical analysis and the evaluation of the petrophysical parameter V_{shale} for reservoir description. *Marine and Petroleum Geology*, **4**, 82–91.

—— 1990. Natural gamma-ray spectrometry in hydrocarbon-bearing sandstones from the Norweigan continental shelf. *In*: HURST, A., LOWELL, M. A. & MORTON, A. C. (eds). *Geological applications of wireline logs*. Geological Society, London, Special Publication, **48**, 211–222.

—— & ARCHER, J. S. 1986. Sandstone reservoir description: an overview of the role of geology and mineralogy. *Clay Minerals*, **21**, 791–809.

IRWIN, H. 1980. Early diagenetic carbonate precipitation and pore fluid migration in the Kimmeridge Clay of Dorset, England. *Sedimentology*, **27**, 577–591.

JOHNSON, H. D., MACKAY, T. A. & STEWART, D. J. 1986. The Fulmar Oil-Field (Central North Sea): geological aspects of its discovery, appraisal and development. *Marine and Petroleum Geology*, **3**, 99–125.

KNOX, R. W. O'B., MORTON, A. C. & LOTT, G. K. 1982. Petrology of the Bridport Sands in the Winterborne Kingston borehole, Dorset. *In*: RHYS, G. H., LOTT, G. K. & CALVER, M. A. (eds). *The Winterborne Kingston borehole, Dorset, England,*, Report of the Institute of Geological Sciences, **81/3**, 107–121.

LEVER, A. & DAWE, R. A. 1987. Clay migration and entrapment in synthetic porous media. *Marine and Petroleum Geology*, **4**, 112–118.

LIEVAART, L. & DAVIES, D. R. 1987. The role of fines during acidizing treatments. *Marine and Petroleum Geology*, **4**, 127–131.

MORTON, A. C. & HUMPHREYS, B. 1983. The petrology of the Middle Jurassic sandstones from the Murchison Field, North Sea. *Journal of Petroleum Geology*, **5**, 245–260.

MUECKE, T. W. 1979. Formation fines and factors controlling their movement in porous media. *Journal of Petroleum Technology*, **31**, 144–150.

NYBERG, Ø., LINDBERG, P. A., LIEN, K. & SMISTAD, J. K. 1978. Mineral composition, an aid in classical log analysis used in Jurassic sandstones of the northern North Sea. *19th Annual Logging Symposium Transactions (Society of Professional Well Log Analysts), Houston*, Paper M, 1–35.

PALLATT, N., WILSON, M. J. & MC HARDY, W. J. 1984. The relationship between permeability and the morphology of diagenetic illite in reservoir rocks. *Journal of Petroleum Technology*, **36**, 2225–2227.

PEVERARO, R. C. A. & RUSSELL, K. J. 1984. Interpretation of wireline log and core data from a mid-Jurassic sand/shale sequence. *Clay Minerals*, **19**, 483–505.

QUIREIN, J. A., BALDWIN, J. L., TERRY, R. L. & HENDRIX, M. 1981. Estimation of clay types and volumes from well log data — an extension of the global method. *Proceedings of 8th Formation Evaluation Symposium Canadian Well Logging Society, Calgary*, Paper 8.

—— GARDNER, J. S. & WATSON, J. T. 1982. Combined natural gamma ray spectral/litho-density measurements applied to complex lithologies. *Society of Petroleum Engineers of AIME*. 11143, 1–14.

RIDER, M. H. 1986. *The geological interpretation of well logs*. John Wiley, New York.

RUHOVETS, N. & FERTL, W. H. 1982. Volumes, types and distribution of clay minerals in reservoir rocks based on well logs. *Proceedings of the SPE/DOE unconventional gas recovery symposium. SPE/DOE paper no. 10796*. Society of Petroleum Engineers, Dallas, 67–82.

STALDER, P. J. 1973. Influence of crystallographic habit and aggregate structure of authigenic clay minerals on sandstone permeability. *Geologie en Mijnbouw*, **52**, 217–219.

SUAU, J. & SPURLIN, J. 1982. Interpretation of micaceous sandstones in the North Sea. *23rd Annual Logging Symposium Transactions (Society of Professional Well Log Analysts), Houston*. Paper G, 1–32.

THOMAS, E. C. 1976. The determination of Qv from membrane potential measurements on shaly sands. *Journal of Petroleum Technology*, **28**, 1087–1096.

THROCKMORTON, H. C. & AL–SHAIEB, Z. 1986. Core-calibrated logs utilization in recognition of depositional facies and reservoir rock of the Henryhouse Formation (Silurian), Anadarko Basin. *27th Annual Logging Symposium Transaction (Society of Professional Well Log Analysts)*, Paper L, 1–18.

Identification of source rocks from wireline logs

A. E. STOCKS & S. R. LAWRENCE
Exploration Consultants Limited, Henley-on-Thames, Oxfordshire, UK

Abstract: This paper reviews the properties of organic-rich rocks which constitute petroleum source rocks and investigates petrophysical methods which can be used for their assessment. A review is made of the common log data and their response to an appreciable organic content in shale formations. Associations and effects which might lead to an erroneous assessment are also discussed. Previous researchers have highlighted relationships between organic content and various log parameters or values. These rely on geochemical control and tend to be stratigraphically and regionally specific. In this paper we show how standard petrophysical processing can be applied for qualitative and semi-quantitative evaluation of organic content and therefore source rock potential. Recommendations are made for a reform in geochemical sampling which will enable better quantitation methods for more widespread application.

The use of petrophysical techniques arises out of the need to assess source rocks for basin studies in various geological settings, where the database is restricted to basic log data and little or no geochemical information, and to provide an integrated 'running' assessment of source rock capability for volumetric determinations.

Source rocks are mostly formed by fine grained sediments such as mudstones and shales, or micritic limestones (Tissot & Welte 1984). The important constituent of source rocks is organic matter (OM). A rule of thumb used by petroleum geochemists is that rocks sourcing commercial amounts of hydrocarbons normally have a TOC (total organic carbon) value of more than 1% (by weight) although their ultimate commerciality depends on other factors such as organic matter type and expulsion/drainage efficiency. Typically, TOCs vary from less than 1% in poor source rocks to over 20% in rich so-called 'oil shales'.

The macroscopic anatomy of a source rock is rarely studied in the subsurface since they are seldom cored. One of the best studied sections however, is that of the Upper Jurassic Kimmeridge Clay in the cliffs at Kimmeridge Bay. The broad variations in organic content have been plotted by Williams (1986) (Fig. 1). The whole section is relatively organic-rich and can be considered a potential source rock, although in this case it is thermally immature. The 'background' illitic mudstone has a TOC content varying up to 10%. Bituminous shale bands with TOCs up to 30% about 2 m thick occur, and 'oil shale' bands with TOCs up to 70% are fairly common but usually less than 10 cm thick.

The microscopic anatomy of source rocks is probably even more poorly understood.

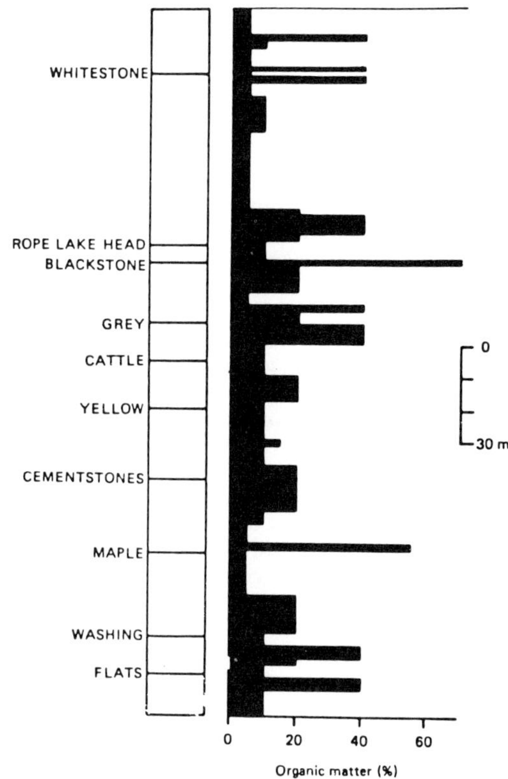

Fig. 1. Distribution of organic matter — Kimmeridge Bay Section Dorset (from Williams 1986).

Organic-rich rocks are commonly laminated in hand specimen (on a scale of mm) and microscopically exhibit concentrations of darker carbonaceous material, light and less organic clay material, and in some cases silty fractions.

From HURST, A., LOVELL, M. A. & MORTON, A. C. (eds), 1990, *Geological Applications of Wireline Logs* Geological Society Special Publication No. 48, pp. 241–252

Log characteristics of organic-rich rocks

A basic log suite is presented which combines gamma-ray, resistivity, acoustic, neutron and density devices (Fig. 2). This example is taken from well 17/9–1 in the Norwegian-Danish Basin and shows the log response of a known source rock in the Tau and Sauda formations (broadly equivalent to the Kimmeridge Clay of England). The expected responses of these common logs to increasing organic content can be summarized as follows:

Gamma ray. Increases in gamma-ray values can be expected where uranium is associated with OM although this is not an exclusive relationship. Logging tool measurements of total gamma radiation are sensitive to small-order variations in U and major diversions may be recorded across organic-rich formations. Perhaps the most notable example is the Upper Jurassic Kimmeridge Clay in the North Sea, the so-called 'hot shale' where gamma-ray values may increase by well over 200 API units across organic-rich intervals.

Direct measurements of U concentration from the natural gamma spectrometry tool (NGT) are potentially a more direct indicator of organic matter because this tool isolates the influences of Th and K associated with mica and clay minerals.

Bulk density. Specific gravities of OM are estimated to be in the range of 0.95 to 1.05 g cm^{-3}. This is in line with the specific gravity of freshwater and contrasts significantly to common rock matrix densities of 2.6 to 2.9 g cm^{-3}. As the OM is considered to form part of the matrix in a source rock, reductions in bulk formation density can be expected. The level of density reduction from organic content is directly comparable (in vol %) with that caused by porosity.

Acoustic transit time. Average transit times of 180 ms ft^{-1} are suggested for OM (Mendelson & Toksoz 1985). As with the density log the acoustic log response to OM is comparable to that of porosity, i.e. increasing the transit time. However, the distribution of the organic matter may also be influential. OM distributed as platelets may have a greater effect on transit time than OM more evenly distributed in the matrix (Mendelson & Toksoz 1985).

Neutron porosity. The neutron tool measures hydrogen concentration which is present as water and hydrocarbons in source rocks. Water has a neutron porosity response of 100 porosity units whilst hydrocarbon may range down to near zero for dry methane. The neutron response to OM is estimated to average 67 porosity units (Mendelson & Toksoz 1985) whereas matrix responses are typically close to zero.

It is realistic to expect neutron porosities to increase in organic-rich rocks. However, there may be some counterbalancing of this effect in mature gas-generating source rocks where hydrogen concentrations may be reduced (gas effect).

Resistivity. OM is considered to be essentially non-conductive. Therefore, increases in resistivity are expected where OM is present at the expense of conductive clays. It can, however, be postulated that the most significant increases in resistivity are likely once hydrocarbon generation has commenced and free hydrocarbons have displaced part of the conductive pore water. Source rocks are characteristically thinly bedded and laminated. Thus, log theory predicts that they will exhibit electrical anistropy, resistance being typically greater vertically. Higher resistivity measurements may result from this cause alone.

In all logs the anticipated effects of OM can be negated or exaggerated by other influences. Washed-out boreholes can reduce apparent formation densities, increase acoustic transit times and neutron porosities but reduce resistivities and gamma ray levels. The presence of pyrite, if widely disseminated, could negate completely the density reductions caused by the presence of OM. The occurrence of radioactive minerals such as micas can cause both gamma-ray and formation density increases. Overpressure in fine-grained rocks leads to an increase in apparent porosity in the same way as the presence of OM.

It is unlikely that any of these additional influences will 'mimic' the effect of OM in all logs. For this reason it is always important to take into account as wide a spread of log data as possible in identifying potential source rocks from logs.

Log evaluation techniques

A number of techniques have been developed that have demonstrated the use of well logs in determining variations and absolute quantities of OM.

(A) In a number of papers Schmoker (e.g. 1979 and 1981) described direct relationships

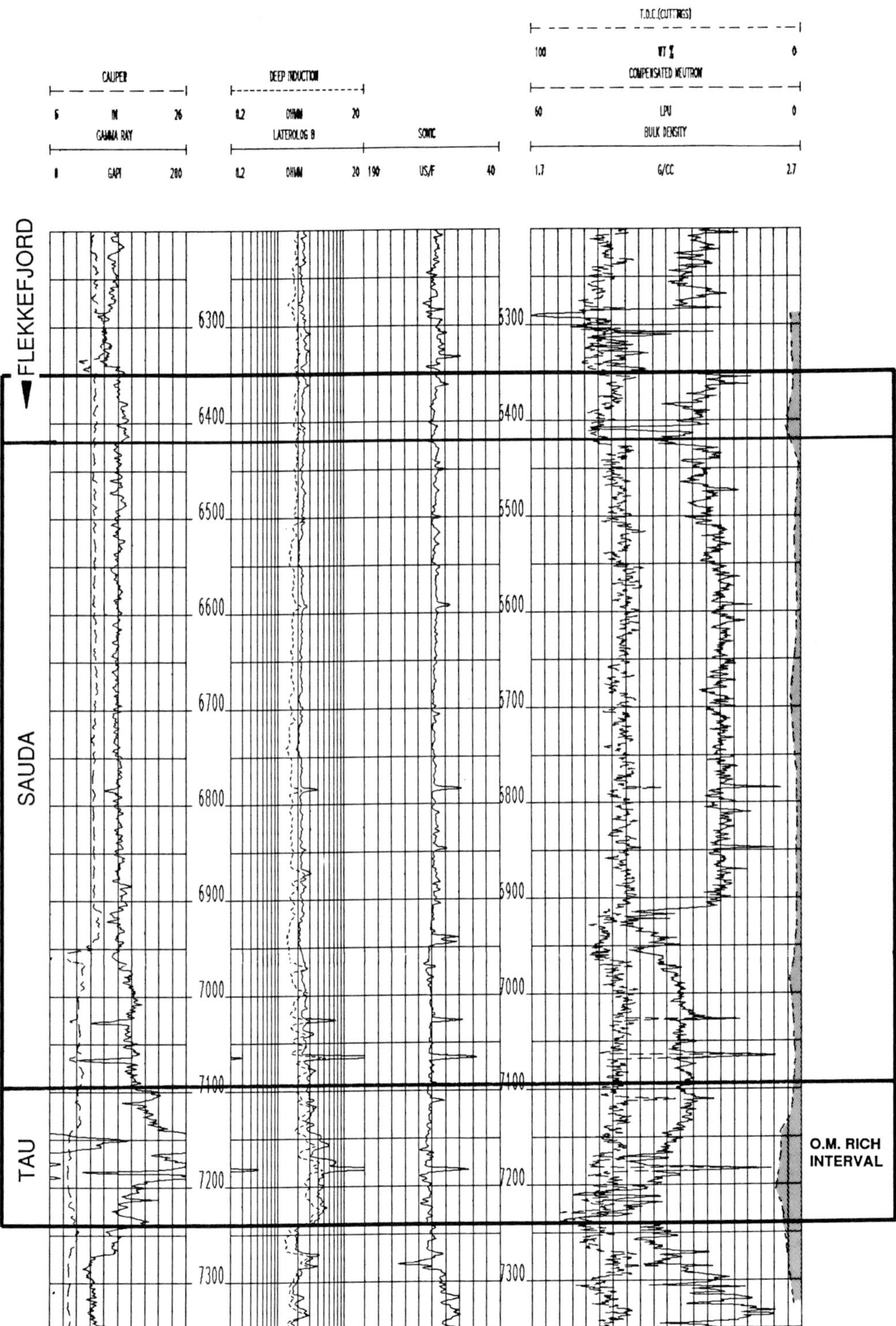

Fig. 2 Basic log suite over organic-rich intervals. Well 17/9–1, Norwegian-Danish Basin.

between gamma ray and TOC from cores, and between density and TOC. (N.B. His algorithms included adjustments for pyrite content which was found to increase with OM content). His studies concentrated on the Appalachian Devonian shales where he concludes that a single relationship between OM content and a log-measured parameter is only valid within a localized area and cannot be applied generally.

(B) Autric & Dumesnil (1985) and Dellenbach *et al.* (1983) concentrated on the use of the gamma ray, sonic and deep resistivity on the basis that these logs are least effected by borehole rugosity. In barren shales transit time and resistivity variations are caused primarily by changes in bound- and pore-water volumes, such that a straight line relationship should be defined (according to Archie principles) (Fig. 3). Deviations from this 'water' line can be expected where OM is present. Similar trends were observed on plots of gamma ray against resistivity.

(C) Meyer & Nederlof (1984) examined sonic/resistivity and density/resistivity relationships. In this case, resistivities were corrected to standard temperature. This method is more qualitative since it identifies source rocks with points lying on the more resistive side of the barren shale or 'water' line (Fig. 4).

(D) Mendelson & Toksoz (1985) examined the relationships between TOC from cores and the sonic, neutron, density and gamma ray logs. In their data set from the North Sea they found definable relationships of varying quality. They also modelled expected log responses in source rocks assuming a simple formation model of matrix, OM and porosity. In calculating a TOC value from logs they used multiple regression analysis to define relationships between TOC and the various responses. Their results are satisfactory but they found differences between wells such that a single algorithm could not be applied regionally with any confidence.

(E) Mann *et al.* (1986) analysed data from a shallow borehole in the Lower Jurassic Posidonia shale in N.W. Germany. They found little relationship between core-measured TOCs and the density, sonic and resistivity levels. Density was influenced by pyrite content, the sonic response by borehole rugosity and the resistivity measurement lacked the necessary resolution. On NGT logs the proportion of U showed the most promising relationship with TOC but this varied according to lithology and the occurrence of fractures.

(F) Herron (1986) used data from the Gamma Spectroscopy Tool (GST) to determine volumes of organic carbon independent of core data. The tool gives elemental proportions of Ca, Cl, Si, Fe and S when run in capture mode and C and O in inelastic mode. Analysis of these data, particularly the C, O and Ca, enabled Herron to produce an estimate of organic carbon. Results from a study on a well in Oklahoma showed a good agreement with core data in estimating organic and inorganic carbon. The data are described as being sensitive to low levels of organic carbon in the 0.5 to 2.0 wt% range. However, the recordings are made at a number

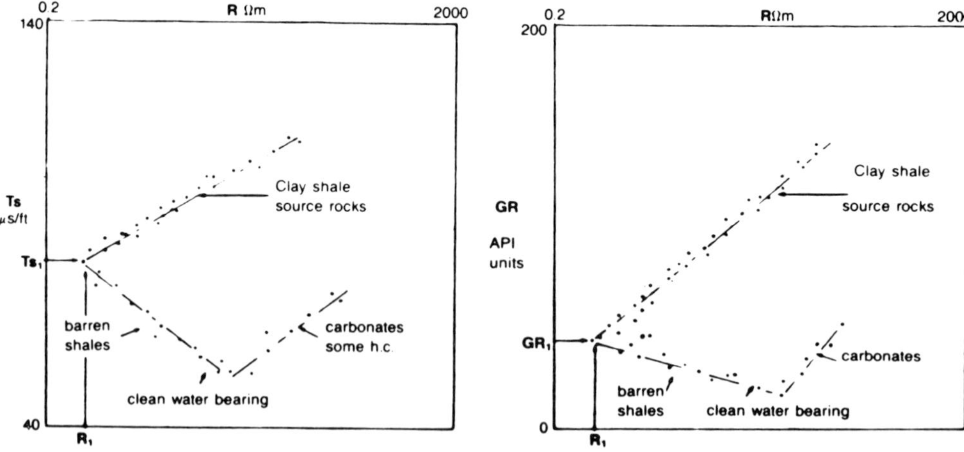

Fig. 3. Characterisation of source rocks on sonic — resistivity and gamma-ray resistivity crossplots (from Autric & Dumesnil 1985).

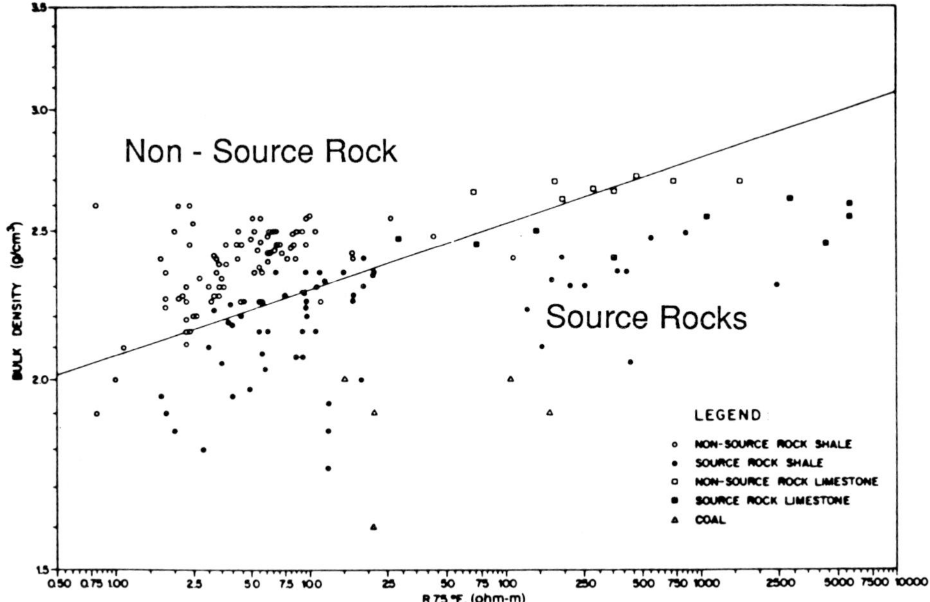

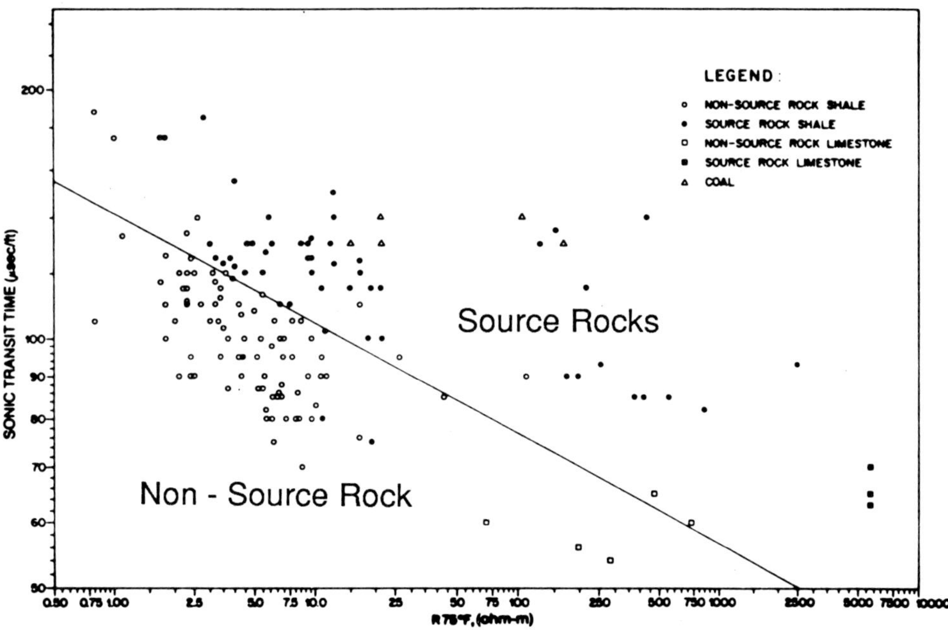

Fig. 4. Identification of source rocks from density–resistivity and sonic resistivity crossplots (from Meyer & Nederlof 1984).

of stationary positions. Continuous recording with a moving sonde may reduce the sensitivity.

Limitations in quantitative appraisal

The extent and nature of geochemical sampling provides the greatest limitation to quantitative appraisal of organic content. There are large discrepancies between the sample intervals of geochemical data and that of log data. Geochemical cuttings samples are usually collected over an interval of about 30 m, or at best 10 m. They represent an 'averaged' assessment of organic content for that interval. In the Kimmeridge Clay section at Kimmeridge Bay organic content can vary from a few per cent to over 60% within a few metres (Fig. 1 and Williams 1986). Such variations may be within the resolution of the conventionally recorded log data but can only be identified by intensive geochemical sampling.

A further limitation is provided by the microscopic textural nature of source rocks. Knowledge of log responses will show that there are evidently significant variations on a laminar- and micro-scale that will be beyond the resolution of the log measurements.

The results of research so far into log evaluation of source rocks have adequately highlighted the limitations inherent in quantitative determination. Methods that involve direct calibration with core data are only reliable within a localized area. In addition, cored source

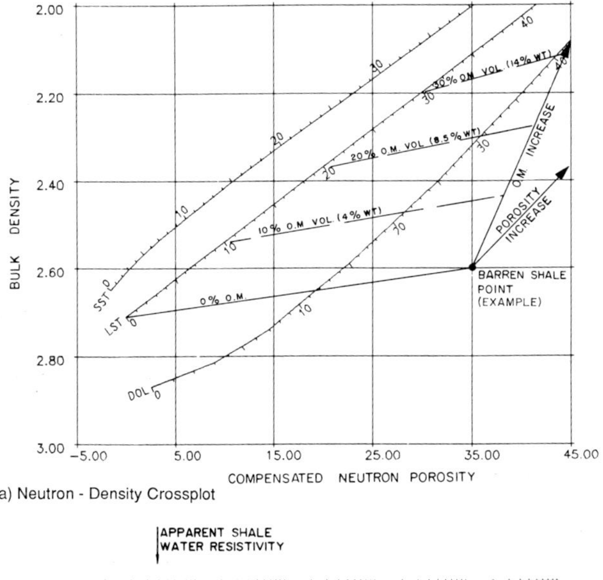

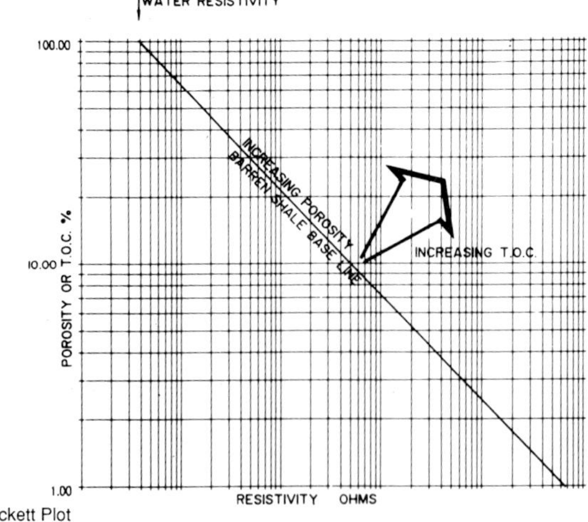

Fig. 5. Expected effects of TOC on crossplots: (a) neutron density crossplot; (b) Pickett plot.

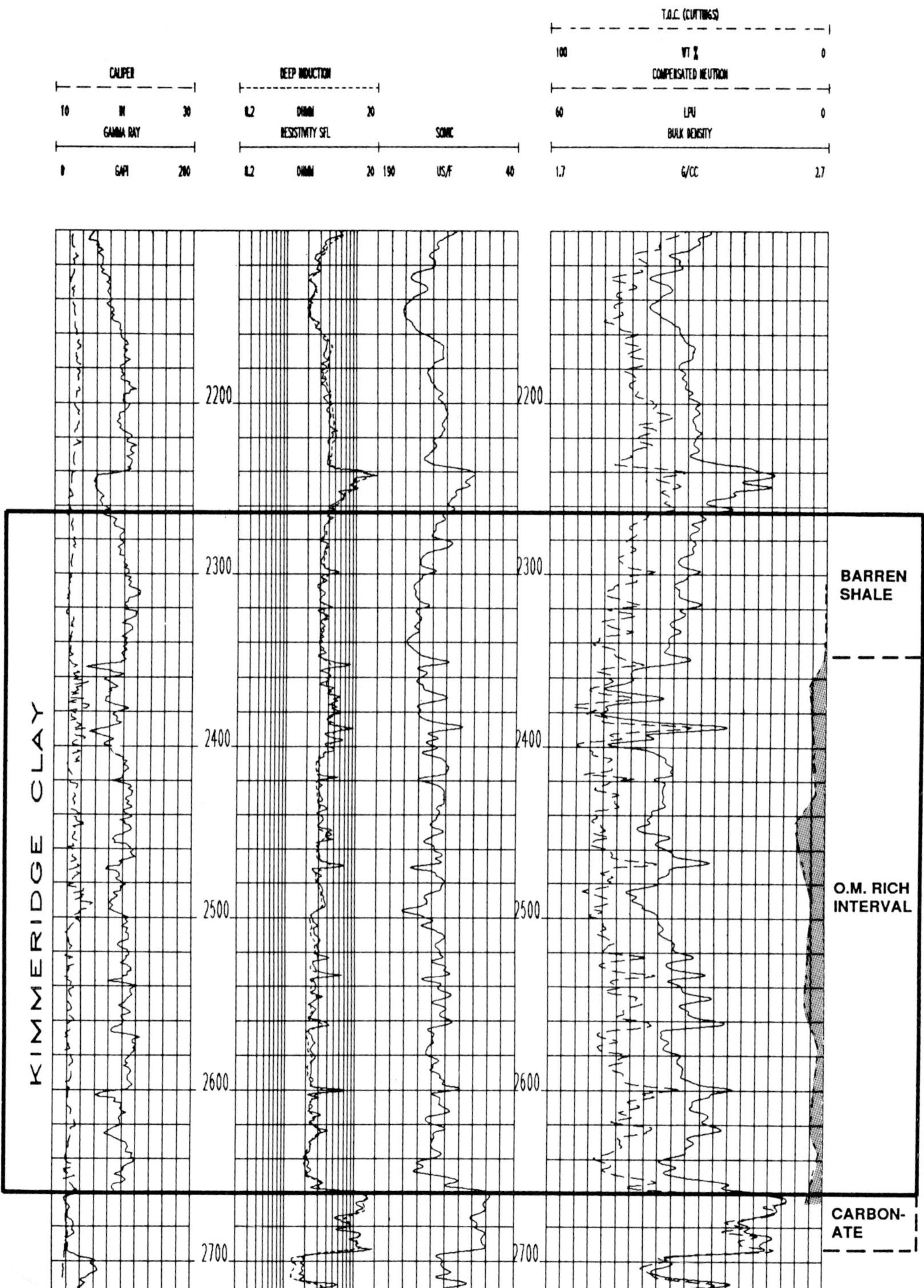

Fig. 6. Basic log suite over Kimmeridge Clay. Well Marchwood-1, S England.

rocks with full geochemical analyses are virtually non-existent. Nevertheless log evaluation techniques may have an application as qualitative indicators until core-based geochemical data are available to enable better calibrations.

Application of conventional petrophysical methods

In a basin evaluation study the researcher is generally in the position of having to accept the existing and generally limited data. Basic log suites will be the norm and geochemical data restricted to composite cuttings samples over long (20 m or more) intervals.

Qualitative visual examinations of log responses may highlight classic 'hot' shale intervals but for quantitative or semi-quantitative evaluation further qualifications are required of the log data. Increases of gamma ray and resistivity may be the result of increased OM but this is not necessarily always the case. There are no absolute values definable for the natural gamma radiation, U levels or resistivity of OM. Such definable responses are theoretically only practical with the density, neutron and sonic logs.

Two basic log analysis techniques have applications in quantifying organic content from well logs. The neutron-density crossplot (Fig. 5a) is a principle display for defining matrix and shale properties. Trends of increasing organic matter are potentially definable away from barren or 'warm' shale points of zero or relatively low (background) organic content. The overall trends, however, may be confused either with a genuine increase of porosity or lithological changes. To distinguish these, a second crossplot of resistivity versus apparent TOC plus porosity (modified Pickett Plot) can be

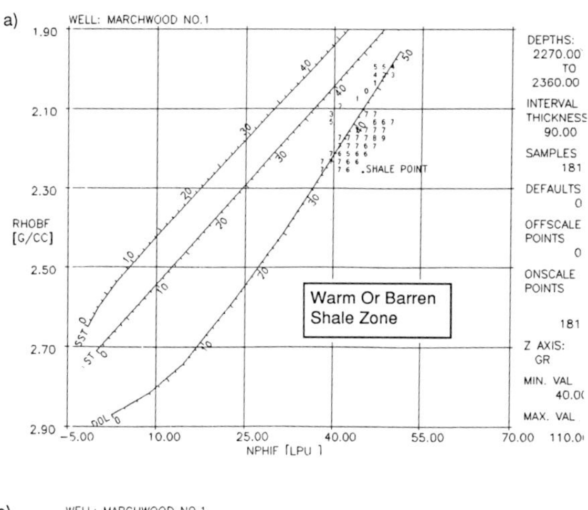

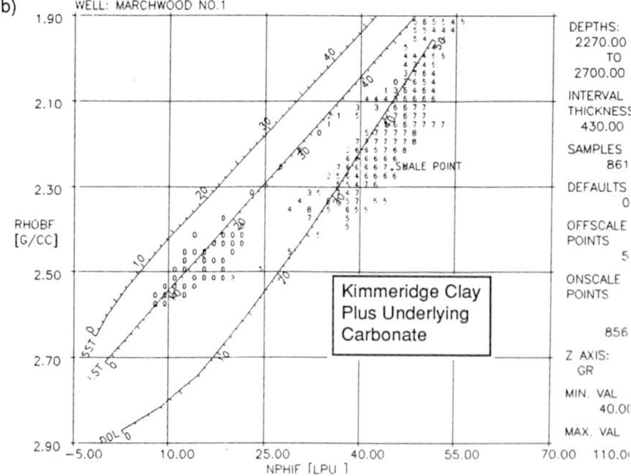

Fig. 7. Neutron density gamma ray z plots. Marchwood-1.

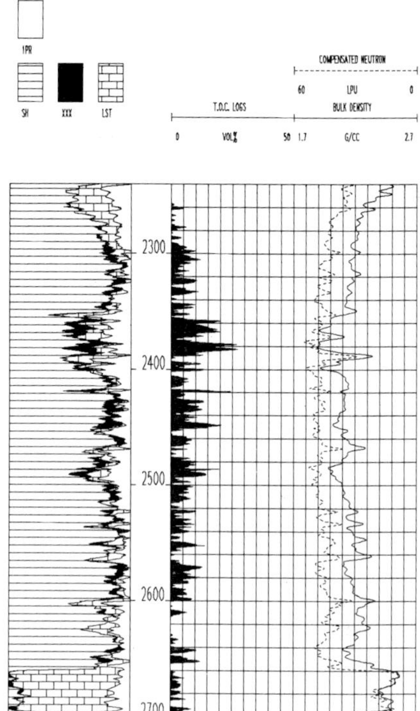

Fig. 8. Interpreted lithology and TOC from logs. Marchwood-1.

used (Fig. 5b). As in standard log analysis techniques barren shale intervals can be used to define a shale 'water' line above which points probably contain organic matter and/or free hydrocarbons.

At its simplest level the porosity log crossplot is only capable of resolving TOC, matrix and shale volumes. It is therefore open to under-or over-estimation of TOC due to the effect of pyrite on the density log or actual changes in porosity and matrix. However, by using multiple crossplots and additional logs such as the sonic, photo-electric factor (PEF, from litho density tool) and, after simple local calibration, the gamma ray, it may be possible to use computer-based deterministic analysis techniques to resolve additional variables such as porosity, and pyrite.

An example is shown for the Kimmeridge Clay section of the Marchwood-1 well, in Dorset, Southern England where a basic log suite was recovered (Fig. 6) and some TOC levels were measured in cuttings samples. Although this penetrated a stratigraphically equivalent formation to that of the 17/9−1 well (Fig. 2) and has similar levels of TOC from cuttings, the gamma ray shows little evidence of the high radioactivity sometimes associated with an organic-rich source rock.

Two neutron-density-gamma ray Z-plots are presented (Fig. 7). The upper plot is from 2270 to 2360 ft where logs show little indication of high organic content and TOCs from cuttings are consistently around about 1%. The zone is considered a 'warm' or relatively barren shale within the unit although it could still constitute a source rock. The lower crossplot extends across the interval 2270 to 2700 ft and includes the section where the geochemical data indicate the presence of OM and maximum TOCs are close to 10% by weight. The trend on the crossplot (Fig. 7) towards low density and high neutron is clearly seen.

Using deterministic formation analysis techniques it is possible to quantify the volumes of OM, shale, matrix and porosity that would be required to give the recorded log responses (Fig. 8). Input logs used are density, neutron and gamma ray. The non-shale matrix is assumed to be carbonate and standard properties for OM were assumed (for density, sonic velocity etc.). For comparison the log-derived TOCs are converted from a vol% to wt% and displayed with the laboratory TOC measurements (Fig. 9). The main anomalies between the two results are associated with intervals of irregular borehole size. However, this form of petrophysical analysis appears able to detect broad qualitative variations in OM, but the validity of the results can only be fully assessed by comparison with laboratory data taken at an equivalent sample frequency.

For further analysis apparent TOCs derived from logs have been substituted for porosity on the Pickett Plot (Fig. 10). Resistivities are expected to increase in source rocks and most significantly in the presence of free hydrocarbons generated by thermal maturation. It can be seen from the plots that in the 'warm'/immature shales resistivity is inversely related to the apparent TOC and decreases as TOC increases (Fig. 8a). Over such a zone the resistivities do not support the evidence of varying TOCs interpreted from the density and neutron logs. A more reasonable explanation may be provided by bound water, porosity and lithological variations or even borehole effects. In the more organic-rich intervals (Fig. 8b) resistivities increase and deviate from the 'warm' shale line. This is interpreted to be an indication of source-rock potential and would also distinguish apparent high TOCs caused by washouts (density log effect) from those caused by genuine high OM content.

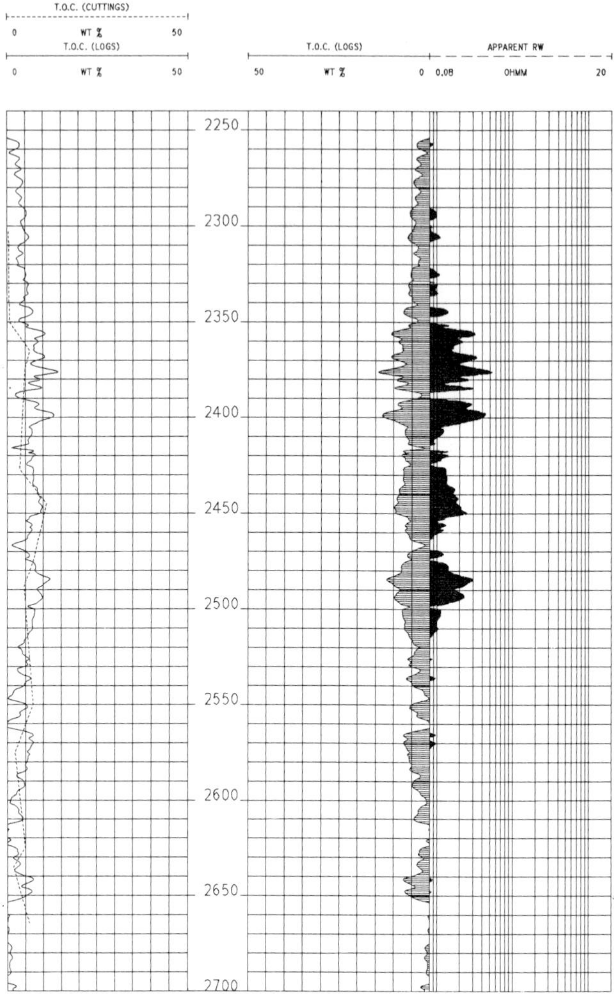

Fig. 9. Log cuttings comparison of TOC indications. Marchwood-1.

In addition, the Pickett Plot results can be interpreted in a semi-quantitative manner. The apparent water resistivity, R_wa, in the shale can be extrapolated from the 'warm' shale line. By assuming the gradient of the 'warm' shale line to be constant for all points, R_wa values for every data point can then be generated and used to identify relative changes in organic richness. The results (Fig. 9) show the computed R_wa values for the Marchwood well and an obvious increase in the interval 2350 to 2500 ft.

The Pickett Plot is a standard petrophysical technique and can help eliminate the effects of changing formation, bound- and pore-water content that will also effect resistivity and perhaps mask the presence of potential source rocks.

Conclusions, recommendations and future directions

In this paper we have attempted to review the petrophysical nature of organic-rich rocks, and their associated components. The changes they undergo with burial and thermal transformation and the resulting variations are the main complications in quantitative evaluation of source rocks. However, it is the same variations, and prediction of their trends, that provide encouragement for future log-based applications.

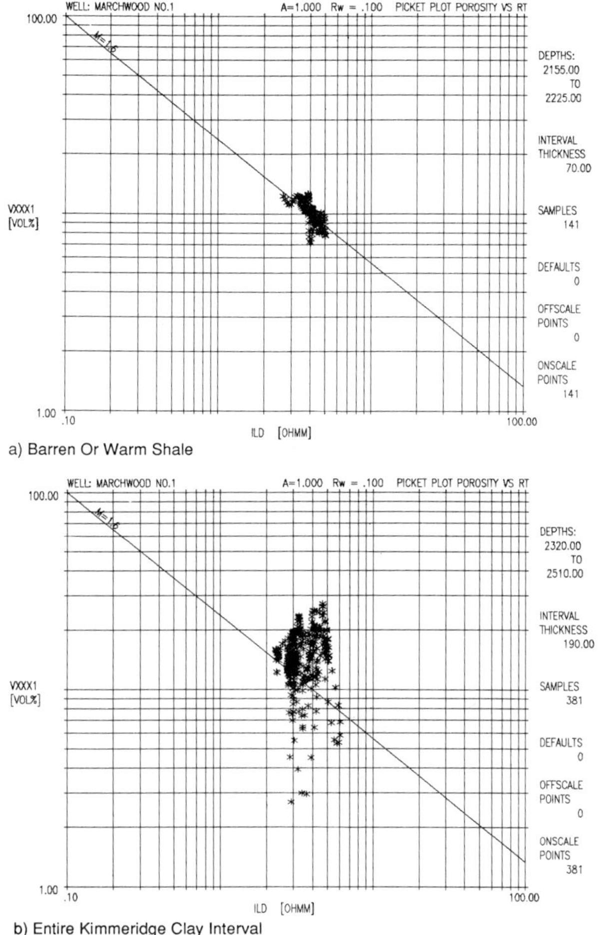

Fig. 10. Modified Pickett plots of TOC against deep resistivity. Marchwood-1. (a) barren or warm shale; (b) entire Kimmeridge clay interval.

Further information on the nature of the OM and its disposition, and physicochemical transformation to hydrocarbons in specific source rocks, will allow a better correlation and calibration of the petrophysical trends, and ultimately a more closely-controlled quantitative interpretation of log data.

This review of techniques clearly shows that logs can be used to identify organic-rich formations and thus potential or actual source rocks. Many techniques to date have started with core data and attempted to calibrate the log responses to it. However, as demonstrated, conventional petrophysical techniques, commonly applied to reservoir intervals, may prove to be useful, at least qualitatively, for source rock evaluation. Realistically, however, OM contents less than about 3% may not always be detectable from conventional open hole logs. On this basis some known commercial source rocks may not have been recognised.

To enable log-based methods to be fully evaluated and calibrated for quantitative determination of OM content, a much more frequent sampling of cuttings and core data for laboratory geochemical analysis is required. The current normally adopted sampling intervals are inadequate to enable calibration. Thus, if advances are to be made, geochemical sampling programmes should be re-appraised with a view to dramatically increasing the frequency of sampling from cuttings and cores. We would in particular recommend the more extensive use of wireline core sample recovery. This would give a true 'spot' sample with a definable OM content at least on a 'macroscopic' scale. This would enable a better absolute calibration for log evaluation at one point, although microscopic variations are beyond the resolution of log data.

We make a broader plea, however, to explorationists for a greater consideration of non-reservoir rocks. Logging programmes could be designed to account for the future evaluation of potential source rock intervals in exploration wells. These would include extending the logged intervals of surveys to include possible source rock intervals and the consideration of running specialized geochemical tools such as the natural gamma (NGT) and gamma spectroscopy tool (GST) specifically for evaluating source rocks.

The authors gratefully acknowledge the assistance provided by Chris Cornford and Jan Bodenhausen who reviewed the manuscript and offered much advice and guidance.

References

AUTRIC, A. & DUMESNIL, P. 1985. Resistivity, radioactivity and sonic transit time logs to evaluate the organic content of low permeability rocks. *The Log Analyst, May-June 1985*, **XXVI, 3**, 36.

DELLENBACH, J., ESPITALIE, J. & LEBRETON, F. 1983. Source rock logging. *SPWLA 8th European Formation Evaluation Symposium*, Paper D.

HERRON, S. L. 1986. A total organic carbon log for source rock evaluation. *SPWLA 27th Annual Logging Symposium, June 1986*, Paper H.H.

MANN, U., LEYTHAEUSER, D. & MULLER, P. J. 1986. Relation between source rock properties and wireline log parameters: An example from Lower Jurassic Posidonia shale, N.W. Germany. *Advances in Organic Geochemistry*, **10**, 1105–1112.

MENDELSON, J. D. & TOKSOZ, M. H. 1985. Source rock characterisation using multivariate analysis of log data. *SPWLA 26th Annual Logging Symposium, June 1985*, Paper U.U.

MEYER, B. L. & NEDERLOF, M. H. 1984. Identification of source rocks on wireline logs by density/resistivity and sonic transit time/resistivity crossplots. *Bulletin of the American Association of Petroleum Geologists*, **68**, 121–129.

SCHMOKER, J. W. 1979. Determination of organic content of Appalachian Devonian Shales from Formation — Density logs. *Bulletin of the American Association of Petroleum Geologists*, **63**, 121–1509.

—— 1981. Determination of organic matter content of Appalachian Devonian shales from gamma ray logs. *Bulletin of the American Association of Petroleum Geologists*, **65**, 1285–1298.

TISSOT, B. P. & WELTE, D. H. 1984. *Petroleum Formation and Occurrence*. Springer, Berlin, 2nd Edn.

WILLIAMS, P. F. V. 1986. Petroleum geochemistry of the Kimmeridge Clay of Onshore Southern and Eastern England. *Marine and Petroleum Geology*, **3**, 258–281.

Identification of faults and fractures

Circumferential acoustic logs detect natural fractures and determine their orientation

W. H. FERTL

Atlas Wireline Services, Western Atlas International. Inc., Houston, TX 77251, USA

Abstract: As the direction of fractures shifts from horizontal/near-horizontal toward the vertical plane with respect to the intersecting wellbore, conventional acoustic logging devices become unreliable for fracture detection purposes. Under these circumstances, circumferential acoustic logging technology has already proven to be an excellent technique to determine vertical and/or near-vertical open fracture systems that intersect a wellbore and to measure the directional subsurface trends of these detected fractures. In addition to visual 'quick-look' interpretation of the recorded acoustic waveform data, novel digital interpretive techniques enhance the fracture evaluation.

For a considerable length of time, conventional acoustic logging devices have been used to locate natural fractures that intersect wellbores in horizontal or near-horizontal planes. However, as the direction of fractures shifts towards the vertical plane with respect to the borehole, conventional logs become unreliable for fracture detection. The circumferential acoustic (CAC) logging system, its theory and interpretation having been described in detail previously (Guy *et al.* 1986, 1987; Noblett *et al.* 1986), overcomes that constraint.

Opposing pairs of acoustic transmitters and receivers are mounted on pad assemblies that are pressed against the borehole wall (Fig. 1). The transmitter operating frequency is approximately 120 kHz. An important feature of the logging instrument is a baffle system, which effectively attenuates the propagation and reception of the direct fluid wave. Hence, the baffle allows the Rayleigh and guided fluid waves to be more easily recognized.

Major directional trends of subsurface fracture systems can be determined via an orientation section incorporated into the logging instrumentation. Azimuth, relative bearing, and deviation are measured, with both the azimuth and relative bearing being referenced to one of the receivers, thus allowing the fracture direction to be determined to within 90°. As the logging instrument rotates while moving up the borehole, additional azimuth readings in the fractured intervals results in a more precise determination of fracture direction. Repeat runs over the interval will further enhance this fracture direction capability. Furthermore, two independent caliper pairs allow two independent borehole-size measurements.

Fractured intervals are characterized by (1) an indication of thick mudcake build-up, or (2)

Fig. 1. Circumferential acoustic logging instrument.

From Hurst, A., Lovell, M. A. & Morton, A. C. (eds), 1990, *Geological Applications of Wireline Logs* Geological Society Special Publication No. 48, pp. 255–262

borehole cross sectional elongation, preferentially observed in the main direction of in situ fracture systems. Such fracture-related 'washout/breakout' indications result from 'chipping' action during drilling.

Acoustic data representing 360° coverage of the borehole wall is recorded in four quadrants and plotted in a Variable Density (VDL) format or signature waveform presentation (Fig. 2). Correlation and lithology identification is provided by the gamma-ray curve. Simultaneously measured orientation data include (1) azimuth, which indicates the direction relative to magnetic north of the reference acoustic pad, and (2) borehole deviation. Acoustic quadrant 1 originates at the reference pad and extends in a clockwise direction for 90°. Quadrants 2, 3, and 4 follow in a clockwise direction, each being displaced by 90°.

Interpretive concepts

'Quick-Look' wellsite analysis is possible using the VDL (or signature) presentation for the attenuation (or lack) of the Rayleigh-type boundary wave. Several formation and borehole conditions such as lithology, high porosity, vugs, permeability, washouts, etc., attenuate this first arrival wave. As a result, attenuation may be observed in one, two, three, or in all four measurement quadrants. Field experience has shown that attenuation in opposite or in any two adjacent quadrants best defines vertical fractures and the direction of subsurface stress planes. Attenuation in one or three quadrants may be attributable to fracturing, but is primarily due to lithologic changes or significant permeability variations. Attenuation observed in all four quadrants usually is the result of variations in high porosity, permeability, and/or lithology.

After zones of such Rayleigh wave attenuation have been pinpointed on the VDL presentation, this information is then transposed onto the orientation presentation. The two quadrants indicative of attenuation are plotted in their respective azimuth-oriented chart sectors for each interval of interest. The general trend direction is determined by drawing a vertical line of intersection through each and every zone of attenuation plotted on the azimuth curve between the top and bottom of the interval of interest. As recently described in detail by Noblett *et al.* (1987), the resulting trends can be defined quantitatively in degrees or simply by general directions such as N–S, NE–SW, etc.

Digital interpretation techniques are based on acoustic time and amplitude processing techniques (Guy *et al.* 1987). Using user-controlled selection criteria for amplitude and frequency, the travel time of first acoustic arrival in each quadrant is determined. This initial arrival time defines either the Rayleigh or the guided fluid wave. Initial arrival time of the guided fluid

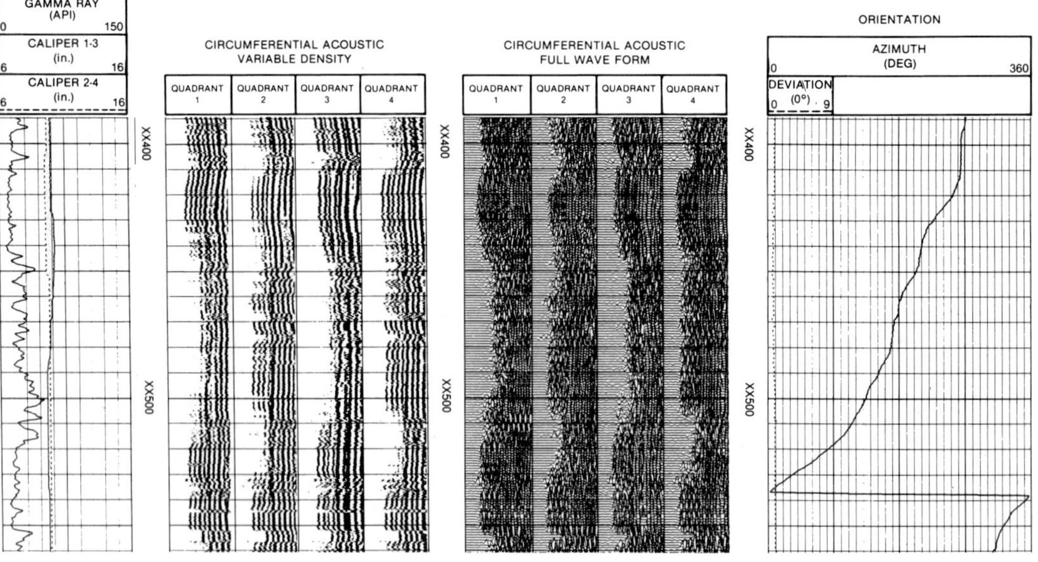

Fig. 2. Acoustic data in variable density (VDL) and signature waveform presentation.

wave is also independently determined for each quadrant. Basically, amplitude information is obtained in a time window positioned ahead of the initial guided fluid wave arrival in each quadrant. Root-mean-square (RMS) amplitude is computed for the portion of the acoustic wave train straddled by the time window. The computed results represent a relative energy level for the Rayleigh wave. The value will approach zero if the Rayleigh wave is absent. In other words, a comparison of time and amplitude characteristics between quadrants is used to define zones of higher fracture probability. A fracture index is assigned to each of the four quadrants based on these comparisons. While a fracture-index value of 0 represents an absence of fracturing, a fracture index of 1.0 represents a maximum probability of fracturing.

Test pit response

Results in the Sierra White Granite test pit of the United States Geological Survey (USGS), located in Colorado, are shown in Figs 3A and B. This test pit contains two granite blocks, 3.05 m thick, stacked one on top of the other. The lower block has a vertical saw cut, intersecting the 25.4 cm borehole, and extending 1.22 m up from the bottom of the block. A 45° saw cut intersects the borehole in the upper block.

The absence of a Rayleigh wave arrival in quadrants I and III in the bottom 1.22 m correlates with the vertical saw cut in the block. Less evident, but still detectable, is the loss of the Rayleigh wave in all four quadrants in the upper block where the 45° saw cut intersects the wellbore.

Figure 3C shows the log response in the Atlas test pit located in Alberta, Canada. The lower section (A) of the log is recorded in the tailpipe, located below the test pit blocks, and shows strong, early first arrival signals. Section B is the high porosity interval with severe attenuation on both pseudo-Rayleigh and guided fluid waves noticeable. The middle porosity section (C) is characterized by a few pseudo-Rayleigh waves preceding the fluid waves. The top portion of the test pit run shows four full oscillations (cycles) of pseudo-Rayleigh waves preceding the guided fluid waves.

The increase in pseudo-Rayleigh energy and velocity is evident in both sections C and D. However, the velocity of guided fluid waves is less dependent on porosity than is their energy. Individual 30.48 cm blocks are easily identified by the abrupt discontinuity of acoustic wave energy. This portrays the ability of the CAC instrument to identify very thin and small acoustic anomalies, such as fractures and vugs.

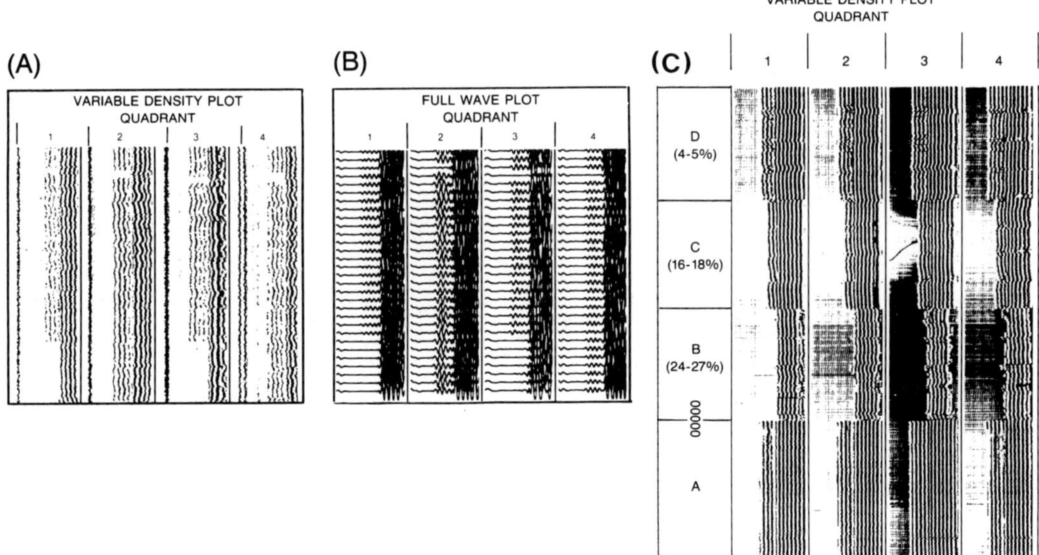

Fig. 3. Circumferential acoustic instrument response in USGS test pit, Colorado (A, B) and Atlas test pit, Canada (C).

Field Examples

Nisku Carbonates. Alberta, Canada

The Nisku Formation is predominantly dolomite with some anhydrite and calcite cement (Guy *et al.* 1986). In the subject well, the intercrystalline porosity development is minor, with most of the Nisku porosity comprising of vugs (Fig. 4). Significant acoustic attenuation is observed in localized zones, which is indicative of vugs. The degree of acoustic attenuation depends on vug shape, size, type, frequency, and distribution.

Cotton Valley Sandstones, East Texas

These sandstones range from high-porosity, high-permeability blanket sands to low-porosity and low-permeability, fine grained, undifferentiated sandstones (Mann & Thomas 1964). Target zones in the example well are low-porosity, low-permeability sands, which normally require massive flow stimulation efforts in order to increase productivity (Hunt *et al.* 1981).

Locating natural fractures is important in the selection of target intervals for stimulation. Furthermore, identification of natural fracture orientation trends will also assist in pinpointing future well locations.

Several CAC logs were run and analyzed without knowledge of local geology and location of possible faults. The results are superimposed on the structure map where two major faults are present (Fig. 5) and which coincide with subsurface fracture and stress plane direction.

Spraberry Formation, West Texas

This complex formation is found in the Leonard series of the Permian age within the Midland basin of west Texas. The reservoir rock consists of interbedded siltstones, fine-grained sandstone, limestone, and dolomite interbedded with shales. Shales are either black and brittle, or very silty.

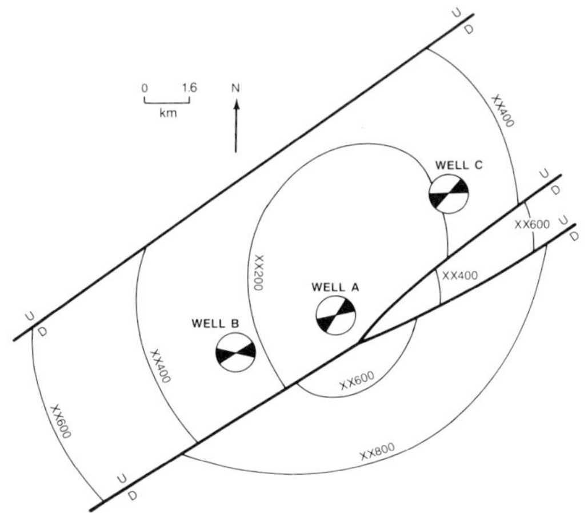

Fig. 5. Structure map and CAC-derived fracture orientation trend in Cotton Valley sandstones, east Texas. U = upthrown side of fault, D = downthrown side of fault.

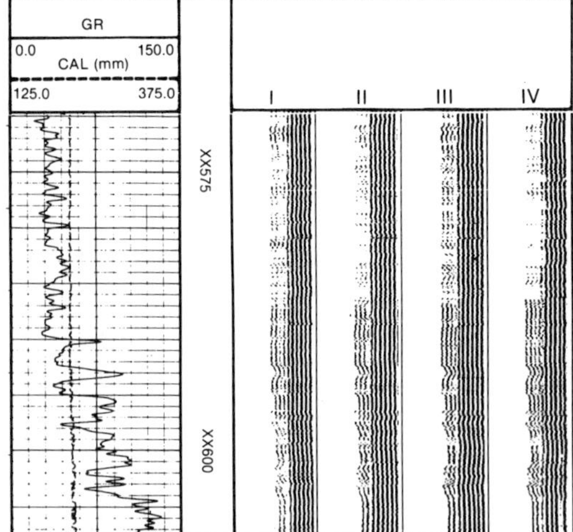

Fig. 4. Circumferential acoustic log response and core with vuggular porosity in Nisku carbonate formation, Alberta, Canada.

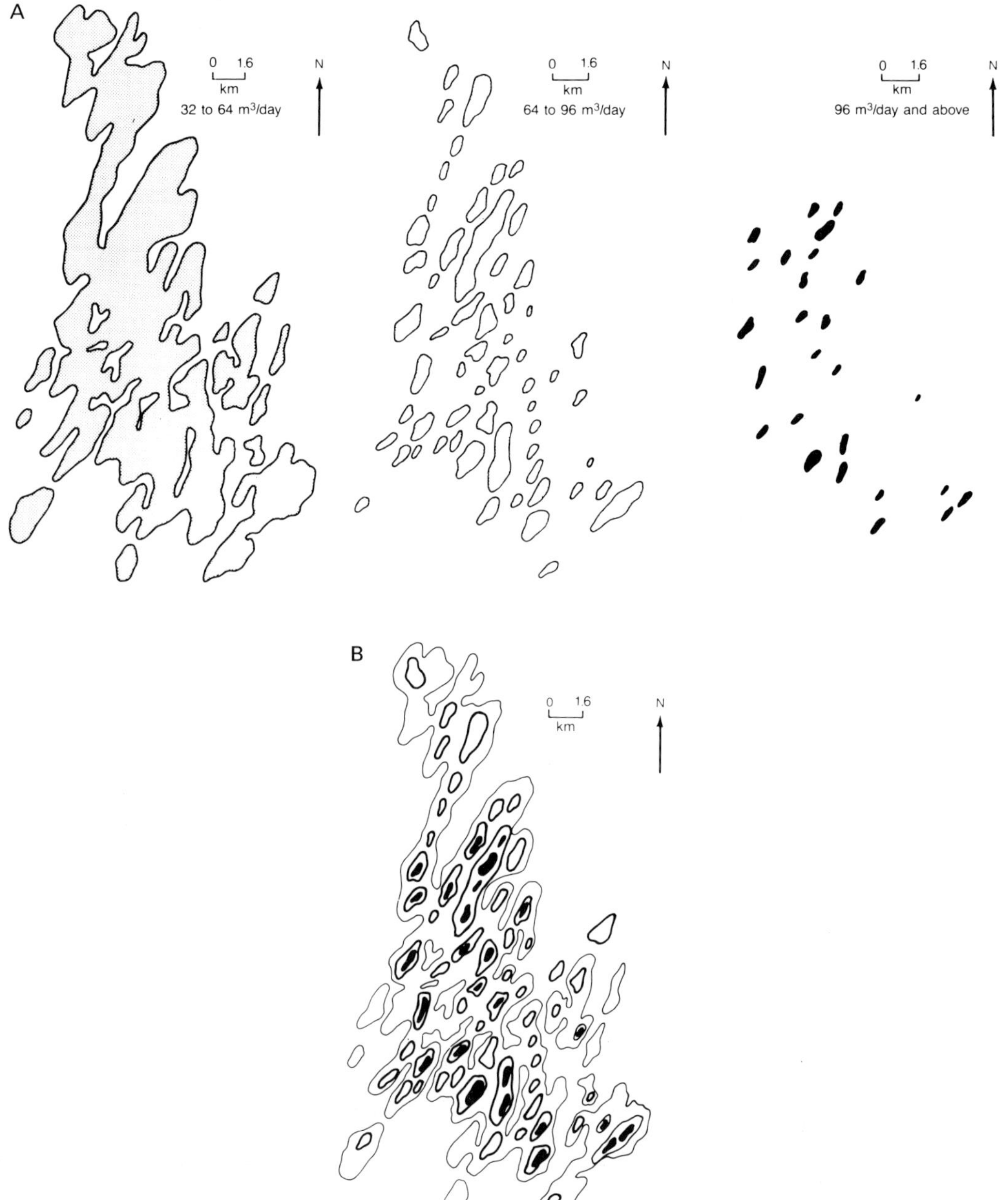

Fig. 6. Isopotential maps of Spraberry formation, west Texas. (after data by Wilkinson 1953). (B) Combined isopotential map of Spraberry formation, west Texas. (Wilkinson 1953. (C) Digital interpretation of the CAC data and VDL plot in one of the Spraberry wells, west Texas.

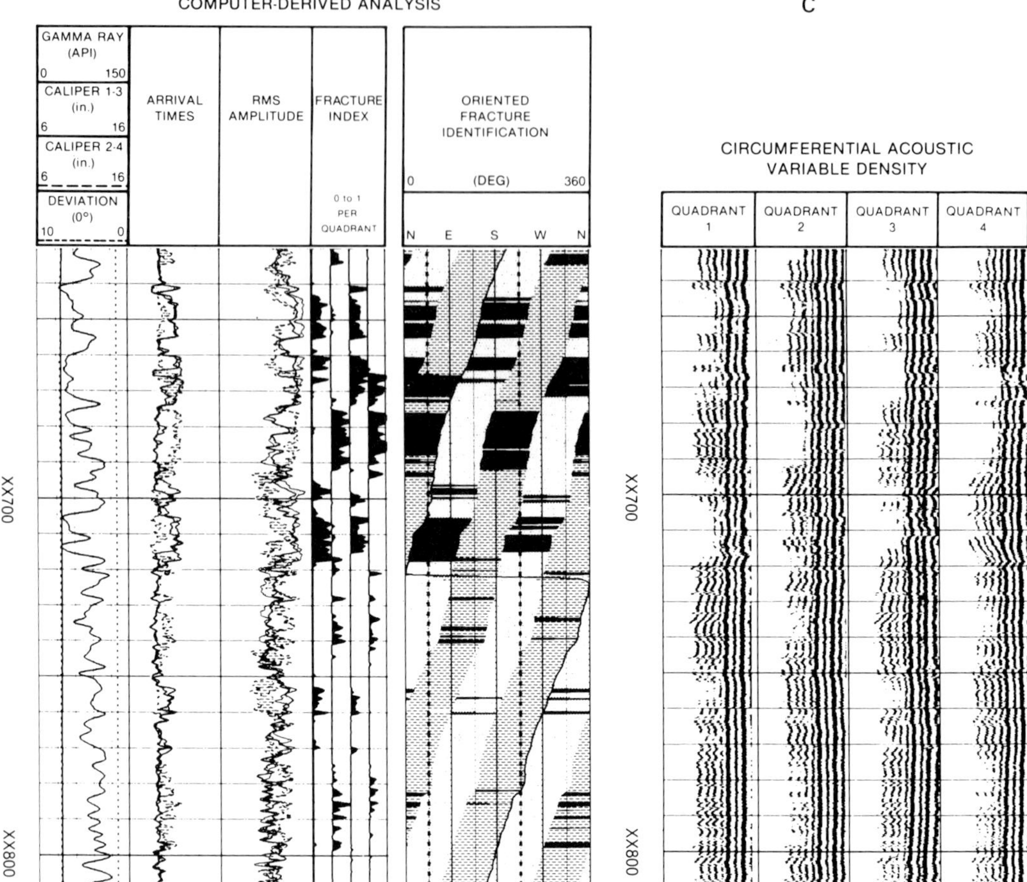

Fig. 6. contd.

Average porosities of the oil-bearing rocks range from 5 to 8%. Intergranular permeability is only 0.5 millidarcy and the reservoirs are typically underpressured. The key to production from the Spraberry is a significant increase in effective permeability due to near vertical fracturing that facilitates draining the tight reservoir (Wilkinson, 1953). The lateral directional fracture trend is clearly reflected in the isopotential maps (Figs 6A and B), which contour the initial or calculated daily rate of production.

The digital interpretation of the CAC data plus the VDL plot in one of these Spraberry wells is shown in Fig. 6C. The analysis indicates no apparent fracturing from the bottom of the interval up to XX800. Between XX718 and XX800, fracturing occurs in several isolated thin, lenticular sands, whereas an extensive fracture system begins at XX718 and extends up to XX642. An almost continuous fracture system is evident in quadrants 1 and 3 between XX718 and XX706. Between XX675 and XX690, a fracture system is evident in quadrants 2 and 4, and between XX644 and XX654, a fracture system is evident, but in quadrants 1 and 3.

A single fracture system with a common directional trend is interpreted to be present. The quadrant change due to instrument rotation occurs between 40 and 55° (40 to 55° east of north) and 220 to 235° (40 to 55° west of south). This interpretation agrees well with published directional trends for the Spraberry.

Mission Canyon Limestone, North Dakota

The carbonates of the Madison Group of lower Mississippian age, located in the Williston basin, are by far the most prolific source of oil produc-

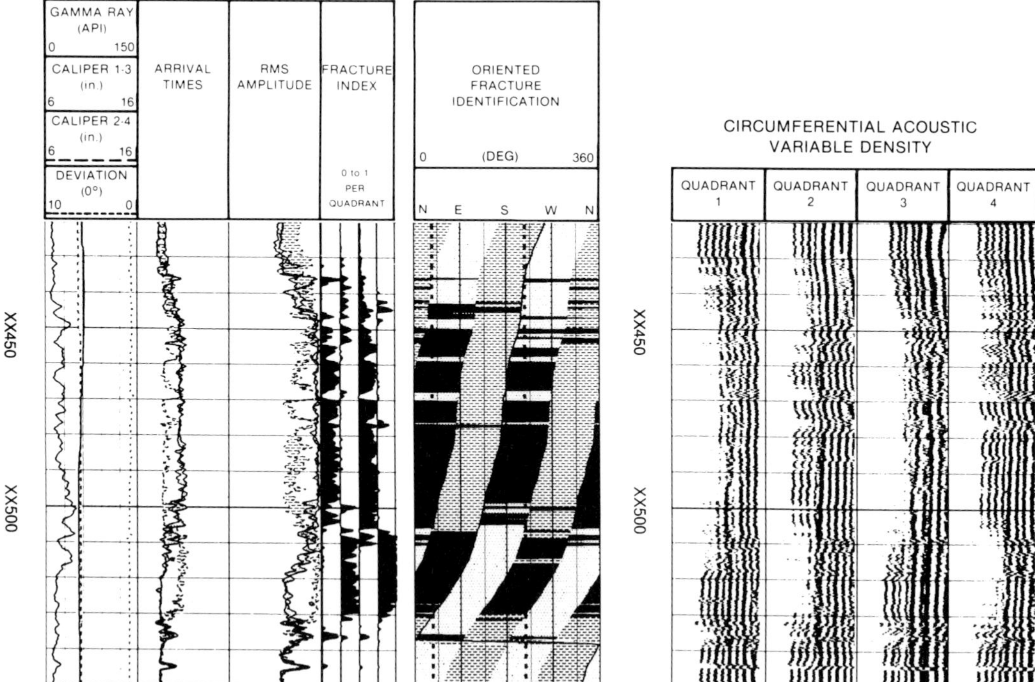

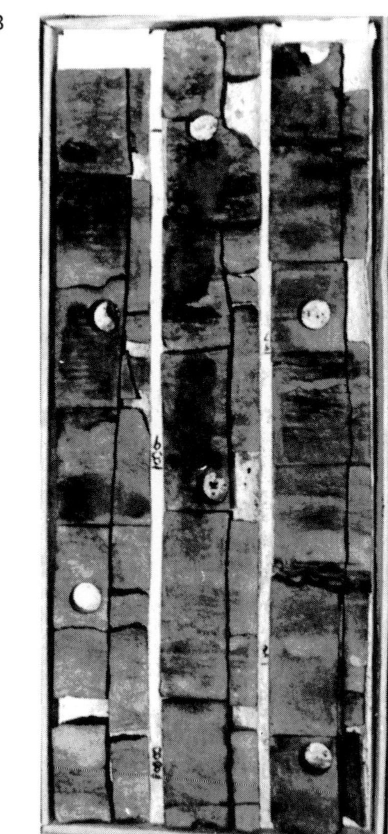

Fig. 7. (A)Digital interpretation of the CAC data and VDL plot in Mission Canyon well, North Dakota. (B) Whole core-defined interval exhibiting vertical fracture in Mission Canyon limestone, North Dakota.

tion in North Dakota (Harris *et al.* 1966). Basically, the Madison Group consists of a thick evaporite section, followed by a massive bedded carbonate interval and a thin-bedded, cherty argillaceous limestone. Reservoir rocks are fine-to medium-grained fragmental and dolomitic limestone. As many as three porous reservoir intervals may be encountered, each from 90 to 150 m thick. Frequently, the porous zones are interconnected via fractures occurring in the denser, finer-grained carbonate matrix.

For example, to the north and east of the major Nesson anticline fields, oil is encountered in the upper Madison, where anhydrite beds act as the seal for hydrocarbons in the porous, interbedded carbonate layers. This stratigraphic section is known as the Mission Canyon Formation.

In a recently drilled well, the CAC log identified vertical fractures (Fig. 7A), confirmed by the whole-core defined fractured interval (Fig. 7B). Furthermore, log-derived orientation data show the identified fracture trend to be in a NNE–SSW direction.

Conclusions

Circumferential acoustic logging devices successfully detect the presence of vertical and near — vertical fractures in open wellbores.

Subsurface fracture trends and regional stress patterns can be determined.

References

GUY, J. O., CORLEY, B., FERTL, W. H. & FROST, E. 1987. Oriented Circumferential Acoustic Logging Evaluates Vertically Fractured Reservoirs. *Transactions of the SPWLA 28th Annual Logging Symposium*, paper QQ.

—— FERTL, W. H. & OLIVER, D. W. 1986. The Use of Circumferentially Propagated Acoustic Waves in Well Logging. *SPE* 15484.

HARRIS, H. H., LAND, C. B., & MCKEEVER, T. H. 1966. Relation of Mission Canyon Stratigraphy to Oil Production in North-Central North Dakota. *Bulletin of the American Association of Petroleum Geologists*, **50**, 2269–2276.

HUNT, E. R., RAYMOND, D. H., HASKETT, C. E., & PIRIE, R. G. 1981. Application of New Well Logs and Geology to Fracturing and Producibility in Tight Gas Sands, Cotton Valley Group. SPE/DOE Low Permeability Symposium, Denver, Colorado.

MANN, C. J. & THOMAS, W. A. 1964. Cotton Valley (Jurassic) Nomenclature, Louisiana and Arkansas, *Transactions of the Gulf Coast Association of Geological Societies*, **14**, 143–152.

NOBLETT, B. R., FERTL, W. H., & GUY, J. O. 1987. Recent Advances in Fracture Evaluation. *SPE* 16226.

WILKINSON, W. M. 1953. Fracturing in Spraberry Reservoir, West Texas. *Bulletin of the American Association of Petroleum Geologists*, **37**, 250–265.

Fracture detection from logs of North Sea chalk

KARL AUDUN LEHNE

Statoil, Stavanger, Norway

Abstract: Depending on the rock type, logs are variously affected by fractures. Only field experience permits us to define logging methods which give the best results locally. This article presents the results of different fracture detection methods based on logs from two wells in the Tommeliten Gamma Field in the North Sea. Cores, dipmeter, formation microscanner, circumferential acoustic, digital sonic (amplitude, spike and criss cross analysis and Stoneley reflection) and echolog (Stoneley reflection) have been used, which allows a comparison of almost all the tools that are available on the market for fracture detection. The Borehole Televiewer was not used because of the high mud weight. The two wells are both deviated, which affects some of the fracture logs. Conventional logs (caliper, resistivity, density-neutron etc.) were of no significant value in fracture detection. Well tests and production logs will be run later to improve this fracture evaluation.

Chalk fields in the North Sea are normally moderately fractured. In the Tommeliten Fields, fracture detection from logs and cores is used to derive a more precise knowledge of fracture intensity, orientation and distribution. The network of fractures will enhance the production characteristics of the reservoir. For each production well, knowledge of the fault and the fracture system is used to select perforation intervals to optimize drainage of the reservoir. The orientation of fractures is also used for in situ stress evaluation of the field and to orientate the fracture system relative to the local structural geology. All the wells in the Gamma Field are used in the field study. This paper presents the results of fracture detection from several log types and from cores from two of the wells. Recommendation concerning the type of fracture logs to be used in deviated wells drilled through chalk reservoirs is included.

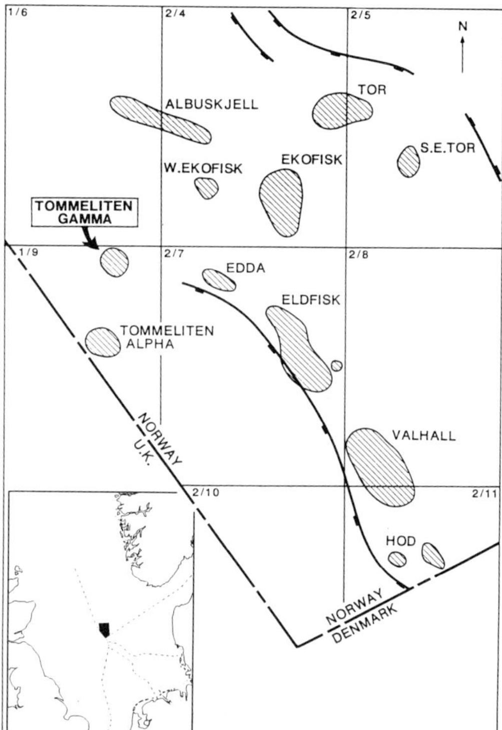

Fig. 1. Location map, Tommeliten Gamma Field.

Summary of the Tommeliten Fields

The Tommeliten Alpha and Gamma Fields are located in the southern part of the Norwegian sector of the North Sea (Fig. 1). They consist of two salt-induced structures 10 km apart which contain hydrocarbons in Upper Cretaceous–Lower Palaeocene chalk reservoirs.

It has been shown that there are two main types of Norwegian chalk sequences, the high productivity, high porosity overpressured reservoirs and the lower productivity, normally pressured, low porosity reservoirs (Sørensen *et al.* 1986). The Tommeliten Fields are overpressured with intermediate porosity.

Fracture detection in the Tommeliten Gamma Field is presented for the main hydrocarbon-bearing units, the Ekofisk and Tor formations. The reservoir chalk of this field accumulated as pelagic and re-sedimented chalks. The re-sedimented chalks have higher porosity and less fractures compared to the pelagic chalks. Some of the zones contain both pelagic and re-sedimented material (hemipelagic chalk). Both the Ekofisk and Tor formations have highly

From Hurst, A., Lovell, M. A. & Morton, A. C. (eds), 1990, *Geological Applications of Wireline Logs* Geological Society Special Publication No. 48, pp. 263–271

variable depositional modes which give large variations in porosity (Figs 2 & 3).

Fracture system, tommeliten gamma

Articles dealing with deformation and fracture systems in chalk overlaying a salt ridge or salt dome include Brewster *et al.* (1986) and Watts (1983), which are mainly based on cores, and Koestler & Ehrmann (1987), which is a study of a surface structure. For Tommeliten Gamma the field-wide deformation of the reservoir by fracturing is believed to be the result of post-diapiric salt movement, faulting and differential compaction. The fracture intensity varies greatly both vertically and areally. The highest intensity is on the top of the structure with a decrease on the flanks. Fracture intensity is also affected by depositional mode, with pelagic and hemipelagic chalks having a higher fracture intensity than the re-sedimented chalk. Almost no fractures can be seen in the re-sedimented lower part of the Tor Formation (Figs 2 & 3).

The fracturing has a significant influence on the effective permeability. The porosity resulting from fracturing is extremely low and does not exceed 0.1 p.u. even in the most fractured intervals.

Four main types of fractures have been identified from cores and logs (Fig. 4).

(i) Fault zones with a complex pattern of conjugate fracture sets and with a displacement of bedding in the range of meters. Fig. 5 is an example of logs from a highly fractured fault zone.

(ii) Joints (fractures without visible displacement) occur as planar, curved and strongly irregular features. They are predominantly sub-vertical and occur as single fractures (Fig. 6) or parallel sets of fractures (Fig. 7) with a fracture spacing as low as 10 cm.

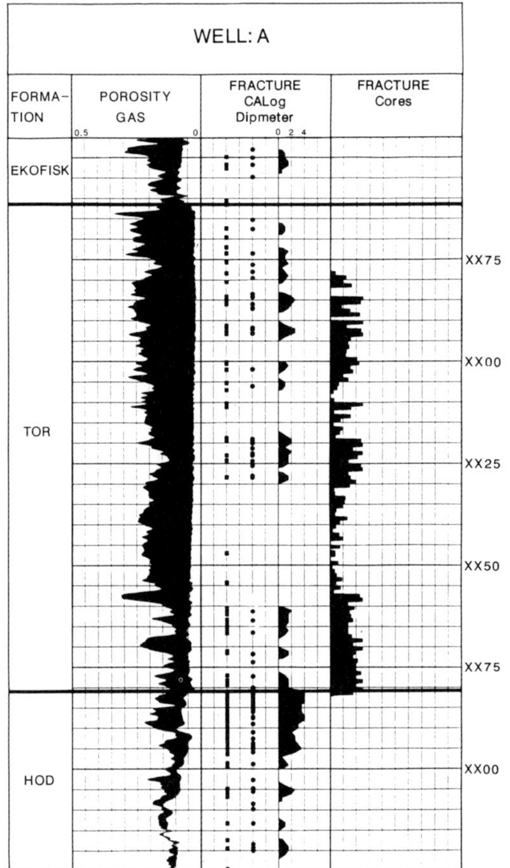

Fig. 2. Fracture detection well A. Track 1: porosity. Track 2: each fracture detected from CALog (■) and dipmeter (●) with the average fracture intensity/2 m to the right. If only one of the logs shows fractures the intensity is zero. Track 3: the variation of fracture intensity from cores.

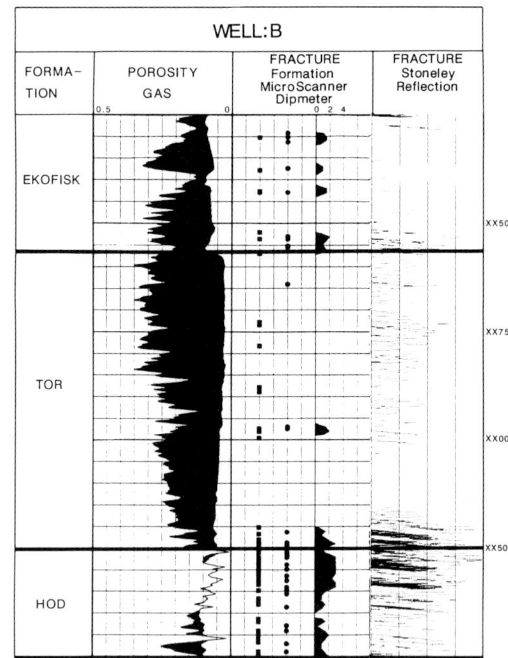

Fig. 3. Fracture detection well B. Track 1: porosity. Track 2: each fracture detected from Formation MicroScanner (■) and dipmeter (●) with the average fracture intensity/2 m to the right. If only one of the logs shows fractures, the intensity is zero. Track 3: Stonely reflection.

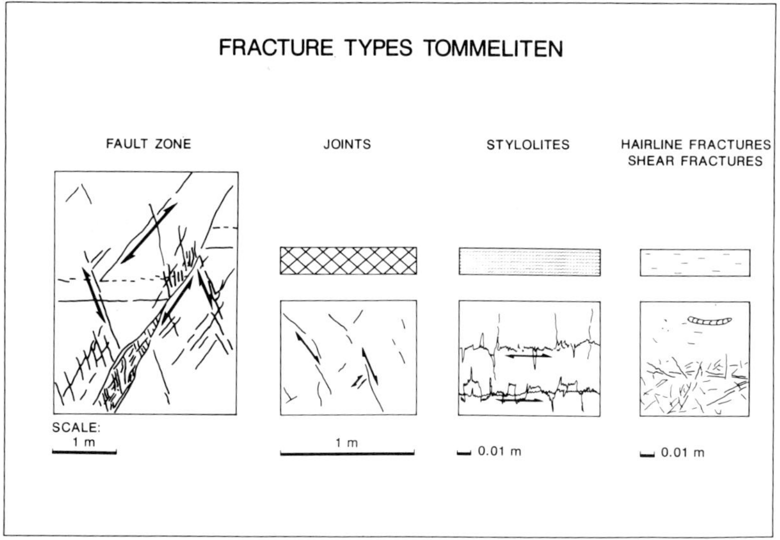

Fig. 4. The four main fracture types.

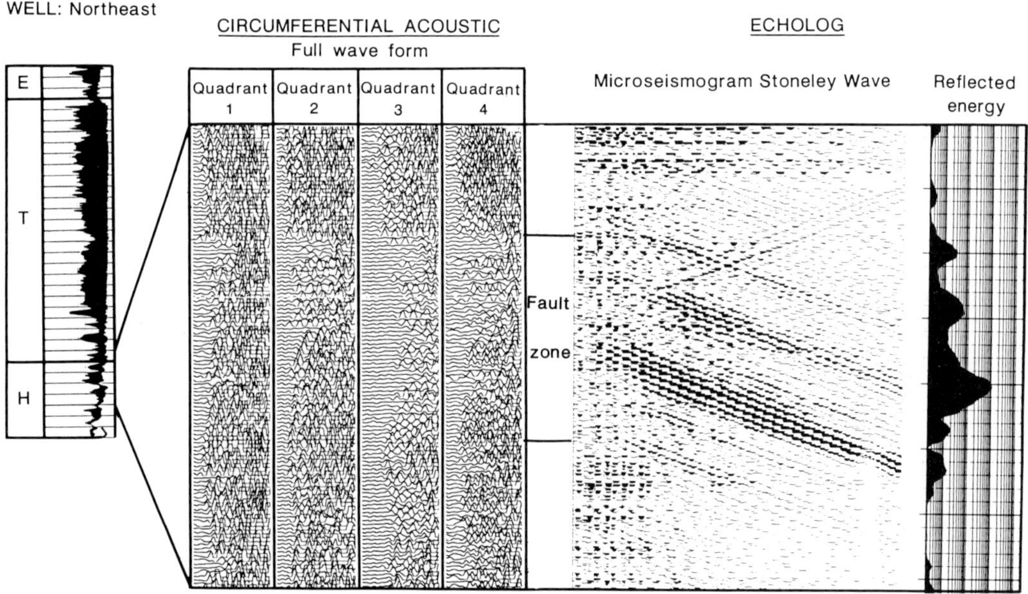

Fig. 5. Full wave plot (CALog and microseismogram with reflected energy (Echolog).

(iii) Stylolite-associated fractures may occur individually in zones with low fracture intensity or as a network of small fractures. They are statistically near-vertical with a fracture length of 3–5 cm. These microfractures terminate at the stylolite surface and never cross them. They are only seen in cores.

(iv) Small hairline and shear fractures are also seen in cores. They are interpreted to have no significant influence on reservoir permeability.

Fig. 8 shows examples of joints, stylolites, thin highly fractured zones and zones without any fractures. Until recently, the only way to analyse the geometry of fractures intersected by

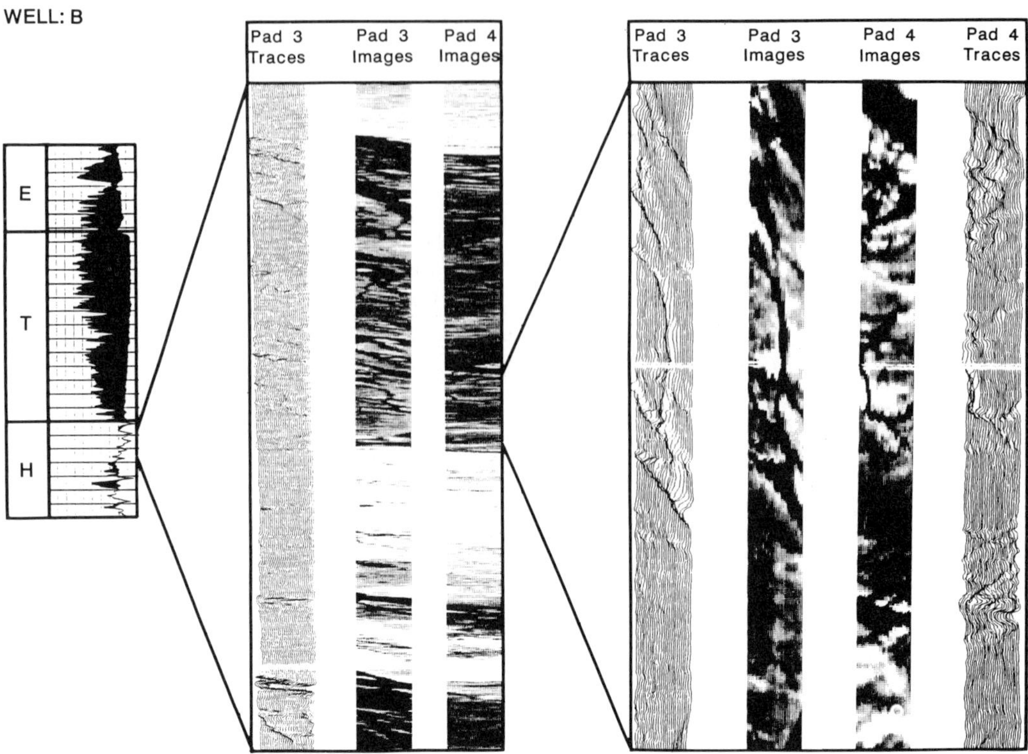

Fig. 6. Image processing (FMS) – irregular natural fracture.

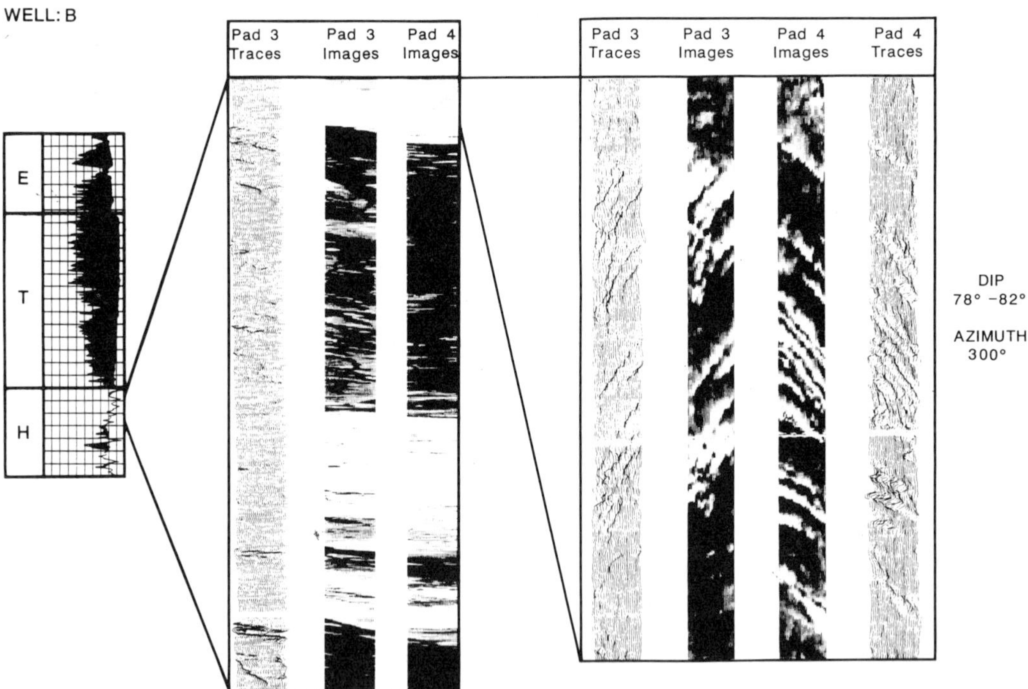

Fig. 7. Image processing (FMS) – parallel planar natural fractures.

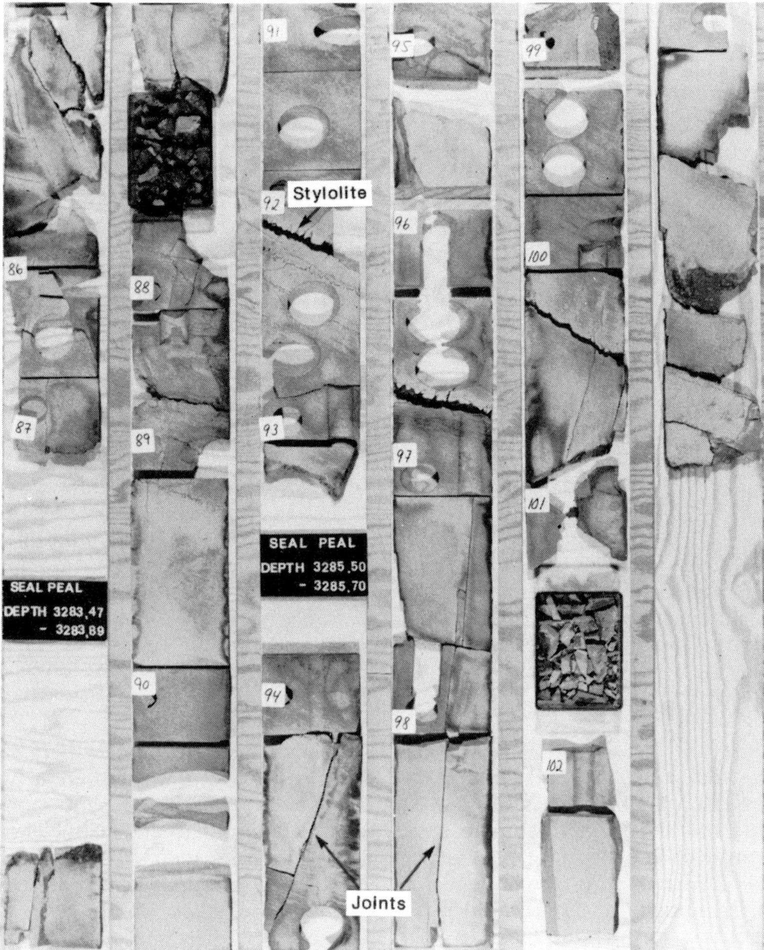

Fig. 8. Core photo with vertical joints, styloite-associated fractures, thin highly fractured zones and zones without any fractures.

the well has been from core analysis. In this study, logs have also been used for fracture typing and orientation. Logs give an in situ characterization of relatively undisturbed fractures, while undamaged core materials of fractured rock are difficult to recover with conventional drilling and coring operations.

Fracture logs

The following fracture logs have been used for fracture detection.

(1) Dipmeter (Diplog,SHDT).
(2) Formation MicroScanner (FMS).
(3) Acoustic Log/Full Waveform Recording.
(3.1) Circumferential Acoustic Wave (CALog).
(3.2) Stoneley Wave Reflection Echolog, SDT).
(3.3) Amplitude, Spike and Criss Cross Analysis (SDT).

Weak indications of fractures were recognised from the conventional logs but are not reported in this paper. The Borehole Televiewer could not be used because of the high mud weights.

Two wells have been chosen for this presentation (Fig. 9), production well A, which has extensive cores and several fracture logs (Diplog, CALog, Echolog), and production well B, which has a different suite of fracture logs

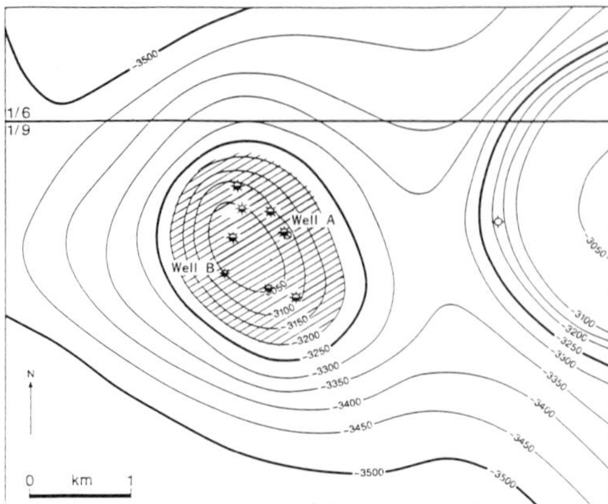

Fig. 9. Structure map, Tommeliten Gamma Field. Well A and well B have been chosen for this presentation.

(SHDT, FMS, SDT). Well B is not cored. From these two wells we have complementary sets of fracture logs, allowing comparison of almost all the tools that are available on the market.

Dipmeter

The fracture detection method using the dipmeter is based on the detection of resistivity reductions due to fluid-filled fracture planes, which are seen by each of the four pads of the dipmeter. The degree of reduction depends on the local matrix porosity and pore fluid resistivity. Low resistivity of the borehole fluid gives the highest reduction in effective resistivity within fractured zones.

On the Tommeliten Gamma Field a computer-based method (for SHDT) was difficult to use because the joint system has planar, curved and irregular features. With the high dip-angle, it is difficult to follow the low resistivity anomalies from pad to pad. For a fracture crossing the well, there can be differences in resistivity curves on opposing sides of the whole (Frisinger & Gyllensten, 1986). Here, it is impossible to correlate the fractures based on the characteristics of the resistivity curves. Fortunately, the borehole ovality (both wells are deviated) gave no rotation of the tools which is also unfavourable for fracture detection.

The best results for fracture detection from dipmeter were obtained by comparing the four dipmeter curves (four pairs, SHDT) using a 1/25 log scale (the FIL log). The fractures were identified as low-resistivity anomalies on one or two of the pads. In some cases fractures parallel to the well were identifiable as a rather long anomaly on one of the pads or possibly on two pads. The problem with this detection method is that the resistivity curves from dipmeter are also sensitive to borehole effects (washouts) and to conductive mineral effects (e.g. pyrite).

Figs 2 and 3 show the results of fracture detection using the dipmeter. Fig. 2 shows the correlation to the CALog and bulk fractures identified from cores. The highly fractured fault zone in the upper part of the Hod Formation is clearly seen on all the fracture logs. For the lower part of Tor Formation (xx55 to xx81), the logs identify fractures in the low-porosity zones and no fractures for the high-porosity zones. The cores also identify fractures in the high porosity zones. These fractures may be due to post-coring dessication. The zone above (xx30 to xx55) is a re-sedimented zone with little or no detection of fractures from cores and logs. For the rest of the Tor Formation, the logs identify marked differences in fracture intensity from zone to zone. The same change in intensity is not observed in cores (Fig. 2).

Formation MicroScanner

The Formation MicroScanner (FMS) maps the microresistivity of the borehole wall from 54 button electrodes (two pads of 27 electrodes each) (Ekstrom *et al.* 1986). The two-dimensional microresistivity data are then converted into image formats which represent changes in formation resistivity presented as a spectrum of 16 grey levels. The darker the image, the higher the conductivity of the formation.

FMS has been run in well B and the results are presented on the FMS images plotted at a scale of 1/10. As with the other fracture logs, FMS gave good fracture detection in the low porosity zones, especially the Hod Formation. From this zone the orientations of the planar natural fractures have been detected. Fig. 7 shows examples of parallel sets of fractures with a fracture spacing as low as 10 cm. Fig. 6 shows an example of an irregular natural fracture seen on the image plot. As shown on the azimuth frequency plot (Fig. 10), the dominant fracture azimuths in the Hod Formation are NW and ENE. The plot is based on orientation of 114 natural planar fractures with high dip angles (70° to 90°). Orientation of the natural fractures in this well are both radial and concentric to the salt dome structure, and are interpreted to have formed in response to the period of diapirism, not the present day in situ stress regime. Evaluation of the present stress regime can be ob-

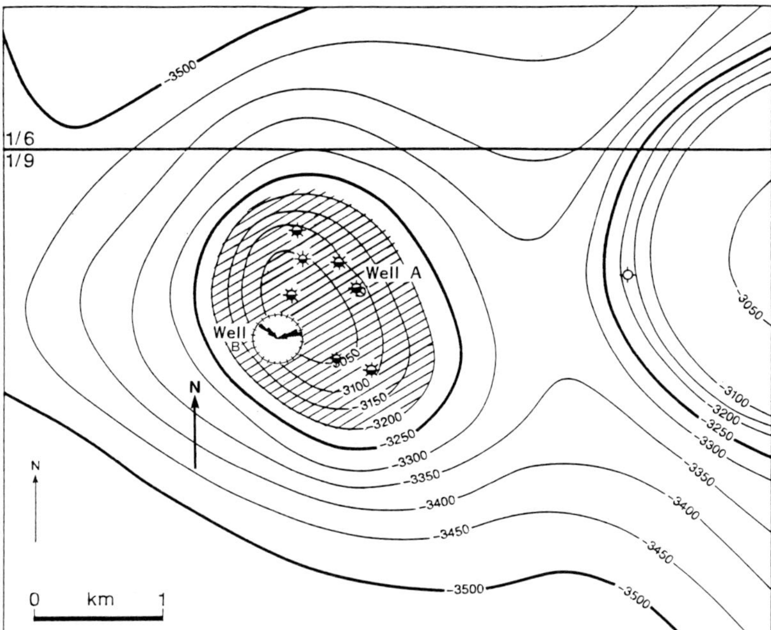

Fig. 10. Orientation of the planar fractures from FMS, well B.

tained from measurement of the orientation of induced fractures.

In the high porosity zones in the Ekofisk and Tor formations the FMS gave no indication of natural fractures. The fractures detected in these zones by the FMS are interpreted as induced fissures. They are identified as irregular vertical lines on the FMS images, sometimes with secondary fissures branching out of the main fissure. The other fracture logs identify fractures in these high-porosity zones, but it is impossible to distinguish between natural and induced fractures from the logs. Fracture detection in the other wells indicates natural fractures in the high porosity zones. One reason for the low detection of fractures in high porosity zones with the FMS is the low contrast between the resistivity of the high-porosity invaded zones and the fractures filled with mud filtrate.

Acoustic logs/full waveform recording

CALog (Guy & Fertl 1987). Circumferential acoustic waves have been used for detection of vertical and subvertical fractures by detection of the decrease in amplitude of the circumferential Rayleigh wave. Since there is one wave measurement in each quadrant every 5 cm, the tool gives a good vertical resolution. Especially, in the laminated zone in the lower part of Ekofisk Formation (Fig. 2) fractures are detected in thin zones (10–20 cm) without any being recognized in the zone above and below. For this deviated well, the attenuation of the Rayleigh wave is different from one borehole quadrant to another (Fig. 5). The receiver on the lower side of the borehole has better contact with the wall than the receiver on the upper side. Since the tool does not rotate because of borehole ovality, it is difficult to determine the orientation of the fractures (Guy *et al.* 1987). The results of the fracture detection from CALog are shown in Figs 2 and 5. The highly fractured fault zone in upper part of Hod Formation is observed in all four quadrants (Fig. 5). For the rest of the well the fractures identified from CALog and dipmeter compare very well. The attenuation of the Rayleigh waves are mainly seen in quadrant 3 and 4 (Fig. 5). Attenuation in one quadrant may also be due to lithologic changes and/or permeability variation. Attenuation and resistivity change from dipmeter used together make fracture detection more reliable.

Echolog, SDT. Detection of the Stoneley Wave reflection is made using both the Echolog (Husten *et al.* 1984) (well A) and the digital sonic (Mathieu *et al.* 1985) (well B). On the

Echolog the transmitter/receiver spacing is zero, and by using low frequency (below 1kHz) transmitter signals the reflected Stoneley Wave is obtained. Fig. 5 shows a strong reflected Stoneley Wave with high reflection energy for the highly fractured fault zone. Using an eight-receiver array sonic tool (SDT) the direct and reflected Stoneley Wave were recorded. The main separation was made using a log pass frequency filter (2kHz-6kHz) and a velocity filter. When the Stoneley (or tube) Wave passes a fracture, some fluid flows into the fracture and gives a pressure drop in the borehole. The pressure drop causes the direct Stoneley Wave to attenuate and in turn generates a reflected Stoneley Wave. As the major causes of Stoneley Wave reflection in a borehole are washouts and permeable fractures and not bed boundaries or other lithological changes (Hsu *et al.* 1987), the effects of washouts can be identified from the caliper and the remaining reflections attributed to fractures.

For well B (Fig. 3) the quality of the reflected Stoneley Wave was good because most of the energy of the signals is concentrated at low frequencies (2kHz–6kHz). Since there was no change in the well diameters recorded by the four arm caliper, the reflections are attributable to natural or induced fractures (Fig. 3).

Amplitude, spike and criss cross. The analyses of the digital sonic have so far shown little or no agreement with the other fracture logs. The attenuation of the energy of the compressional and shear waves crossing a fracture depends to a large extent on the efficiency of mode conversions at the fracture interface. The compressional and shear waves must be converted to a fluid wave at the first fracture interface and then converted back again at the second. Obviously, therefore, the dip of the fracture is important. The attenuation is also affected by the lithology. Furthermore, separating the shear energy from other transmission mode is difficult. The amplitude analysis for fracture detection was of no use.

Spike analysis is a method of fracture detection based on the occurrence of positive spikes in the compressional amplitude. The spikes are sharp increases of amplitude of roughly a wavelength in width which arise as a result of enhanced acoustic coupling between the borehole fluid and the formation in the vicinity of a fracture.

Criss Cross patterns occur when a fracture, and to a lesser extent a bed boundary, occurs between the transmitter and receiver. As a result of mode conversion by wave at the irregularity, a second wave is generated. This wave creates a criss cross interference pattern with the first wave at the receiver.

The original reference (Cheung *et al.* 1984) describes the spike and criss cross analysis used data from the long spacing (8', 10' and 12') sonic tool. Here in Tommeliten Gamma, the SDT was used, which has a much lower-frequency spectrum than the long spacing sonic. Perhaps the decrease in high-frequency energy is the reason why the spike and criss cross analysis did not work.

Conclusions

The intention of this paper has been to provide a comparison of several types of fracture logs for fracture identification in chalk.

(i) Both Dipmeter, Formation MicroScanner, CALog, Echolog and Digital Sonic (Stoneley Wave Reflection) identify fractures.

(ii) The Formation MicroScanner gave the best identification although interpretation of the signal is often difficult in the high porosity zones. CALog and Stoneley Wave reflection also gave good identification but the data quality of CALog is dependent on the deviation of the well.

(iii) Fracture detection from Amplitude, Spike and Criss Cross analysis from SDT did not give results comparable with the other fracture logs.

(iv) Orientation of fractures was only possible from Formation MicroScanner. Fig. 10 shows both radial and concentric orientation to the salt dome of the natural fractures which mean that the local stress from the rising diapir was the dominating in situ stress.

I wish to thank E. B. Nielsen and T. A. Knai from Statoil and J. Spurlin from Schlumberger for their support on this project. I also wish to thank Agip, Fina and Statoil for permission to publish this paper. A version of this paper was presented at the SPWLA 11th European Symposium, Oslo, September 1988.

References

BREWSTER, J., DANGERFIELD, J. & FARRELL, H. 1986 The Geology and Geophysics of the Ekofisk Field Waterflood *Marine and Petroleum Geology*, **3**, 139–169.

CHEUNG, P. S. Y. 1984. Fracture Detection Using the Sonic Tool, *Proceedings of the 8th International Symposium of Societe pour l'avancement de l'interpretation des diagraphies, Paris, October, 1984*.

EKSTROM, M. P., DAHAN, C. A., CHEN, Y., LLOYD, P. M. & ROSSI, D. J. 1986. Formation Imaging with Micro-Electrical Scanning Arrays, *SPWLA 27th Annual Logging Symposium, Houston, June, 1986*.

FRISINGER, M. R. & GYLLENSTEN, A. 1986. Fracture Detection in North Sea reservoires, *SPWLA 10th European Symposium, Aberdeen, April, 1986*.

GUY, J. O. & FERTL, W. H. 1987. Circumferential Acoustic Measurements. The CALog, *SAID 7th European Logging Symposium, Paris, October, 1987*.

—— CORLEY, B., FERTL, W. H. & FROST, E. 1987. Oriented Circumferential Acoustic Logging Evaluates Vertically Fractured Reservoirs, *SPWLA 28th Annual Logging Symposium, London, June/July, 1987*.

HSU, K., BRIE, A. & PLUMB, R. A. 1987. A New Method for Fracture Identification Using Array Sonic Tools, *Journal of Petroleum Technology*, June, 1987.

HUSTEN, P., FREI, W. & BUKOWIECKI, S. 1984 Tube Wave Logging – The ECHOLOG, *SPWLA*.

KOESTLER, A. G. & EHRMANN, W. U. 1987. Fractured Chalk Overburden of a Salt Diapir Laegerdorf, NW Germany – Exposed Example of Possible Hydrocarbon Reservoir In: LERCHE, I. & O'BRIEN, J. (eds) *Dynamics of salt diapirs*, Academic, New York.

MATHIEU, G., YVER, J., CHEUNG, P., BRIE, A., HORNBY, B. & HSU, K. 1985. Applying Sonic Data: Fracture Detection, *The Technical Review*, **33**, 69–79.

SØRENSEN, S., JONES, M. HARDMAN, R. F. P., LEUTZ, W. K., SCHWARZ, P. H. 1986. Reservoir Characteristics of High- and Low-productivity Chalk from the Central North Sea *In*: SPENCER, A. M. *et al.* (eds). *Habitat of Hydrocarbons on the Norwegian Continental Shelf* Graham and Trotman, London, 91–110.

WATTS, N. L. 1983. Microfractures in Chalks of Albuskjell Field, Norwegian Sector, North Sea: Possible Origin and Distribution, *Bulletin of the American Association of Petroleum Geologists*, **67**, 201–234.

Structural interpretation of dipmeter results in the Gullfaks Field

T. M. RØNNINGSLAND

Norsk Hydro Exploration and Production, P.O. Box 200, N-1321 Stabekk, Norway

Abstract: Dipmeter interpretation has been carried out on seven wells from the Gullfaks Field. Orientation and style of major and associated minor faults have been investigated using dipmeter results. Major faults, as inferred from dipmeter data, are in fairly good agreement with faults defined from seismic interpretation. Several less significant dipmeter-defined faults were below the resolution of seismic data. The orientation of these faults probably corresponds with that expected of antithetic and synthetic faults associated with the main fault system. Analyses of dipmeter results indicate little or no influence of syn-sedimentary tectonism. Thickness variations of the different formations can be explained by post-depositional faulting (fault cut-out) and erosion. The differences in structural dip between the seven wells support the interpretation that they are located on separate fault blocks. The density of detected faults, one per 140 m (from dipmeter results), proves the presence of an intensely faulted area. This study therefore emphasises the importance of using dipmeter results as a supplement to seismic for the purpose of structural interpretation.

A structural dipmeter interpretation has been performed on seven wells from the Gullfaks Field (Fig. 1), a major oil field in the Norwegian part of the North Sea where hydrocarbons have been found in Lower and Middle Jurassic sandstones (Erichsen *et al.* 1987).

Resolution of seismic is, under optimal conditions, 20 to 30 m. This represents a considerable uncertainty regarding interpretation and mapping of structural features. Dipmeter data were applied to obtain more accurate interpretation at the well location, and to clarify the importance of smaller scale structural features beyond seismic resolution. The interpretation is based on dipmeter data used together with conventional logs and core information.

Dipmeter data

The dipmeter data are of the SHDT type for well 34/10-19, otherwise HDT-data were used. During processing, a correlation length/step of 1 m/0.5 m was used. Processing of data from 34/10-19 and 34/10-14 used the MSD-programme and INCLINE-programme respectively.

The other five wells were processed by the CLUSTER program which also takes into account adjacent results (Hepp & Dumestre 1975). The program tends to single out those curve pair displacements (out of all possible) which result in the most consistent dips over several mutually overlapping levels. At each level the chosen displacements are those which contribute dip to, or is near to, the predominating cluster (concentration of possible dips). This is done separately over short intervals or zones (maximum 14 dip levels), such that any long-range bias is avoided. If no contribution is found from a level close enough to any of the defined clusters, this level is computed as a 'no result' interval.

The selection by CLUSTER is not made so restrictive that it will prevent the build up of upwards increasing or decreasing dip trends. However, scattered dips will generally not be plotted.

Structural setting

The Gullfaks Field (Figs 2 & 3) is situated in the East Shetland Basin of the northern North Sea, just to the west of the Viking Graben. Gullfaks is a positive structural element and comprises a major fault block tilted to the west. As a result of extensive block faulting in late Middle Jurassic to early Cretaceous time, there are several individual rotated fault blocks associated with N–S to NE–SW trending, easterly dipping fault zones. Associated with these fault systems are adjustment faults (Fossen 1989) that strike approximately E–W.

The Gullfaks structure is generally considered to reflect extensional tectonics (Sealand & Simpson 1982). There is, however, evidence of compressional movements which may explain the anomalously high structural position of the field relative to other fields in the area. Indications of reverse fault movements are present and the Gullfaks structure may have undergone

From HURST, A., LOVELL, M. A. & MORTON, A. C. (eds), 1990, *Geological Applications of Wireline Logs* Geological Society Special Publication No. 48, pp. 273–286

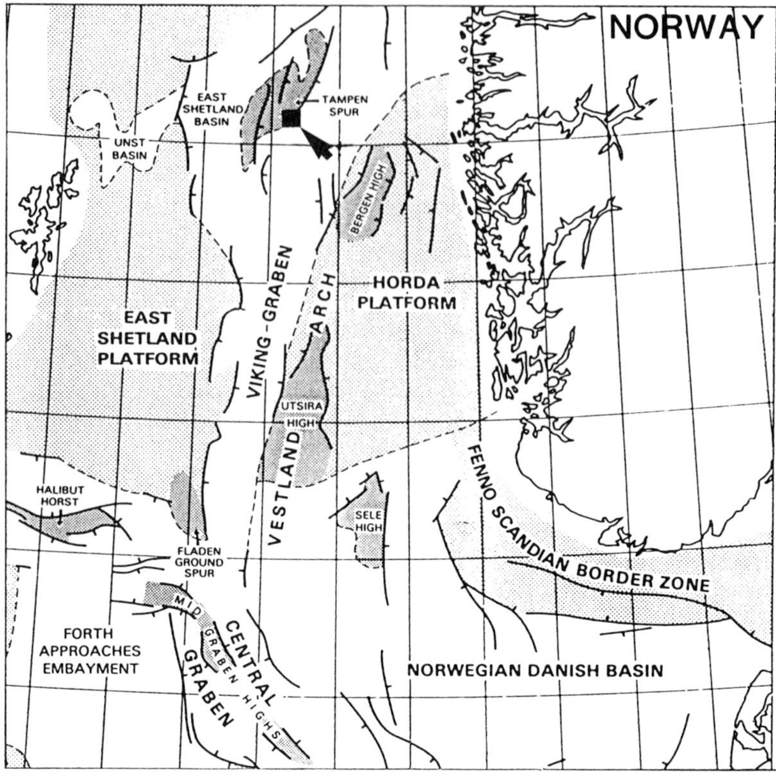

Fig. 1. Location map with the main structural elements in the central and northern North Sea (After Day *et al.* 1981).

E–W compression after a major extensional phase associated with Kimmerian tectonic movements (Fossen 1989). The inversion of the Gullfaks structure (and the Tampen Spur) probably post-dates deposition of the Middle Jurassic Brent sediments. Compressional movements (post-Middle Jurassic), however, predate the Upper Cretaceous sediments that are draping the structure. Upper Jurassic to Lower Cretaceous rocks are rather thin and incomplete due to uplift and erosion. They are best preserved downflank on the fault blocks.

Stratigraphy

The stratigraphy of the northern North Sea is shown in Fig. 4. This study examines the Late Triassic Lunde Formation and the Jurassic sequence comprising the Statfjord Formation and the Dunlin, Brent and Viking Groups.

The Lunde Formation consists of continental sandstones and shales. The overlying Statfjord Formation is of fluvial origin, and dominated by sandstones. The Amundsen and Burton Formations comprise silty shales deposited in a marine environment. The Cook Formation is a coarsening-upwards sequence, ranging from silty shale near the base to shallow marine sandstone in the upper part, partly developed as megaripples (P. Kloumann, pers. comm.). The unit is overlain by the extensive marine, silty shales of the Drake Formation.

The Middle Jurassic Brent Group represents a prograding deltaic sequence. The lowermost Broom Formation is interpreted to be a prodelta deposit of alternating sandstone and shale. The Rannoch Formation is interpreted to be a delta front/lower shoreface deposit, dominated

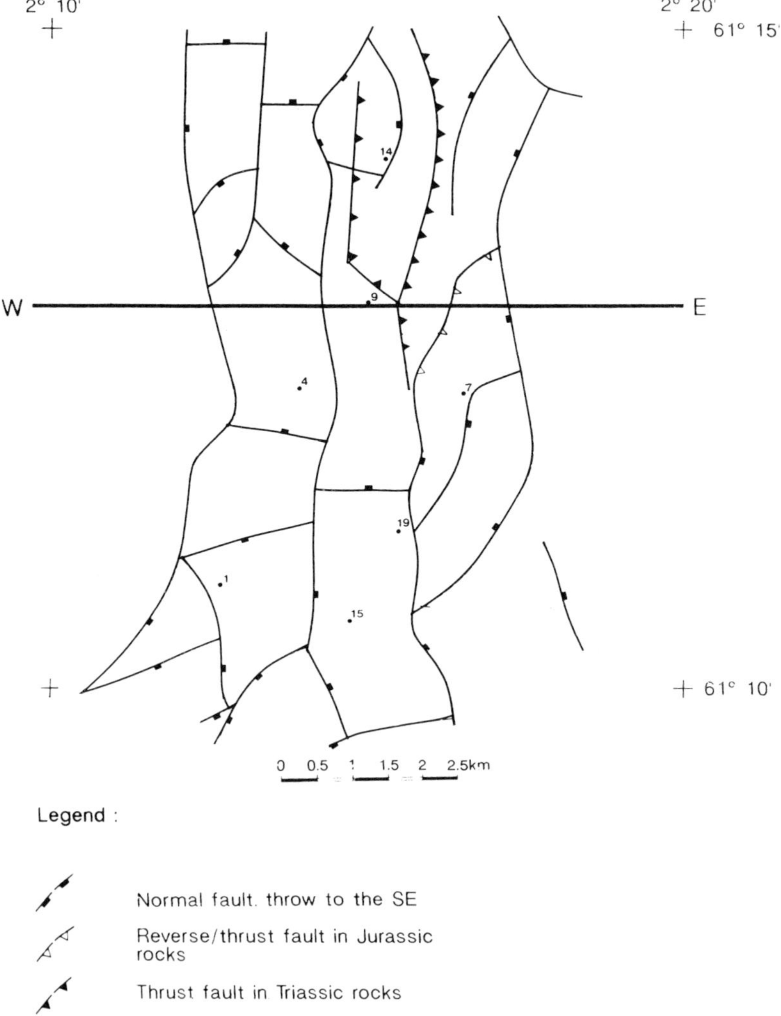

Fig. 2. Structural elements in the Gullfaks area, top Dunlin Group. The main thrust faults are at the Triassic (Carnian) level. (After Fossen 1989).

by fine-grained sandstones. The overlying coarser sandstones of the Etive Formation are generally of upper shoreface or mouth bar origin. These sandstones are overlain by coastal plain shales/siltstones, sandstones and coals of the Ness Formation. The uppermost Tarbert Formation of the Brent Group comprises deltaic to shallow marine sandstones. The overlying Viking Group is represented by marine, silty shales.

Structural interpretation from well data: general

Structural interpretation from well data is primarily based on dipmeter data. Lithostratigraphical correlation is used to identify missing and repeated sections; the former may be due to unconformities or normal faults (fault cutouts) and the latter to reverse faults.

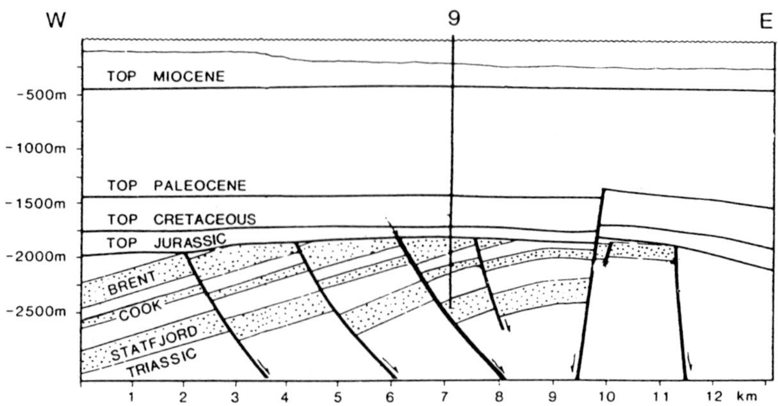

Fig. 3. W–E profile illustrating the structural development of the Gullfaks Field. For location of the profile see Fig. 2. (After Erichsen *et al.* 1987).

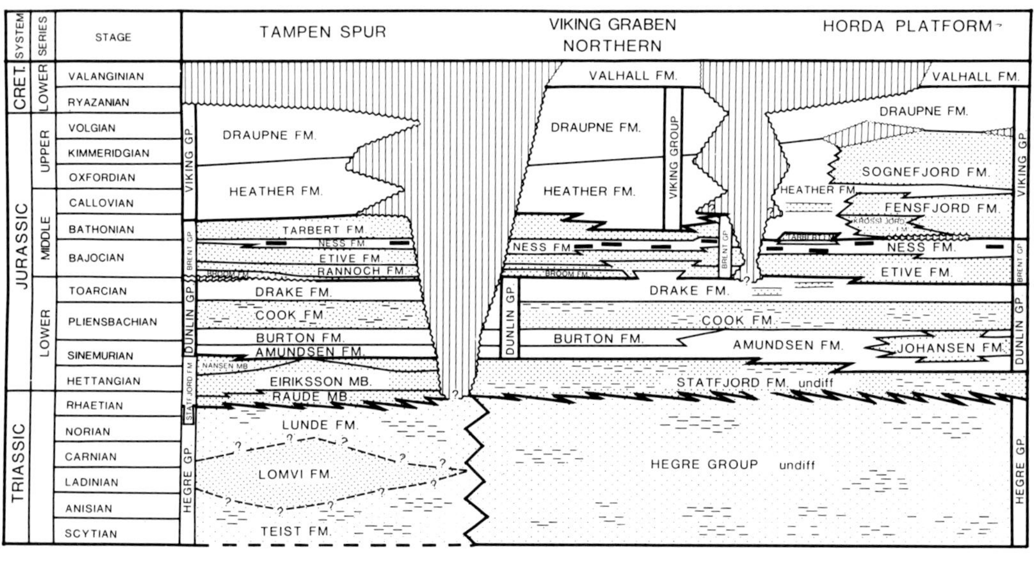

Fig. 4. Triassic and Jurassic lithostratigraphy of the Nothern North Sea (after Vollset & Doré 1984).

Fine-grained sediments normally have near-horizontal bedding. Therefore, present-day dips reflect the degree of structural dip, indicating post-depositional tilting, syn-sedimentary movements, or folding. Post-depositional tilting will show up as a constant dip trend, while syn-sedimentary movements may cause an upwards decrease in dip (Fig. 5).

Faulting is indicated by abrupt changes in structural dip, by dip patterns that might reflect deformation or drag close to a fault plane, and missing or repeated intervals. The latter is based

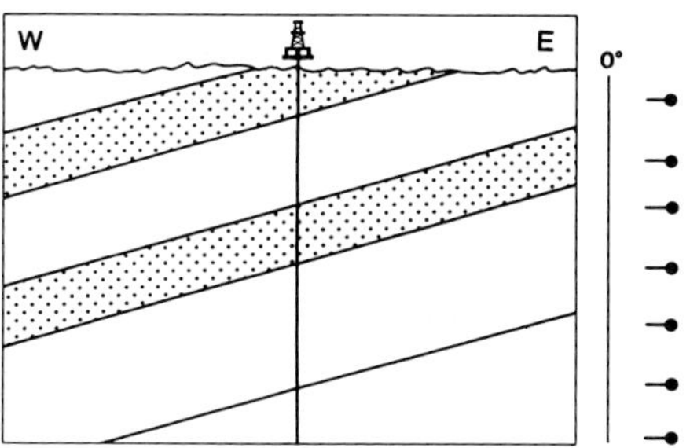

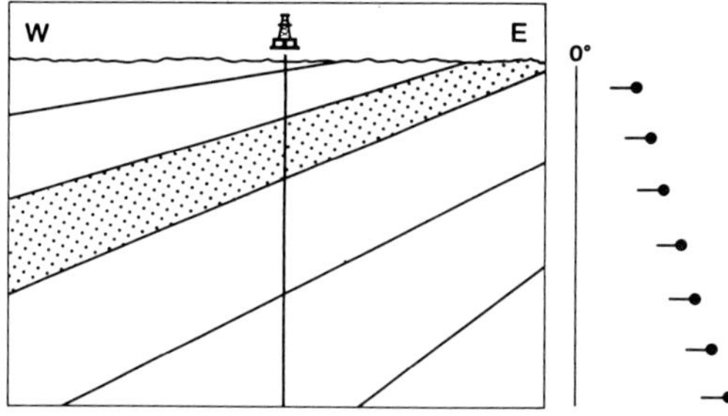

Fig. 5. Dip patterns of tilted bedding and downdip thickening of sediments (after Schlumberger 1981). The first represents post-depositional movements, while the latter can be caused by synsedimentary movements.

on lithostratigraphic correlations. Figs 6 & 7 show examples of dip patterns associated with normal and reverse faults as well as determination of strike from dipmeters. This is done by plotting the dip results from the zone of fault drag onto a stereogram, thereby constructing a drag fold axis that indicates the strike of the fault plane itself.

Identification of unconformities is based on changes in dip and azimuth at a certain level; therefore, only angular unconformities will show up. Additional geological evidence is required to distinguish faults from angular unconformities.

In order to evaluate the uncertainties in interpretation of faults, they have been ranked into three groups (Fig. 8): major faults, faults and possible or minor faults. Major faults have, by acquired definition, a missing section of at least 50 m. The second group represents faults which are fairly easily documented, while minor or possible faults are more questionable. Sketches of interpreted fault configurations are often tentative.

Structural interpretation of the Gullfaks Field

Structural interpretation of each well is described below. In the text, a structural dip will be shown as 10°/270° giving the magnitude and azimuth in degrees. The abbreviations IU and DU are used to denote increasing upwards dip (cf. Schlumberger 'blue pattern') and decreasing

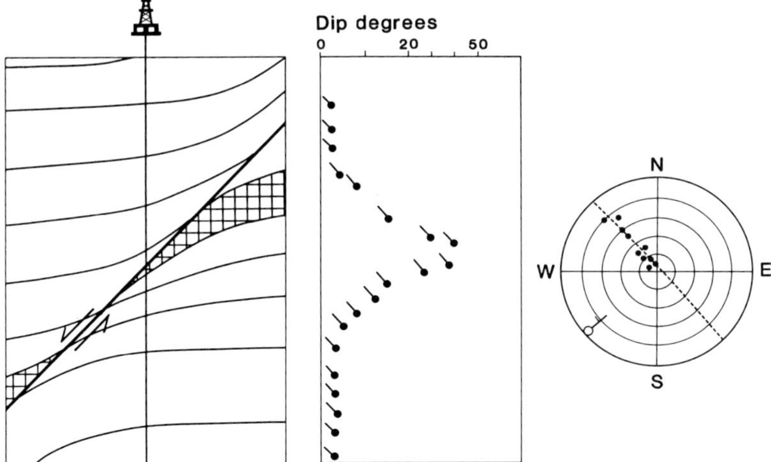

Fig. 6. Dip pattern of drag associated with a normal fault (after Dresser Industries Inc. 1987). Representative dips are plotted onto a stereogram (Wulff), i.e. poles of planes projected on the upper hemisphere, dip at periphery is 50°. Associated structural axis is given where dip at periphery is opposite or 0°. Strike of the axis defines the strike of the fault plane itself.

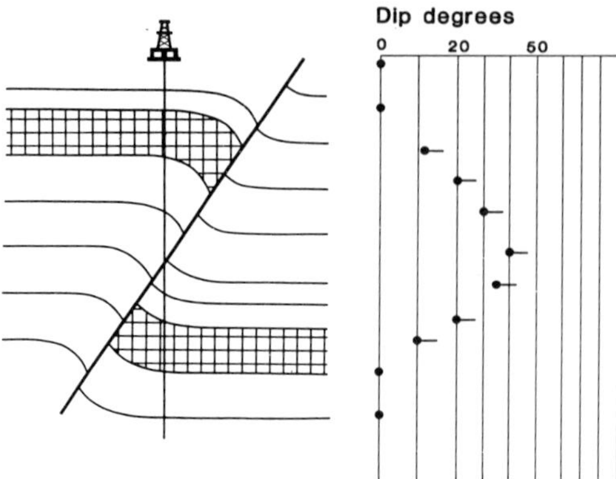

Fig. 7. Dip pattern of drag associated with a reverse fault.

upwards dip (cf. Schlumberger 'red pattern') respectively (Schlumberger 1981). 'Blank' zones characterize intervals with almost no computed dips. The dipmeter results are summarized on the seven-well correlation scheme (Fig. 8). Finally, several faults have been projected to top Dunlin level and are shown together with the seismically defined fault system (Fig. 9). The structural dip below top Dunlin is marked in for each well.

Well 34/10-1

Well 34/10-1 is located on a westerly dipping fault block bounded by N–S trending faults. Additional WSW–ENE trending faults in the area are interpreted as transfer or adjustment faults (Fossen 1989). Structural dips in the Lunde, Statfjord and Amundsen Formations are westerly throughout, although changing from 20° dip at Lunde level to 15° in the Amundsen Formation. In addition to a few minor faults in this lower section, a major fault with an estimated fault cut-out of approximately 90 m is observed around the Lunde/Statfjord boundary (Fig. 8).

A major fault zone (2020–2105 m), affecting the whole of the Cook Formation, comprises two faults, at 2082 m and 2020 m, with a rotated 'block' in between (Fig. 10a). The throw on the lowermost fault is significant and a fault cut-out of 70 m is estimated. The upper fault may be represented by a missing section of only 10–15 m, with no development of drag patterns. The faults are interpreted as normal, and the geometrical configuration assumed to be associated with this fault zone is shown in Fig. 10b. Some apparent reverse drag against the lower fault plane is observed, but this may be explained by the presence of additional minor faults, with associated rotations which produce

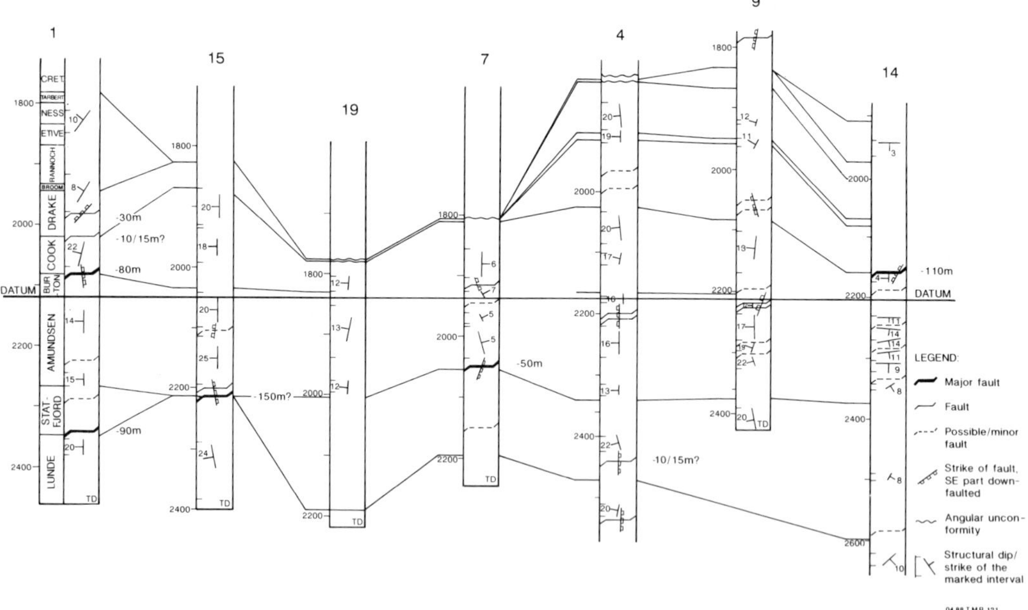

Fig. 8. Dipmeter interpretation results together with lithostratigraphical correlations. The results are suggesting that thickness variations are mainly controlled by post Brent faulting and erosion.

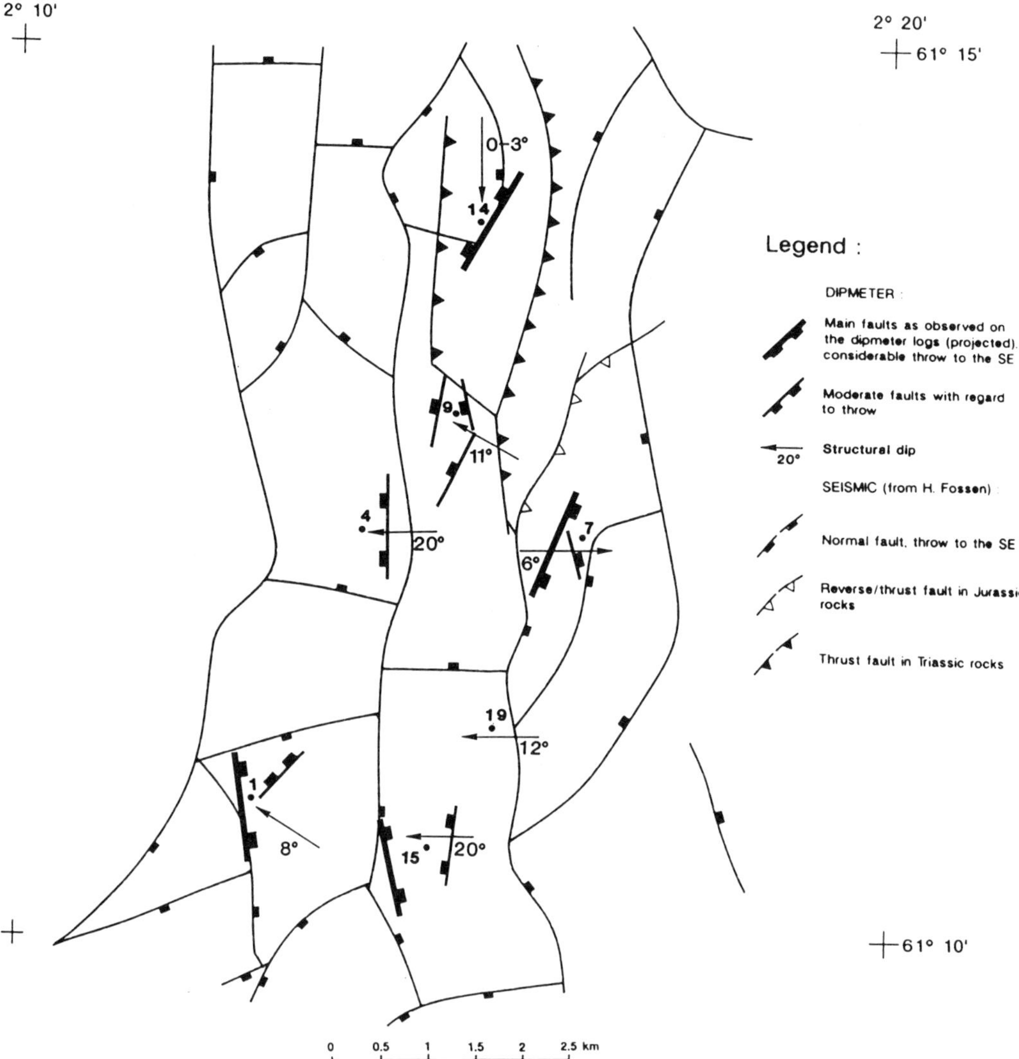

Fig. 9. Structure map where faults interpreted from dipmeter are shown together with seismically defined faults (Fossen 1989). Structural dip from dipmeter is marked in for each well. A fairly good agreement can be seen between major faults from dipmeter and seismic results.

an 'apparent' reverse drag. Below the fault plane there is considerable drag with reversal of the dips close to the fault plane (Fig. 10a). The fault zone is interpreted to correspond with the fault zone determined from seismic.

Structural dip is 8°–10°/310° in the upper part of the Drake Formation and the Brent Group, presumably due to a SW–NE oriented axis of rotation. The movements seem to be associated with the upper fault at 2020 m and another normal fault at 1984 m, where a fault cut-out of approximately 30 m is indicated.

Well 34/10–4

This well is also situated on a westerly dipping fault block, bounded by N-S striking main faults (Fig. 9). It is located a considerable distance from the seismically defined faults, and no major 'seismic' fault seems to intersect the well. This is supported by well data. Structural dips are fairly constant from the Lunde Formation to the top of the Brent Group (16–22° westwards).

The faults observed in this well seem to strike N–S, dipping to the east in the Statfjord and

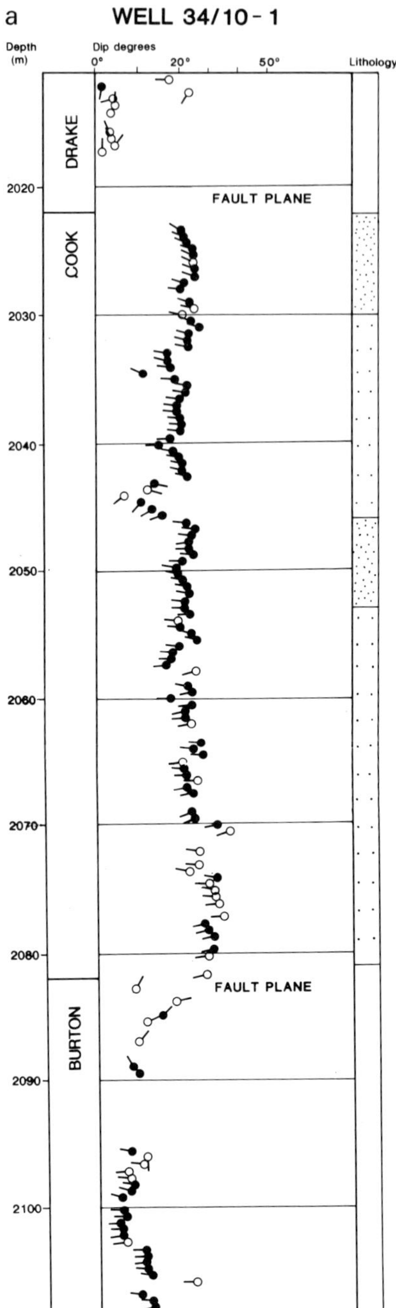

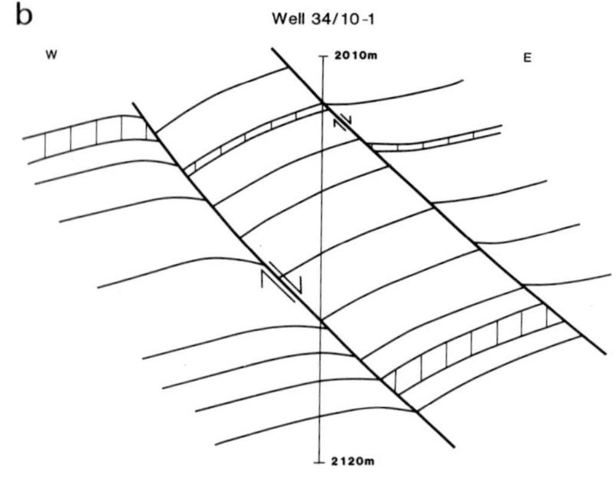

Fig. 10. A major ENE dipping fault zone comprising two normal faults and a rotated 'block' in between. (a) shows the arrow plot and (b) the interpreted fault configuration. A fault cut-out of 70 m (Cook) is estimated at the lower fault, while only 10–15 m seems to be missing at the upper fault. Apparent reverse drag against the lower fault plane could be due to additional minor faults and rotation.

Lunde Formations and to the west in the Amundsen Formation (Fig. 8). The faults have little throw but have well developed drag patterns; none are identified on seismic. They are probably of synthetic and antithetic types associated with the main system of easterly dipping faults. The variation in structural dip in the Amundsen and Statfjord formations may be indicative of additional minor faults.

The suggested antithetic fault-zone in the Amundsen Formation is illustrated in Fig. 11. A fault couple with a rotated fault slice in between is recognized. Drag is interpreted below the lower and above the upper fault respectively. The deformation affects a fairly thick interval (2190–2215 m) considering that an insignificant amount of section seems to be faulted out. In contrast the deformation involves quite small dip angles (computed dips not exceeding 30°). The fault zone is thus suggested to represent several displacements with fairly little throw.

Well 34/10–7

This well has been drilled on a slightly easterly dipping fault block. Reverse fault movements are identified in Jurassic rocks along the main fault to the west of the well (Fig. 9). An ESE-dipping major fault zone (2037–2053 m) is intersected by the well close to the base of the Amundsen Formation. About 50 m of the lower part of the formation is faulted out. However, very little drag pattern is seen (Fig. 12a). The lack of computed dips in the central part of the fault zone are indicative of its heavily fractured or brecciated nature.

Two small groups of dips, above and below the 'blank' zone, are interesting as they cannot be explained by normal drag. Both the lower group, with dips to the WSW, and the upper group, with E dips, form an IU pattern, with a dip magnitude of less than 20°. A tentative and simplified sketch (Fig. 12b) illustrates a possible fault configuration, in which normal faulting was followed by a phase with some reverse movements. A normal fault is interpreted at this level based on seismic evidence, with later reverse movement along the same fault (T. Lilloe-Olsen, pers. comm.).

A change in structural dip from 5°/60–70° to 6°/90° is observed in the lower part of the Cook Formation. This is interpreted as due to a fault (1902–1922 m) where rotation (pivotal) has occurred between two fault blocks.

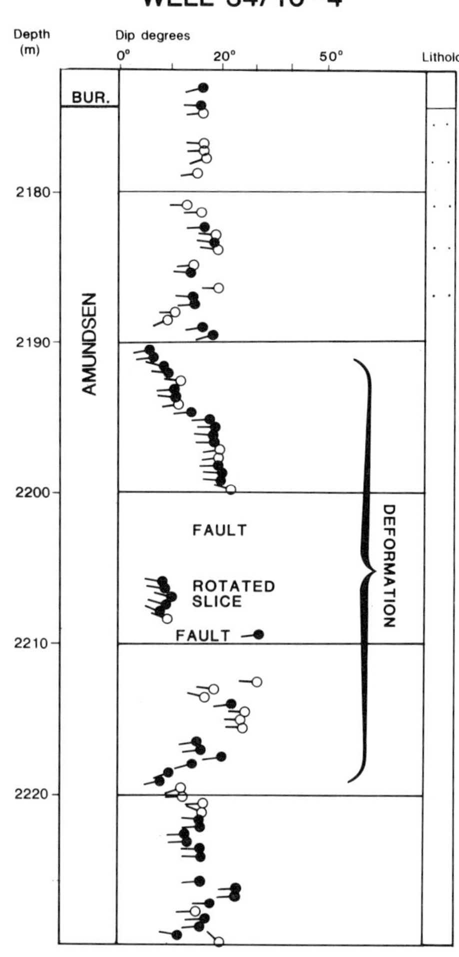

Well 34/10-4

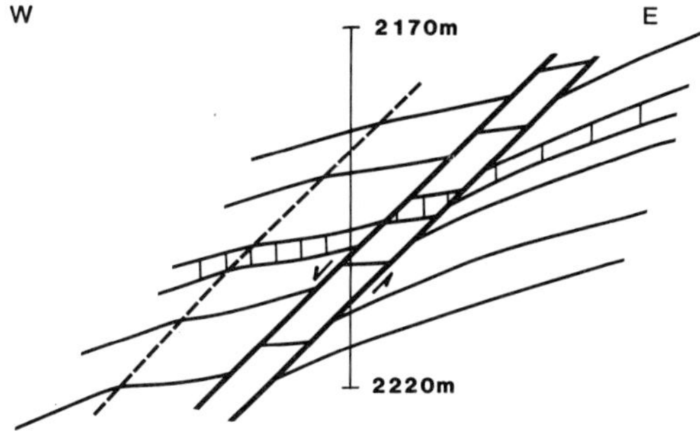

Fig. 11. Fault zone interpreted as a fault couple (normal) with a thin rotated slice in between. Drag is observed above and below the faults. Missing section seems to be insignificant, and the fault zone is suggested to be antithetic.

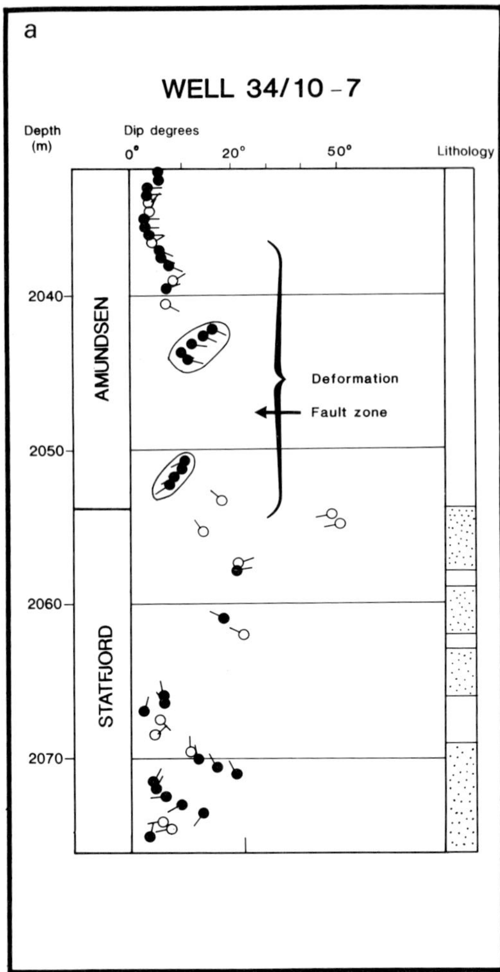

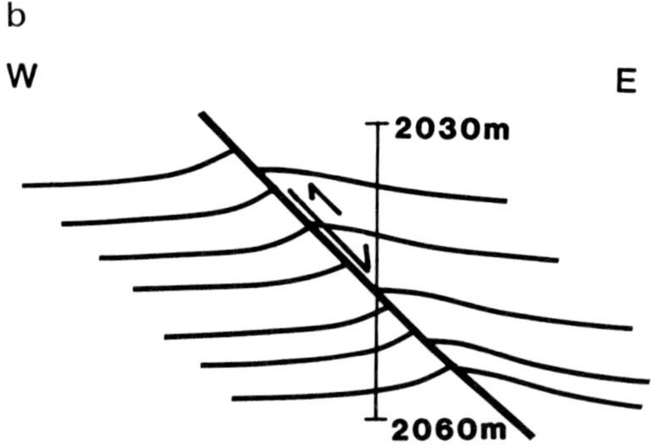

Fig. 12. A major ESE dipping fault zone intersecting the well close to the base of the Amundsen formation. Fault cut-out is estimated to 50 m. The fault is interpreted as normal, followed by a later phase of reverse movement along the same fault.

Well 34/10–9

No major faults are intersected by well 34/10–9. The most significant fault zone (not seen on the seismic) is found near the top of the Amundsen Formation, where deformation/drag affects the interval 2205–2235 m, but produces little dip variation (Fig. 13a). Two WNW-dipping fault planes are inferred, with a slightly rotated slice in between (Fig. 13b). The change in structural dip, from 17–20°/260–270° in the Statfjord/Amundsen to 11–13°/275–300° in the Cook, Drake and Brent, was mainly caused by this fault couple. Fault cut-outs seem not to be significant. However, there is little indication of vertical movement. Two minor faults higher up in the well also dip approximately westwards. The faults are probably antithetic.

Well 34/10–14

This well is located in a structurally complex area as shown by the structure map (Fig. 9). Structural dips of about 8–10°/120–150° are measured in the Lunde, Statfjord and the lowermost part of the Amundsen formations. Small variations in structural dip are observed further up throughout the Amundsen Formation (9–14°/180°), probably caused by minor faults.

A major fault is seen at 2159–2160 m (Fig. 14a). The suggested fault plane is indicated on the microresistivity curves as a conductivity anomaly and is computed to dip 45–50°/300°. About 75 m of the Cook Formation and 35 m of the Drake Formation are estimated to be missing due to this fault. Both dipmeter and seismic data show a major fault at this level.

A rotated fragment of the Cook Formation is seen between 2163 m and 2173 m; this may suggest another fault below. A sketch of the fault zone is shown in Fig. 14b.

Above this fault zone, structural dip is 0–3°/180°, up to the top of the Brent Group.

Well 34/10–15 and Well 34/10–19

The wells are both located on the same W-dipping fault block according to seismic interpretation (Fig. 9). In well 34/10–15 the whole Statfjord Formation has been faulted out, and 150 m of missing section is estimated. The main fault intersects the well at approximately 2213 m, although the interval between 2195 and 2220 m seems to have been affected (Fig 15a). There may be another fault at approximately 2200 m. The fault zone (dipping ENE) has also been verified by seismic interpretation (T.

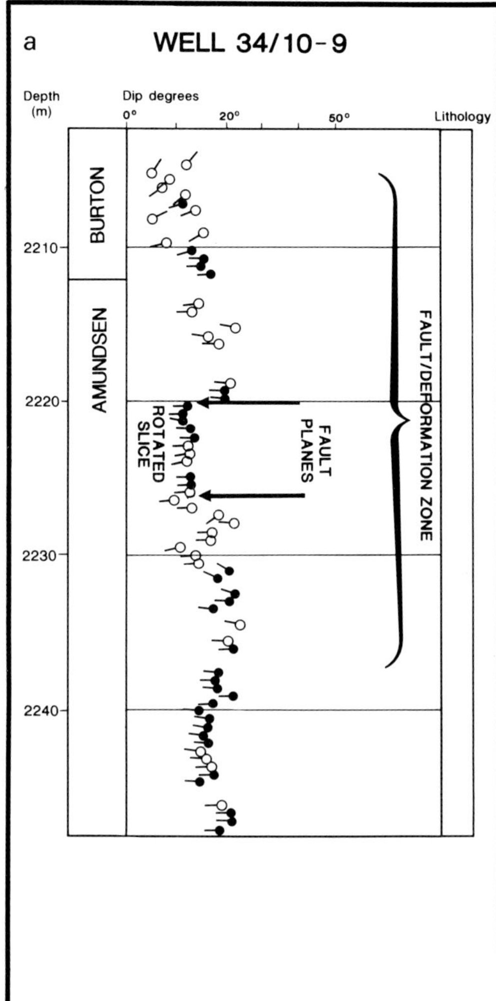

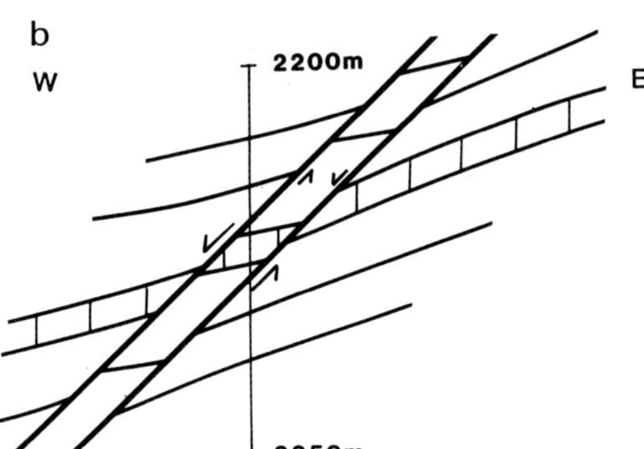

Fig. 13. Fault zone comprising two faults (normal) with a rotated slice in between. Slight drag is seen above and below the faults. No significant fault cut-out is observed. The fault zone is considered as antithetic.

Lilloe-Olsen, pers. comm.), and is assumed to correspond with the main seismically defined fault to the W of this well (Fig. 15b).

The dip results show no clear evidence of faulting, but a certain reduction in dip magnitude towards the fault zone is observed (footwall and hanging wall), possibly reflecting some normal drag. The lack of 'good fault patterns' may be explained by heavy fracturing/brecciation, a hypothesis which is substantiated by examination of core recovered from around these depths, where brecciated zones and slickensides are observed. Structural dip is estimated to be approximately 20°/270° through the whole Triassic/Jurassic section.

Well 34/10−19 is located in a relatively stable area as no faulting is recognized, and the structural dip is around 12°/270 throughout. The Cook Formation is directly overlain by Upper Cretaceous rocks with a distinct, angular unconformable contact.

The marked difference in structural dip between 34/10−15 and 19 implies that the two wells are located on separate fault blocks. An additional fault is thus inferred.

Results and discussion

The interpretations of the well data are summarized in Fig. 8. Variation in structural dip in each well is generally moderate. When changes are significant, they appear quite abruptly, and seem to be associated with faulting. No vertical trends in the structural dip that may be indicative of syn-sedimentary tectonic movements have been observed. Structural dip is almost constant between the faults, a consistency that only locally could be caused by the CLUSTER processing. The variation in structural dip is therefore ascribed to tilting, during post-Middle Jurassic faulting. Variations in structural dip between the different wells is considerable, suggesting that they are all situated on different fault blocks.

Faults are generally thought to be normal. Some evidence for reverse movement is found in well 34/10−7. Faulting often appears as deformed zones rather than single fault planes. In several cases two faults have been identified with a rotated slice in between. Faults may also show a rather 'blank' expression, especially faults of considerable throw. The 'blank' expression may be due to intensive brecciation of the rocks producing high dip scatter which is subsequently removed by the CLUSTER processing. The average density of faults is estimated at one per 140 m vertically (Fig. 8), as

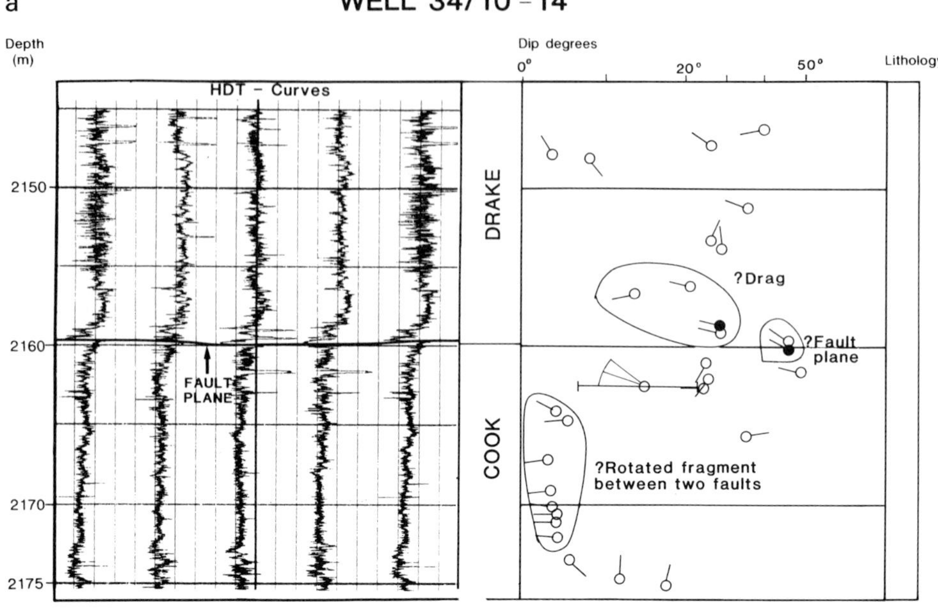

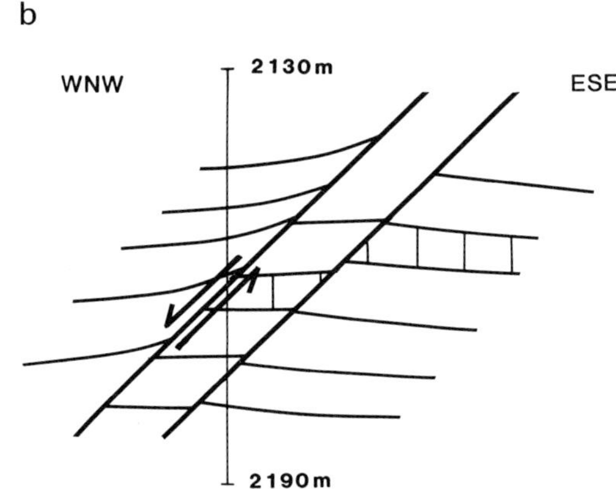

Fig. 14. A major fault at 2159–2160 m indicated on the micro-resistivity curves as a conductivity anomaly is computed to dip 45–50°/WNW. 110 m missing section is indicated. A rotated fragment suggest another fault below.

seen in the wells. Most thickness variations can be explained by faulting (fault cut-out) and/or by post-Brent erosion.

The structure map (Fig. 9) shows faults interpreted from dipmeter together with seismically defined faults. There is reasonably good agreement between the two interpretations, although perfect match between the two types of data is not expected. HDT data, for instance, may give quite inaccurate azimuths, with up 20° uncertainty. The projection of faults at top Dunlin is based on assumed 45° dip of the fault planes,

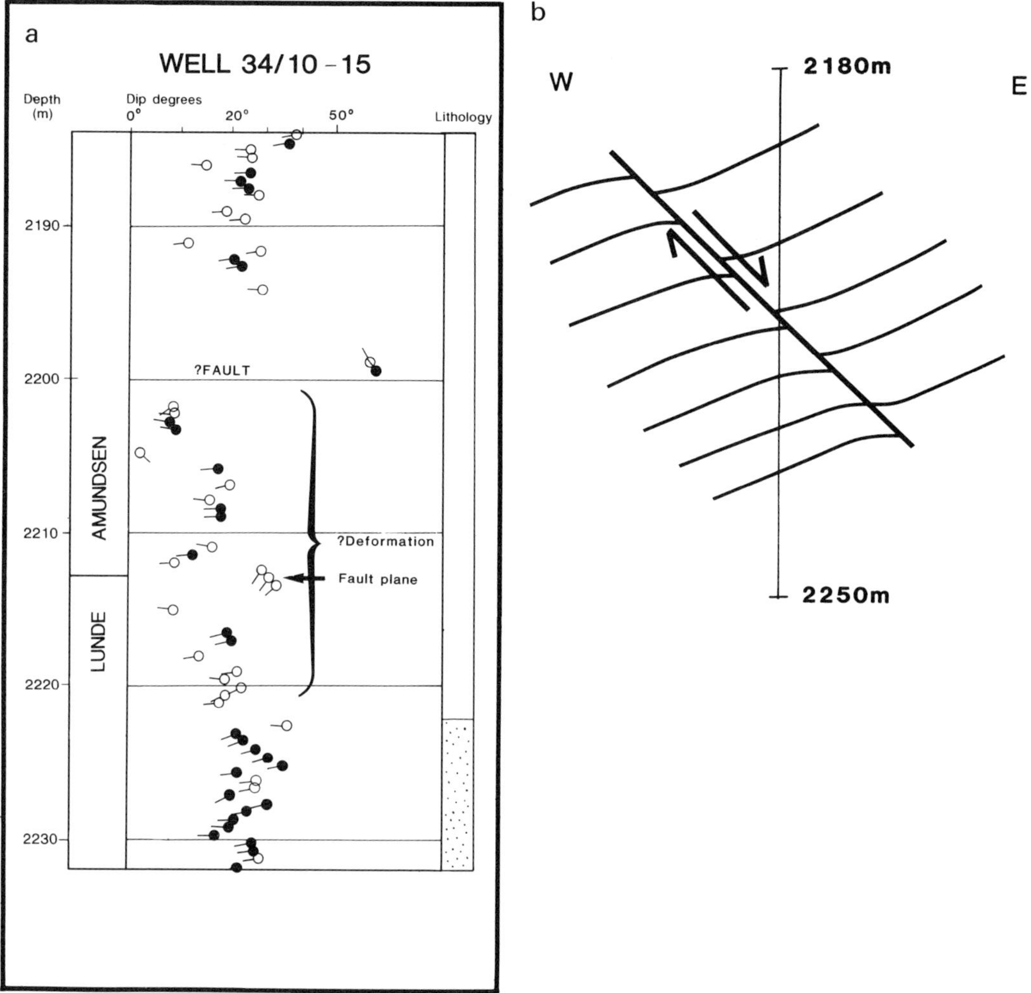

Fig. 15. A major ENE dipping fault where 150 m, i.e. the whole Statfjord Formation, is faulted out. The fault is poorly defined from dipmeter, though some normal drag is suggested.

which may introduce a considerable error when faults are projected over great vertical distances. Seismically derived faults are presented as planar features, although they actually represent zones.

Other less important faults presented in Fig. 8 are not detected on the seismic. They are subparallel to the main fault system, partly westerly and partly easterly dipping. They are interpreted as mainly representing synthetic and antithetic faults associated with the main fault system.

The advantage of dipmeter, compared to seismic, is that additional details may be obtained and detection of structural features beyond seismic resolution can be found. The lack of well developed fault patterns often make interpretation of even major faults uncertain. The identification of faults depends much on interpreted fault cut-outs. High dips from cross-bedded sandstones may complicate the structural interpretation. Despite these limitations the dipmeter data can give useful additional information, thus supplementing seismic interpretation.

The author would like to thank B. Mair for valuable discussions, R. Færseth for comments during the preparation of this manuscript, T. Lilloe-Olsen for providing seismic interpretation results, and P. Kloumann for carrying out much of the editing.

References

DAY, G. A., COOPER, B. A., ANDERSEN, C., BURGERS, W. F. J., RφNNEVIK, H. C. & SCHφNEICH, H. 1981. Regional Seismic Structure Maps of the North Sea. *In*: MING, L. V. & MOBSON, G. D. (eds). *Petroleum Geology of the Continental Shelf of North-West Europe*, Weyden, London, 76–84.

DRESSER INDUSTRIES INC. 1987. *Fundamentals of Diplog Analysis*.

ERICHSEN, T., HELLE, M., HENDEN, J. & ROGNEBAKKE, A. 1987. Gullfaks. *In*: A. M. SPENCER *et al.* (eds). *Geology of the Norwegian Oil and Gas Fields*. Graham & Trotman, London, 273–286.

FOSSEN, H. 1989. Indication of transpressional tectonics in the Gullfaks oil-field, Northern North Sea. *Marine and Petroleum Geology*, **6**, 22–30.

HEPP, V. & DUMESTRE, A. C. 1975. CLUSTER, — a method for selecting the most probable dip results from dipmeter surveys. *SPE of AIME, 50th Annual Fall Meeting, Dallas*, SPE Paper 5543.

SÆLAND, G. T. & SIMPSON, G. S. 1982. Interpretation of 3-D data in delineating a subconformity trap in block 34/10, Norwegian North Sea. *Memoir of the American Association of Petroleum Geologists*, **32**, 217–236.

SCHLUMBERGER, 1981. *Dipmeter Interpretation*. Vol. 1.

VOLLSET, J. & DORE, A. G. 1984. A revised Triassic and Jurassic lithostratigraphic nomenclature for the Norwegian North Sea. *Bulletin of the Norwegian Petroleum Directorate*, **3**.

Example of fault identification using dipmeter data

M. C. DEVILLIERS & PH. WERNER
Total CFP, Paris, France

Abstract: Dip results obtained by conventional processing techniques are not always suitable for the identification of faults. Dipmeter identification of faults is usually based on the recognition of increasing angles of dip associated with drag zones adjacent to fault zones. Several case studies confirm this relationship. Interpreting faults in this way is, however, risky, and can lead to erroneous data being integrated into structural interpretation. By reviewing original log quality and processing methods, much erroneous information can be eliminated. The influence of faulting on raw dipmeter curves is studied, and actual fault geometries from outcrops are compared with dipmeter patterns. Outcrop data are taken from several tectonic contexts in order to provide a significant amount of information, in spite of the restricted vertical perspective of the outcropping faults. Reliable detailed fault interpretation from dipmeter data is enhanced by reconstruction of dip profiles from field examples, in particular where similar data are available from other wells and local structural geological knowledge is available.

Faults and faulted zones are frequently encountered when drilling a well, particularly in areas of intense deformation. Dipmeter logs provide useful geometrical information from the subsurface, but interpretation of effect of faulting on the raw dipmeter data is often neglected. Typically, fault interpretation is made simply by the dipmeter pattern method (Schlumberger 1981).

In order to determine what kind of fault has been detected with dipmeter data, the geologist needs a thorough knowledge of the main geometrical features associated with faulting, and correlation of geological data with dipmeter response, using both raw data and processed results.

Characterization of faults with dipmeter data

Fault interpretation from dipmeter data is often based on pattern identification from arrow-plot results. Increasing angle of dip due to drag effects along fault planes is the characteristic most often used to identify faults from dipmeter data. However, great care must be taken in using dipmeter results in this way, because the increasing dip pattern, which is attributed to drag, is not observed systematically on arrow-plots. Faults, however, may not have a drag zone, or there may be insufficient variations in the resistivity of the formation to permit dipmeter detection of fault-induced structures.

Additionally, choice of processing parameters may be the cause of 'missing a fault' and thus should be carefully selected. The search angle often treated as 'standard' (Hepp & Dumestre 1975), 30° × 2 or 35° × 2, should in general be replaced by 70° × 1. A 35° × 2 search angle makes an initial search within the first 35° of inclinations. If no correlation if found, the program then resumes the search in the same interval with an angle of 35° to 70°. In this case, identification of low structural dips can thus hinder the detection of the high dips, commonly associated with faults. The correlation interval of 1 m (feet) in our experience is adequate for fault analysis.

Examination of the raw dipmeter data is essential for confirmation of arrow-plot interpretations. The detailed analysis of resistivities, plotted at an adequate scale (1/20 or 1/40), gives a qualitative assessment of the validity of automatically computed dips. The effect of faulting on the dipmeter resistivity curves may be minor and focused (Fig. 1), but faulted zones are often marked by conductivity anomalies which are readily discernible at a 1/200 scale (Fig. 2). Reliable interpretation of the raw data requires that effects such as tool rotation or borehole ovality are not responsible for producing the resistivity anomalies.

Faults images from dipmeters and outcrops

Field observations (CSRPPG 1976) of faults are very helpful for understanding the relationships between the type of fault, the geometrical variations of beds adjacent to the fault plane, the magnitude of throw, and of the distortion of bedding planes. Occurrence of drag folding of bedding planes provides the theoretical basis for detection of faults using dipmeter data.

However, the drag associated with normal

From HURST, A., LOVELL, M. A. & MORTON, A. C. (eds), 1990, *Geological Applications of Wireline Logs* Geological Society Special Publication No. 48, pp. 287–295

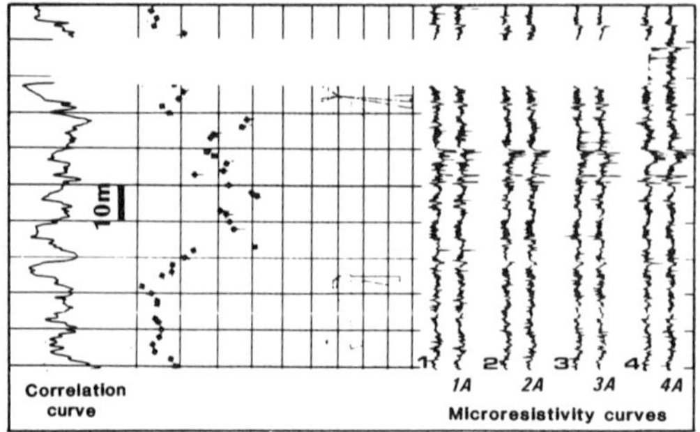

Fig. 1. Effects of a fault on micro-resistivity curves. A major normal fault is shown with a throw of approximately 85 m.

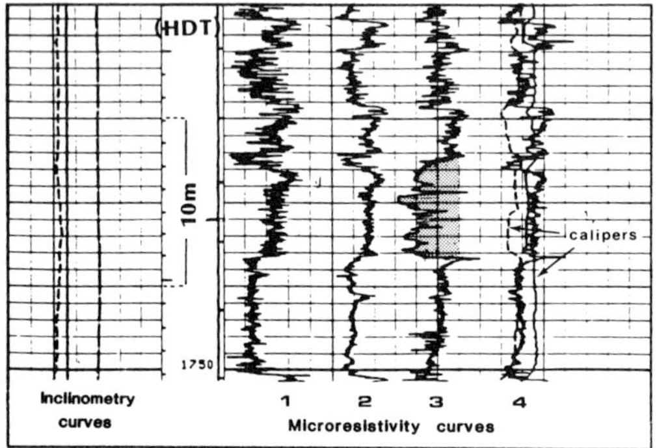

Fig. 2. A fault zone marked by a conductivity anomally marked as the shaded interval in track 3. Throw is estimated to be 75 m.

faulting does not necessarily create concave folds in the downthrown block and convex folds in the upthrown block, as represented traditionally in dipmeter publications (Schlumberger 1981). Our field obsevations prove that drag in the footwall block of normal faults is very often limited.

Azimuth variations may only appear in the hanging wall block (Fig. 3). In the example, evidence of rotational movement is identified in the hanging wall block with respect to the adjacent footwall block. Rotational effects are in our experience common and provide the best conditions for interpreting faults from dipmeter data.

In an example of extensional strain, faulted blocks bounded by several sub-parallel faults (Fig. 4) are clearly defined by dip data, due to the tilting effect. An intermediate tilted block (B) is located between two major normal faults F1 and F2. The dips within block B tilt down to the east, as shown by the arrow-plot (Fig. 4b). A similar pattern, which is interpreted as an intermediate tilted block, is obtained from borehole data (Fig. 4c).

In grabens, the drag effect is often limited to very narrow zones, too narrow to be identifiable on arrow-plots (Fig. 5). Detailed field examination of the Cingoli anticline reveals an absence of drag effects along the fault planes. Construction of arrow plots, based on three vertical sections through the anticline, illustrate the absence of identifiable drag (A to D, Fig. 5b). No horizontal dip is measurable due to the northward plunge of the fold. The rotational movement appears abruptly on the E–F arrow plot and corresponds to the change of dip and azimuth over the western fault plane.

In compressive zones, complex geology caused by low angle reserve faults is common (Boyer & Elliot 1982), (Chapman & Williams 1985). Faulting may often be associated with folding, which produces well-marked changes of dip and azimuth, both of bedding planes and faults. Overturning of folds often results in the development of shear planes which rupture the fold and form reverse faults. This demonstrates the common relationship between reverse faults and overturned folds (Fig. 6). The minor thrust (throw $\sim$ 20 m) is characterized by a brecciated zone which in vertical section ranges from 1 to 2 m thick. The overturned section is about 1.5 m thick, and the synclinal fold in the footwall block creates a decreasing dip pattern in a 1 m thick section.

Similar fold-thrust structures, including major faults, may also be revealed by the dipmeter (Fig. 7). In this example, hand-made correlations confirm the dip results in the faulted zone. The overturned flank and very high dips between 1170 m and 1180 m deduced from these correlations, provide an argument in favour of a southwest dipping thrust plane near 1165 m.

Gentle folding may, however, be sufficiently significant to confirm the presence of a reverse fault. Such a structure is identifiable on the arrow plot between 400 and 550 m in Fig. 8. The dips have a clear change of azimuth from west ot east corresponding to an anticlinal fold. The change in azimuth is associated with a minor reverse fault which is clearly marked by an increasing dip pattern at 480 m. Repetition of the sequence gives an apparent throw of 25 m over the reverse fault.

The detailed analysis of dips is complex when successive tectonic phases are superimposed (Vialon *et al.* 1976). A rollover shape identified in the hanging wall block (Fig. 9) is due to the

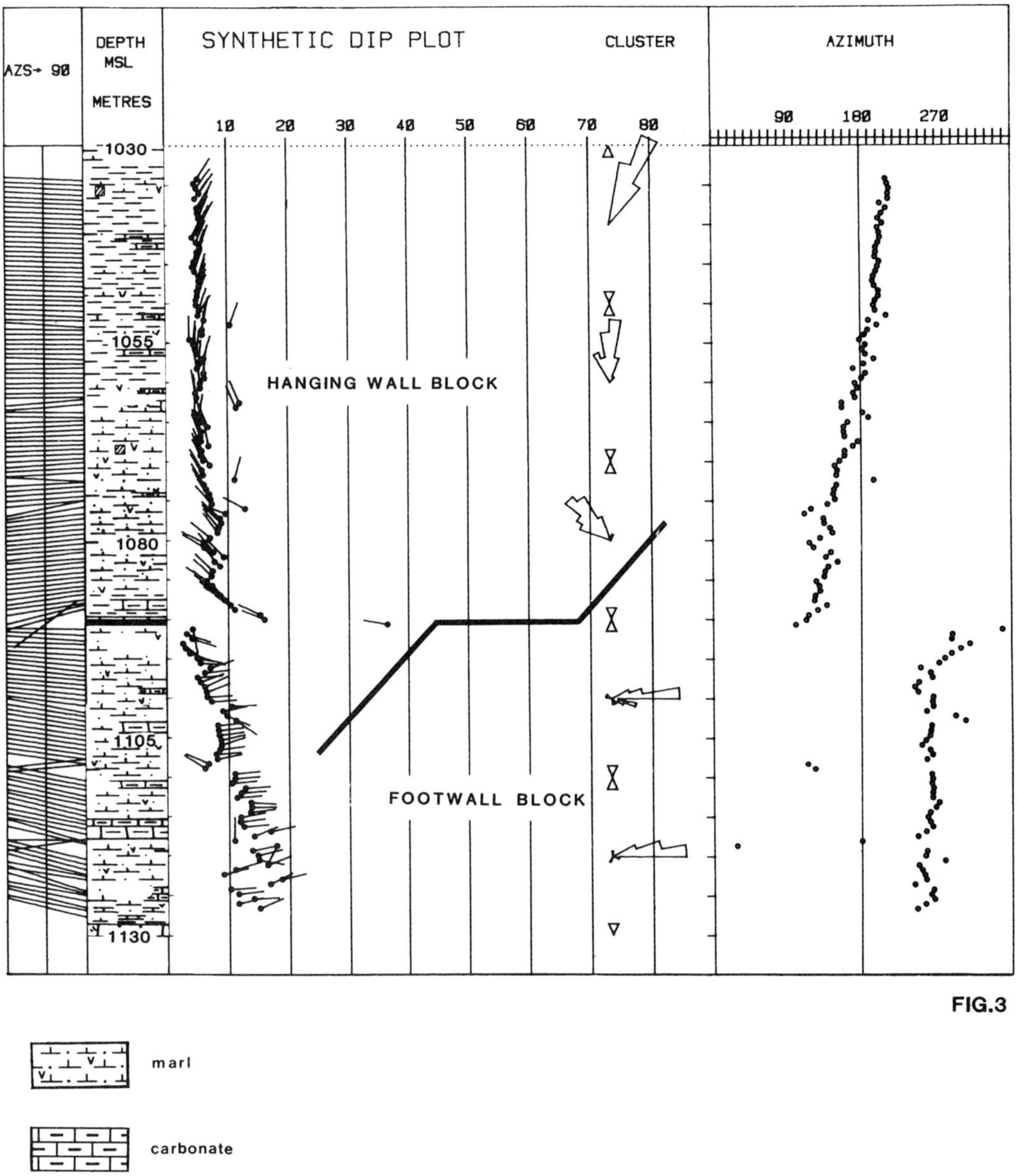

FIG.3

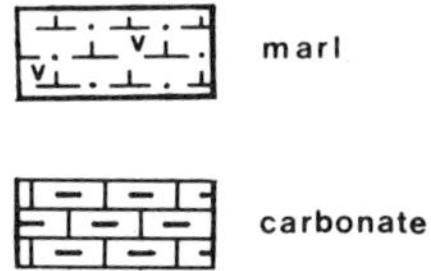

Fig. 3. Location of a major normal fault at 1090 m. Throw is estimated to be 250 m. A progressive rotation of azimuths is seen in the hanging wall block as the fault plane is approached. Low dip (< 40°) of the fault plane, is indicated by the thick interval in which azimuth rotation is recorded.

rejuvenation of a normal fault by a compressive phase, which involved a strike-slip movement along the fault. Proof of the strike-slip movement comes from field observations of slickensides on the fault plane. If no high dips are computed on the schematic arrow-plot, the geometrical description of the fault remains ambiguous.

The distortion of beds in the footwall block of a normal fault will often be limited to the major fault plane itself. If this assumption is correct, the fault patterns on Fig. 10 are supportive of

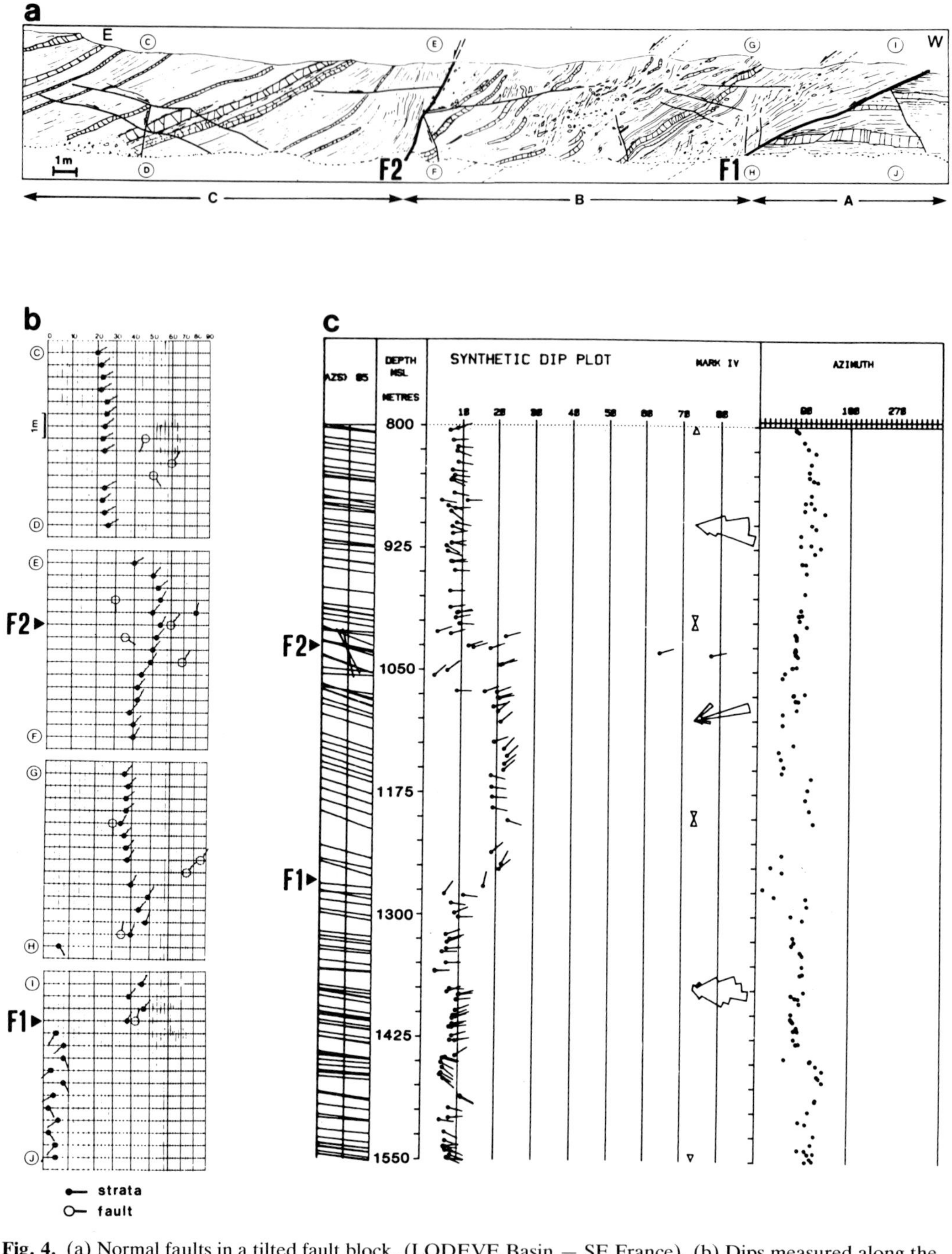

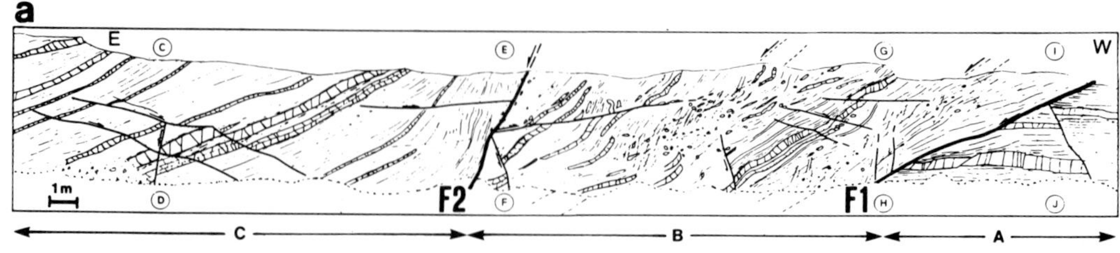

Fig. 4. (a) Normal faults in a tilted fault block. (LODEVE Basin — SE France). (b) Dips measured along the vertical sections C to J on the outcrop shown in (a). (c) Borehole data which have similar dip patterns to the field example.

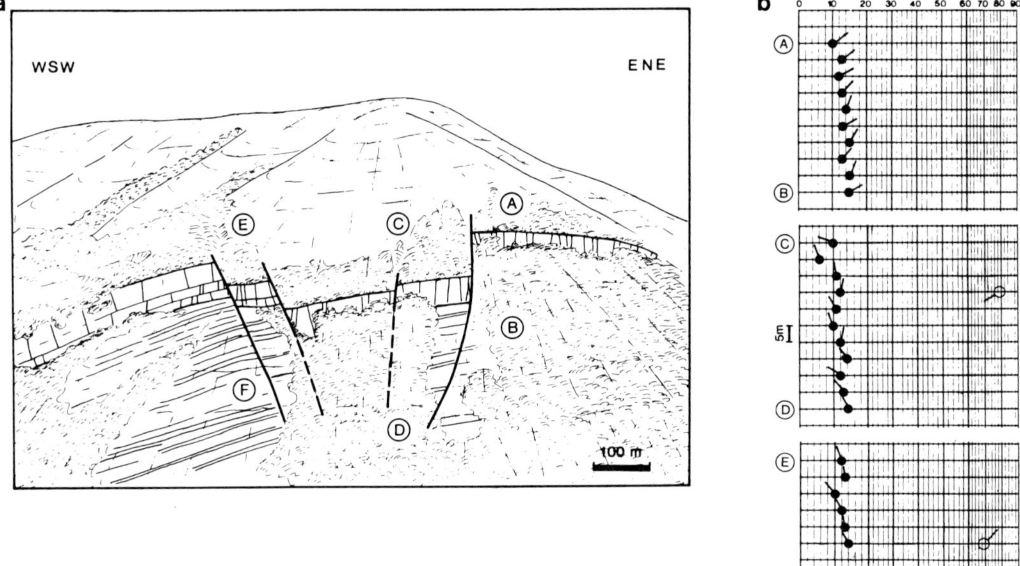

Fig. 5. (a) Normal faults in the Cingoli anticline, Ancona, in the northern Appennines. (b) Dips measured along the vertical profiles A to F. Dip symbols as for Fig. 4.

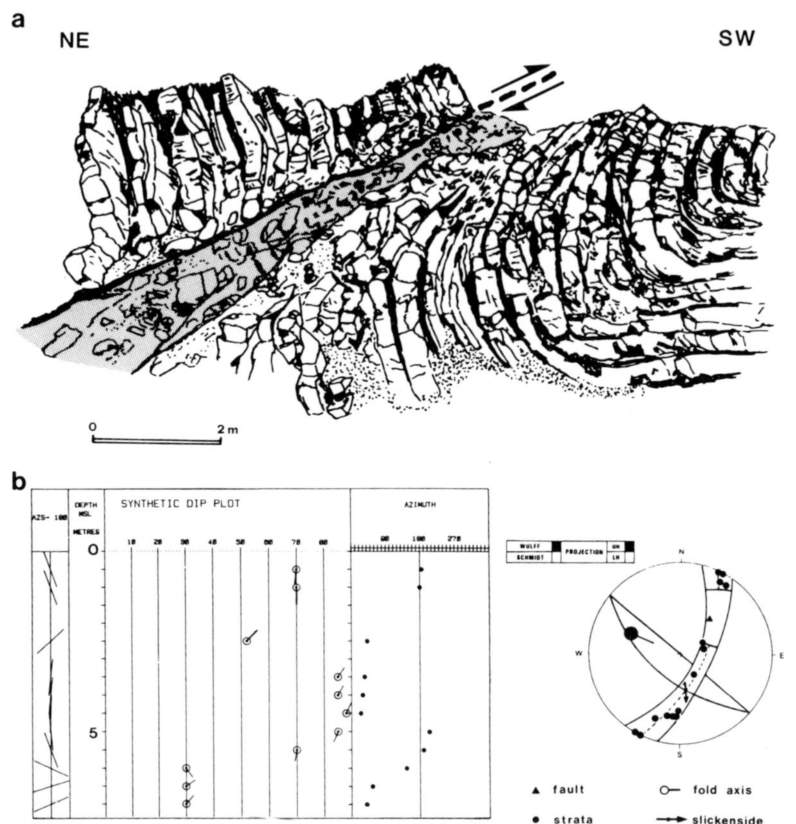

Fig. 6. (a) Reverse faulting in Hauterivian limestones of the Castillon anticline (SE France). (b) Schematic arrow plot reconstructed from the outcrop shown in (a). Throw is estimated to be tens of metres.

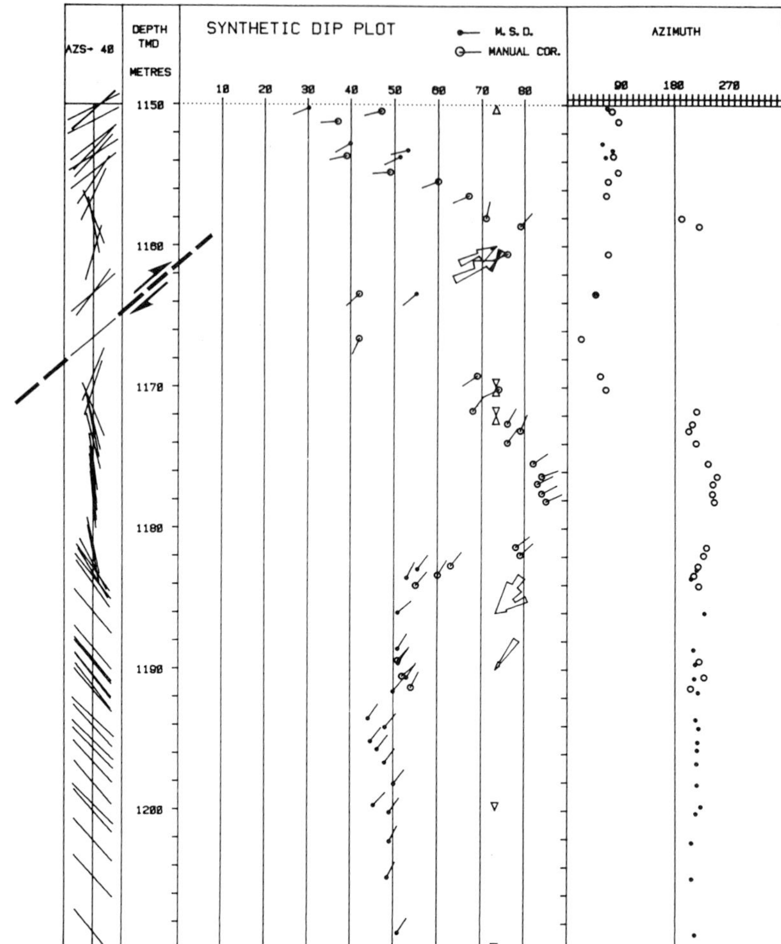

Fig. 7. Borehole data interpreted to show the presence of a major reverse fault. Compare with Fig. 6: throw is hundreds of metres.

fault plane movement, caused by compressive stress (a possible strike-slip fault), which has created a drag fold in the footwall block. No distortion of bedding planes is discernible in the hanging wall block. A strike-slip interpretation in this example is also supported by the regional tectonic data.

Growth faults, which are often listric (Ellis & McClay 1988), are reasonably straightforward to define from dipmeter data, particularly when roll-overs and divergent bedding planes are present in hanging wall block. Growth faults are also characterized by the preservation of a greater thickness of sediments in the hanging wall block. The dip patterns associated with these structures (Fig. 11) are summarized as follows.

(i) Divergent bedding is clearly marked by increasing dips from 2200 to 2850 m. The dips do not increase gradually downward. Some minor unconformities possibly reflect variations of fault movement during growth.

(ii) Drag zones are identifiable by decreasing dips, for example between 2850 m and 2975 m which corresponds to the hanging wall syncline.

(iii) no distortion of bedding planes appear in the footwall block.

Conclusions

Dipmeter interpretation of faults can be made with confidence subject to the following conditions.

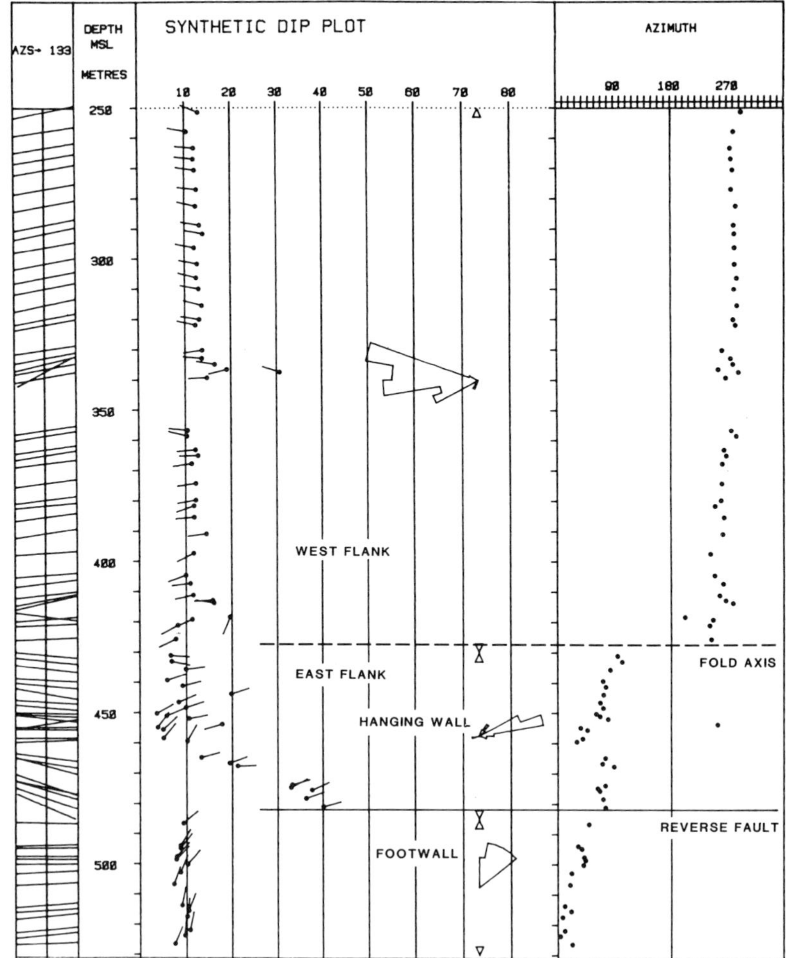

Fig. 8. Borehole data from a sequence of folded and reverse-faulted rocks.

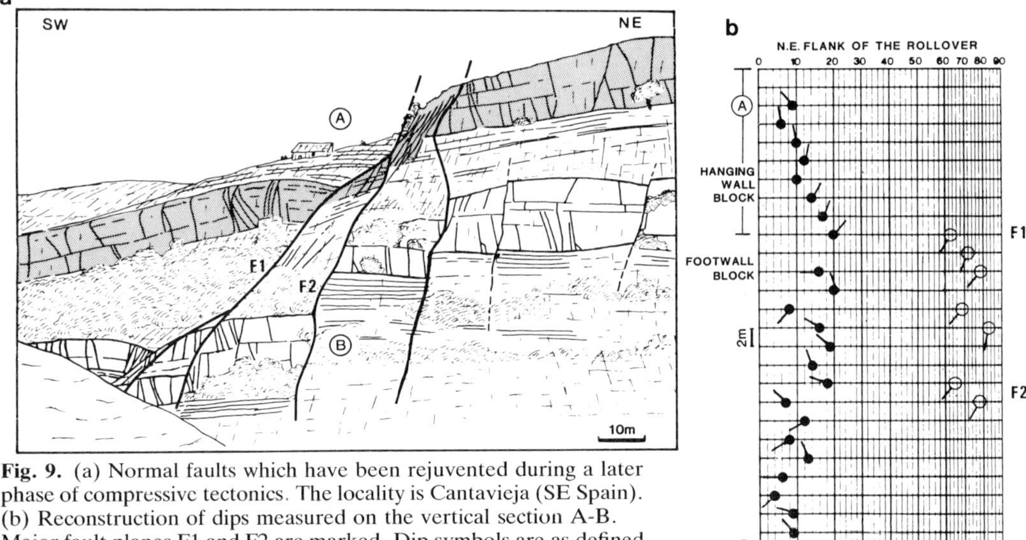

Fig. 9. (a) Normal faults which have been rejuvented during a later phase of compressive tectonics. The locality is Cantavieja (SE Spain). (b) Reconstruction of dips measured on the vertical section A-B. Major fault planes F1 and F2 are marked. Dip symbols are as defined in Fig. 4.

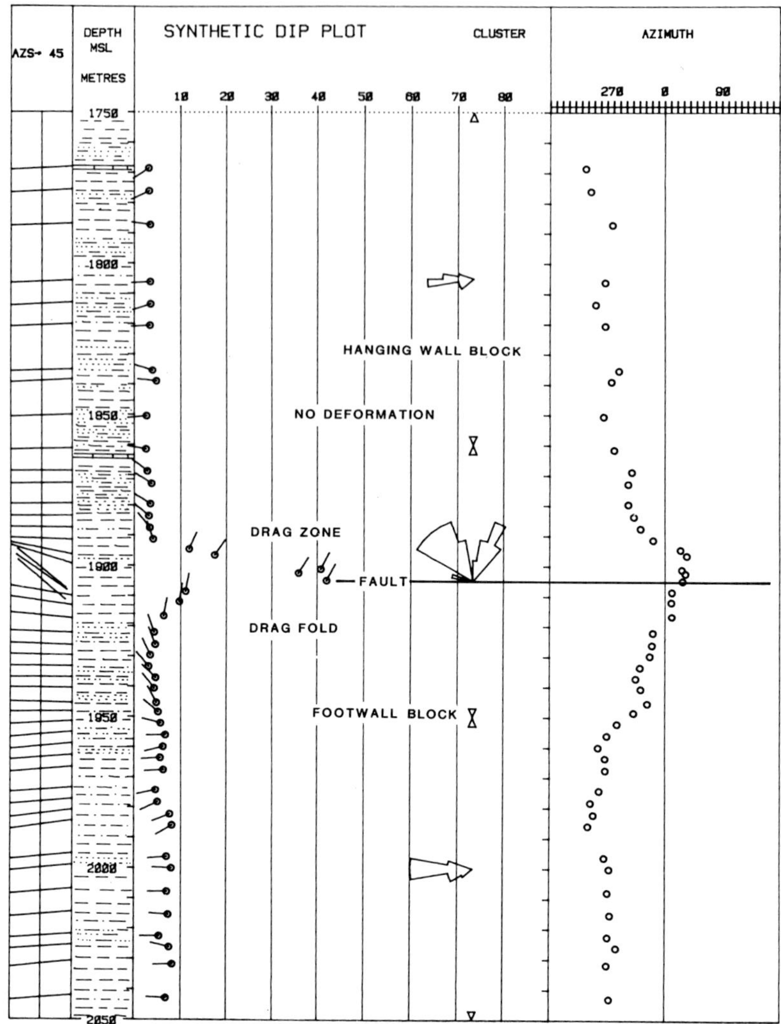

Fig. 10. Borehole which has penetrated a sequence containing a possible strike-slip fault, created by compressive movements. Compare this example with the outcrop data shown in Fig. 9.

(i) The integration of independent information, i.e. regional geological data and seismic interpretation.

(ii) Fault identification from dip results is not made without detailed analysis of resistivity curves, including the processing parametres.

(iii) The pattern method commonly used for fault detection is only helpful if a good knowledge of the geometrical and geological character of the faults is already available.

(iv) Study of outcropping faults and faulted zones affords optimal conditions for fault characterization (geometry, movement, size) and provides valuable analogue data for comparison with borehole data.

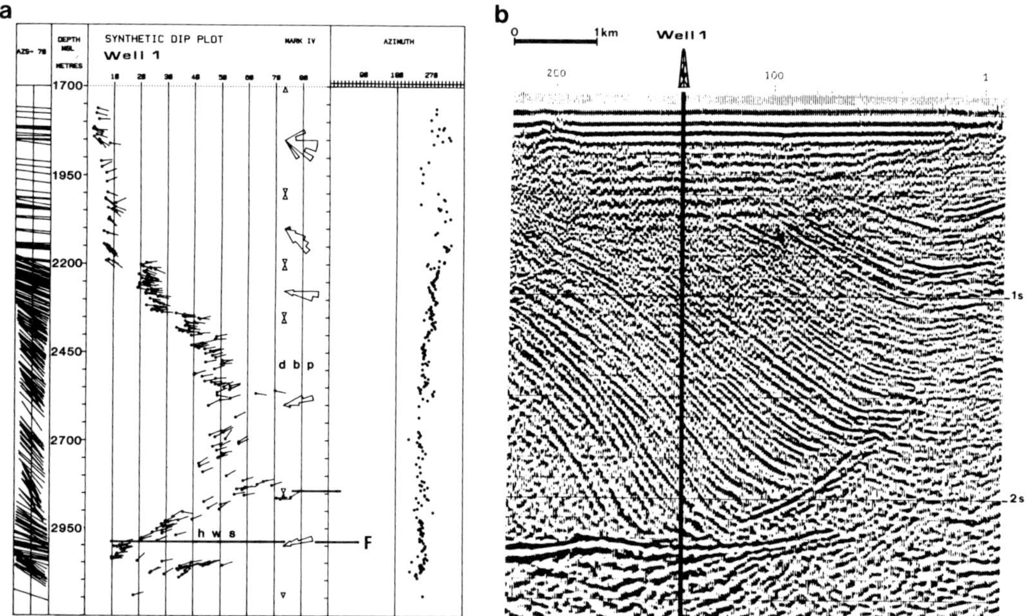

Fig. 11. (a) A borehole section through a growth fault: F, fault plane; hws, hanging wall syncline; dbp, divergent bedding planes. (b) Seismic section through the listric fault penetrated by the well shown in (a).

References

Boyer, S. E & Elliott, D. 1982. Thrust systems. *AAPG Bulletin*, **66**, 1196–1230.

Chambre Syndicale de la Recherche et de la Production du Pétrole et du Gaz naturel, Comité des Techniciens 1976. Méthodes modernes de géologie de terrain. 2b — Manuel d'analyse structurale, Traitement des données. Editions Technip, Paris.

Chapman, T. J. & Williams, G. D. 1985. Strains developed in the hanging walls of thrusts due to their slip/propagation rate: a dislocation model. *Journal of Structural Geology*, **7**, 755–762.

Ellis, P. G. & McClay, K. R. 1988. Listric extensional fault systems results of analogue model experiments. *Basin Research* **1**, 55–70.

Hepp, V. & Dumestre, A. C. 1975. Cluster: A method for selecting the most probable dip results from dipmeter survey. *SPE of AIME, 50th Annual Fall Meeting*, Dallas Paper SPE 5543.

Hobbs, B. E., Means, W. D & Williams, P. F. 1976. *An outline of Structural Geology*. John Wiley, New York.

Ramsay, J. G. & Huber, M. I. & 1987 *The Techniques of modern structural geology*, **1 & 2**, Academic, New York.

Schlumberger, Ltd, 1981. *Dipmeter interpretation*, vol. 1.

Vialon, P., Ruhland, M. & Grolier, J. 1976. *Eléments de tectonique analytique*. Masson ad., Paris.

A correlation of hydraulic conductivity from pulse tests with sonic log amplitudes

D. GOLDBERG[1], D. SPEED[2], C. WILKINSON[1] & E. SCHOLZ[3]
[1]*Lamont-Doherty Geological Observatory of Columbia University, Borehole Research Group, Palisades, NY 10964, USA*
[2]*IBM Corp., Hopewell Jct., NY 12533, USA*
[3]*Lamont-Doherty Geological Observatory of Columbia University, Borehole Research Group, Palisades, NY 10964, USA (now at Stanford University, Dept. of Geophysics, Stanford CA 93002)*

Abstract: A 100 m water resource well penetrating shaly dolostone in East Fishkill, NY was tested to identify fracturing and to relate in situ hydraulic conductivity to acoustic properties derived from sonic waveform logs. The vertical fracture distribution was determined by borehole televiewer imaging. Sonic full waveforms were also recorded in the well and were analysed for compressional and Stoneley amplitudes, and differential energy logs were calculated. Hydraulic pulse tests were conducted in 1.5 m intervals isolated using a re-inflatable borehole packer system, and the measured conductivity in unfractured zones ranged from 10^{-9} to 10^{-5} cm s^{-1}. Our preliminary comparison of results show that hydraulic conductivity is not related to Stoneley amplitude and weakly related to compressional amplitude and Stoneley differential energy. However, hydraulic conductivity may be reduced by partially closed or clay-filled fractures and volumetric differences in the packed interval which simultaneously increase attenuation. Further analysis by hydraulic flow tests in highly permeable zones are expected to improve the correlation. In principle, independent estimation of hydraulic conductivity from sonic attenuation measurements alone could be extended to untested intervals in nearby wells.

Theoretical studies support the hypothesis that formation permeability is related to sonic attenuation due to frictional shear losses between the rock matrix and pore fluids that move in association with a propagating wave (Biot 1956). Early empirical studies in a variety of porous rocks indicated that compressional attenuation varied directly with formation permeability (Lebreton *et al.* 1978). Recently, the Stoneley wave has been favoured as the most effective indicator of fracture and formation permeability since it propagates on the borehole interface and depends strongly on shear wave attenuation in the formation (Rosenbaum 1974; Burns & Cheng 1986). However, Stoneley waves are most sensitive to fluid and interface effects at frequencies lower than those used with standard logging tools (Schmidt *et al.* 1987). Several investigators have nevertheless concluded that Stoneley wave attenuation in sonic logs is related to formation permeability as measured from core samples (Williams *et al.* 1984; Castagna *et al.* 1987; Cheruvier & Winkler 1987). Paillet *et al.* (1987) correlated Stoneley attenuation with the fracture permeability in crystalline rock using in situ flow tests.

In this paper, we present our efforts in one well to relate in situ pulse test measurements in unfractured rock intervals to compressional and Stoneley attenuation and to differential energy calculations derived from sonic logs. Hydraulic pulse tests were used to measure in situ conductivity, an improvement over core plug measurements of permeability for comparison with logging data. Hydraulic conductivity is directly related to formation permeability in a packed interval by the fluid density and viscosity. At the time of this report, only selected intervals in one well located within the Wappingers Formation dolomite in East Fishkill, N.Y. were tested. Hydraulic flow testing in fractured zones and additional waveform analysis will be necessary to quantify the conductivity-attenuation relationship as a continuous function of depth.

Acoustic borehole televiewer

The acoustic borehole televiewer enables the determination of location and orientation of features intersecting a wellbore, for example, sedimentary layering and fractures, and accurate measurement of the diameter, surface roughness and ellipticity of the borehole. The borehole televiewer is an acoustic device which scans the wall of a borehole producing an image

From HURST, A., LOVELL, M. A. & MORTON, A. C. (eds), 1990, *Geological Applications of Wireline Logs* Geological Society Special Publication No. 48, pp. 297–302

of the reflectivity of the surrounding rock as a function of depth and azimuth. A piezoelectric transducer mounted on a central shaft rotates three times per second and transmits and receives a high-frequency acoustic pulse 600 times per revolution. The source pulse is transmitted at 1.3 MHz through the borehole fluid, reflected from the wall of the borehole, and received again at the transducer. The amplitude of the returning signal is recorded and displayed to produce an azimuthal section of the borehole unwrapped about its vertical axis.

The geometrical technique to obtain the orientation of planar features intersecting the wellbore involves peak-to-trough analysis of the sinusoidal trace of features on the unwrapped televiewer image. The strike of a feature is given by the azimuthal orientation of the midpoint between peak and trough; the dip is calculated by the arc tangent of the height (of the sinusoid) divided by the borehole diameter. Analysis of the borehole televiewer log in this well indicates a NW orientation of the poles normal to fracture planes which dip approximately 20° SE. However, the televiewer cannot recognize features which do not affect either the roughness, reflectivity or radius of the borehole, and cannot differentiate between features which affect borehole reflectivity in a similar manner. Typically, the televiewer image can resolve 0.5 cm open fractures in a water-filled borehole.

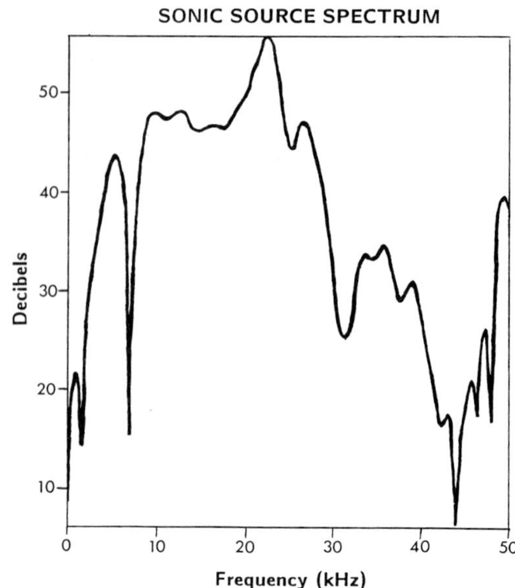

Fig. 1. Sonic source spectrum in a water bath from the 12-channel sonic tool. Peak power occurs in a narrow band centred on 22 kHz, and generates compressional, shear and Stoneley waves.

Multichannel sonic log

The multichannel sonic log is configured with a single 22 kHz source spaced 2 m above an array of 12 equally spaced receivers spanning 1.65 m. Fig. 1 shows the spectral signature of the sonic source in a water bath. Peak power occurs in a narrow band at about 22 kHz, but a significant peak below 10 kHz is also present. The tool geometry is similar to the geometry of a surface refraction survey. The tool is centred in the borehole by means of bowspring centralizers and the data are recorded on magnetic tape as the tool is pulled uphole at a constant rate of about 3 m per minute. The energy that arrives at each receiver travels as a compressional pulse in the borehole fluid and converts at the borehole wall into a series of body and guided wave modes. Typical guided wave modes produced in the borehole environment are controlled by the properties of both the formation and fluid-filled borehole. These guided waves include the pseudo-Rayleigh and Stoneley modes, which in addition to the compressional mode, enable complete characterization of the elastic properties of the formation (Williams *et al.* 1984).

In Fig. 2, sonic waveform data from a single receiver in the lower interval of this 7-inch diameter well are presented in variable density format. The waveforms were analysed to yield compressional and shear velocities across the receiver spread using a modified semblance calculation (Goldberg & Gant 1988). In the past, semblance has been applied extensively to seismic reflection data to obtain stacking velocities and was applied here as measure of the coherence between channels to calculate the compressional and shear velocities. In addition, variations in amplitudes of compressional and Stoneley wave modes within a 200 μs window were determined by Fourier transformation of the signals.

Differential energy logs were calculated according to the methods outlined by Brie *et al.* (1988). First, the peak amplitude from the Fourier transformed signal is selected for each receiver and each depth. These are referred to as energy logs. The logs are depth-shifted so that the recorded amplitudes at any receiver in the array refer to the same formation interval (common-receiver mode). For each pair of receivers at 1-, 2-, or 3-foot relative spacing, the difference in amplitude is divided by the receiver spacing. This differential energy value is re-

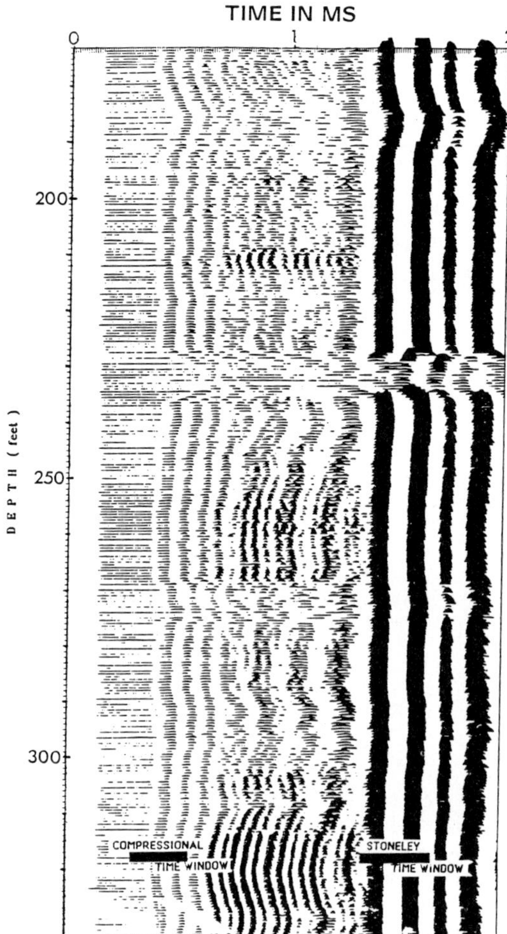

Fig. 2. Single-channel sonic waveforms at a 2 m transmitter–receiver spacing from 333 ft to 173 ft in an 18 cm diameter borehole in shaly dolostone.

ferred to the midpoint of each receiver pair, and the energy logs are similarly differentiated in a common-transmitter mode. Average differential energy logs for each of the three receiver spacings in both common-receiver and common-transmitter modes are calculated. This is referred to as borehole-compensated differential energies since it averages the energy loss between receiver pairs for energy travelling both up (common-transmitter) and down (common-receiver) the borehole, and thus is free from both receiver and transmitter coupling effects.

Hydraulic testing

A re-inflatable straddle packer system was designed to isolate 1.5 m intervals of the borehole for hydraulic conductivity testing (Fig. 3). Tests are generally conducted by measuring the decay over time of an instantaneous pressure impulse (pulse tests) or by injecting fluid under a constant head and measuring the decay of flow rate over time (flow tests). The hydraulic head and system geometry are described by a partial differential equation and a set of boundary conditions which produce 'type-curve' solutions (Cooper *et al.* 1967; Papadopulos *et al.* 1973; Bredehoeft & Papadopulos 1980). The data were matched against these type-curves to obtain the hydraulic conductivity and storage coefficient (see final section).

While it is desirable to measure a continuous profile of hydraulic conductivity with depth, successful tests can be accomplished only where the packers can obtain a good seal with the borehole wall. The televiewer log was used to locate suitable zones for packer placement, where borehole wall roughness and fracturing were relatively small. Upon positioning, the packers were inflated to a consistent pressure above the hydrostat, a downhole vent valve was closed, and the pressure was monitored above, below, and within the interval to ensure that the packers sealed and the pressure stabilized prior to commencing the pulse tests. A minimum of

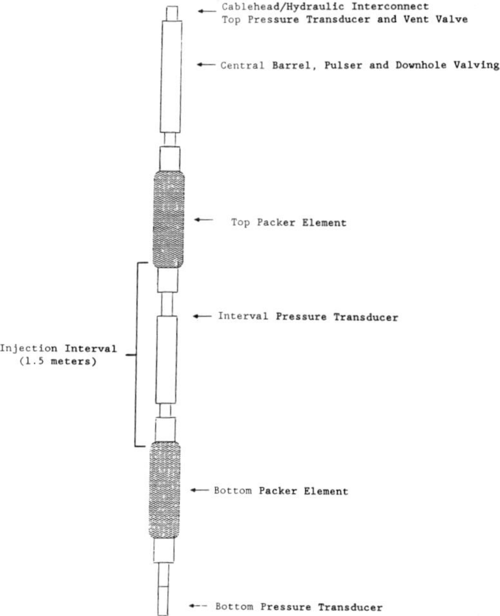

Fig. 3. Schematic of straddle packer tool configuration. Top and bottom packer elements inflate to seal off a 1.5 m interval for hydraulic testing.

three pressure pulses, created by the activation of a hydraulic ram, were recorded in each interval using absolute-reading transducers at a 5 Hz data acquisition rate.

Results and discussion

The caliper, gamma-ray, compressional (V_p) and shear (V_s) velocities and V_p/V_s ratio, compressional and Stoneley amplitudes, conductivity, and interpreted televiewer logs are displayed in Fig. 4. In this well, packed intervals were located over both fractured and unfractured zones, although during testing in fractured zones, the pressure pulse technique could not be used to measure conductivity due to a rapid decay of the pulse. Consequently, only the low conductivity data are presented. From the data shown in Fig. 4, it is difficult to correlate between the elastic and physical property measurements, the V_p/V_s ratio, fracturing and clay content (gamma-ray). This may be due to the effect of partially open or clay-filled fractures which cause similar decreases of both V_p and V_s.

The amplitudes of sonic waves, which are dominated by guided-wave modes, are, however, more sensitive than velocities to fracturing and enviromental changes in the well. Poor transmission of energy across bedding planes, fractures and laminations, as well as fluid motions, cause losses of amplitude in both compressional and Stoneley waves. In order that variations in amplitude be attributed uniquely to formation effects, the borehole size must be constant or the amplitudes must be borehole compensated by receiver differencing (Anderson & Castagna 1984; Goldberg *et al.* 1984; Brie *et al.* 1988). Borehole sizer (caliper log) is relatively constant in this well, except for washout zones near 51 and 225 ft.

Stoneley wave amplitudes are particularly influenced by borehole size changes, the character of the borehole wall, and the physical properties of the formation (Burns & Cheng 1986). In this well, Stoneley amplitudes are low only in the washed-out zones and do not correlate with the presence of fracturing or variations in conductivity. The greater variation in compressional amplitude is indicative of conductivity variations, but the correlation is limited by the narrow range of conductivities measured using the pulse technique.

Fig. 4 also shows the correlation between Stoneley differential energy and hole size, as well as V_p/V_s versus depth. These suggest that variations in borehole size are responsible for the largest changes in velocities and differential energies. New data from constant flow tests in highly permeable intervals are expected to improve the overall correlation of low amplitude

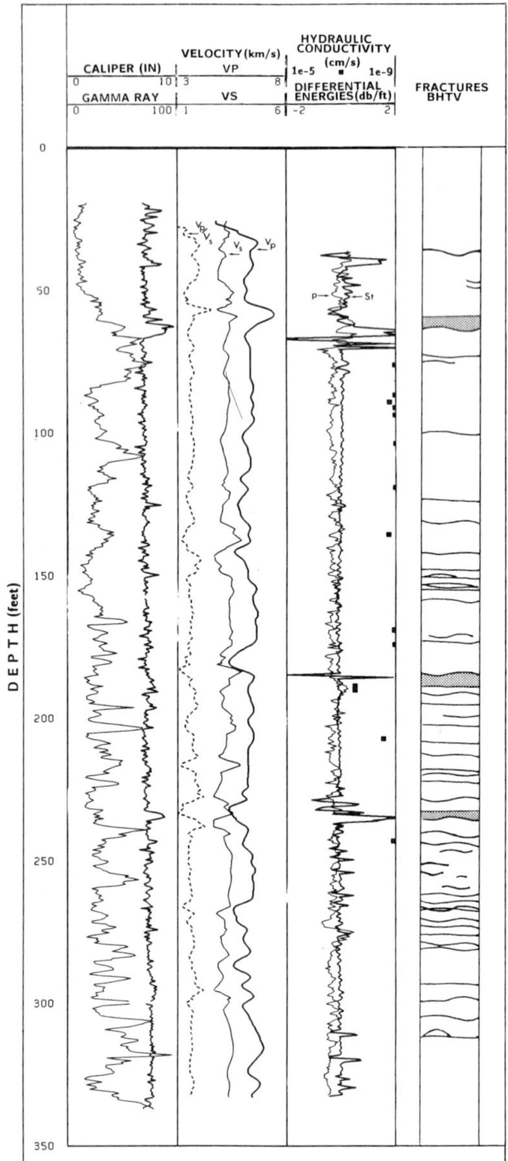

Fig. 4. Gamma-ray, Caliper, V_p, V_s, and V_p/V_s ratio, Stoneley and compressional differential energy (1 ft spacing), hydraulic conductivity, and interpreted BHTV logs. Stoneley attenuation (amplitude loss) is relatively insensitive to fracturing or clay content (gamma-ray increase), but greatly affected by borehole washout (caliper increase).

with high conductivity such that an empirical relationship can be derived.

Estimating packed interval volume and storage coefficient

The initial pressure pulse for hydraulic testing is often found to deviate from that anticipated on the basis of the calibration. In permeable zones, rapid pressure decay during the initial pulse generation may attenuate the resulting pulse height. However, in many cases, attenuated pulse heights do not correlate with rapid decay rates, and based on successive calibration tests, it appears that the hydraulic ram delivered consistent pressure pulses. A likely cause of pulse attenuation is the variation of the packed interval volume due to fractures and/or irregularities in the borehole wall. Conversely, deviations in pulse height may be used to estimate the packed interval volume.

Our analysis of the hydraulic head, h, in a well having finite storage volume is based on an analytical solution of

$$2\pi r_w \times T \frac{\partial h}{\partial r} = S_w \frac{\partial h}{\partial t}$$

for radial flow from the well into the formation (Cooper *et al.* 1967; Bredehoeft & Papadopulos 1980). For a finite volume, the boundary condition to the solution yields the hydraulic conductivity from the transmissivity T and the storage capacity $S_w = \rho g V_w C_p$.

Experimental determination of the compliance by calibration is based upon the measured pressure impulse $\triangle P$ due a known volume change $\triangle V$ within an isolated interval of specified volume V_o (e.g. fluid-filled casing). The system compressibility $C_p = \triangle V/V_o \triangle P$ includes both the effects of the compressibility of water and compliance of the equipment (Neuzil 1982). Successive tests determined that $C_p = 1.4 \times 10^{-9}$ Pa^{-1}, which compares well with values of 2.1×10^{-9} and 2.7×10^{-9} determined for the equipment used by Hsieh *et al.* (1983) and Neuzil (1982), respectively.

We have modified the total system compressibility to accomodate the compressibility of any additional volume V_2 in open test intervals. We assume that the solid matrix is incompressible and that V_2 is determined solely by the compressibility of water C_w. Thus,

$$\triangle P = \frac{\triangle V}{V_o C_p + V_2 C_w}$$

and solving for V_2, each added increment of interval volume may be estimated by

$$V_2 = \frac{(\triangle V/\triangle P) - V_o C_p}{C_w}.$$

The well storage capacity S_s must also be modified to accomodate for V_2 in the solution

$$S_s = \rho g (V_o C_p + V_2 C_w).$$

References

Anderson, J. & Castagna, J. 1984. Analysis of sonic log compressional wave amplitudes using borehole compensation techniques: *Transactions of the Society of Professional Well Log Analysts Logging Symposium.*, paper K.

Biot, M. A. 1956. Theory of propagation of elastic waves in fluid saturated porous solid. I. Low-frequency range, II. High-frequency range. *Journal of the Acoustic Society of America*, **28**, 168–178.

Brie, A., Hsu, K. & Eckersley, C. 1988. Using the Stoneley Normalized Differential Energies for Fractured Reservoir Evaluation: *Transactions of the Society of Professional Well Log Analysts Logging Symposium.*, paper XX.

Burns, D. R. & Cheng, C. H. 1986. Determination of in situ permeability from tube wave velocity and attenuation. *Transactions of the Society Professional Well Log Analysts Logging Symposium.*, paper KK.

Bredehoeft, J. D. & Papadopulos, I. S. 1980. A method for determining the hydraulic properties of tight formations. *Water Resources Research*, **16**, 233–238.

Castagna, J., Zucker, S. & Shoberg, T. 1987. Permeability indication with conventional sonic waveforms: *Transactions of the Society Professional Well Log Analysts Logging Symposium.*, paper MM.

Cheruvier, E. & Winkler, K. 1987. A field example of in-situ permeability indication from acoustic full wavetrains: *Transactions of the Society of Professional Well Log Analysts Logging Symposium.*, paper NN.

Cooper, H. H., Bredehoeft, J. D. & Papadopulos, I. S. 1967. Response of a finite diameter well to an instantaneous charge of water. *Water Resources Research*, **3**, 263–269.

Goldberg, D., Castagna, J. & Kan, T. K. 1984. Attenuation measurements from sonic log waveforms: *Transactions of the Society of Professional Well Log Analysts Logging Symposium.*, paper NN.

Goldberg, D. & Gant, W. 1988. Shear-wave processing of sonic log waveforms in a limestome reservoir. *Geophysics*, **53**, 668–676.

HSIEH, P. A., NEUMAN, S. P. & SIMPSON, E. S. 1983. Pressure Testing of fractured rocks — a methodology employing three-dimensional cross-hole tests NUREG/CR-3213, Washington, D. C.

LEBRETON, F., SARDA, J. P. TROCQUEME, F. & MORLIER, P. 1978. Logging tests in porous media to evaluate the influence of their permeability on acoustic waveforms: *Transactions of the Society of Professional Well Log Analysts Logging Symposium.*, paper O.

NUEZIL, C. P. 1982. On conducting the modified "slug" test in tight formations. *Water Resources Research*, **18**, 439–441.

PAILLET, F. L., HSIEH, P. & CHENG, C. H. 1987. Experimental verification of acoustic waveform and vertical seismic profile measurements of fracture permeability: *Transactions of the Society of Professional Well Log Analysts Logging Symposium.*, paper OO.

PAPADOPULOS, I. S., BREDEHOEFT, J. D. & COOPER, H. H. 1973. On the analysis of slug test data: *Water Resources Research*, **9**, 1087–1089.

ROSENBAUM, J. H. 1974. Synthetic microseismograms: logging in porous formations. *Geophysics*, **39**, 14–32.

SCHMIDT, D. P, CHENG, C. H. & TOKSOZ, M. N. 1987. Shear wave logging in saturated porous formations, *Transactions of the Society of Professional Well Log Analysts Logging Symposium*, paper OO.

WILLIAMS, D. M., ZEMANEK, J., AANGONA, F. A., DENNIS, C. L., CALDWELL, R. L. 1984. The long spaced acoustic logging tool: *Transactions of the Society of Professional Well Log Analysts Logging Symposium.*, paper T.

Physical properties

Investigating stress regimes in sedimentary basins using information from oil industry wireline logs and drilling records

J. S. BELL

Geological Survey of Canada, Atlantic Geoscience Centre, Bedford Institute of Oceanography, PO Box 1006, Dartmouth, Nova Scotia, B2Y 4A2, Canada (Present address: Geological Survey of Canada, Institute of Sedimentary and Petroleum Geology, 3303 33rd St. NW, Calgary, Alb T2L 2A7, Canada)

Abstract: Estimates of the in situ stress tensor in sedimentary basins can be inferred from information gathered during the drilling and logging of hydrocarbon exploration wells. Principal stress magnitudes can be obtained from information provided by density logs, drilling histories, leak-off test results and pore pressure measurements. Horizontal principal stress'orientations are indicated by borehole breakouts recorded on four-arm dipmeter logs. The stress magnitude estimates are comparable with values obtained from hydraulic fracturing and the orientation data provide a level of detail hitherto unobtainable. The information obtained assists in designing hydraulic fractures, in secondary recovery programmes and in obtaining optimum drill bit performance. It also contributes to understanding lithospheric plate movements, crustal subsidence and tectonic deformation.

The stress acting at any point in the Earth's crust can be described by three mutually perpendicular vectors σ_1, σ_2 and σ_3, which refer to the maximum, intermediate and minimum principal stresses, respectively. At all but the shallowest depths, these stresses are compressive and, in most locations, of different magnitudes (Gay 1980). To describe the stress tensor at a point the orientations and magnitudes of the three principal stresses are required. With certain simplifying assumptions, enough data are often gathered while drilling wells to do so.

Since no shear stresses can exist in the plane of the Earth's surface, one of the principal stresses in the upper crust should be oriented approximately perpendicular to it (Hubbert & Willis 1957). McGarr & Gay (1978) found that one principal stress is generally oriented within 20° of vertical except beneath very rough topography. Sedimentary basins are often flat-lying regions so, within them, one principal stress is likely to be close to vertical and aligned approximately parallel to most boreholes. Consequently, the other two principal stresses will be oriented more or less horizontally, parallel to the ground surface, and will act approximately perpendicular to wellbores.

Stress orientations around a wellbore can be determined by hydraulic fracturing provided the azimuth of the fracture is recorded. Hydraulic fractures generally extend away from boreholes as planar features oriented perpendicular to σ_3, the least principal stress (Hubbert & Willis 1957; Haimson & Fairhurst, 1970). Fracture orientations can be recorded on inflatable packers (Raleigh *et al.* 1972) or mapped at the surface by tiltmeter arrays (Holzhausen *et al.* 1980) and microseismic techniques (Majer & Doe 1986). These methods are, however, time-consuming and expensive and do not form part of normal exploration drilling programs.

Due to a most fortunate phenomenon, the rocks lining borehole walls can 'feel' the stresses applied to them, and will frequently respond in a manner that enables the orientations and relative magnitudes of the principal stresses to be determined. Moreover, the response can be measured by wireline tools designed and run for different purposes. The features that provide these insights are well bore breakouts (Babcock 1978).

In situ stress orientations

Wellbore breakouts

In recent years it has been recognized that the walls of many boreholes cave or spall so as to produce intervals with ellipsoidal cross sections known as breakouts (Cox 1970; Babcock 1978). They were identified first on four-arm dipmeter records. This tool has extendible pads which record the long and short diameters of a borehole and their azimuths (Fig. 1). Cox (1970) noted that certain intervals in the boreholes had become elongated through spalling of opposed faces of their walls, and that the long axes of these elongated, semi-elliptical zones, were commonly aligned. Cox identified 31 breakouts zones in 17 wells in Alberta, Western Canada in rocks ranging from Cretaceous shales to

From HURST, A., LOVELL, M. A. & MORTON, A. C. (eds), 1990, *Geological Applications of Wireline Logs* Geological Society Special Publication No. 48, pp. 305–325

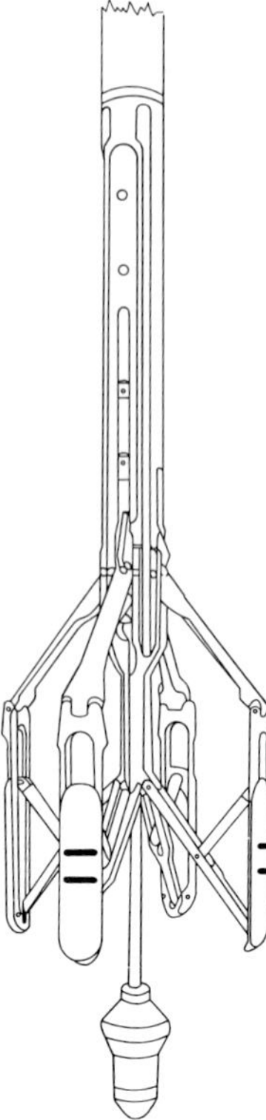

Fig. 1. The HDT-E four-arm dipmeter tool of Schlumberger which has been used to identify breakouts. The four extendable pads are hydraulically actuated and their extension is recorded as they are pressed against the wall of a borehole. When closed, the tool diameter is 12.1 cm, the pads are 5.7 cm wide and 40.6 cm long; they will extend to a diameter of 54 cm.

Devonian carbonates that exhibited dips to 30°. The directions of hole elongations bore little or no relation to the dip azimuths (Cox 1970).

Babcock (1978) extended Cox's observations and described the chief diagnostic criteria for breakouts. The four-arm dipmeter tool rotates as it is pulled up a circular wellbore but, when a breakout is encountered, one pair of electrodes on opposed caliper arms becomes trapped near the major axis and rotation ceases (Fig. 2). Meanwhile, the other pair of calipers records the shorter hole diameter. Babcock (1978) analysed 246 breakouts in 23 wells and confirmed Cox's observations that their direction was consistent regardless of depth or lithology and showed a consistent preferred NW–SE orientation throughout much of the Western Canadian Basin. Within any given well, he found that the breakouts were discrete; they began and ended abruptly and were separated by uncaved beds. Breakouts were also independent of bedding dip and hole deviation, and occurred in shales, siltstones, sandstones, limestones and dolostones. After considering several possible explanations Babcock concluded that the breakouts resulted from the drill encountering steeply-dipping fractures or fracture zones. He noted that 'the parallelism between azimuths of elongation and regional joint sets in Alberta and the consistency of the azimuths of elongation within a well, between wells, and in varying lithologies, all indicate an origin caused by the drill encountering a fracture or zone of fractures' (Babcock 1978).

Bell & Gough (1979) found this interpretation difficult to accept. There were four well defined concentrations of surface joint directions in Alberta (NW, NE, N and E) but only one significant concentration of sub-surface breakout directions (NW). Furthermore, the breakout zones affected hundreds of feet of section in many wells and it seemed unlikely that fractures or fracture zones would exactly coincide with the path of a well so frequently. It was also appreciated that, for four-arm dipmeter calipers to measure breakouts, at least one pad 2.25 inches (5.7 cm) wide and 16 inches (40.6 cm) long would have to penetrate into the borehole wall. Fractures so wide and continuous are extremely rare in outcrop and it appeared unlikely that they were widespread at depths of several thousand feet (Secor & Pollard 1975).

As an alternative explanation, Bell & Gough (1979) suggested that the Western Canadian Basin was stressed anisotropically, that the holes themselves could concentrate and amplify the stresses, and that the breakout long axes were aligned parallel to the lesser horizontal principal stress, $S_{H_{min}}$.

Theory of breakout generation

The generation of breakouts can be understood by considering an infinitely large plate with a small circular hole in it which is subjected to a uniaxial compressive stress, S. As Kirsch (1898) demonstrated, the stress components at the point (r, θ) are given by the equations

$$\sigma_r = 1/2S\left(1 - \frac{a^2}{r^2}\right) + 1/2S\left(1 + \frac{3a^4}{r^4} - \frac{4a^2}{r^2}\right)\cos 2\theta \quad (1)$$

$$\sigma_\theta = 1/2S\left(1 + \frac{a^2}{r^2}\right) - 1/2S\left(1 + \frac{3a^4}{r^4}\right)\cos 2\theta$$

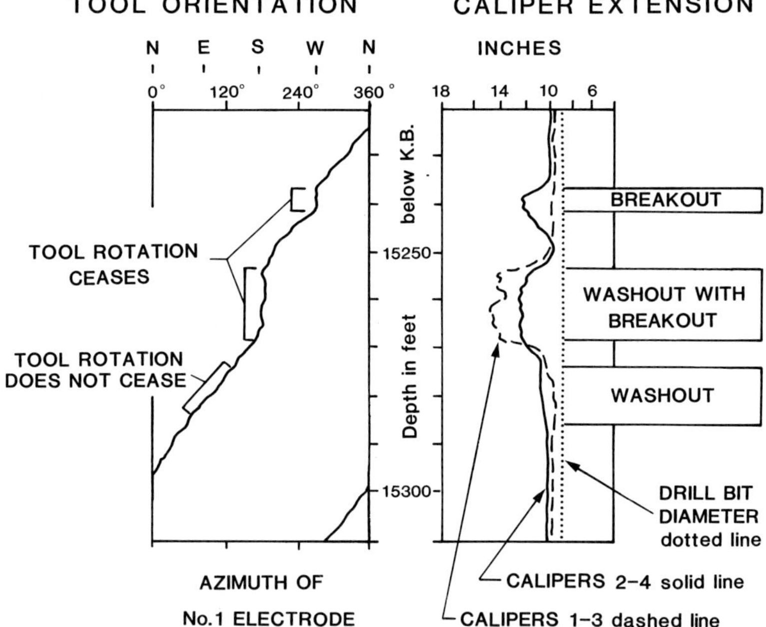

Fig. 2. Simplified representation of an unprocessed four arm dipmeter log showing how breakouts are identified by coincident cessation of tool rotation and differential caliper extension. Note that washouts alone do not affect tool rotation.

$$\tau_{r\theta} = 1/2S\left(1 - \frac{3a^4}{r^4} + \frac{2a^2}{r^2}\right)\sin 2\theta$$

where a is the radius of the hole, r the distance from the centre of the hole and compressive stress is positive. The angle θ is measured from the direction of S and σ_r, σ_θ and $\tau_r\theta$ are the radial, tangential and shear stresses respectively. At the hole wall where $r = a$, σ_r and $\tau_{r\theta}$ become zero

$$\sigma_\theta = S - 2S\cos 2\theta. \quad (2)$$

It has a maximum value of $3S$ when θ=90° or 270° which occurs at the ends of the diameter perpendicular to S. At the ends of the diameter aligned with the uniaxial compressive stress, σ_θ has a minimal value of $-S$ and is tensile. In this idealized uniaxially stressed situation, S is amplified around the hole perimeter and stress is concentrated due to the realignment of stress trajectories near the hole (Fig. 3).

In the sub-surface, rocks in the Earth's crust will be subjected to triaxial stresses, most probably of different magnitudes. One principal stress will be approximately vertical and its magnitude will increase with depth, so that it remains close to the overburden pressure. The orientation of this principal stress will coincide approximately with the axis of a semi-vertical borehole. Thus to a first approximation, a vertical borehole will be subjected to plane biaxial stress specified by the two horizontal principal stresses, $S_{H_{max}}$ and $S_{H_{min}}$ where $S_{H_{max}} > S_{H_{min}}$. As before, the radial and shear stresses, σ_r and $\tau_{r\theta}$, will have values of zero at the borehole wall and the tangential stress will be

$$\sigma_\theta = S_{H_{max}} + S_{H_{min}} - 2(S_{H_{max}} - S_{H_{min}})\cos 2\theta. \quad (3)$$

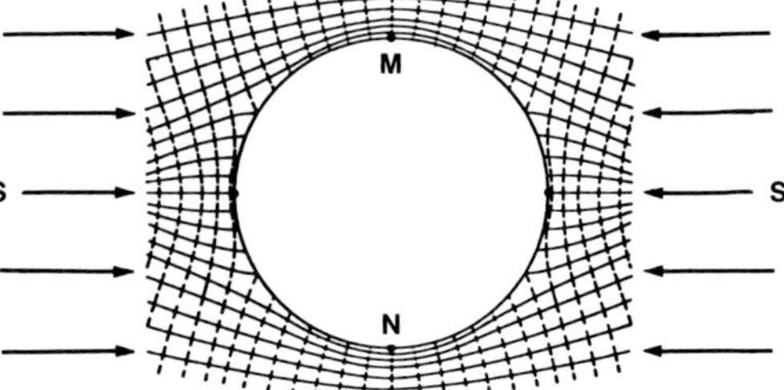

Fig. 3. Stress trajectories near a circular hole in an infinite plate which is subjected to uniaxial compression (after Jaeger 1962). Note the "clustering" of trajectories near the hole wall prependicular to the applied stress around points M and N.

When $\theta = 90°$ or $270°$, σ_θ reaches its maximum value of

$$\sigma_\theta = 3S_{H_{max}} - S_{H_{min}}. \quad (4)$$

The stress amplification is very local and the stress difference approaches $(S_{H_{max}} - S_{H_{min}})$ one radius from the wall (Jaegar & Cook 1976).

In an open hole, where no pressure is exerted against the borehole walls by a mud column, the degree of stress amplification and the zones affected by compressional and tangential stress are determined by the relative magnitudes of the stresses (Fig. 4). The stress amplification decreases as the contrast in principal horizontal stress magnitudes becomes less, and tensionally stressed zones will only exist on borehole walls when the ratio of principal horizontal stresses is greater than 3:1. In practice, this degree of horizontal stress anisotropy is unlikely to be found, except in exceptional near-surface situations. If the principal horizontal stresses are of equal magnitude, the applied stress will be doubled all around the borehole wall (equation (3)).

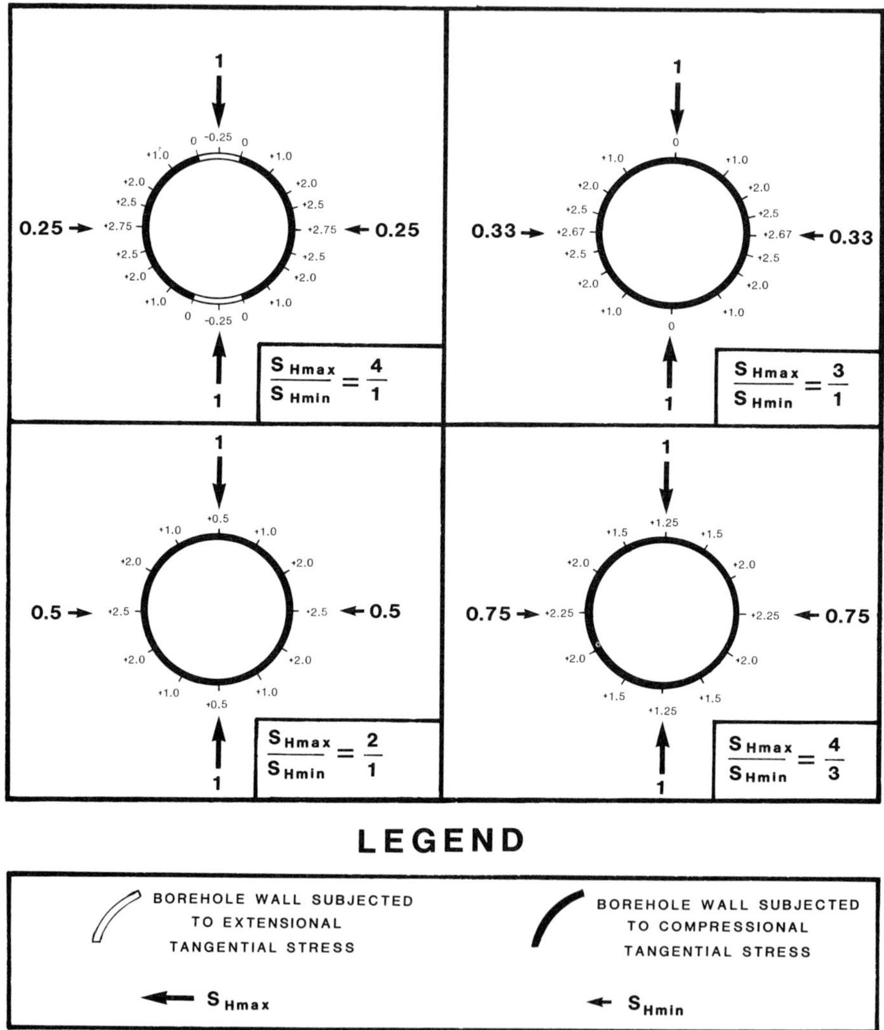

Fig. 4. Stress amplification on borehole walls for wells drilled through rocks subjected to varying horizontal principal stresses. The numbers around the borehole walls indicate the magnitudes of the acting tangential stresses as multiples of the larger horizontal principal stress. Note that the stress amplification decreases as the contrast in principal horizontal stress magnitudes becomes less.

To explain the origin of breakouts in vertical boreholes, consider the case where the horizontal principal stresses are unequal. The region of greatest stress concentration occurs on the part of the hole wall that is oriented at right angles to the greater horizontal principal stress. In cross section, there are two lunate-shaped regions around the hole wall (Hoek & Brown 1980) where the tangential stresses may be amplified to levels for which the shear strength of the surrounding rock is exceeded (Fig. 5). Shear fractures can then form within the lunate-shaped regions and rock chips fall away from the hole wall to produce breakouts (Kirsch 1898, Zoback *et al.* 1985).

Zoback *et al.* (1985) have shown how this can occur. They take account of fluid formation pressures in the rocks around the wellbore and pressures exerted by drilling mud as the well is drilled. These pressures do not affect the orientation of fractures propagating to form breakout zones, but they do alter the magnitudes of the stress differences around the borehole walls, as the modified Kirsch equations show:

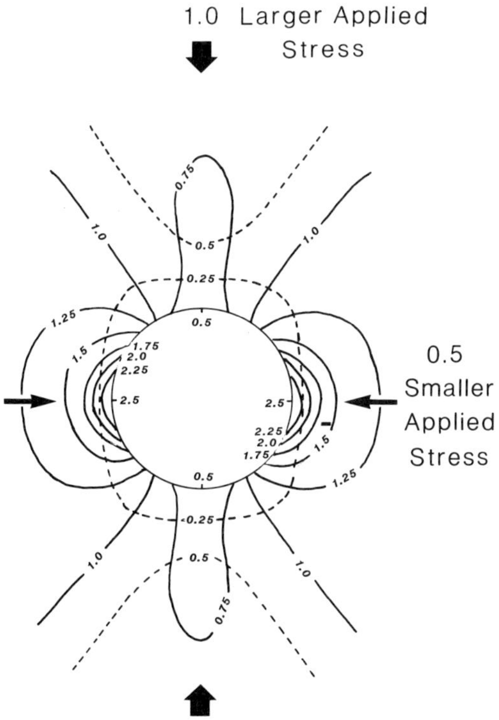

$$\sigma_r = 1/2\,(S_{H_{max}} + S_{H_{min}})\left(1 - \frac{a^2}{r^2}\right) + 1/2\,(S_{H_{max}} - S_{H_{min}})\left(1 - \frac{4a^2}{r^2} + \frac{3a^4}{r^4}\right)\cos 2\theta + \frac{a^2}{r^2}(P_m - P_f) \quad (5)$$

$$\sigma_\theta = 1/2\,(S_{H_{max}} + S_{H_{min}})\left(1 + \frac{a^2}{r^2}\right) - 1/2\,(S_{H_{max}} - S_{H_{min}})\left(1 + \frac{3a^4}{r^4}\right)\cos 2\theta - \frac{a^2}{r^2}(P_m - P_f) \quad (6)$$

$$\tau_{r\theta} = 1/2\,(S_{H_{max}} + S_{H_{min}})\left(1 + \frac{2a^2}{r^2} - \frac{3a^4}{r^4}\right)\sin 2\theta \quad (7)$$

where the notation is the same as used previously. $P_m - P_f$ is the difference between the fluid pressure in the borehole and that in the formation. Excess pressure in the borehole is positive.

According to Zoback *et al.* (1985), the region around a wellbore that will fail can be predicted from the magnitudes of the horizontal principal stresses, the shear strength of the rocks and their coefficient of internal friction. They employ the extended Griffith criterion of McClintock & Walsh (1962), which considers the extension of closed cracks with a finite frictional strength in a biaxial stress field. The analysis is equivalent is zones of compressive stress to the Coulomb criterion in which the failure envelope has a slope equal to the coefficient of internal friction, μ, and an intercept, τ_o, equal to the shear strength of the rock (Paterson 1978; Jaeger & Cook 1976). The region around the wellbore that will fail can be computed from Mohr's circle when the radius of the circle

$$\left[\left(\frac{\sigma_\theta - \sigma_r}{2}\right)^2 + \tau_{r\theta}^2\right]^{1/2}$$

is greater than or equal to the distance from the centre of the circle to the failure line given by

$$\frac{\mu}{(1 + \mu^2)^{1/2}}\left(\frac{\sigma_\theta + \sigma_r}{2}\right).$$

1.0 Larger Applied Stress

0.5 Smaller Applied Stress

Fig. 5. Principal stress contours around a circular hold in an infinite plate under biaxial stress where applied stress ratio is 2 : 1 (after Hoek & Brown, 1980). Note the two lunate shaped zones adjacent to the hole walls where the larger local principal stress is greater than twice the far field applied stress. The solid contours give the ratio of the larger principal stress to the larger far field applied stress. The dashed contours indicate the smaller principal stress.

The above expressions can be arranged to obtain the maximum value of shear strength at which the rocks will fail given by

$$\tau_o = (1 + \mu^2)^{1/2} \left(\left[\frac{\sigma_\theta - \sigma_r}{2} \right]^2 + \tau_{r\theta}^2 \right)^{1/2} - \mu \left(\frac{\sigma_\theta + \sigma_r}{2} \right). \tag{8}$$

For most rocks μ varies between 0.6 and 1.0 (Byerlee 1978), and τ_o can vary from a few MPa to a few tens of megapascals (Handin 1966). Zoback *et al.*, (1985) substitute appropriate values into equations 5–8 to predict the size of the initial region in which the ratio of shear to normal stress is great enough to cause failure on potential shear surfaces. The shear surfaces are curved (Fig. 6a) and failure can be expected in arcuate lunate zones resembling breakouts (Fig. 6b). Breakouts have been observed with rough surfaces defined by a similar series of intersecting fractures (Fig. 7).

According to Zoback *et al.* (1985) ideal breakouts will be broad and flat-bottomed (Figs 8 & 9). The lower the shear strength of the rock, the deeper and wider will be the breakout

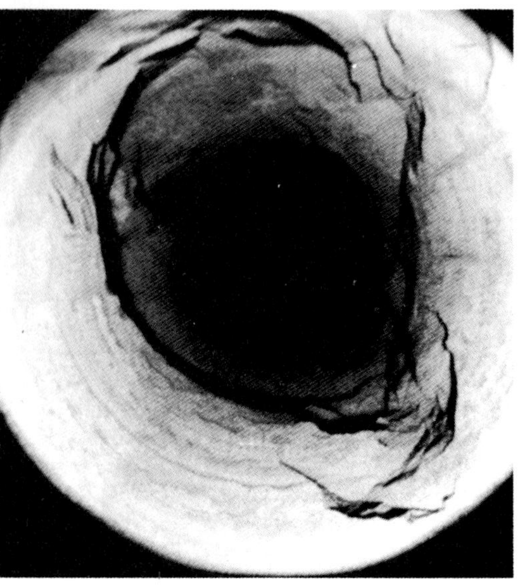

Fig. 7. A breakout which occurs in a borehole drilled in volcanic tuff beneath the Nevada Test Site. Note the rugged surface of the breakout, which is consistent with an origin via spalling of wall rock following growth of intersecting fractures (Photograph provided by J. E. Springer).

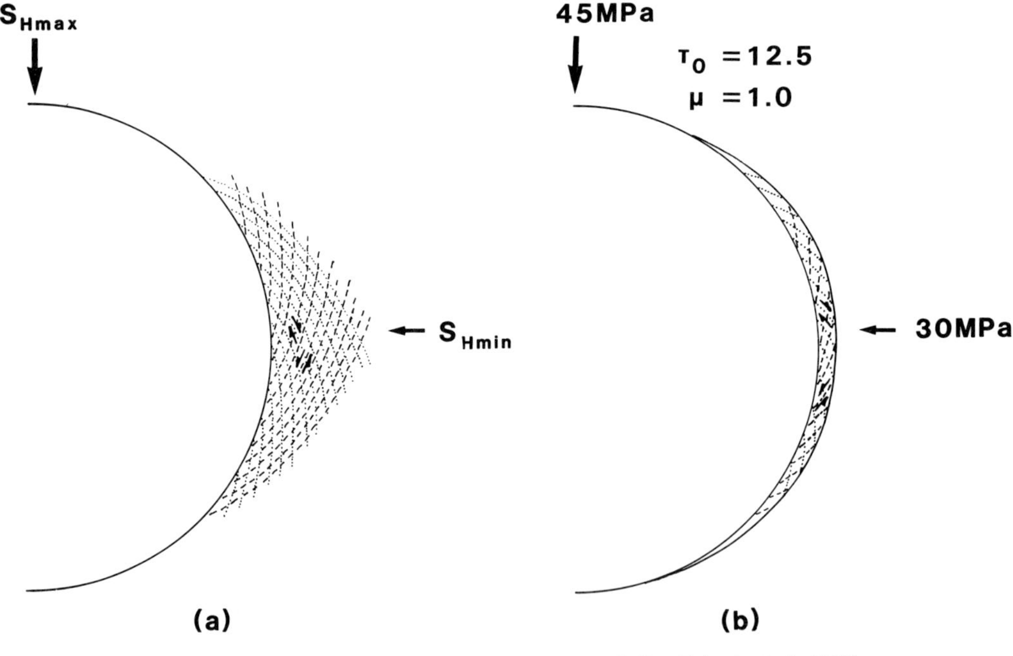

Fig. 6. (a) Potential shear failure surfaces adjacent to a borehole drilled through biaxially stressed rock. (b) Lunate-shaped failure zone of borehole wall predicted for a hole drilled through isotopic rock with compressive shear strength, t_0, of 12.5 MPa and μ equal to 1.0, when subjected to biaxial stresses of 45 and 30 MPa. Stress magnitudes of this order exist at depths of 1500–2000 metres in many basins.

(Fig. 8). Increasing the stress contrast increases the likelihood of forming breakouts, since rocks with increasingly high shear strengths will fail. It also implies that breakouts will be larger for given values of μ and τ_o (Fig. 8). Varying the mud pressure will also effect breakout growth (Fig. 9).

Zoback *et al.* (1985) suggest that, for isotropic rocks, the ratio between the horizontal principal stress magnitudes can be determined from breakout geometry. Unfortunately, the relationship is particularly sensitive to the maximum radius of the breakout. This value is not only difficult to measure but, when obtained, there is presently no means of establishing whether the breakout's shape has evolved to a stable 'ideal' stage or whether it is only an incipiently caved elongation. There is the further problem that excavated breakouts give rise to similar stress concentration to that which initially generated them, so that they will, to a degree, be self-propagating (Woodland 1988) and the Zoback *et al.* (1985) simulation only considers a single cycle of shear failures.

In spite of these qualifications, the theoretical analysis outlined above predicts the evolution of breakouts, their shape and their orientation with respect to the directions of the horizontal principal stresses prevailing in the rocks involved. The stress anisotropy and magnitudes required are known to be widely present in the upper crust (McGarr & Gay 1978; Gay 1980; Zoback & Zoback 1980; Gough & Gough 1987). Most importantly, aligned breakouts have been reported in many areas and the horizontal principal stress azimuths deduced from them agree with stress orientations obtained by other methods (see references in Evans & Brereton, 1990).

Identification of breakouts in wells

The best tools for identifying breakouts in boreholes are imaging devices that record breakout shape and orientation (Fig. 7). As downhole cine cameras cannot be used in drilling mud, borehole images in oil industry wells are obtained from acoustic borehole televiewers (Wiley 1981) or resistivity-measuring formation microscanners (Ekstrom *et al.* 1987).

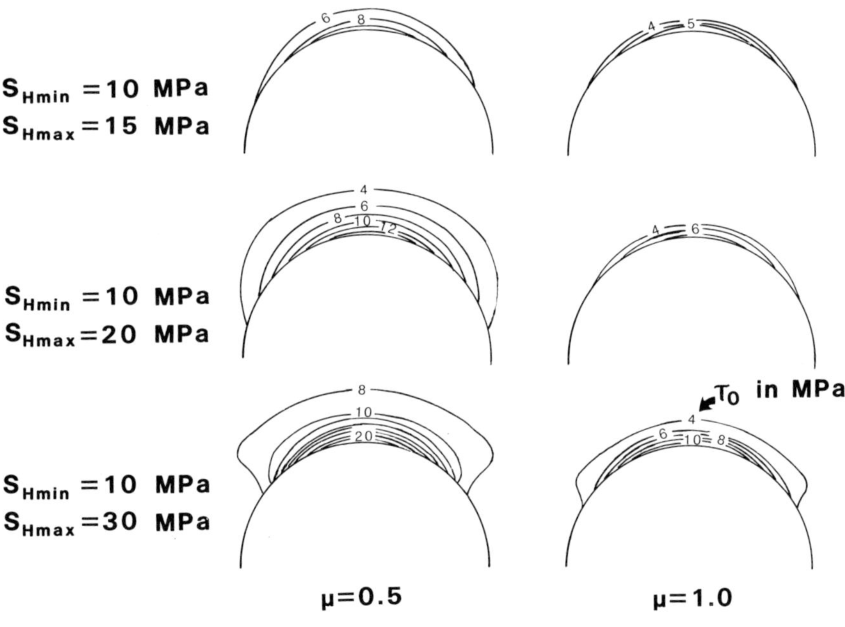

Fig. 8. Variation in the theorectical size of the zones in which the compressive shear strength, t_0, of isotropic rock is exceeded by amplified stresses close to the borehole wall (from Zoback *et al.* 1985). The zone boundary 'contours' are labelled with the applicable values of t_0 in MPa. Note that rocks with low shear strengths will develop deep and wide breakouts. As the biaxial stress contrast incrases, stronger rocks will begin to fail and form breakouts.

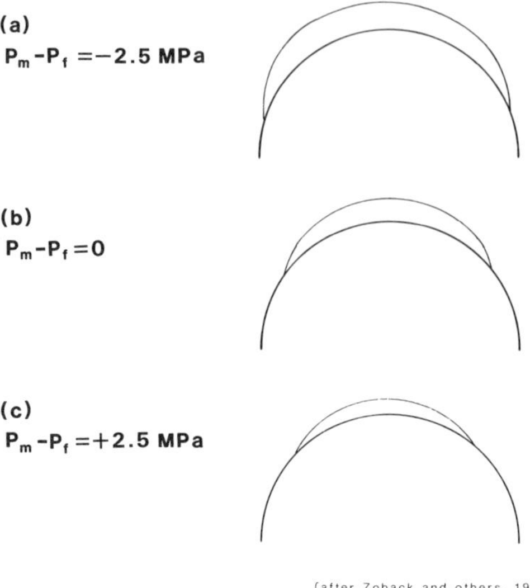

Fig. 9. The effect of wellbore fluid pressure (mudweight) on the size of breakouts which would theoretically form in a borehole drilled through isotropic rock for which τ_o is 7.5 MPa and μ is 0.6 in a stress regime where $S_{H_{max}}$ is 22 MPa and $S_{H_{min}}$ is 11 MPa. (a) Wellbore pressure, P_m is 2.5 MPa less than formation pressure, P_f (b) Wellbore pressure equals formation pressure; (c) Wellbore pressure is 2.5 MPa greater than formation pressure.

Borehole televiewers provide spectacularly detailed records of breakouts (Paillet & Kim, 1987; Fig. 10). Televiewer records can be processed to show borehole shape (Plumb & Hickman, 1985; Newmark *et al.* 1984; Fig. 11) and can be displayed as 'moulds' of the inside hole shape (R. N. Anderson, pers. comm. 1987). Similar images can be obtained from formation microscanner records (Ekstrom *et al.* 1987). Borehole televiewers are time consuming and expensive to run and are rarely used to log oil wells. The formation microscanner is much faster but has only recently been developed so few logs are yet available. For most breakout investigations, one is therefore obliged to rely on four-arm dipmeter tools. Although the design of these devices varies somewhat, they all contain electrode pads mounted on hydraulically extendible caliper arms (Fig. 1) that monitor hole diameter and caliper orientation as the tool is drawn up a borehole. At the same time, the tool's orientation is recorded by an inclinometer and a flux gate magnetometer. Four microresistivity curves are recorded and correlated to compute dips. They will also identify any directional resistivity anomalies. Four-arm dipmeter tools are raised up boreholes at approximately 10 m min^{-1} and rotate clockwise due to torque on the logging cable.

Logging companies present four-arm dipmeter resistivity, caliper extension and tool orientation data in several standard formats from which can be derived breakout orientations, resistivity anisotopy, hole deviation angles and hole deviation directions (Bell & Gough 1983; Cox 1982). Measurements can be made directly from the logs or, alternatively, by computer analysis of digital log data (Blümling *et al.* 1983; Evans & Brereton 1990). The latter approach is almost mandatory for examining the logs of four-arm dipmeters fitted with high angle inclinometers.

Researchers have summarized and presented breakout data in various ways. Blümling *et al.* (1983) and Zhonghuai & Ajia (1987) prepared composite borehole profiles for each well by superimposing images of caliper extensions measured at close intervals. The mean azimuth for each well could be observed visually and calculated statistically.

Borehole shape can be influenced by phenomena other than breakouts. The drill string can erode an elongated wellbore if the drill deviates significantly (Babcock 1978), asymmetric mudcaking can distort a hole's shape (Bell & Babcock 1986) and washouts (Fig. 2) will enlarge a hole in all directions so that both calipers are extended significantly beyond bit diameter (Cox 1983). Breakouts can be superimposed on washouts so that, although the hole becomes elongated in all directions, its long axis coincides with that of classical breakouts in the same well (Springer *et al.* 1984). Intersecting fractures may pose a more serious problem (Babcock 1978; Brown 1979) since, theoretically, they can give rise to features very similar to breakouts. Fractures are indicated by differences in conductivity between adjacent or op-

Table 1. *Criteria for recognizing breakouts on four-arm dipmeter logs*

1. The tool must stop rotating.
2. Significant changes in conductivity should be recorded by all four pads.
3. There should be no conductivity anomaly recorded by one or two pads.
4. The larger caliper curve should exceed hole gauge. The smaller should not be less than hole gauge and its trace should be straight.
5. The larger diameter of hole elongation should not coincide with azimuth of hole deviation.
6. The direction of the hole elongation should not coincide with azimuth of hole deviation.

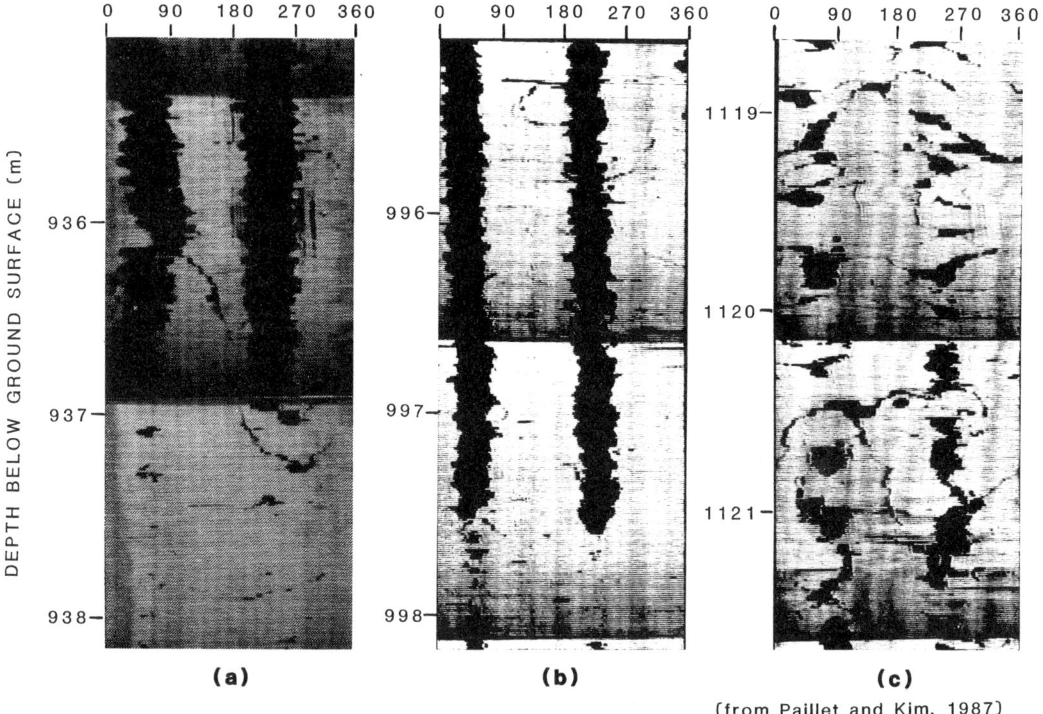

Fig. 10. Borehole televiewer images of (a,b) continuous and (c) discountinuous wellbore breakouts published by Paillet and Kim (1987). The breakouts occur at the indicated depths within Columbian River basalt flows in hole RRL-6 at the Hanford Site in southern Washington State, U.S.A. Image (a) also contains a sinusoidal trace of a steeply dipping natural fracture which crosses the wellbore at right angles to the long axis of the breakout (Photographs provided by F.L. Paillet).

posing microresistivity curves so, if differential caliper extension and cessation of tool rotation are associated with a large lateral conductivity anomaly, it may be unwise to identify the interval as a breakout (Brown 1979). The situation is not clearcut, however. Breakouts form as a consequence of borehole fragments breaking away along intersecting fractures generated soon after the hole has been drilled. The distal ends of such fractures may be present in the hole walls behind breakout excavations, and could cause the dipmeter pads to record resistivity anisotropy. With these considerations in mind, Fordjor *et al.* (1983) and Springer (1987) have proposed breakout recognition criteria for dipmeter log analysis (Table 1).

In situ stress magnitudes

Established methods

Measuring reliable in-situ stress magnitudes at depth in sedimentary basins is not easy. Three dimensional information can be obtained by strain relief methods involving overcoring behind mine openings (McGarr & Gay 1978), but the measurements frequently appear to be influenced by excavation-induced stresses (Zoback 1987). Holographic analysis of sidewall borehole deformation holds promise of providing reliable vertical and horizontal principal stress magnitudes at depth in wells (Bass *et al.* 1986), but suitable instrumentation has not yet been developed. Horizontal principal stress magnitudes should be obtainable by measuring core expansion through differential strain analysis (Simmons *et al.* 1974), although results have not always been consistent (Montgomery & Ren 1983).

The most reliable measurements to date have been derived from hydraulic fracturing (Hubbert & Willis 1957; Haimson & Fairhurst 1970). The method involves isolating a section of a borehole over a known depth interval with inflatable packers, and then increasing the

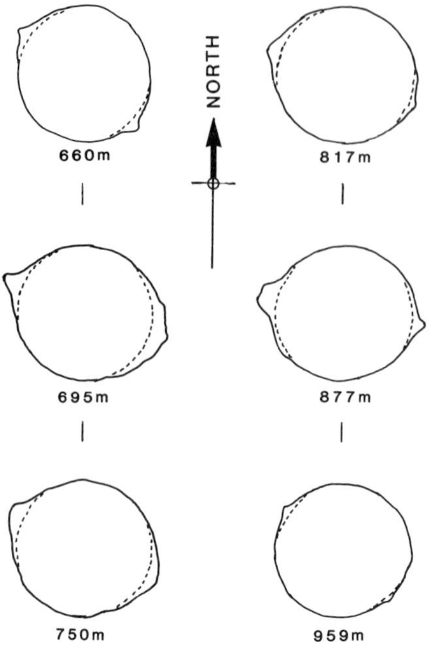

Fig. 11. Cross sectional images of borehole breakouts in basalts identified in DSDP hole 504B by the borehole televiewer (Newmark *et al.* 1984).

pressure within the interval by pumping in fluids until they induce a fracture in the surrounding rocks. Pressure is then reduced to allow the fracture to close. The pressure at which this occurs, the 'instantaneous shut-in pressure', is equated with the least principal stress. It is believed that fracture closure corresponds to the pressure at the inflection point on the pressure decay slope of the pressure-time record (Fig. 12) (Gronseth & Kry 1983; Aamodt & Kuriyagawa 1983). In controlled experiments, a series of opening and closing cycles may be run so that the instantaneous shut-in pressure for a particular fracture may be determined precisely (Haimson & Fairhurst 1970; Zoback *et al.* 1977). The vertical principal stress, S_V, at a given depth is equated with the overburden pressure and obtained by integrating formation density logs or otherwise estimating the pressure exerted by overlying rocks.

Most controlled hydraulic fractures in sedimentary basins have yielded least principal stress values smaller than the corresponding overburden stress, indicating that $S_{H_{min}}$, the lesser horizontal principal stress, had been measured. In such cases $S_{H_{max}}$, the larger horizontal principal stress, can be determined from

$$P_b = T + 3S_{H_{min}} - S_{H_{max}} - P_0 \qquad (9)$$

where P_b is the breakdown pressure, T is the tensile strength of the rock, $S_{H_{min}}$ is the smaller horizontal principal stress, and P_0 is the static pore pressure in the rock surrounding the borehole (Hubbert & Willis 1957). The equation describes the first cycle of pumping during a hydraulic fracture experiment when an initially intact interval of isotropic rock is broken (Fig. 12). In later cycles this initial rupture will not recur, so that the tensile strength, T, will be equal to the difference in breakdown pressure between the first and later pumping cycles. Therefore, for later pumping cycles, or in a first cycle when no breakdown pressure can be measured, equation (9) can be modified to

$$P_r = 3S_{H_{min}} - S_{H_{max}} - P_0 \qquad (10)$$

where P_r is the fracture opening pressure (Bredehoeft *et al.* 1976).

In many well-explored basins, the records of industrial hydraulic fracture programs will yield instantaneous shut-in pressures (Nolte 1979). If available, these data should be utilized to investigate the stress regime.

Leak-off tests

Hydraulic fracturing is expensive and is not performed during routine oil industry exploration drilling. However, leak-off tests, or formation integrity tests, are often carried out while drilling exploration wells, particularly if the wells are located in areas where abnormal formation pressures are anticipated. Leak-off tests resemble the first part of a hydraulic fracture experiment. They are run to measure the strength of rock units immediately below a casing shoe to indicate the maximum mud column pressure or formation pressure that the open wellbore can withstand (Dickey 1981). A well is drilled to a predetermined depth; steel casing is lowered to the base of the hole and cement is injected so that it sets between the outside of the casing pipe and the borehole wall. After the cement has hardened, it is drilled out at the bottom of the well and the hole is deepened several metres beneath the base of the casing. At this point drilling ceases and borehole fluid pressures are increased by pumping a small quantity of additional drilling mud into the well. The drilling mud begins to 'leak-off' when a fracture is initiated in the borehole wall in the open interval below casing.

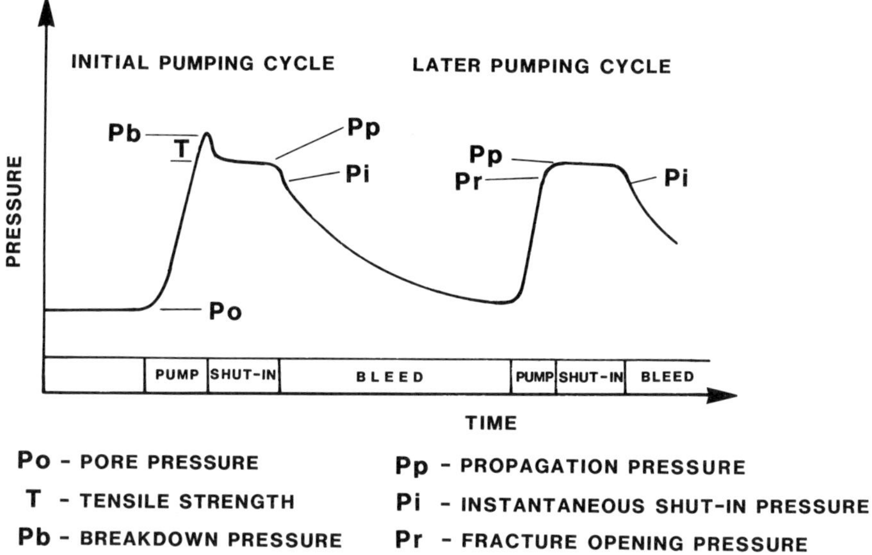

Fig. 12. Schematic pressure–time record of a hydraulic fracture experiment.

As soon as the pressure record indicates that leak-off has occurred, pumping is halted and the borehole fluid pressures are allowed to decay back to normal mudweight levels (Fig. 13).

Unlike a complete hydraulic fracture test, only a small volume of fluid is injected into the fracture and no attempt is made to propagate it. Only fracture initiation pressures are recorded and no data is collected to measure fracture closure pressures, so that leak-off tests will not give a direct measurement of the least principal stress at depth. Fracture initiation involves overcoming the amplified tangential stresses acting on a borehole wall that were referred to earlier in the discussion of breakouts (Breckels & Van Eekelen 1981) as well as any tensile strength the rocks may possess (Haimson & Fairhurst 1970) so that leak-off test pressures are likely to be higher than the least principal stress affecting the surrounding rocks.

Leak-off and instantaneous shut-in pressures were obtained over the same test intervals in 14 wells in Brunei (Breckels & Van Eekelen 1981), and the leak-off values exceeded the ISIPs by an average of 11%. On the other hand, more than 300 leak-off pressures and ISIPs from Gulf Coast wells were plotted against depth and a correspondence demonstrated between the lower bound values from both data sets (Breckels & Van Eekelen 1981). This suggests that the relatively lower leak-off pressures in a large data set can give a reasonable estimate of the least principal stress in a basin.

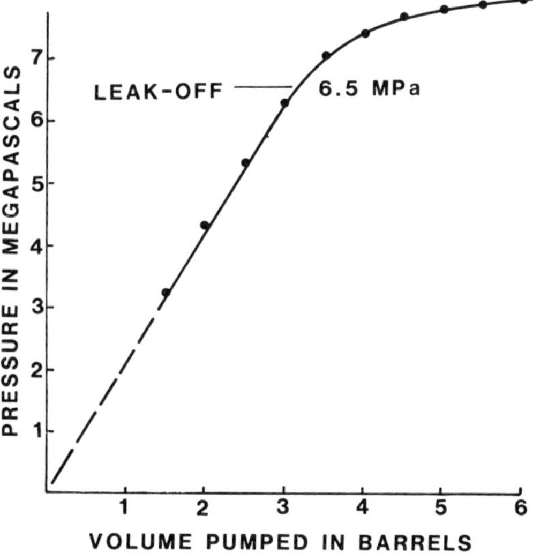

Fig. 13. A typical leak-off test pressure/mud volume record. Pressure increase ceased to be linear after approximately 3 barrels (0.5 m^3) of mud had been pumped into the well, at which point pump pressure had reached 6.5 MPa. Adding this pressure to the weight to the column of mud in the well gives the leak-off pressure at which a small fracture began to be propagated from the well into the surrounding rock.

In several wells drilled in offshore eastern Canada, a series of multiple leak-off tests were run beneath the same casing shoes and successively lower pressures recorded as fractures were initiated and then reopened (Ervine & Bell 1987). The initial fracturing in the first leak-off test would reduce the wall rock's tensile strength to zero and modify the tangential stresses acting on the borehole wall, with the result that the subsequent lower pressure leak-off tests resembled second or later fracturing cycles in conventional hydraulic fracture experiments (Fig. 16). From the above data, Ervine & Bell (1987) concluded that the lower leak-off tests pressures of the data set were likely to be close to fracture opening pressures and that, since the latter pressures were only marginally greater than instantaneous shut-in pressures in many hydraulic fracture experiments (Fig. 12), such leak-off pressures gave good estimates of the least principal stress (which appeared to be horizontal on the Scotian Shelf). In other words

Lower bound

$$\text{Leak-off pressure} \approx P_r \approx \text{ISIP} = S_{H_{min}}. \quad (11)$$

This equivalence allowed them to reduce equation (10) to

$$S_{H_{max}} = 2P_e - P_0 \quad (12)$$

where P_e is the formation leak-off pressure and P_0 is the static formation fluid pressure (Ervine & Bell 1987). Formation fluid pressure values can be inferred from drilling mud weight records and are measured directly during formation testing.

As before, the vertical principal stress, S_v, at a specified depth in a well can be equated with the pressure generated by the weight of the overlying rocks and expressed as

$$S_v = \int_o^z \rho(z) g dz \quad (13)$$

where $\rho(z)$ is the density function of a rock at depth z and g is the acceleration due to gravity. Strain-relief measurements of vertical stresses at depth are generally consistent with this assumption (McGarr & Gay 1978; Hoek & Brown 1980). Lithostatic stress information can be obtained from density logs by integrating the depth/density record.

Applications of in situ stress information in basins

Hydraulic fracture orientations

If the directions and relative magnitudes of the principal stresses at a sub-surface location are known, the orientation of induced hydraulic fractures can be predicted (Hubbert & Willis 1957). This can be advantageous in a number of situations that arise during oil industry exploration and production programs. If an exploration well has missed a small target such as a pinnacle reef (Fig. 14), and the target's precise location is known, it may prove feasible to induce a fracture that will connect the well to the productive reservoir in the reef (Bell & Babcock 1986), provided the well and the reef are aligned parallel to $S_{H_{max}}$ and the distance between them is only a few hundred metres (Smith *et al.* 1978). This is likely to be a cheaper option than directional drilling. Another possible application is blow-out control (Hassan 1982). When a relief well is being drilled to pump heavy mud into a well that has blown out, the relief well should be located so that hydraulic fractures, if any are induced, will propagate towards the blowout well, thus increasing the chances of establishing fluid communication between the two wells.

In many cases optimum recovery of oil by waterflooding requires that no inadvertent fracturing occur. In the Norman Wells field in northern Canada (Kempthorne & Irish 1981; Gronseth & Kry 1987), oil occurs in a Devonian limestone reef complex that dips to the southwest and at depths between 300 and 600 m. The limestone is naturally fractured and the fractures do not penetrate the overlying shales. The fractures trend NE–SW and the reservoir exhibits directional permeability with maximum permeability occurring parallel to the fracture trend (Kempthorne & Irish 1981). Breakouts do not

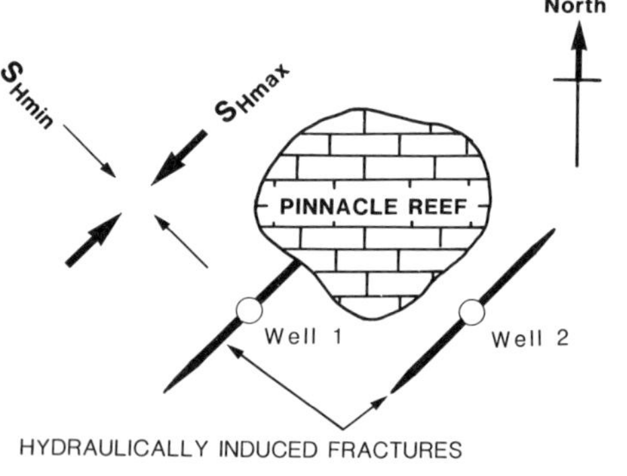

Fig. 14. Map view of a subsurface oil-bearing pinnacle reef. Well 1 can be connected to the reef oil reservoir by hydraulic fracturing since the local stress regime causes fractures to propagate northeastwards and southwestwards. Fractures originating in well 2 will miss the target.

occur in the fractured limestone (Gough & Bell 1982), but are present in the overlying shales and indicate that $S_{H_{max}}$ is oriented NE–SW.

The original depletion plan involved water flooding from the start, perforating new wells in the lower part of the reservoir and hydraulically fracturing them. This strategy was based on the assumption that $S_{H_{min}}$ was horizontal and oriented NW–SE, so that the induced hydraulic fractures would be vertical and extend parallel to the natural fracture direction. It was also assumed that the fractures would be confined to the reservoir limestone. However, an extensive in situ stress determination program involving controlled generation of small hydraulic fractures and measurements of core relaxation showed that, even at 600 m, the minimum principal stress was oriented vertically (Gronseth & Kry 1987). Induced hydraulic fractures would be horizontal and vertical conformance could not be achieved by augmenting the natural fracture system with hydraulic fractures. As a result, the final completion plans were changed to include multiple perforated intervals in wells based on natural permeability variations in the reservoir and the proposed waterflood injection pressures were reduced to prevent induced hydraulic fracturing from occurring (Gronseth & Kry 1987).

In the Pembina oil field in western Canada, most of the oil is trapped stratigraphically in Cretaceous sandstones at depths of 1500 to 2000 m (Nielsen 1957). Water-flooding has been responsible for 80% of total production to date and has been employed since 1956, three years after the field's discovery (Nielsen & Porter 1984). Some of the early waterfloods were injected at pressures well above the minimum principal stress, which is horizontal and oriented NW–SE in the Pembina field (Gough & Bell 1981). The result was that some injection wells induced NE and SW propagating hydraulic fractures, which connected with adjacent production wells that were offset in these directions (McLeod 1977; Hassan 1982). Initially, the production rate of oil increased markedly at wells adjacent to and southwest or northeast of the injection wells. However, within four months, the increased flow of oil was replaced almost completely by water as water-filled propagating fractures, which initially had swept oil ahead of them, intersected the production wells. Once these fractures were established, they induced a preferred permeability in the sandstones and prevented the waterflood from achieving optimum recovery. Now that this behaviour is understood, waterflood secondary recovery configurations can be designed to take advantage of induced hydraulic fractures (Fig. 15).

The gas-bearing sandstones of the Wattenberg Field in the western United States are tight and require massive hydraulic fracturing to be productive. The induced fractures exhibit a preferred orientation of 340° (Smith *et al.* 1978) and simulation studies indicate an optimum fracture length of 1200 m (Roberts 1981). Field monitoring has documented gas drainage from axially aligned ellipsoidal zones around the fractures and the most profitable wells are those located to exploit the optimum fracture length and orientation (Smith *et al.* 1979).

Both gas and oil production from tight reservoirs can be enhanced by inducing multiple fractures in a single well. If the hydrocarbon-bearing section is thick, multi-fractured inclined wells could be effective, but the well must be drilled in a direction close to $S_{H_{min}}$ (Fig. 16). Knowledge of the stress tensor is needed to orient such a well.

An experimental inclined production well was drilled through low permeability chalk in the Caddo-Pine Island oil field of Louisiana in the southern United States (Strubhar *et al*, 1975). It was drilled normal to $S_{H_{min}}$ at an angle of 52° from the vertical. By packing-off successive intervals four hydraulic fractures were induced at depths between 469 and 518 m. This multi-fractured well flowed two to three times faster than a conventional single-fracture vertically-drilled well in the field.

Borehole stability

One very practical application of in situ stress information is related to minimizing caving while drilling. As discussed earlier, caving is a function of the shear strength of the rock, the magnitude of the regional stresses affecting the borehole and their orientation with respect to it. Cooling and increasing the density of the drilling mud can suppress growth of fractures in the borehole wall (Zoback *et al.* 1985; Woodland 1988), but such measures are not always cost effective.

Active faulting

Induced hydraulic fracturing occurs as a result of artificially increasing the pore fluid pressure in a rock body. This can occur naturally and give rise to faulting. Listric normal faulting has occurred between the Cretaceous and late Tertiary in the Sable Basin offshore of eastern

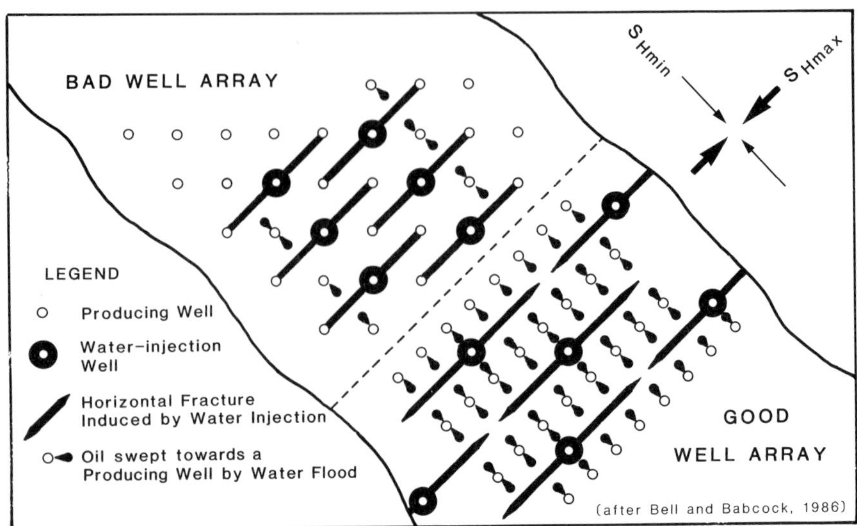

Fig. 15. Map of well arrays in an oilfield where water flooding promotes hydraulic fracturing. In the 'bad' array, little oil is recovered because induced fractures connect the water injection wells with intended production wells. Water flows between them but does not effectively sweep oil towards many well bores. In the 'good' array, injection wells are spaced so that induced fractures do not connect to potential production wells but distribute the water so that it drives oil towards man production wells.

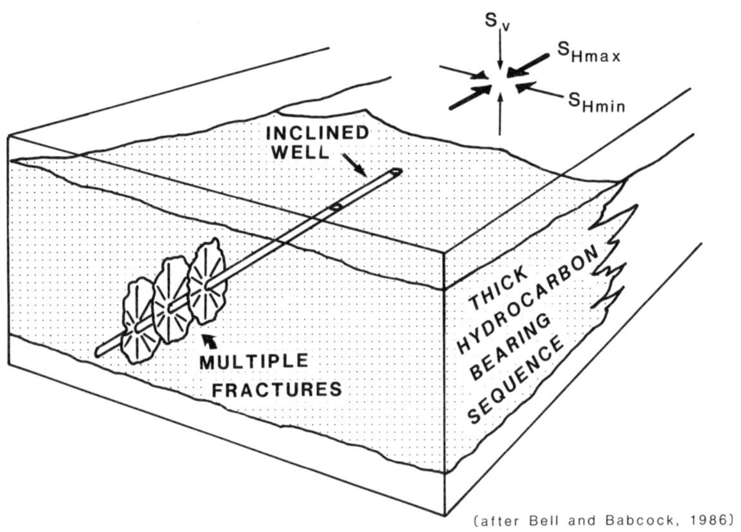

Fig. 16. Schematic block diagram showing a multiply fractured inclined well drilled into a relatively thick hydrocarbon reservoir sequence in a direction trending parallel, and at approximately 45° to $S_{H_{min}}$. Multiple fractures propagated from such an inclined well would provide greater drainage and higher flow rates than could be obtained from a vertical well with a single fracture feeding it.

Canada (Jansa & Wade 1975). The stress regime in the basin (Fig. 17a) has been established from breakout orientations (Podrouzek & Bell 1985) and log and drill record analyses (Ervine & Bell 1987). Above approximately 500 m below sealevel, $S_V > S_{H_{max}} > S_{H_{min}}$, from 500 m to approximately 6000 m $S_{H_{max}} > S_V > S_{H_{min}}$. Below approximately 6000 m, it appears likely that $S_{H_{max}} > S_{H_{min}} > S_V$.

The formation fluid pressures are hydrostatic

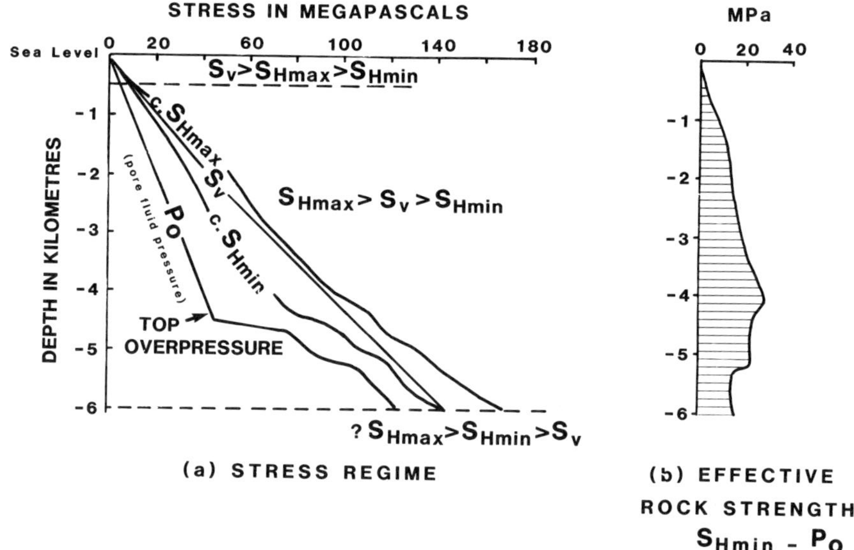

Fig. 17. (a) Variation in principal stress magnitudes with depth and pore fluid pressure with depth in the Sable Basin, offshore eastern Canada. (b) Effective rock strength versus depth for the Sable Basin. The values shown represent the additional pressures required to initiate hydraulic fracturing of the rocks. Note the decrease in effective strength below 5 k.

down to approximately 4500 m. Below this they increase rapidly and the section becomes overpressured (Drummond 1986; Wade *et al.* 1989). The increase in fluid pressure required to initiate a hydraulic fracture is the difference between the formation fluid pressure and the fracture pressure or formation leak-off pressure which, in Figure 17a, is equated to $S_{H_{min}}$. In essence, this is the effective horizontal stress (Hubbert & Rubey 1959), but here the difference between these two values is described informally as the effective strength of the rock section (Fig. 17b). The nature of faulting which would occur in the rocks (Anderson 1951), once fractured, assuming they were free to become displaced is defined by the relative magnitudes of the principal stresses (Fig. 18a). Also shown in Fig. 18a is a profile of effective rock strength and the orientations along which induced hydraulic fractures would form, as predicted by the relative magnitudes of the principal stresses (Hubbert & Willis 1957). The rocks are weakest near the surface where they tend to fracture vertically. At intermediate depths, effective rock strength increases, but it drops significantly within the overpressured zone and any increase in fluid pressures below approximately 6000 m will tend to generate horizontal fractures. If the rock body is slightly tilted and/or free to undergo extension, large listric normal faults and associated antithetic faults could form (Fig. 18b). It is suspected that many such faults in sedimentary basins form in response to similar dynamic regimes.

Finite element modelling has demonstrated that stress trajectories may be deflected around active faults (Chinnery 1966). Regional stress trajectories are deflected close to the San Andreas fault zone and it is suggested that it may be acting as an 'almost free surface' (Mount & Suppe 1987).

A similar phenomenon is found in the Jeanne d'Arc Basin of offshore eastern Canada (Bell *et al.* in preparation). Here the regional orientation of $S_{H_{max}}$ is NE–SW (Adams 1987), but several wells that were drilled on downthrown blocks adjacent to the Murre and Mercury Faults exhibit E–W S_{Hmax} directions (Fig. 19). These faults are basin bounding listric normal features which offset seismic reflections to within 500 milliseconds of the seafloor (Enachescu 1987). They have been active in latest Tertiary time and may still be (A. Edwards, personal communication, 1987). Movement of the hanging wall blocks could compress the rocks therein about an E–W axis, and so reorient $S_{H_{max}}$ in that local area. Alternatively, the regional horizontal principal stresses may be being deflected by the Murre and Mercury Faults, which are acting as free surfaces. No major earthquakes

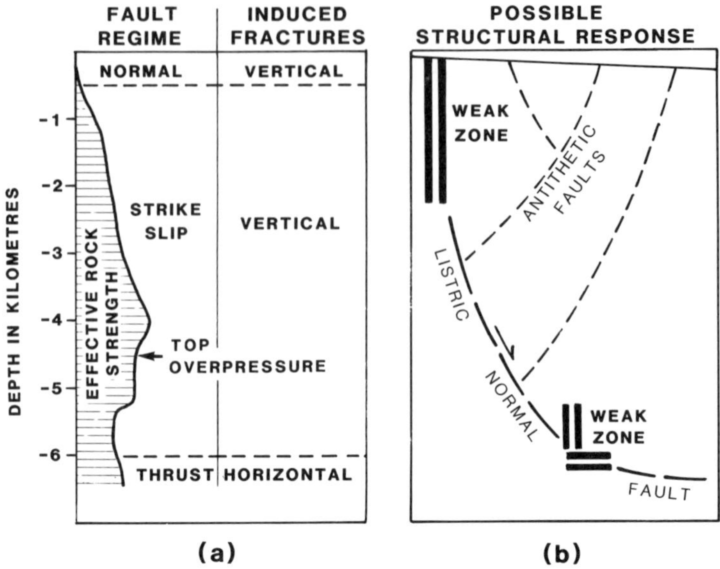

Fig. 18. (a) Relationship between orientations of induced fractures, faulting tendencies and effective rock strength versus depth in the Sable Basin. (b) Possible structural response if the section was free to undergo extension.

involving these two faults have been recorded since 1929 (Basham & Adams 1982; Adams, 1986), but that does not necessarily mean they have ceased moving. Looking ahead, locally anomalous principal stress orientations may help identify potentially active seismic zones that have been relatively quiescent during historic time.

Plate Tectonics and Seismic Stratigraphy

Two large-scale applications of in situ stress data should be mentioned but will not be discussed at length here.

As more and more horizontal principal stress orientations are becoming available, stress provinces which exhibit common axes of compression or extension are being recognized (Zoback & Zoback 1980). The stress provinces appear to be a manifestation of the tractions responsible for lithospheric plate movement and their configurations constrain hypotheses about how and why this occurs (Gough 1984). Work is currently underway to prepare a horizontal stress orientation map of the world in the expectation that this will help elucidate global tectonic driving mechanisms (Zoback 1987).

The seismostratigraphic record at basin margins, which has been documented extensively by Vail *et al.* (1977) and later workers, may result in part from the variation of intra-plate horizontal stress magnitudes through time (Cloetingh 1986). If stress variation proves to be a controlling cause of the high-frequency coastal onlap and offlap cycles, this could have major implications for understanding Mesozoic continental margin sedimentation. One way of testing the model's validity is to compare $S_{H_{max}}$ magnitudes at comparable depths in different regions to see if variations of the required order of magnitude exist today. Like the World Stress Map, this is a long term endeavour, but one that could provide some exciting insights in the years ahead.

Conclusions

It has been demonstrated that information gathered during logging and drilling exploration wells can yield enough information to infer principal in situ stress magnitudes and orientations at depth around the well bore. Suitable wells are those which have been logged with density logs and dipmeters and in which both leak-off and drillstem tests have been run. A detailed drilling history report is also desirable. In many jurisdictions such information is collected by regulatory agencies and released for public examination some years after the wells have been drilled. At the present time, numer-

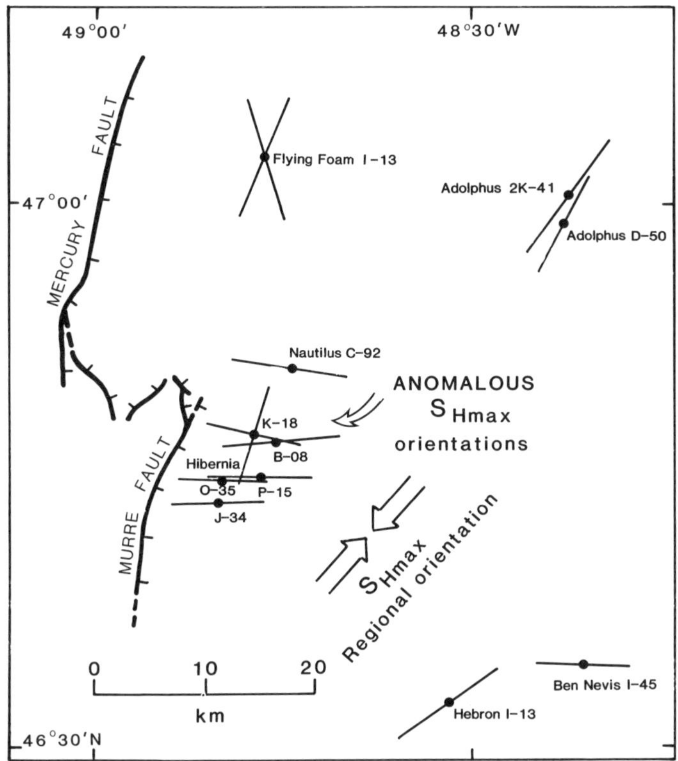

Fig. 19. Map of the western part of the Jeanne d'Arc Basin showing $S_{H_{max}}$ orientations derived from breakout populations. The regional orientation of $S_{H_{max}}$ is NE–SW but anomalous directions are given by several wells most of which are located close to the Murre and Mercury Faults. Note how $S_{H_{max}}$ aximuths in wells close to the Murre Fault are oriented approximately normal to its trace. This deflection may be due to the fault zone acting as a free surface, or be caused by local E–W compression associated with recent slop on the fault.

ous records of breakouts are being analysed for indications of horizontal principal stress directions. The results are proving to be remarkably consistent and are augmenting our understanding of stress configurations in the lithosphere (Zoback 1987). Leak-off test data have not yet been used for stress magnitude estimation to the same extent, so their utility is not so firmly established.

Appendix

Notes on breakout analysis

Standard format logs for dipmeters fitted with low angle inclinometers are discussed elsewhere (Cox 1970; Babcock 1978; Bell & Gough 1983; Bell & Babcock 1986) and data retrieval methods for other formats are identical, except that recovery of Caliper-1 azimuths may vary. Digital processing or manual extraction can be used and sampling intervals specified. Gough & Bell (1981) recorded azimuth data at 2 ft intervals within breakouts and separated populations in a given well with histograms.

The mean breakout azimuth of a population is the resultant of the unit vectors (Mardia 1972) given by

$$\cos^{-1} C/R \quad \text{or} \quad \sin^{-1} S/R \tag{14}$$

where

$$C = \frac{1}{n} \Sigma f \cos\theta$$

$$S = \frac{1}{n} \Sigma f \sin\theta$$

$$R = (C^2 + S^2)^{1/2}.$$

In these equations n is the total number of

Table 2 *Leak-off test information recored in Olympia A-12 well history report*

A	B	C	D	(B×C) + D
Drilled depth 1531.5	Depth of base of casing 1510.5 m	Specific gravity of drilling mud 1.126 g/cm^{-2}	Surface pressure at leak-off 1265 kg cm^{-2}	Total leak-off pressure 2965 kg cm^{-2}

measurements in a population, f is the number of readings of a specific orientation and θ refers to breakout long axis directions corrected for magnetic declination.

Mardia (1972) derives the standard deviation of the mean azimuth as $360/2\pi(-2 \log_e R)^{1/2}$.

Example of the measurement of stress magnitudes from well data

The Olympia A-12 well in offshore eastern Canada was leak-off tested below intermediate casing (Table 2).

The vertical principal stress, S_v at 1510.5 m was obtained by integrating corrected density log measurements, extrapolating readings to the seafloor and adding the water column weight to give a magnitude of 3009 kg m^{-2}. As this exceeded the leak-off pressure, the latter value was equated to $S_{H_{min}}$, the smaller horizontal principal stress. The larger horizontal principal stress was then estimated from equation 12:

$$S_{H_{max}} = 2(\text{Leak-off pressure}) - \text{pore pressure}.$$

Pore pressure at 1510.5 m was estimated to be 1621 kg cm^{-2} because the drilling mud used uphole had varied in specific gravity from 1.030 to 1.126 g cm^{-2} indicating that a normal seawater formation pressure gradient (1073 kg m^{-3}) existed. This assumption yielded an $S_{H_{max}}$ magnitude of 4309 kg cm^{-2}. These principal stress magnitude estimates indicate that $S_{H_{max}} > S_v > S_{H_{min}}$ unless $S_{H_{min}}$ has been significantly overestimated.

The writer is most grateful to his many colleagues and collaborators with whom most of the contents of this paper have been generated or discussed. These include: N. Anderson, R. N. Anderson, E. A. Babcock, M. W. Barr, J. W. Cox, M. B. Dusseault, A. Edwards. W. B. Ervine, C. K. Fordjor, D. I. Gough, A. C. Grant, J. M. Gronseth, P. K. Kry, K. Y. Lo, A. J. Podrouzek, G. S. Stockmal, J. A. Wade, R. E. Wyman, M. D. Zoback and M. L. Zoback. The photographs for Figs 7 & 10 were loaned by J. E. Springer and F. L. Paillet respectively. N. H. Koziel, E. Kun, C. Thompson and P. Pheby are thanked for their assistance. In situ stress studies have been supported by the Geological Survey of Canada's Frontier Geoscience Program in Project 870039. The manuscript has been compacted and improved by the suggestions of A. Hurst, A. M. Jessop, M. S. King, G. S. Stockmal and P. Talwani. This is Geological Survey of Canada Contribution number 55588.

References

AAMODT, R. L. & KURIYAGAWA, M. 1983. Measurements of instantaneous shut-in pressure in crystalline rock. *In*: ZOBACK M. D. & HAIMSON B. C. (eds), *Hydraulic Fracturing Stress Measurements*. National Academy Press, Washington, D. C., 139–142.

ADAMS, J. 1986. Changing assessment of seismic hazard along the southeastern Canadian margin, *3rd Canadian Conference on Marine Geotechnical Engineering* St. John's, Newfoundland, **1**, 42–53.

—— 1987. Canadian crustal stress data — a compilation. *Geological Survey of Canada, Open File Report*, 1622.

ANDERSON, E. M., 1951. *The dynamics of faulting:* 2nd edn. Oliver and Boyd, Edinburgh.

BABCOCK, E. A., 1978. Measurement of subsurface fractures from dipmeter logs. *Bulletin of the American Association of Petroleum Geologists*, **62**, 1111–1126.

BASHAM, P. W. & ADAMS, J. 1982. Earthquake hazards to offshore development on the Eastern Canadian continental shelves. *2nd Canadian Conference on Marine Geotechnical Engineering*. Halifax, Nova Scotia.

BASS, J. D., SCHMITT, D. R. & AHRENS, T. J. 1986. Holographic in-situ stress measurements. *Geophysical Journal of the Royal Astronomical Society*, **85**, 13–41.

BELL, J. S. & BABCOCK, E. A. 1986. The stress regime of the Western Canadian Basin and implications for hydrocarbon production. *Bulletin of Canadian Petroleum Geology*, **34**, 364–378.

—— & GOUGH, D. I. 1979. Northeast–southwest compressive stress in Alberta: evidence from oil wells. *Earth and Planetary Science Letters*, **45**, 475–482.

—— & —— 1983. The use of borehole breakouts in the study of crustal stress. *In*: ZOBACK, M. D. & HAIMSON, B. C. (eds). *Hydraulic Fracturing Stress Measurements*. National Academy Press, Washington, D. C. 201–209.

—— PODROUZEK, A. J. & ERVINE, W. B., in preparation. The stress regime of offshore eastern Canada.

BLÜMLING, P., FUCHS, K., & SCHNEIDER, T. 1983. Orientation of the stress field from breakouts in a crystalline well in a seismic active area. *Physics of the Earth and Planetary Interiors*, **33**, 250–254.

BRECKELS, I. M., & VAN EEKELEN, H. A. M. 1981. Relationship between horizontal stress and depth in sedimentary basins. Paper SPE 10336, *56th Annual Fall Technical Conference, Society of Petroleum Engineers of AIME*, San Antonio, Texas, October 5–7, 1981.

BREDEHOEFT, J. D., WOLFF, R. G., KEYS, W. S. & SHUTER, E. 1976. Hydraulic fracturing to determine the regional in-situ stress field, Piceance Basin, Colorado. *Bulletin of the Geological Society of America*, **87**, 250–258.

BROWN, R. O. 1979. Fracture identification log use in Cretaceous of N. Louisiana, Mississippi. *Oil and Gas Journal*, **77**, 350–355.

BYERLEE, J. D. 1978. Friction of rocks. *Pure and Applied Geophysics*, **116**, 615–626.

CHINNERY, M. A. 1966. Secondary faulting. *Canadian Journal of Earth Sciences*, **3**, 163–190.

CLOETINGH, S. 1986. Intraplate stresses: a new tectonic mechanism for fluctuations of relative sea level. *Geology*, **14**, 617–620.

COX, J. W., 1970. The high resolution dipmeter reveals dip-related borehole and formation characteristics. *Transactions of the SPWLA 11th Annual Logging Symposium*, Los Angeles, May 3–6, 1970.

—— 1983. Long axis orientation in elongated boreholes and its correlation with rock stress, data. *Transactions of the SPWLA 24th Annual Logging Symposium*, Calgary, Canada, June 27–30, 1983.

DICKEY, P. A. 1981. *Petroleum development geology*, 2nd edn. Penwell Books, Oklahoma.

DRUMMOND, K. J. 1986. Geopressure in Venture and the Sable Island area. *Canadian Society of Petroleum Geologists Convention*, Reserves 21, June 21–24, 1986, Calgary, Alberta.

EKSTROM, M. P., DAHAN, C. A., CHEN, M. Y., LLOYD, P. M. & ROSSI, D. J. 1987. Formation Imaging with Microelectrical Scanning Arrays. *Log Analyst*, **28**, 294–306.

ENACHESCU, M. E. 1987. The tectonic and structural framework of the northeast Newfoundland continental margin. *In*: BEAUMONT, C. & TANKARD, A. J. (eds). *Sedimentary Basins, and Basin Forming Mechanisms*. Canadian Society of Petroleum Geologists, Memoir **12**, 117–146.

ERVINE, W. B. & BELL, J. S. 1990. Subsurface in-situ stress magnitudes from oil well drilling records: an example from the Venture area, offshore eastern Canada. *Canadian Journal of Earth Sciences*, **24**, 1748–1759.

EVANS, C. J. & BRERETON, N. R. 1990. In situ crustal stress in the United Kingdom from borehole breakouts. *In*: HURST, A., LOVELL, M. A. & MORTON, A. C. (eds). *Geological applications of wireline logs*. Geological Society, London, Special Publication **48**, 327–337.

FORDJOR, C. K., BELL, J. S. & GOUGH, D. I. 1983. Breakouts in Alberta and stress in the North American plate. *Canadian Journal of Earth Sciences*, **20**, 1445–1455.

GAY, N. C. 1980. The state of stress in the plates, *In*: BALLY, A. W. Penden, P. L., MCGETCHIN, T. R. & WALCOTT, R. I. (eds). *Dynamics of Plate Interiors. American Geophysical Union/ Geological Society of America*, 145–154.

GOUGH, D. I. 1984. Mantle upflow under North America and plate dynamics. *Nature*, **311**, 428–433.

—— & BELL, J. S. 1981. Stress orientations from oil well fractures in Alberta and Texas. *Canadian Journal of Earth Sciences*, **18**, 638–645.

—— & —— 1982. Stress orientations from borehole wall fractures with examples from Colorado, east Texas and northern Canada. *Canadian Journal of Earth Sciences*, **19**, 1358–1370.

—— & GOUGH, W. I. 1987. Stress near the surface of the Earth. *Annual Review of Earth and Planetary Sciences*, **15**, 545–566.

GRONSETH, J. M. & KRY, P. R. 1983. Instantaneous shut-in pressure and its relationship to the minimum in-situ stress. *In*: ZOBACK, M. D. & Haimson, B. C. (eds). *Hydraulic Fracturing Stress Measurements*. National Academy Press, Washington, D. C. 55–60.

—— & ——, 1987. In-situ stresses and the Norman Wells expansion project. Paper 87–38–57, *Petroleum Society of the Canadian Institute of Mining and Metallurgy, 38th Annual technical Meeting*, Calgary, June 7–10, 1987.

HAIMSON, B. C. & FAIRHURST, C. 1970. In-situ stress determination at great depth by means of hydraulic fracturing, *In*: SOMERTON, W. (ed.), *Rock Mechanics – Theory and Practice*. Proceedings of the 11th Symposium on Rock Mechanics, AIME, New York, 559–584.

HANDIN, J. 1966. Strength and ductility. *In*: CLARK, S. P. Jr. (ed.) *Handbook of Physical Constants*. Geological Society of America, Memoir **97**, 223–291.

HASSAN, D. 1982. A method for predicting hydraulic fracture azimuth and the implication thereof to improve hydrocarbon recovery. Paper 82–33–19, *33rd Annual Technical Meeting of the Petroleum Society of the Canadian Institute of Mining and Metallurgy*, Calgary, June 6–9, 1982.

HOEK, E. & BROWN, E. T. 1980. *Underground Excavations in Rock*. Institute of Mining and Metallurgy, London.

HOLZHAUSEN, G. R., WOOD, M. D., RAISBECK, J. M. & CARD, C. C. 1980. Results of deformation monitoring during steam stimulation in a single-well test. *Applied Oil Sands Geoscience Conference*, Edmonton.

—— & RUBEY, W. W. 1959. Role of fluid pressure in mechanics of overthrust faulting. *Bulletin of the Geological Society of America*, **70**, 115–166.

—— & Willis, D. G. 1957., Mechanics of hydraulic fracturing. *AIME Petroleum Transactions*, **210**, 153–166.

Jaeger, J. C. 1962. *Elasticity, fracture and flow*. 2nd edn. Meutheun, London.

—— & Cook, N. G. W. 1976. *Fundamentals of rock mechanics*. 2nd edn, Chapman and Hall, London.

Jansa, L. F. & Wade, J. A. 1975. Geology of the continental margin off Nova Scotia and Newfoundland. *In*: *Offshore Geology of Eastern Canada*. Geological Survey of Canada, Paper 74–30, 51–105.

Kempthorne, R. H. & Irish, J. P. R. 1981. Norman Wells — a new look at one of Canada's largest oil fields. *Journal of Petroleum Technology*, **33**, 985–991.

Kirsch, G. 1898. Die Theorie der Elastizität un die Bedürfnisse der Festigkeitslehre. *Zeitschrift des Vereines Deutscher Ingenieure*, **42**, 797–807.

Majer, E. L. & Doe, T. W. 1986. Studying hydrofractures by high frequency seismic monitoring. *International Journal of Rock Mechanics and Mining Sciences*, **23**, 185–199.

Mardia, K. V. 1972. *Statistics of directional data: probability and mathematical statistics*. Academic Press, London and New York.

McClintock, F. A. & Walsh. J. B. 1962. Friction on Griffith cracks under pressure. *4th U. S. National Congress of Applied Mechanics Proceedings*, 1015–1021.

McGarr, A. and Gay, N. C. 1978. State of stress in the Earth's crust. *Annual Review of Earth and Planetary Sciences*, **6**, 405–436.

McLeod, J. G. F. 1977. Successful injection pattern alteration, Pembina J Lease, Alberta. *Paper presented at the 28th Annual Meeting of Petroleum Society of Canadian Institute Mining*, Edmonton.

Montgomery, C. T. & Ren, N-K. 1983. Differential strain curve analysis: does it work? *In*: Zoback, M. D. and Haimson, B. C. (eds). *Hydraulic Fracture Stress Measurements*. National Academy Press, Washington, D. C. 239–245.

Mount, V. S. & Suppe, J. 1987. State of stress near the San Andreas fault: Implications for wrench tectonics. *Geology*, **15**, 1143–1146.

Newmark, R. L., Zoback, M. D. & Anderson, R. N. 1984. Orientation of *in situ* stresses in the oceanic crust. *Nature*, **311**, 424–428.

Nielsen, A. R. 1957. Cardium stratigraphy of the Pembina field. *Alberta Society of Petroleum Geologists Bulletin*, **5**, 64–72.

—— & Porter, J. W. 1984. Pembina oil field — in retrospect, *In*: Scott, D. F. & Glass, D. J. (eds). *The Mesozoic of Middle North America*. Canadian Society of Petroleum Geologists, Memoir 9, 1–14.

Nolte, K. G. 1979. Determination of fracture parameters from fracturing pressure decline. *54th Annual Technical Conference and Exhibition of Society of Petroleum Engineers*, Las Vegas, Nevada.

Paillet, P. L. & Kim, K. 1987. Character and Distribution of Borehole Breakouts and Their Basalts. *Journal of Geophysical Research*, **92**, 6223–6234.

Paterson, M. S. 1978. *Experimental rock deformation: the brittle field*. Springer, New York.

Plumb, R. A. & Hickman, S. H. 1985. Stress-induced borehole elongation: a comparison between the four-arm dipmeter and the borehole televiewer in the Auburn geothermal well. *Journal of Geophysical Research*, **90**, 5513–5522.

Podrouzek, A. J. & Bell, J. S. 1985. Stress orientations from wellbore breakouts on the Scotian Shelf, eastern Canada. *In*: *Current Research, Part B*. Geological Survey of Canada, Paper 85-1B, p. 59–62.

Raleigh, C. B., Healy, J. H. & Bredehoeft, J. D. 1972. Faulting and crustal stress at Rangely, Colorado. *In*: *Flow and Fracture of Rocks*. American Geophysical Union Monograph, **16**, 275–284.

Roberts, C. N. 1981. Fracture optimization in a tight gas play: Muddy "J" Formation, Wattenberg field, Colorado. Paper SPE-DOE 9851, *Low Permeability symposium*, Denver, Colorado, May 27–29, 1981.

Secor, D. T. & Pollard, D. D. 1975. On the stability of open hydraulic fractures in the Earths crust. *Geophysical Research Letters*, **2**, 510–513.

Simmons, G., Siegfried, R. W. & Feves, M. 1974. Differential strain analysis: a new method for examining cracks is rocks. *Journal of Geophysical Research*, **79**, 4382–4385.

Smith, M. B. 1979. Effect of fracture azimuth on production with application to the Wattenberg gas field. Paper SPE 8298, *54th Annual Fall Technical Conference of the Society of Petroleum Engineers of AIME*, Las Vegas, September 23–26, 1979.

——, Holman, G. B., Fast, C. R. & Corlin, R. J. 1978. The azimuth of deep, penetrating fractures in the Wattenberg field. *Journal of Petroleum Technology*, **30**, 185–193.

Springer, J. E. 1987. Stress orientations from the well bore breakouts in the Coalinga region. *Tectonics*, **6**, 667–676.

——, Thorpe, R. K. & McKague, H. L. 1984. *Borehole elongation and its relation to tectonic stress at the Nevada Test Site*. Lawrence Livermore National Laboratory Report UCRL 53528.

Strubhar, M. K., Fitch, J. L. & Glen, E. E. 1975. Multiple, vertical fractures from an inclined well bore — a field experiment. *Journal of Petroleum Technology*, **27**, 641–647.

Vail, P. R., Mitchum, R. M. Jr. and Thompson, S. III. 1977. Global cycles of relative changes of sealevel. *American Association of Petroleum Geologists, Memoir* **26**, 83–97.

Wade, J. A., Campbell, G. R., Procter, R. M. & Taylor, G. C. in press. Petroleum resources of the Scotian Shelf. Geological Survey of Canada, Paper.

Wiley, R. 1981. *The borehole televiewers: An update on application*. World Oil, March 1981, 47–53.

WOODLAND, D. C. 1988. Borehole instability in the Western Canadian Overthrust Belt. Paper SPE 17508, *Rocky Mountain Regional Meeting, Society of Petroleum Engineers of AIME*, Casper, Wyoming, May 11–13, 1988.

ZHONGHUAI, X. & AJIA, G. 1987. Oil-well breakouts is Sichuan Basin of China. *International Union of Geodesy and Geophysics XIX General Assembly Vancouver, Canada. Abstracts*, **3**, 1116.

ZOBACK, M. D., MOOS, D., MASTIN, L. & ANDERSON, R. N. 1985. Wellbore breakouts and in-situ stress. *Journal of Geophysical Research*, **90**, 5523–5530.

——, RUMMEL, F., JUNG, R. & RALEIGH, C. B. 1977. Laboratory hydraulic experiment in intact and prefractured rock. *International Journal of Rock Mechanics and Mining Science*, **14**, 49–58.

ZOBACK, M. L. 1987. State of stress in North America. *International Union of Geodesy and Geophysics XIX General Assembly, Vancouver, Canada. Abstracts*, **3**, 1114.

—— and ZOBACK, M. D. 1980. State of stress in the conterminous United States. *Journal of Geophysical Research*, **85**, 6113–6156.

In situ crustal stress in the United Kingdom from borehole breakouts

C. J. EVANS & N. R. BRERETON

British Geological Survey, Keyworth, Nottingham, NG12 5GG, UK

Abstract: Until recently very little in situ stress data have been available from the UK crust. Literature reviews list only a limited number of measurements which include information on the direction of maximum horizontal compression. Unequal horizontal crustal stresses imposed on vertical boreholes will often cause localized rock spalling along the direction of least compressive stress. The resultant elongations of the borehole cross section, called breakouts, can be identified from four arm caliper records collected routinely with conventional geophysical dipmetre logs. Eighty-one boreholes in the United Kingdom have been analysed for breakouts using criteria for identification of breakouts unique to this paper. The majority of borehole breakouts in southern Britain are aligned NE–SW, indicating a maximum horizontal stress in a NW–SE direction. There are insufficient data to confidently assess a dominant borehole breakout direction in Northern Britain, although the four boreholes analysed indicate possible variations in the crustal stress field.

Stresses in the Earth's crust are speculated to arise from many sources: plate driving forces, temperature changes (cooling and heating), gravitational loading by volcanic and sedimentary processes, and from poorly understood processes of mid-plate tectonics (Solomon *et al.* 1980). The magnitude and direction of the principal stresses in the brittle crust influence the propagation of faults, the opening and closing of fractures, and pore water circulation.

The databank of stress measurements in the earth's crust remains inadequate because methods of measurement are in general expensive. Research has demonstrated that stress measurements cannot be reliably extrapolated any distance from the measurement location (Hyett *et al.* 1986). Therefore, our knowledge of regional or local stress levels remains poor due to a lack of measurements both of a qualitative and quantitative nature.

Theoretical and laboratory studies carried out in recent years have shown that unequal horizontal crustal stresses imposed on near vertical boreholes will often cause localized rock spalling along the direction of least compressive stress (Zoback *et al.* 1985, Haimson & Herrick 1986). The resultant elongations of the borehole cross section have been referred to as breakouts. The measurement of the azimuthal frequency of breakouts, using data from existing boreholes, presents a unique opportunity to expand the regional stress direction data base and to study the distribution and mechanisms of stress fields in more detail than has been possible hitherto.

Borehole breakouts can be identified from the four arm caliper records collected routinely with conventional geophysical dipmeter logs. The majority of boreholes drilled for hydrocarbon exploration purposes are logged for dipmeter and so the number of boreholes potentially available for breakout analysis in the UK is large (Brereton & Evans 1987).

A review of stress measurements in Britain and NW Europe

Until comparatively recently, very few stress measurements had been made in Britain and knowledge of the state of stress in British rocks was limited. Klein & Brown (1983) carried out a comprehensive review and found only six sets of measurements which included complete information on the state of stress (Fig. 1). Even so, the authors cast doubt on the reliability of many of these data.

Until recently the only high-quality horizontal stress direction and magnitude determinations came from work at Rosemanowes Quarry in the Carnmenellis Granite, Cornwall (Batchelor & Pine 1986). Sixteen hydraulic fracture tests were attempted from near-surface to a maxium depth of 2550 m. Although the tests were successful in achieving 'breakdowns', only in one case was a hydraulic fracture identified. Therefore, although the minimum stress results are considered to be reliable, the maximum stress and the orientation of the stress fields remain questionable. A series of overcoring tests in South Crofty Mine, 10 km from Rosemanowes, plus hydrofracture testing and overcoring (Cooling and Hudson 1986) from Carwynnen, also in the Carmenellis Granite, are in good agreement with the work of Batchelor & Pine (1986) and confirm a regional

From Hurst, A., Lovell, M. A. & Morton, A. C. (eds), 1990, *Geological Applications of Wireline Logs* Geological Society Special Publication No. 48, pp. 327–338

Fig. 1. Directions of maximum horizontal compression in the British Isles (after Klein & Brown 1983).

NW–SE maximum stress orientation and a very high ratio of maximum to minimum stress in the granite.

Cox (1983) was the first to publish breakout results from western Europe when he included data from three boreholes in the North Sea. This was followed by Blümling *et al.* (1983) who reported results from the Urach-3 crystalline rock geothermal borehole in southern Germany. European breakout analyses have subsequently been reported for a further twelve boreholes in Northern Germany and Holland (Draxler & Edwards 1984) and for five additional boreholes in Switzerland (Becker *et al.* 1984, 1987).

Klein & Barr (1986) extended the data set for Britain by presenting borehole breakout azimuth plots for eleven boreholes from the North Sea, two offshore to the north of Scotland and a further six from onshore Britain (mostly in the East Midlands). The boreholes penetrated a variety of generally near horizontally-bedded sedimentary rocks including, sandstones, siltstones and mudstones. Borehole penetration ranged in depth from 240 to 3970 m. Stress magnitude data have recently been further augmented by three hydrofracturing measurements at Morley Quarry and Rempstone in the East Midlands and at Wray in northern England (Evans 1987). The results from these hydrofracturing tests are in broad agreement with those reported for the Cornish granite.

The more recent stress measurements, along with compilations of British and western European data from in situ stress measurements, earthquake focal plane solutions and geological stress indicators (Klein & Barr 1986), all provide evidence which support a generalized NW–SE maximum stress direction for much of Northwest Europe and the British Isles. It has been argued that the predominant orientation is largely determined by plate tectonic boundary forces (Bott & Kusznir 1984; Klein & Barr 1986).

Comparison of British stress magnitudes with European results is difficult because of a lack of deep data in Europe. However, a global review and theoretical analysis of the stress pattern in the upper crust (Rummel 1986) cited the Cornish results as an extreme case of high deviatoric stress. It is thus not surprising, that all three fault plane mechanism solutions available from the United Kingdom demonstrate strike-slip motion within the middle–upper crust (0–20 km). These earthquakes, and recent stress measurements (Evans 1987), confirm a high shear stress regime for Britain.

The breakout hypothesis

Babcock (1978) introduced the term breakouts following a study in which he examined the relationship between borehole elongations and in situ stress regimes. Although Babcock (1978) was somewhat sceptical of the relationship between the orientation of breakouts and in situ stress, Bell & Gough (1979) argued that boreholes drilled through rocks bearing unequal horizontal stresses could concentrate the stresses so as to produce sub-surface breakouts parallel to the minimum stress direction. In a subsequent paper, Gough & Bell (1981) suggested that breakouts first propagate in the borehole very soon after the passage of the drill bit and continue to develop with time both along the axis of the borehole and into the formation. In a series of subsequent studies by Bell & Gough and their co-workers, it became increasingly clear that breakouts are caused by localized compressive shear fracturing of the borehole walls in a direction normal to that of the maximum horizontal principal stress (Podrouzek & Bell 1985).

The application of breakout studies has expanded rapidly covering topics from tectonics of the oceanic crust (Newmark *et al.* 1984) to geothermal energy (Plumb & Hickman 1985).

A comprehensive test of the breakout hypothesis in a single USGS – Schlumberger borehole experiment demonstrated successfully that dipmeter-derived borehole elongations can be used to infer directions of minimum horizontal in situ stress and that these directions are in excellent agreement with those determined independently from borehole televiewer and hydraulic fracturing tests (Plumb & Hickman 1985; Hickman *et al.* 1985; Zoback *et al.* 1985). Plumb & Cox (1987) corroborated the findings of the USGS – Schlumberger experiment by analysing borehole elongations (breakouts) for 47 boreholes in eastern North America. The stress directions they determined were in agreement with published interpretations from hydraulic fracture testing and earthquake fault plane solutions. They found that breakouts are systematically aligned perpendicular to the direction of hydraulic fractures, NE-striking natural fractures and centreline fractures (near-vertical, tensile fractures which form during coring). They further concluded that the uniform breakout pattern is not caused by boreholes intersecting a set of NW-striking natural fractures but, is caused by stress-induced borehole failure, so confirming the work of Hickman *et al.* (1985).

Experimental and theoretical studies

With the advent of the borehole televiewer it is possible to determine not only the depth and azimuth of the breakouts but also their span and shape. This has led to a quest to develop theoretical relationships between breakout dimensions and principal stress magnitudes.

Bell & Gough (1979) presented a theoretical hypothesis based upon the Kirsch (1898) equations to support their thesis that breakouts are caused by the symmetrical spalling of regions of the borehole wall where the stress concentrations exceed the shear strength of the rock. Gough & Bell (1981) and Bell & Gough (1982) predicted that the localized stress concentrations cause a region of failure, triangular in cross section and enclosed by flat conjugate shear planes. Extension of the theoretical analysis of breakout formation mechanisms (Zoback *et al.* 1985) has shown that many breakouts can be rather broad and flat-bottomed, unlike the triangular or 'dog-ear' shapes predicted by the work of Gough & Bell (1981).

Paillet & Kim (1987) found that breakouts appear to have little effect on refracted compressional and shear waveforms, and that velocities calculated from the waveform arrivals also showed no obvious changes. On the basis of their observations, they concluded that the effects of spalling are confined to the immediate vicinity of the borehole wall and do not extend to depths equivalent to a significant portion of one acoustic wavelength (about 25 cm). The validity of these observations is, however, debatable, since the exact path of the seismic wave is unknown with regard to the restricted zone of fracturing caused by the breakouts.

Zoback *et al.* (1985) used a plane strain model to calculate the radial, tangential and shear stress. These were then compared with the stresses required to cause shear failure using the linear Mohr–Coulomb failure criterion. Near the borehole, the stress concentration produces markedly curved potential shear failure surfaces brought about by rotation of the azimuths of the principal horizontal stresses near the free surface of the cylindrical borehole (Fig. 2). Zoback *et al*'s theory was able to predict the general shapes of observed breakouts and demonstrated that, as the stress ratio increases, the depth, though not necessarily the width, of the breakouts will increase. It was also demonstrated that the breakout shape is strongly influenced by the difference between the fluid pressure in the borehole and that in the formation. A high fluid pressure in the borehole diminishes substantially the size of the breakouts while a lower pressure promotes breakout

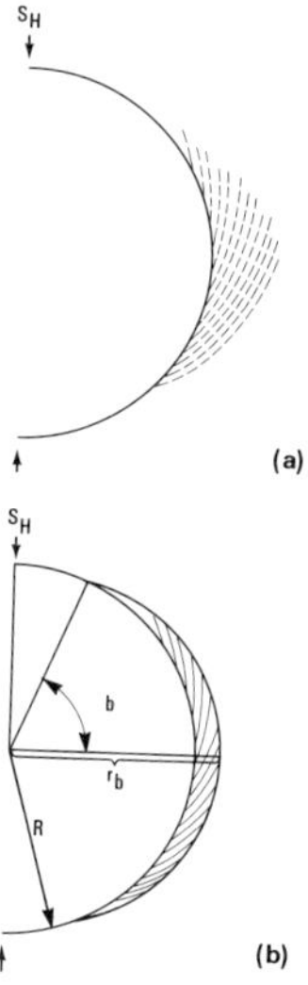

Fig. 2. (a) Orientation of potential shear failure surfaces adjacent to a borehole wall. (b) Areas in which failure to expected (after Zoback *et al.* 1985).

development. This influence is due to the change in normal stress on failure planes near the borehole wall. The cohesive strength of the rock was also shown to be important, no breakouts developing in rocks with high cohesive strength under certain conditions of differential stress and frictional sliding (Zoback *et al.* 1985). However, for rocks with low cohesive strength the breakouts would be so large as to nearly extend around the borehole. The latter situation could appear to be similar to that often described as a washout (Cox 1983).

One of the limitations of the theory described by Zoback *et al.* (1985), is that it can only accommodate the formation of breakouts in initially cylindrical boreholes (Detournay & Roegiers 1986). Although the theory explains satisfactorily the broad flat-bottomed breakouts, it is not able to explain the deeper V-shaped breakouts which are very likely influenced by anelastic deformation during failure and time-dependent effects related to subcritical fracture growth.

If the maximum to minimum stress ratio is greater than three it is postulated by Blümling *et al.* (1983) that compressional stress at the borehole wall in the maximum stress direction is reduced, and may even become tensile. Therefore, the drilling of a circular borehole may cause tensile fracturing parallel to the maximum stress direction in addition to breakout elongations parallel to the minimum stress direction. The mechanism of tensile failure is a possible explanation for the centre-line fracturing observed in cores (Plumb & Cox 1987). Furthermore, Blümling *et al.* (1983) suggested that the determination of the horizontal compression axis from breakouts, in conjunction with fault plane solutions, provides a means not only of determining the internal friction of the rock at depth but also of defining the active fault plane.

During laboratory tests Haimson & Herrick (1986) observed four stages of breakout. Firstly, short vertical hairline cracks appeared which resulted from grain splitting. The hairline cracks occurred at the position where the minimum stress axis intersects the borehole wall, but the cracks themselves were aligned roughly with the maximum stress axis. Secondly, the segment defined by the hairline cracks remained approximately constant, but as the magnitude of the maximum stress increased the cracks tended to link up to form long winding cracks. Thirdly, at higher values of maximum stress magnitude, thin slabs of rock began to spall away from the borehole wall while others were detached from behind. Finally, a maximum stress magnitude was reached at which spalling ceased and the breakout appeared to stabilize. The experiments confirmed that when the state of in situ stress is the only controlling factor, breakouts concentrate along two diametrically opposed arcs of the borehole cross section and the centres of the arcs are aligned with the direction of minimum stress. Confirmation of the third spalling stage is known from the recovery of two sandstone cavings, from a depth of about 3000 m, which depicted both the original concave borehole wall and a typical convex compressive shear failure surface (Klein & Barr 1986).

The most important result from the experimental work of Haimson & Herrick (1986) is the demonstration of the clear correlation between breakout dimensions and in situ stress magnitude. Theoretical curves, derived from the work of Zoback *et al.* (1985), which relate the depth and span of the breakouts to the minimum stress magnitude, show good qualitative agreement with the experimental results. However, the quantitative agreement is not so good, implying that refinements to the theoretical model are still needed.

Breakout recognition criteria

The four arm dipmeter measures the strike and dip of bedding planes by cross-correlation of electrical microresistivity traces recorded by each of four orthogonal pads pressed against the borehole wall. One of the pads is magnetically oriented (usually referred to as pad 1) and the two opposite pairs of pads (pads 1–3 and 2–4) provide two independent caliper measurements of the borehole diameter. The tool is also equipped to measure the angle of deviation from vertical of the borehole. As the dipmeter tool is winched up on the wireline it rotates and when one pair of the caliper arms encounters an elongation, the pads tend to 'lock' into the elongation and rotation ceases.

Interpretations of characteristic dipmeter caliper logs from out-of-gauge boreholes are described in terms of breakouts, washouts or key seats (Fig. 3). Plumb & Hickman (1985) extended breakout recognition criteria devised by Fordjor *et al.* (1983) as follows:

(1) the tool rotation stops in the zone of elongation;
(2) the difference between the two calipers is greater than 0.6cm (0.24 inches);
(3) the smaller of the caliper readings is close to bit size, or if the smaller caliper reading is greater than bit size it should exhibit less variation than the larger caliper;

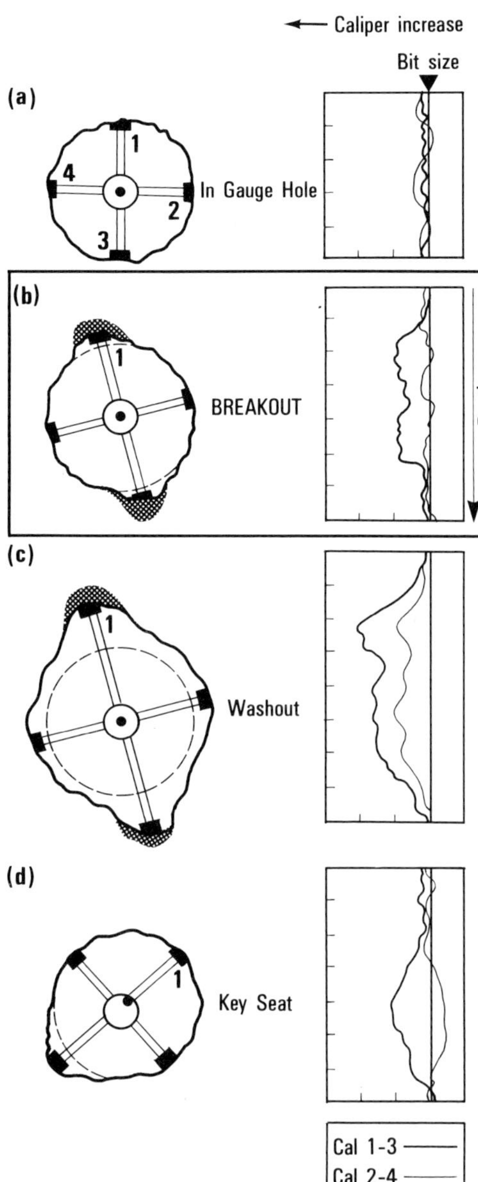

Fig. 3. Examples of dipmeter caliper logs and common interpretations of the borehole geometry (after Plumb & Hickman 1985).

(4) the elongation zone length is greater than 30 cm (11.81 inches);
(5) the direction of elongation should not consistently coincide with the azimuth of the high side of the borehole when the hole deviates from vertical.

Conditions (2) and (4) were introduced in recognition of the limitations imposed by the physical size of the dipmeter tool. On the other hand condition (5) takes account of the fact that as a borehole becomes increasingly deviated from the vertical, drill-pipe wear of the formation causes preferential elongation (key seating) which, combined with the weight of the tool, inhibits tool rotation.

Borehole breakouts in the United Kingdom and methods of analysis

The study has concentrated on two aspects, a critical assessment of the breakout recognition criteria and methods of analysis, and the analysis of data from as many UK onshore boreholes as resources allowed. A statistical approach is adopted where data are analysed in detail from each borehole, without examining the possible relationships between stratigraphy, lithology, fracturing, faulting or other information which may be available for each borehole or each region.

It is felt that the available borehole breakout recognition criteria are somewhat subjective and that their application would result in the rejection of valid data. For example, discontinuous breakouts of the type depicted by Paillet & Kim (1987) from televiewer records would be unlikely to cause the dipmeter tool to cease rotation. Similarly, poorly developed breakouts in rocks of high cohesive strength could not only inhibit the cessation of tool rotation but could also be rejected on the grounds that the difference between the two calipers is not great enough.

To minimize the likelihood of asymmetric elongations due to drill pipe wear (key seats) being confused with breakouts, a borehole deviation upper limit of 10° was set for all the boreholes analysed. A sensitivity analysis was carried out on several boreholes which deviated from less than 1° to greater than 30° and it was found that the 10° limit was well within the range where the effects of the deviation could no longer be detected. The value of 10° is in concurrence with that set by Klein & Barr (1986), although Plumb & Cox (1987) set a limit of only 5°.

In a borehole with a predominant elongation in a particular direction, the tool rotation will result in the pad 1–3 caliper becoming progressively larger and then smaller, while the pad 2–4 caliper will become correspondingly smaller and then larger. Thus, the difference between the two calipers, referred to as the

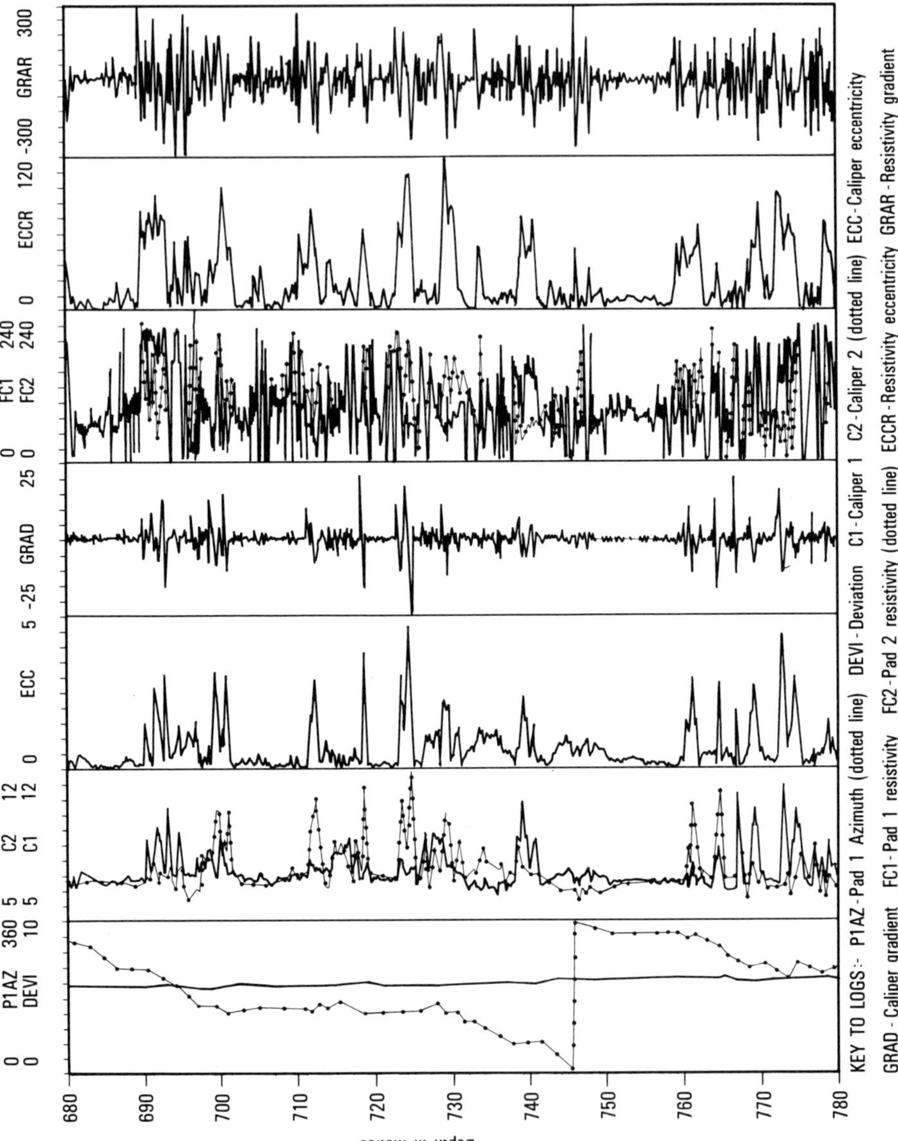

Fig. 4. Section of caliper and resistivity logs from Hobgoblin borehole with associated derived eccentricity and gradient functions.

caliper eccentricity, will oscillate between positive and negative values. Since only the magnitude of the eccentricity is of interest, a computer program is set to assign the larger of the two caliper values to caliper 1 irrespective of whether it is actually pad 1–3 or pad 2–4 caliper. Similarly, as the tool rotates from 0° to 180° the caliper values measured by the pad 1–3 caliper will mimic the values measured by the same caliper as the tool rotates from 180° to 360°. Since the borehole must be assumed to be symmetrical, it is only necessary to consider azimuth values from 0° to 180°. All values from 180° to 360° can be assigned to correspond to 0° to 180°.

The Hobgoblin borehole drilled near Stoke-on-Trent by British Coal is used as an example of the concepts developed as part of the analytical methods. In the borehole there are clear differences between the overlain caliper logs which may be attributed to breakouts (Fig. 4). The azimuth of pad-1 data for Hobgoblin borehole shows a main trend between about 45° and 65° (Fig. 5a). A cross plot of eccentricity against azimuth shows that the highest eccentricity values occur mainly between 40°and 80° (Fig. 5b).

An additional method of presenting the data is to produce a cross sectional view of the borehole by 'stacking' all the caliper information with depth onto one diagram (Fig. 5c). The same 40° to 80° azimuthal data grouping can be seen along with a clearly defined orthogonal grouping which exhibits much smaller caliper values. The most illustrative presentation of azimuthal frequency is the rose diagram (Fig. 6a). This is essentially an alternative to the frequency histogram but it allows a much clearer visual presentation of the azimuthal distribution of the data. Of the various methods of data presentation the rose diagram was chosen as the most suitable for the final presentation of results.

During the initial stages of developing analysis concepts it was recognized that classical breakout occurrences (Fig. 3) usually exhibit a sharp rise in the eccentricity value at the onset of the breakout interval, followed by an equally sharp fall at the end of the breakout interval. However, in some boreholes there can be long depth intervals where there is a continuous and significant caliper difference but with no clear preferential orientation. This behaviour can be seen to a limited degree between 743 and 750 m in the Hobgoblin borehole (Fig. 4). A measure of the rate of change of the eccentricity with depth, referred to as the caliper gradient, should be able to differentiate between these different

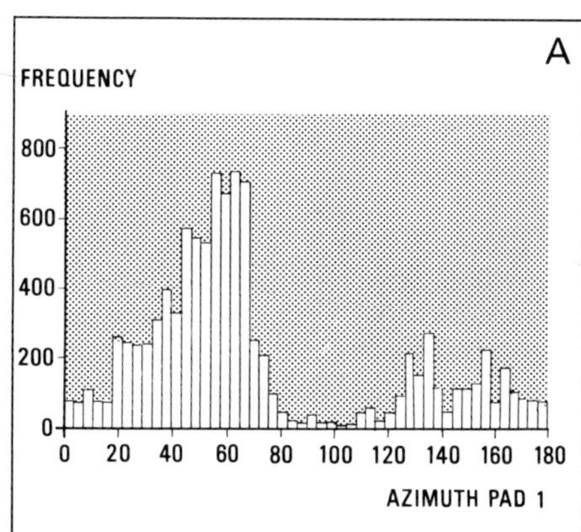

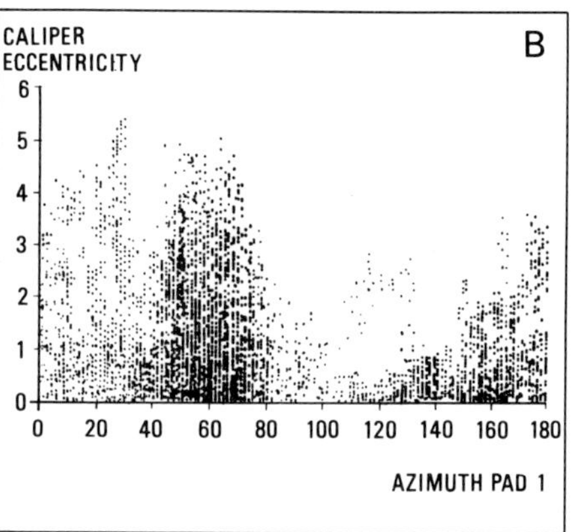

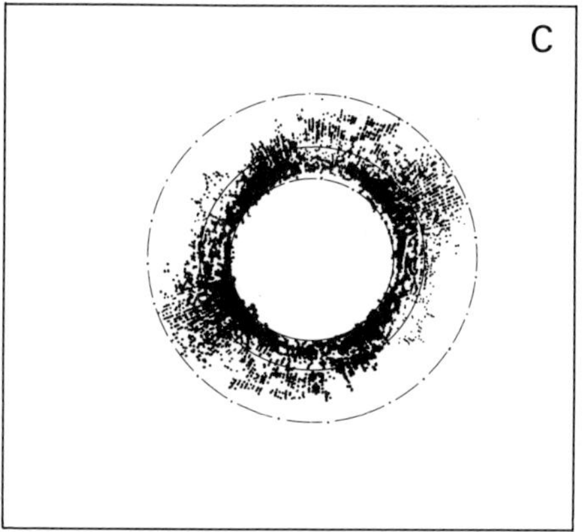

HOBGOBLIN BOREHOLE

Depth range from 500 to 1300 metres

Number of data points: 9843

Fig. 5. (a) Frequency distribution of Pad 1 Azimuth. (b) Crossplot of Pad 1 Azimuth versus caliper eccentricity. (c) Cross section of borehole from caliper data. (Dashed circles represent 6, 8.8 and 12.5 inch diameter boreholes)

types of eccentricity occurrence by filtering out such depth intervals and highlighting the classical type of breakout occurrence (Fig. 4).

The rose diagram presented in Fig. 6a does not take account of the magnitude of either eccentricity or gradient but simply represents the number of caliper 1 (i.e. larger caliper) values within each azimuthal class interval. There are several algorithms which can be adopted to utilize the eccentricity/gradient magnitudes. One method is to apply a cut-off function to remove all the low eccentricity/gradient data (cf. Plumb & Hickman 1985), the main weakness of which is that a subjective decision is needed to identify which cut-off function to apply. The preferred method of data treatment is to weight each azimuthal data point according to the magnitude of the eccentricity or gradient. The simple approach is to choose a linear weighting function whereby the sum of the eccentricity/gradient values within each azimuthal class interval is plotted (Figs 6b and 6c). Comparative studies show that this method produces a similar result to that achieved by the cut-off function, without recourse to some arbitrary choice of cut-off value.

The linear weighting function gives equal prominence to both the very large number of low eccentricity/gradient values and to the much smaller number of higher values. For some boreholes, where there are very few high eccentricity/gradient breakouts, it is sometimes difficult to decide upon a prominent breakout direction. It is also difficult for the computer to locate unequivocally a maximum breakout trend. To find the orientation of minimum stress, the computer program locates the maximum breakout trend and computes the mean azimuthal angle within a $\pm 25°$ window about the trend.

Although the mean breakout orientations have been derived from linearly weighted data, this approach, as applied here, is not statistically rigorous in that the eccentricity/gradient values are not normalised to dimensionless quantities. Therefore, the minimum stress orientations derived from the eccentricity and gradient data, and more particularly their associated errors, are not strictly comparable. Fitting a Gaussian distribution to the data sets would enable a normalized weighting function to be devised based upon the standard deviation about a dimensionless mean. Direct comparability could then be established between the eccentricity and gradient results and between the caliper and resistivity results.

Despite these reservations the mean orientations derived from the eccentricity data and from the gradient are usually very close. Where there is a significant difference it is substantial, and often there are two orientations which are orthogonal, or nearly so.

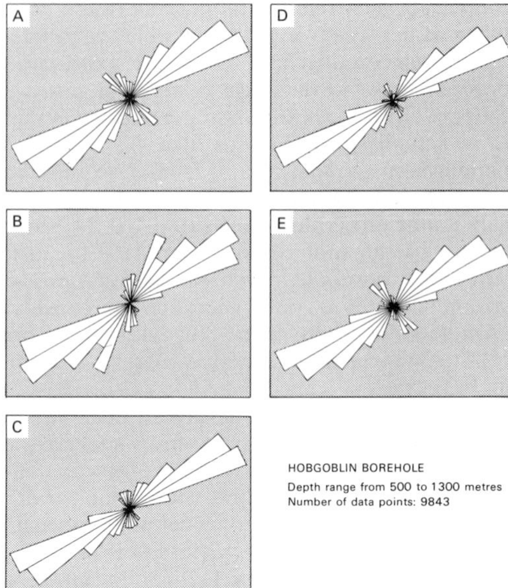

Fig. 6. (a) Rose diagram of caliper 1 direction (where caliper 1 is larger caliper). (b) Rose diagram of linearly weighted caliper eccentricity. (c) Rose diagram of linearly weighted caliper gradient. (d) Rose diagram of linearly weighted resistivity eccentricity. (e) Rose diagram of linearly weighted resistivity gradient.

Stress orientations from resistivity anomalies

Zoback *et al.* (1985) noted that stress concentrations near the borehole result in curved potential shear failure surfaces, and that breakouts occur when the surfaces spall away from the borehole wall (Fig. 2). Microfractures associated with borehole failure result in a localized enhanced formation porosity which is preferentially invaded by drilling fluid (shaded areas on Fig. 3). The localized invasion produces anomalously low electrical microresistivity readings on the dipmeter pads as they traverse the minimum stress orientation. The resistivity information recorded by the dipmeter pads can therefore be used to determine minimum stress orientations in much the same way as the caliper differences. Such an azimuthal correlation between caliper eccentricity and resistivity anomalies confirmed breakout orientations in the Auburn borehole (Plumb & Hickman 1985).

At any given depth, differences are to be expected between the resistivity pads by virtue of the strike and dip of bedding features. Ideally, the effects of the strike and dip should be taken into account before the stress-induced effects are evaluated. However, this would significantly increase the complexity of the computing so a simple averaging is used instead. Thus, the resistivity values recorded by each opposite pair of pads are averaged to give resistivity 1 (pads 1 and 3) and resistivity 2 (pads 2 and 4). The relative values of resistivity 1 and resistivity 2 are directly analogous to caliper 1 and caliper 2 and the subsequent data analysis can therefore be treated in the same way. A direct comparison of the results of the eccentricity and gradient for caliper and resistivity logs show a broad correspondence (Fig. 4). When the resistivity data are used to plot linear weighted eccentricity and gradient rose diagrams (Figs 6d and 6e), the overall characteristics and the mean derived orientations of minimum stress are again very similar to those exhibited by the caliper data.

Results and discussion

Insufficient resistivity data have been analysed to draw any firm conclusion. However, from the nine boreholes with resistivity data there is often, though not always, a good agreement between the caliper-derived and resistivity-derived orientations. The deviation about the mean for the resistivity data may sometimes be higher than for the caliper data, but this may be because the effects of dip and strike have not been accounted for.

The *dominant* eccentricity and gradient directions for all boreholes studied are shown at their appropriate geographical locations (Figs 7a and 7b respectively). On each of the figures is an inset representing the dominant direction of the summation of all breakout orientation data presented. The results presented in the figures have not been normalized in terms of depth or the number of data values used in the analysis, i.e. results from a 300 m borehole will appear to be of equal significance to those from a 3 km borehole.

The majority of the boreholes analysed in southern England and the Midlands give a coherent picture of minimum in situ stress with an average orientation of 54°/234° ± 11. This is in close agreement with previous stress studies (Klein & Brown 1983; Klein & Barr 1986) and also with geological indicators (Bevan & Hancock 1986). The majority of breakouts in one borehole exhibit unimodal orientation, but a significant number are bimodal and exhibit an orthogonal distribution, while others have a unimodal trend which is orthogonal to the average orientation. Orthogonal trends are particularly applicable to the four boreholes analysed from Scotland.

Bimodal orthogonal distributions have been observed in many North American boreholes. Plumb & Hickman (1985) suggested that orthogonal distributions may be due to a transition from elastic to plastic deformation and argued that a borehole drilled into rocks that deformed plastically will tend to have the largest diameter aligned with the maximum stress direction rather than the minimum stress direction. Plastic deformation has been identified in this study. High drilling-mud pressures, which are produced by using dense additives to stabilize boreholes drilled in partially consolidated rocks, not only minimize spalling in the minimum stress direction (Zoback *et al.* 1985) but may initiate localized hydraulic fracturing in the maximum stress direction, which subsequently cave and increase the borehole diameter (Bell & Babcock 1986). As an alternative mechanism, Bell & Babcock (1986) proposed that preferential mudcake build up in breakouts could result in an eccentricity aligned with the maximum stress direction. In addition, Blümling *et al.* (1983) postulated that tensile fracturing in the maximum stress direction was dependent upon the maximum to minimum horizontal stress ratio. Whatever the mechanism it seems that breakout orientations which correlate orthogonally with the regional average breakout orientation may be aligned parallel to the maximum in situ stress direction.

The Sutherland borehole in northern Scotland (northernmost data point Fig. 7a and b) exhibits a clear unimodal breakout orientation which is almost orthogonal to the regional trend and to the minimum stress orientation which would be expected from geological indicators. Clearly, a more detailed study of the available data is needed in such cases before it can be said whether this orientation represents the maximum stress direction derived from tensile fractures or a regional variation in minimum stress direction derived from breakouts.

Hyett *et al.* (1986) stated that the residual stresses, locked into the rock by previous tectonic events, are responsible for a significant component of the scatter in results experienced during most in situ stress measurement programmes. Paillett & Kim (1987) observed a systematic drift in breakout direction shown by

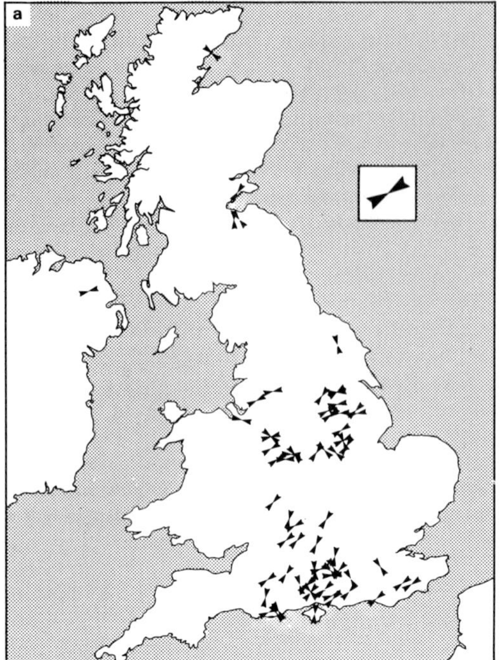

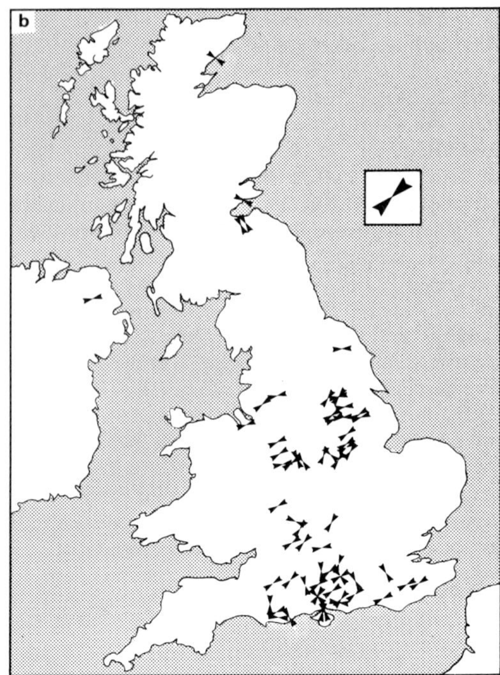

Fig. 7. Dominant breakout orientations for the United Kingdom derived from (a) eccentricity data (b) gradient data. (Inset shows the result of the summation of data from all the boreholes).

a televiewer record which could indicate a real rotation in stress orientation or, as they felt to be more likely, represent the effects of ferromagnetic minerals in the borehole wall on the orienting device. Plumb & Hickman (1985) attributed anomalous orientations in the Auburn borehole to drill pipe wear but they also affirmed the value of the resistivity anomalies to help distinguish between stress-induced and anomalous elongations. Systematic variations in the orientations of anomalous breakout trends could, on the other hand, lend support to the idea of stress decoupling at geological discontinuities (Becker *et al.* 1987; Haimson 1982).

To help interpret the regional significance of stress orientations in western Canada, breakout orientations were used to construct a horizontal stress trajectory map from which stress field rotations were inferred (Bell & Babcock 1986). Stress rotations were then correlated with subsurface basement topography. The application of similar interpretational techniques to the UK and to western Europe could help elucidate some of the detail of finer stress-field orientation. No attempt has been made to advance geological interpretations based upon the results presented here since this requires a wider-ranging integration of information pertaining to deep geological and geophysical structures.

Conclusions

It is clear from previous research that the drilling of a borehole in pre-stressed rock produces a concentration of stress near the free surface of the borehole wall. This in turn may induce shear fractures with associated spalling of material in a direction parallel to the minimum stress orientation, a borehole breakout. The fracturing of the rock also results in anomalously low formation resistivity values which can also be used to determine the minimum stress orientation.

Analytical methods are developed which limit the need for subjective judgements. However, the level of detail that can be achieved during an analysis is such that a range of results is derived for each borehole. Additional refinements of the analytical techniques (i.e. depth, lithological correlation) are proposed which will allow for a closer comparison of results derived from within and between individual boreholes.

In the United Kingdom, breakout orientations have been analysed for 81 onshore boreholes and regional distributions of inferred minimum stress orientations have been plotted. The overall picture is one of an average breakout orientation of 54°/234° ± 11°, which is similar to the previous few measurements of minimum stress directions in Britain and western Europe. In selected areas the results can be a little more variable. In the Midlands there is a bimodal trend at 71°/251° ± 10° and at approximately 45°/225°. In Southern England the orientation is at 52°/232° ± 10°, while two sub-regions of southern England are 45°/225° ± 9° (Hampshire Basin) and 61°/241° ± 8° (Wytch Farm). Individual boreholes exhibit variations about the average trends, some of which are orthogonal and could be explained as indicating the maximum stress orientation. There are also indications of stress rotations or decoupling horizons, however, the fine details encompassed within the results have not yet been fully exploited.

The research demonstrates that borehole breakouts can be used successfully to construct coherent distributions of regional stress orientations. The raw information suitable for breakout analysis has been, and continues to be, widely collected for other purposes in the hydrocarbon and minerals industries. It is clear therefore, that the potential to evaluate detailed regional and local stress systems is considerable.

The authors gratefully acknowledge the Commission of the European Communities (Contract EN3G. 0054.UK (H)) for funding this research. Thanks are given to Mr T. J. Fowkes, Mr C. P. Royles and Mr A. D. Evans who helped with the analysis. We also wish to express our gratitude to British Coal, British Petroleum, Carless Exploration, Taylor Woodrow Energy, and Ultramar Exploration for providing data. This paper is published with the permission of the Director, British Geological Survey (N. E. R. C.).

References

BABCOCK, E. A. 1978. Measurement of subsurface fractures from dipmeter logs. *American Association of Petroleum Geologists Bulletin*, **62**, 1111–1126.

BATCHELOR, A. S. & PINE, R. J. 1986. The results of in situ stress determinations by seven methods to depths of 2500 m in the Carnmenellis granite. *In*: STEPHANSSON, O. (ed). *Proceedings of the International Symposium on Rock Stress and Rock Stress Measurements*. Stockholm, 467–478.

BECKER, A., BLÜMLING, P. & MÜLLER, W. H. 1984. Rezentes Spannungsfeld in der zentralen Nordschweiz. Nationale Genossenschaft Für die Lagerung radioaktives Abfälle, Technischer Bercht, 84–37.

——, —— & ——, 1987. Recent stress field and neotectonics in the Eastern Jura Mountains, Switzerland. *Tectonophysics*, **135**, 277–288.

BELL, J. S. & BABCOCK, E. A. 1986. The stress regime of the Western Canadian Basin and implications for hydrocarbon production. *Bulletin of Canadian Petroleum Geologists*, **34**, 3, 364–378.

—— and GOUGH, D. I. 1979. Northeast–Southwest compressive stress in Alberta: Evidence from oil wells. *Earth and Planetary Science Letters*, **45**, 475–482.

—— & —— 1982. The use of borehole breakouts in the study of crustal stress. *United States Geological Survey Open-file Report,* **82–1075**, 201–209.

BEVAN, T. G. & HANCOCK, P. L. 1986. A late Cenozoic regional mesofracture system in Southern England and Northern France. *Journal of the Geological Society of London*, **143**, 355–362.

BLÜMLING, P., FUCHS, K. & SCHNEIDER, T. 1983. Orientation of the stress field from breakouts in a crystalline well in a seismically active area. *Physics and Earth Planetary Interiors*, **33**, 250–254.

BOTT, M. H. P. & KUSZNIR, N. J. 1984. The origin of tectonic stress in the lithosphere. *Tectonophysics*, **105**, 1–13.

BRERETON, N. R. & EVANS, C. J. 1987. Rock stress orientations in the United Kingdom from borehole breakouts. *Report of British Geological Survey*, RG 87/14.

COOLING, C. M. & HUDSON, J. A. 1986. The importance of in situ rock stress in repository design. *In*: STEPHANSSON, O. (ed). *Proceedings of the International Symposium on Rock Stress and Rock Stress Measurements*. Stockholm, 647–656.

COX, J. W. 1983. Long axis orientation in elongated boreholes and its correlation with rock stress data. *SPWLA 24th Annual Logging Symposium, Calgary, Canada*, paper J.

DETOURNAY, E. & ROEGIERS, T. C. 1986. Comment on 'Well bore breakouts and in situ stress' by Zoback, Moos, Mastin and Anderson. *Journal of Geophysical Research*, **91**, 14161–14162.

DRAXLER, J. K. & EDWARDS, D. P. 1984. Evaluation procedures in the Carboniferous of northern Europe. *Transactions 9th International Formation Evaluation Symposium, Paris*, **27**, 1–12.

EVANS, C. J. 1987. Crustal stress in the United Kingdom. Investigations of the Geothermal Potential of the UK. *Report of the British Geological Survey*.

FORDJOR, C. K., BELL, J. S. & GOUGH, D. I. 1983. Breakouts in Alberta and stress in the North American plate. *Canadian Journal of Earth Sciences*, **20**, 1445–1455.

GOUGH, D. I. & BELL, J. S. 1981. Stress orientations from oil well fractures in Alberta and Texas. *Canadian Journal of Earth Sciences*, **18**, 638–645.

HAIMSON, B. C. 1982. A comparative study of deep hydrofacturing and over-coring stress measurements at six locations with particular interest to the Nevada Test Site. *In*: ZOBACK, M. D. & HAMISON, B. C. (eds). *Proceedings of the XVII Workshop on hydraulic fracturing stress measurements, Volume 1. California*, 276–304.

—— and HERRICK, C. G. 1986. Borehole breakouts – a new tool for estimating in situ stress? *In*: STEPHANSSON, O. (ed). *Proceedings International Symposium on Rock Stress and Rock Stress Measurements*. Stockholm, 271–280.

HICKMAN, S. H., HEALEY, J. H. & ZOBACK, M. 1985. In situ stress, natural fracture distribution and borehole elongation in the Auburn geothermal well, Auburn, New York. *Journal of Geophysical Research*, **90**, 5497–5512.

HYETT, A. J., DYKE, C. G. & HUDSON, J. A. 1986. A critical examination of basic concepts associated with the existence and measurement of in situ stress. *In*: STEPHANSSON, O. (ed). *Proceedings of the International Symposium on Rock Stress and Rock Stress Measurements*. Stockholm, 387–396.

KIRSCH, G. 1898. Die Theorie der Elastizitat und die Beaurforisse der Festigkeitslehre. *Zeitschrift des Vereines Deutsches Ingenieure*. **42**, 707.

KLEIN, R. J. & BARR, M. V. 1986. Regional state of stress in western Europe. *In*: STEPHANSSON, O. (ed). *Proceedings of the International Symposium on Rock Stress and Rock Stress Measurements*. Stockholm, 33–44.

—— & BROWN, E. T. 1983. The state of stress in British rocks. *Report for the Department of the Environment DOE/RW/83.8*, 1–73.

NEWMARK, R. L., ZOBACK, M. D. & ANDERSON, R. N. 1984. Determination of in situ directions in the oceanic crust, DSDP Hole 504B. EOS, *Transactions of the American Geophysical Union* **65**, 219.

PAILLET, F. L. & KIM, K. 1987. Character and distribution of borehole breakouts and their relationship to in situ stresses in deep Columbia River Basalts. *Journal of Geophysical Research*, **92**, 6223–6234.

PLUMB, R. A. & COX, J. W. 1987. Stress distributions in eastern North America determined to 4.5 km from borehole elongation measurements. *Journal of Geophysical Research*, **92**, 4805–4816.

—— & HICKMAN, S. H. 1985. Stress induced borehole elongation: A comparison between the four arm dipmeter and the borehole televiewer in the Auburn Geothermal Well. *Journal of Geophysical Research*, **90**, 5513–5521.

PODROUZEK, A. J. & BELL, J. S. 1985. Stress orientations from well bore breakouts on the Scotian Shelf, eastern Canada. Current Research, Part B, Geological Survey of Canada, Paper 85–1B, 59–62.

RUMMEL, F. 1986. Stresses and tectonics of the Upper Continental Crust; a review. *In*: STEPHENSSON, O. (ed). *Proceedings of the International Symposium on Rock Stress and Rock Stress Measurements*. Stockholm, 177–186.

SOLOMON, S. C., RICHARDSON, R. M. & BERGMAN, E. A. 1980. Tectonic stress: models and magnitudes. *Journal of Geophysical Research*, **85**, 6086–6092.

ZOBACK, M. D. MOOS, D., MASTIN, L., & ANDERSON, R. N. 1985. Well bore breakouts and in situ stress. *Journal of Geophysical Research*, **90**, 5523–5530.

Electrical properties of basalts from DSDP Hole 504B: a key to the evaluation of pore space morphology

M. A. LOVELL[1] & P. A. PEZARD[2]

[1]*Geology Department, University of Nottingham, Nottingham NG7 2RD, UK*

[2]*Borehole Research Group, Lamont Doherty Geological Observatory, Columbia University, Palisades, New York 10964, USA*

Abstract: Archie's Law is a well known empirical relationship between electrical formation factor and porosity:

$$\mathrm{FF} = a\,\phi^{-m}.$$

As such it is widely applied in the analysis and interpretation of downhole measurements. Despite its frequent application, however, the mechanics underlying the relationship that it embodies remain largely a mystery. A brief review of some previously published, but often neglected, data is used here to describe recent results for laboratory and downhole electrical measurements in basaltic ocean crust. The results disagree with many previously published studies of crystalline rocks in that $a \neq 1.0$ and m is not always close to 2.0. Instead a increases with increasing fracturing, while m remains low and constant, though it increases slightly in the altered rock, possibly implying increased 'tortuosity'. The value of a allows discrimination between fractured pillows and massive flows and may be inversely related to the degree of crack connectivity. Cation exchange capacity measurements on core samples show there is potential for surface conduction to occur, but probably only at higher temperatures. The precise nature of this temperature dependence is unclear, as is the interplay of the surface conduction mechanism with that of electrolytic (pore fluid) conduction.

Electrical flow in porous rocks is generally dominated by electrolytic conduction through the pore fluid. The grain fabric is often considered to be non-conducting relative to the electrical conductivity of the pore fluid. Under these circumstances, Sundberg (1932) showed that the resistivity of a fully saturated cohesionless material (R_o) is directly proportional to the resistivity of the saturating fluid. (R_w).

$$R_o = \mathrm{FF} \times R_w. \tag{1}$$

Archie (1942) described the constant of proportionality (FF) as the formation factor. As such, it is an intrinsic quantity which relates to a specific formation and is independent of the salinity (and hence electrical resistivity) of the saturating fluid. Moreover, Archie (1942) noted an empirical relationship between this formation factor and the porosity (ϕ) of the rock resulting in the formulation of Archie's equation or, as is commonly stated, Archie's law:

$$R_o = a\,\phi^{-m}\,R_w \tag{2}$$

or

$$\mathrm{FF} = a\,\phi^{-m} \tag{3}$$

where $a = 1$, and m is a constant. Patnode & Wyllie (1954) showed that for a given piece of rock the formation factor is not always constant, but that for certain formations it varied with the resistivity of the saturating fluid. These deviations in sedimentary rocks have frequently been attributed to additional conductivity related to the presence of clay minerals and may be held partly responsible for transforming Archie's original observations into a much more complex but more generally applicable problem.

The presence of clay minerals in an otherwise 'clean' rock might modify the measured value of the formation factor due to the creation of conduction mechanisms additional to the free-fluid conductivity. The presence of clays thus increases the conductivity of the sample (see Worthington 1982). Consequently, the measured formation factor (or apparent formation factor $\mathrm{FF_a}$) is of lower magnitude than the intrinsic value of FF described by Sundberg (1932).

In defining the relationship between formation factor and porosity, Archie (1942, 1950) took care to emphasize that the equation only represented a trend or statistical average line through a number of measurements made on core samples. In the latter paper he also noted two particularly important features. Firstly, the physical character of a rock centres mainly around the pore size distribution. Secondly, the presence of clay minerals would no doubt be of

From HURST, A., LOVELL, M. A. & MORTON, A. C. (eds), 1990, *Geological Applications of Wireline Logs* Geological Society Special Publication No. 48, pp. 339–345

interest in future studies. These early statements point to developments which have since taken place and which have dominated investigations aimed at interpreting Archie's law.

Deviations from Archie's specific relationship were first documented by Winsauer *et al.* (1952) for a series of brine-saturated sands and a law of the form

$$FF = a\,\phi^{-m} \qquad (4)$$

with $a = 1$

This was followed by the well known Humble relationship (Winsauer *et al.* 1954) for eight sand samples ranging in porosity from 13.6 to 26.8%.

$$FF = 0.62\,\phi^{-2.15}. \qquad (5)$$

Removing the constraint ($a = 1$) from equation (4) generates a peculiar boundary condition in that an FF value of 1.0 corresponds with a porosity of 80%, rather than 100% as in equation (3). This is explained most easily by the invoking of an apparent formation factor FF_a as a consequence of the presence of a conductive element within the rock matrix. Subsequent work on various types of rocks has shown both the multiplier a and the exponent m to vary (for example see Winsauer *et al.* 1954, Carothers & Porter 1970, Lovell & Ogden 1984). The latter parameter m has frequently been described as some measure of the textural characteristics, whilst a is probably dependent on the nature of the pore structure and the conductivity of the matrix relative to that of the pore fluid.

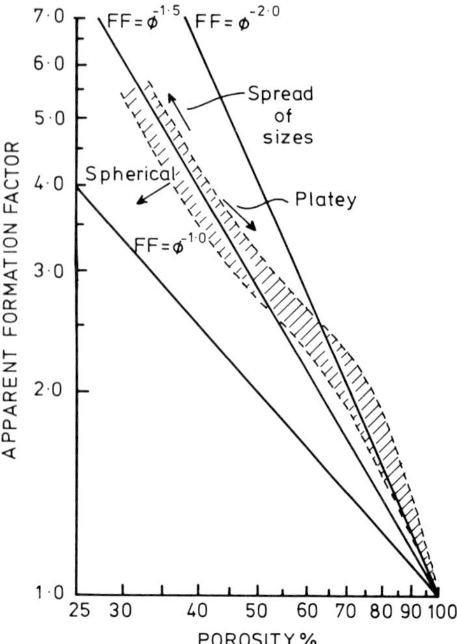

Fig. 1. Electrical formation factor — porosity relationship for clean sands (after Jackson *et al.* 1978).

Characteristics of *m* and *a*

Jackson *et al.* (1978) completed a series of excellent laboratory experiments on cohesionless beach sands of varying shape and size distributions. The results show that each of these clean samples adhere to Archie's law with a value of $a = 1$, and with m having a direct relationship with particle shape. For spherical glass ballotini m is as low as 1.4, increasing to 1.5–1.6 for rounded quartz sands and rising as high as 1.8–2.0 for platelet-shaped carbonate shell sands (Fig. 1).

Measurements on near-surface deep sea clays in the laboratory (Lovell & Ogden 1984, Lovell 1985) showed each particular sample to follow a trend similar to that described by Winsauer *et al.* (1952) with $a \neq 1$. The deviation of a from unity was interpreted to be due to the surface conduction effects of the clay particles. Lovell (1985) concluded from a study of the porosity, electrical flow and fluid flow properties of sands of varying shape and size distributions that the exponent m is a function of pore shape, and the constant a is a function of both pore size and pore shape. The two constants are inevitably inter-related; a decrease in m results in an increase in a or vice versa, for any specific rock sample.

Numerous studies of reservoir rocks and synthetic sedimentary rocks have been conducted (see for example, Carothers 1968, Carothers & Porter 1970, Gomez-Rivero 1976, Lang 1976, Worthington 1982). Whilst these have resulted in innumberable specific formulae along the lines of both Archie (1950) and Winsauer *et al.* (1952), as well as modifications thereof, there has been little fundamental progress. Helander & Cambell (1966) produced a list of eight factors influencing the value of m, covering everything from the nature of the pore space to that of the matrix. Worthington (1982) suggested that data from a wide range of sources points increasingly to Archie's law relating not to particle shape but to pore morphology, a point originally hinted at by Archie (1950) and Winsauer *et al.* (1952).

Attempts to explain the behaviour of 'non-clean' formations through the application of models have frequently been based on the concept of excess or increased conductivity of the rock beyond that attributable to the nature of the electrolyte and its distribution (Winsauer & McCardell 1953, Wyllie & Southwick 1954, Waxman & Smits 1968). More sophisticated models were introduced by Rink & Schopper (1974) and Clavier *et al.* (1977), who considered the concept of free and bound (or 'far' and 'near') water. Bussian (1983), however, suggests this may be a misconception given the present understanding of clay mineralogy. Perez-Rosales (1982) produced an interesting theory based on the concept of an internal geometry consisting of channels which actively participate in the flow of electric current and traps which are regions of stagnation. He concluded that a decreased with increasing complexity of the geometry (pore-shape) while m increased with increasing complexity (stagnation). Introducing a conducting matrix into the system would decrease m and increase a as a consequence of the reduction in the geometric complexity.

Electrical conduction in basalts

The nature of the porosity of crystalline rocks, and particularly basalts, has been discussed at length (Brace 1971, Mathews *et al.* 1983, Becker 1985). Generally there are considered to be three types of pores.

(1) Quasi — equidimensional pores and grain boundaries; these would represent a poor potential current path.
(2) Microcracks, often defined as having a length of about 100 μm and an aspect ratio of 10^{-3}–10^{-5} (Simmons & Richter 1974); these would represent a small scale network of current paths.
(3) Large cracks and voids, either plugged with alteration products or fluid-filled.

Typically, cores exhibit (1) and (2). The presence of (3) emphasizes the need for in situ measurements (see Houtz & Ewing 1976, Newmark *et al.* 1985, Pezard & Anderson 1988).

Studies of crystalline rocks from both cores and logs tended initially to follow the trend set by Archie (see, for example, Brace *et al.* 1965, Brace & Orange 1968, Cann & Von Herzen 1983, Mathews *et al.* 1983, Becker 1985) with the value of m close to, or slightly higher than, 2.0 and a equal to 1.0. It is interesting to consider the philosophy behind using a power law which is constrained by the value of the formation factor at 100% porosity when the samples under test rarely reach 20% porosity and the fresh samples are often characterized by porosities less than 5%.

A recent study by Broglia & Moos (1988) of DSDP Hole 418A used log data which has been carefully calibrated with core data. In the extremely altered upper part of the basement an equation with a = 11.5 and m = 1.85 was derived. The lower and fresher part of the drillhole gave a = 29.5 and m = 1.16. The larger value of m occurs in the altered section.

The influence of surface conductivity on the electrical resistivity of basalts has been studied and discussed by Olhoeft (1981), Drury & Hyndman (1979), and Karato (1985), among others. Analyses show that clays do not generally modify the resistivity of low-porosity samples. Whilst surface conduction at low temperature can be considered as negligible in ocean crust, it might, however, become as important as pore fluid conduction at about 80°C, and eventually become the dominant mechanism (Olhoeft 1981) at higher temperatures. This temperature dependence is obviously important in interpreting results from deep boreholes.

Pezard *et al.* (1988) report results of cation exchange capacity (CEC) measurements for a series of cores from DSDP Hole 504B. This hole penetrates 274.5 m of sediments and 1287.8 m of basaltic ocean crust. The basalts are generally considered to be equivalent to the seismically defined layers 2A, 2B and 2C of the classic ocean crust ophiolite model (see Gass & Smewing 1973, Kidd 1977). The measurements are shown in Fig. 2 and exhibit a clear zonation into those above 4370 m (Layers 2A & 2B) and below 4370 m (Layer 2C). In Layers 2A and 2B the average CEC value is 5.8 meq/100 g, whereas in Layer 2C the average is 2.3 meq/100 g. The high values in the upper part of the hole correlate with the porous nature identified by other downhole measurements (see Becker *et al.* 1988) and with the nature of alteration to the rock (see for example Alt *et al.* 1985). Intensely brecciated regions are identified by particularly high average values of CEC (11.2 meq/100 g). These values may be compared to results for the East Pacific Rise at 13°N on zero age crust which exhibit little alteration of fresh pillows (average 1.1 meq/100 g).

Electrical resistivity — porosity relationships at DSDP Hole 504B

Laboratory measurements of electrical resistivity have been made on samples, from which the ends were trimmed for CEC measurements, at ambient temperature and pressure using a

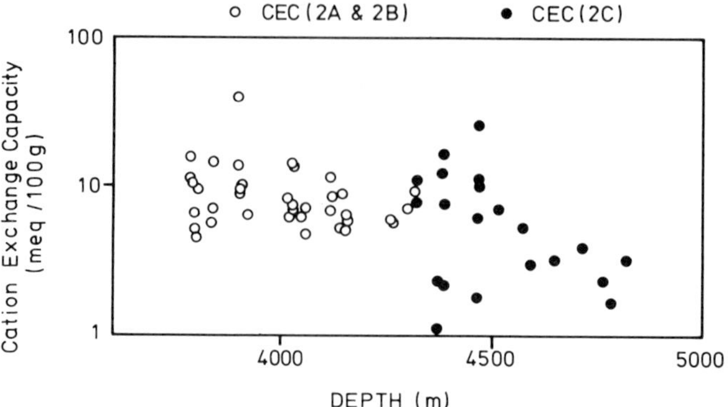

Fig. 2. Cation exchange capacity–depth plot Hole 504B showing the clear zonation between the upper altered section (Layers 2A & 2B) and the lower fresher section (Layer 2C).

10 mV, 50 Hz signal. The results (Fig. 3) show an increasing trend of electrical resistivity with depth for a given porosity. An inverse relationship between the CEC and apparent formation factor is shown in Fig. 4 for Layers 2A and 2B, suggesting that surface conduction may be important at certain temperatures though electrolytic conduction through pore space cannot be ruled out. Layer 2C results exhibit considerable scatter which may be attributed to the large variability of alteration facies at that depth in the hole.

Fig. 3 shows laboratory porosity — resistivity data for cores taken from recovered material at DSDP Hole 504B. Further plots in Fig. 5 show that data are generally constrained between two lines of similar gradients (equations (6) & (7)).

$$FF = 12.1\ \phi^{-1.02} \quad (6)$$

$$FF = 6.2\ \phi^{-1.05}. \quad (7)$$

The fracture-free samples from the massive flows plot on the upper limit of this zone (equation (6)), while the lower limit (equation (7)) is defined by the most fractured, but sealed, samples from the pillow units. A progression from the upper limit to the lower limit corresponds with a decrease in the value of a in Archie's law, while the exponent m remains approximately the same.

The results compare favourably with those of Broglia & Moos (1988) where the lower, fresher part of DSDP Hole 418A gave $a = 29.5$ and $m = 1.16$, while in the extremely altered upper part of the basement $a = 11.5$ and $m = 1.85$. In 418A m is larger in the altered section. This implies increased tortuosity since the relative contribution of the altered products to electrical conduction is likely to be small at low temperatures. In 504B a similar effect is observable although the value of m changes only slightly. These results are in general agreement with the conclusions reached in earlier studies of unconsolidated and consolidated sediments (Jackson *et al.* 1978, Lovell & Ogden 1984, Lovell 1985; see Fig. 5).

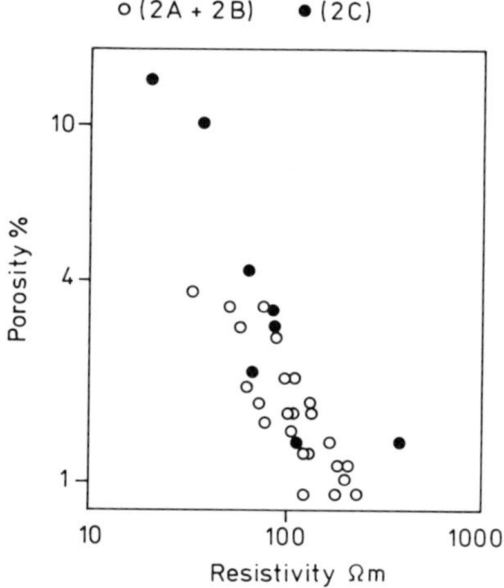

Fig. 3. Porosity–resistivity plot for Hole 504B. (open circles, Layers 2A & 2B; solid dots, Layer 2C).

The results for 504B (cores) and 418A (logs/cores) show similar behaviour of both a and m with respect to increasing alteration. Preliminary analysis of the log responses for 504B indicate that porosity and resistivity formation factor are related in situ by a very low m value, comparable to that determined from laboratory data (Broglia, pers. comm.). Since m remains almost constant from the massive flows through

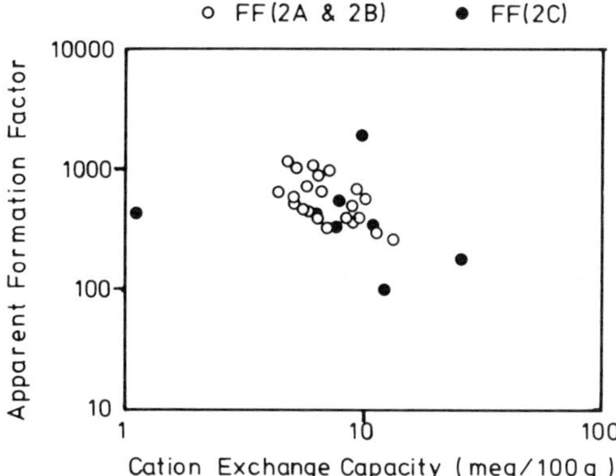

Fig. 4. Cation exchange capacity — apparent formation factor plot for Hole 504B showing the inverse relationship which suggests surface conduction may be important at certain temperatures.

to the pillows of Hole 504B, it is possible that the structure of the microscopic pore space of young mid-ocean ridge basalt does not vary with lithology. The value of a allows discrimination between the more altered pillows and the massive flows, which may reflect the degree to which the cracks are connected.

Conclusions

1. From a review of published results for unconsolidated sediments, sedimentary rocks, and models of electrical flow in porous media it is suggested that in the generalized version of Archie's law (equation 3):
   (i) the exponent m is dependent on the shape of the connected pore structure.
   (ii) the multiplier a is dependent on the shape of the connected pores and the pore size. As a consequence m seems to increase with increasing complexity of geometry, whilst a decreases.
   (iii) if there is a conducting element within the solid matrix of the rock then m is most probably decreased and a consequently increases; this is presumably a consequence of the reduction in the complexity of the pore geometry in terms of potential conduction paths.
2. Previous studies of the electrical properties of basalts using downhole data have tended to use Archie's generalised law with a equal to one, and a value of m at, or above, 2. This assumes surface conduction due to clays is negligible and that it is reasonable to constrain electrical conduction in rocks of very

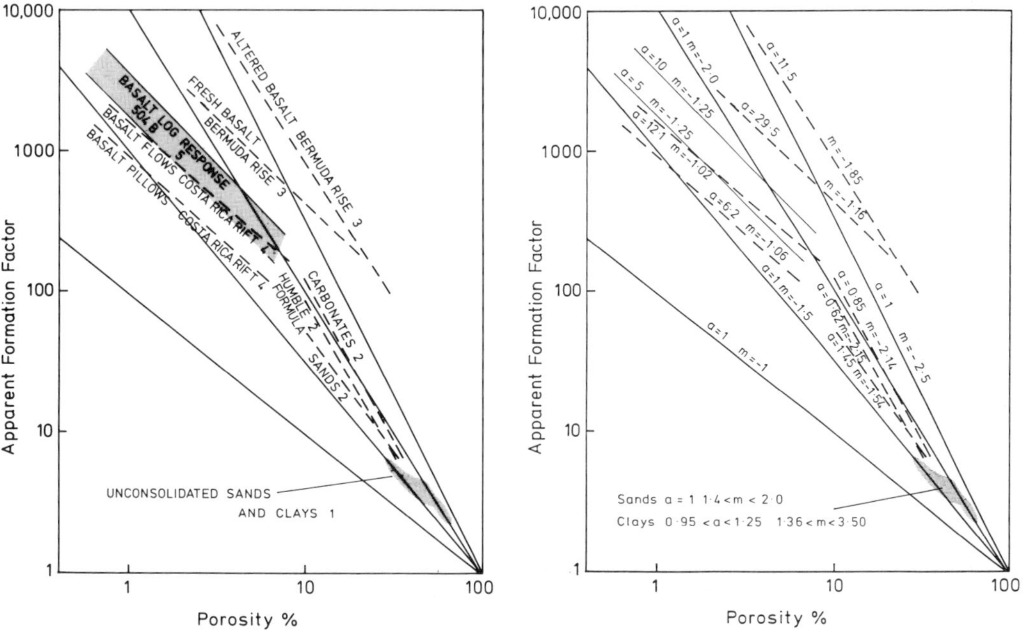

Fig. 5. Apparent formation factor — porosity plots for various rock types. Shaded zone (5) is laboratory data for Hole 504B. Data sources are: 1. Jackson *et al.* (1978), Lovell (1985); 2. Carothers (1968); 3. Broglia & Moos (1988); 4. Broglia, pers. comm.

low porosity within the boundary condition applicable to extremely high porosity rocks (i.e. FF = 1 at 100% porosity).

3. Measurements of the cation exchange capacity of basalt cores from DSDP Hole 504B show substantial surface conduction potential to exist, particularly at shallow depths where the rock is most altered. An inverse relationship is additionally observed between electrical formation factor and CEC from the upper part of the basement, although there is no distinct depth zonation. For both cases m is larger in the altered section, implying increased tortuosity. Surface conduction effects are relatively low here because of the temperatures involved.

4. Electrical formation factor — porosity results for basalt cores from DSDP Hole 504B show a to decrease with increasing fracturing, while m remains low and constant. These results agree with in situ log responses for Hole 418A, and are supported by preliminary analyses of DSDP Hole 504B log responses. It is suggested that the value of a allows discrimination between fractured pillows and massive flows and may be inversely related to the degree of crack connectivity.

5. The temperature dependences of the two conduction mechanisms (electrolytic pore conduction and electrical surface conduction) and how they vary in detail downhole is not clear.

We would like to thank the co-chiefs Keir Becker and Hitoshi Sakai for allowing us to participate in Leg 111 of the Ocean Drilling Program, and Russ Merrill, Gerry Bode, and Andy Adamson for enabling access to samples. This manuscript benefitted greatly from numerous informal discussions with Peter Jackson, Jeffrey Alt, Cristina Broglia, and Roger Anderson. We should also like to thank the two reviewers Helen Steward and Keir Becker for their careful and exhaustive attention to detail. In Nottingham Dave Jones, Josie Wilkinson and Dulcie Bryant greatly assisted in the production of this paper. This research forms part of a programme of work funded by the Natural Environment Research Council (UK) and the National Science Foundation (USA).

References

ALT, J. C., LAVERNE, C. & MUEHLENBACHS, K. 1985. Alteration of the upper oceanic crust: mineralogy and processes in Deep Sea Drilling Project Hole 504B, Leg 83. *In*: ANDERSON, R. N., HONNOREZ, J., BECKER, K., *et al.*, *Initial Reports of the Deep Sea Drilling Project*, **83**, 249–262.

ARCHIE, G. E. 1942. The electrical resistivity log as an aid in determining some reservoir characteristics. *Transactions of the American Institute of Mining and Metallurgical Engineers*, **146**, 54–62.

—— 1950. Introduction to petrophysics of reservoir rocks. *Bulletin of the American Association of Petroleum Geologists*, **34**, 943–961.

BECKER, K. 1985. Large-scale electrical resistivity and bulk porosity of the oceanic crust, DSDP Hole 504B, Costa Rica Rift, *In*: ANDERSON, R. N., HONNOREZ, J., BECKER, K., *et al. Initial Reports of Deep Sea Drilling Project*, **83**, 419–427.

—— SAKAI, H., *et al.* 1988. *Proceedings of the Ocean Drilling Program, Initial Reports (Part A)*, **111**.

BRACE, W. F. 1971. Resistivity of saturated crustal rocks to 40 spars km based on laboratory studies. *In*: HEACOCK, J. G. (ed.), *The structure and physical properties of the Earth's Crust*. American Geological Union, Washington, 243–255.

—— & ORANGE, A. S. 1968. Electrical resistivity changes in saturated rocks during fracture and frictional sliding. *Journal of Geophysical Research*, **73**, 1433–1445.

—— & MADDEN, T. R. 1965. The effect of pressure on the electrical resistivity of water-saturated crystalline rocks, *Journal of Geophysical Research*, **70**, 5669–5678.

BROGLIA, C. & MOOS, D. 1988. In situ structure and properties of 110 M, a crust from geophysical logs in DSDP Hole 418A.

SALISBURY, M. H., SCOTT, J., AUROUX., C. A., *et al. In*: *Proceedings of the Ocean Drilling Program, Initial Reports*, (Part B), **102**.

BUSSIAN, A. E. 1983. A comparison of shaly sand models. *Society of Professional Well Log Analysis 24th Annual Logging Symposium*, Paper E, 1–16.

CANN, J. R. & VON HERZEN, R. P. 1983. Downhole logs at DSDP sites 501, 504, & 505, near the Costa Rica Rift. *In*: CANN, J. R., LANGSETH, M. G., *et al. Initial Reports of the Deep Sea Drilling Project*, **69**, 281–300.

CAROTHERS, J. E. 1968. A statistical study of the formation factor relation. *The Log Analyst*, **9**. October.

—— & PORTER, C. R. 1970. Formation factor-porosity relations from well log data. *Society of Professional Well Log Analysis 11th Annual Logging Symposium*.

CLAVIER, C., COATES, G. & DUMANOIR, J. 1977. The theoretical and experimental bases for the dual water model for the interpretation of shaly sands: *52nd Annual Fall Technology Conference.*, Paper SPE 6859, Denver, 1977.

DRURY, M. J. & HYNDMAN, R. D. 1979. The electrical resistivity of oceanic basalts. *Journal of Geophysical Research*, **84**, 4537–4546.

GASS, I. G. & SMEWING, J. D. 1973. Intrusion, extrusion and metamorphism at constructive margins: evidence from the Troodos Massif, Cyprus.

Nature, **242**, 26–29.

GOMEZ-RIVERO, O. 1976. A practical method for determining cementation exponents and some other parametres as an aid in well log analysis. *The Log Analyst*, September, 8–24.

HELANDER, D. P. & CAMBELL, J. M. 1966. The effect of pore configuration, pressure and temperature on rock resistivity. *Society of Professional Well Log Analysts 7th Annual Logging Symposium*.

HOUTZ, R. & EWING, J. 1976. Upper crustal structure as a function of plate age, *Journal of Geophysical Research*, **81**, 2490–2498.

JACKSON, P. D., TAYLOR SMITH, D. & STANFORD, P. N. 1978. Resistivity-porosity-particle shape relationships for marine sands, *Geophysics*, **43**, 1250–1268.

KARATO, S. I. 1985. Physical properties of basalts from DSDP Hole 504B, Costa Rica Rift. *In*: ANDERSON, R. N., HONNOREZ, J., BECKER, K., *et al. Initial Reports of the Deep Sea Drilling Project*, **83**, 687–695.

KIDD, R. G. W. 1977. A model for the process of formation of upper oceanic crust. *Geophysical Journal of the Royal Astronomical Society*, **50**, 149–183.

LANG, W. H. 1976. Porosity-resistivity cross-plotting. *Society of Professional Well Log Analysts 13th Annual Logging Symposium*, Paper F.

LOVELL, M. A. 1985. Thermal conductivity and permeability assessment by electrical resistivity measurements in marine sediments. *Marine Geotechnology*, **6**, 205–240.

—— & OGDEN, P. 1984. Remote assessment of permeability/thermal diffusivity of consolidated clay sediments. *Report to the European Atomic Energy Commission*, Brussels:EUR-series, No. EN9206.

MATHEWS, M., SALISBURY, M. H. & HYNDMAN, R. D. 1983. Basement logging on the Mid-Atlantic Ridge, DSDP Hole 395A. *In*: HYNDMAN, R. D., SALISBURY, M. H., *et al. Initial Reports of the Deep Sea Drilling Project*, **78**, 717–730.

NEWMARK, R. L., ANDERSON, R. N., MOOS, D. & ZOBACK, M. D. 1985. Sonic and ultrasonic logging of Hole 504B and its implications for the structure, porosity, and stress regime of the upper 1 km of the oceanic crust. *In:* ANDERSON, R. N., HONNOREZ, J., BECKER, K., *et al. Initial Reports of the Deep Sea Drilling Project,* **83**, 479–510.

OLHOEFT, G. R. 1981. Electrical properties of rocks. *In*: TOULOUKIAN, Y. S., JUDD, W. R. & ROY, R. F. (eds). *Physical properties of rocks and minerals*, McGraw-Hill, New York.

PATNODE, H. W. & WYLLIE, M. R. J. 1950. The presence of conductive solids in reservoir rock as a factor in electric log interpretation. *Transactions of the American Institute of Mining and Metallurgical Engineers*, **189**, 47–52.

PEREZ-ROSALES, C. 1982. On the relationship between formation resistivity factor and porosity. *Petroleum Engineering Journal of the American Institute of Mining and Metallurgical Engineers*, 531–536.

PEZARD, P. A. & ANDERSON, R. N. 1988. The physical structure of the ocean crust in Hole 504B; relation between morphology and alteration. *In*: BECKER, K., SAKAI, H. *et al.* 1988. *Proceedings of the Ocean Drilling Program, Initial Reports* (Part B), **111**.

—— HOWARD, J. J. & LOVELL, M. A. 1988. The influence of clays on the electrical conductivity of basalts from Hole 504B and changes in the structure of the pore geometry due to hydrothermal alteration of the oceanic crust. *In*: BECKER, K., SAKAI, H., *et al. Proceedings of the Ocean Drilling Program Initial Report*, (Part B) **111**.

RINK, M. & SCHOPPER, J. R. 1974. Interface conductivity and its implications to electric logging. *Society of Professional Well Log Analysis 15th Annual Logging Symposium Transactions*, Paper J.

SIMMONS, G. & RICHTER, D. 1974. Microcracks in Rocks. *In*: FRANCHETEAU, J. (ed). Processes at mid-ocean ridges. *Tectonophysics*, **55**

SUNDBERG, K. 1932. Effect of impregnating waters on electrical conductivity of soils and rocks. *Transactions of the American Institute of Mining and Metallurgical Engineers*, **79**, 367–391.

WAXMAN, M. H. & SMITS, L. J. M. 1968. Electrical conductivities in oil-bearing shaly sands. *Transactions of the Society of Petroleum Engineers*, **243**, 107–122.

WINSAUER, W. O. & MCCARDELL, W. M. 1953. Ionic double layer conductivity in reservoir rocks. *Petroleum Transactions of the American Institute of Mining and Metallurgical Engineers*, **198**, 129–134.

WINSAUER, W. O., PERKINS, F. M. Jr. & BRANNON, H. R. Jr. 1954. Interrelation of resistivity and potential of shaly reservoir rock. *Transactions of the American Institute of Mining and Metallurgical Engineers,* **201**, 176–183.

—— SHEARIN, A. M., MASSON, P. H. & WILLIAMS, M. 1952. Resistivities of brine saturated sands in relation to pore geometry. *Bulletin of the American Association of Petroleum Geologists*, **36**, 253–277.

WORTHINGTON, P. F. 1982. The influence of shale effects upon the electrical resistivity of reservoir rocks. *Geophysical Prospecting*, **30**, 673–687.

WYLLIE, M. R. J. & SOUTHWICK, P. F. 1954. An experimental investigation of the S. P. and resistivity phenomena in dirty sands. *Transactions of the American Institute of Mining and Metallurgical Engineers*, **201**, 43–56.

An assessment of the use of Christensen's equation for the prediction of shear wave velocity and engineering parameters

D. C. ENTWISLE[1] & D. M. McCANN[2]

[1] *Engineering Geology Research Group, British Geological Survey, Keyworth, Nottingham, NG12 5GG, UK*

[2] *Regional Geophysics Research Group, British Geological Survey, Keyworth, Nottingham, NG12 5GG, UK*

Abstract: The full wave train sonic log is widely used in civil engineering for the determination of formation dynamic elastic moduli. In some instances, the shear wave amplitude is too small to time its arrival accurately. In these situations Christensen's equation can be used to predict the shear wave velocity from the compressional wave velocity and the formation density. In this paper the origins of Christensen's equation are investigated in relation to both field and laboratory studies on a wide range of natural materials. It is suggested that differences between the measured and predicted shear wave velocities are lithologically based. The equation is not valid for uncemented sediments. The dynamic elastic moduli reflect these differences. Bulk modulus is least affected by large differences in shear wave velocity. There is no relationship between the predicted and the measured values for Poisson's ratio. The use of Christensen's equation will be improved if corrections are made locally for lithology.

For the design of many civil engineering constructions, investigation boreholes provide core samples on which a wide range of tests are carried out in the laboratory to provide design parameters and other geotechnical data. These tests, however, give only an approximate representation of the actual in situ engineering properties of the rock mass, since little account can be taken of discontinuities present on anything but a small scale. The effects of most bedding plane partings, jointing and faulting cannot be included in the laboratory measurements and, thus, only the engineering properties of the intact rock are known. These results in themselves may be suspect, since it is difficult to allow for the stress relief caused to the sample when it is extracted from the ground or the disturbance to which it is subjected during the coring procedure.

Geophysical well-logging techniques examine the properties of an annulus of rock round the borehole and, thus, the influence of jointing, fracturing, weathering and even lateral lithological variations will be reflected in the properties of the rock mass determined indirectly from the geophysical logs. Of particular importance is the measurement of the seismic properties of the rock since the dynamic elastic moduli can be derived from the compressional (P) and shear wave (S) velocity values and the formation density.

The dynamic elastic moduli are used in engineering design where structures are subjected to dynamic loading effects. Examples that can be cited are the effects of earthquakes on the foundation of a sensitive installation, such as a nuclear power station, or the dynamic loading of marine oil platforms by wave motion. By carrying out standard geophysical logging of a borehole it is possible to use the dynamic elastic values in the following ways:

(a) the engineering properties can be examined to the full depth of the investigation borehole;
(b) the engineering properties of the lithological units present in the borehole can be determined and compared with static and dynamic tests carried out in the laboratory on core samples.

Determination of the dynamic elastic moduli

The compressional and shear wave velocities can be obtained from the full wave train sonic log, where use is made of the P-wave energy that is converted to S- wave energy at the borehole wall by mode conversion; this is later reconverted to P- wave energy and detected at the receiver (Fig. 1a). A typical sonic wavetrain observed at the receiver is shown in Fig. 1b.

The results are often displayed as a variable density log (Fig. 2) where the shear wave arrival is relatively easy to pick out. This, however, is not always the case since very often there is no

From HURST, A., LOVELL, M. A. & MORTON, A. C. (eds), 1990, *Geological Applications of Wireline Logs* Geological Society Special Publication No. 48, pp. 347–354

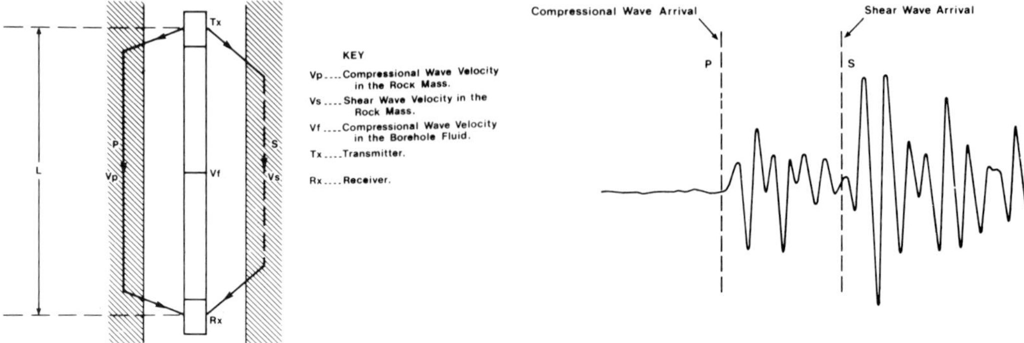

Fig. 1. (a) Full wave train sonic tool. (b) Received waveform form full wave sonic log.

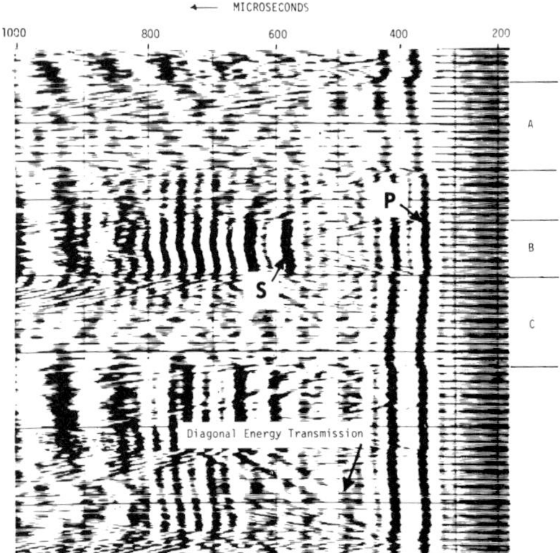

Fig. 2. Variable density log in granite (after Geye & Myung 1970).

distinct break at the onset of the shear wave in the sonic wave train or the shear wave is highly attenuated and lost in the compressional wave. In addition, shear wave arrivals are not present in the sonic wave train when the shear wave velocity is less than the compressional wave velocity in the borehole fluid. The tube wave or Stonely wave can be used to calculate the shear wave velocity following the theoretical approach of White (1965), provided his theory is extended to the higher frequencies used in the sonic log tool and the effects of velocity dispersion are included; this is considered by Cheng & Toksoz (1983).

Using the value of formation density computed from the gamma/gamma log together with the values of the compressional and shear wave velocities, the dynamic elastic moduli can be computed as follows.

(a) Bulk modulus (K)

$$K = \rho\ (V_p^2 - 4/3\ V_s^2) \qquad (1)$$

where ρ is the formation density, V_p is the compressional wave velocity and V_s is the shear wave velocity.

(b) Shear modulus (G)

$$G = \rho\ V_s^2. \qquad (2)$$

(c) Poisson's ratio (δ)

$$\delta = \frac{0.5\ (V_p/V_s)^2}{2[(V_p/V_s)^2 - 1]}. \qquad (3)$$

(d) Young's modulus (E)

$$E = 2\rho\ V_s^2\ (1 - \delta). \qquad (4)$$

Thus, provided V_p and V_s can be accurately determined from the full wave train sonic log it is possible to obtain a continuous measurement of the dynamic elastic moduli with depth using these three parameters. Early work in this field was carried out by Christensen (1964), who made an in-depth study of the performance of the full wave train sonic tool (the 3D sonic log) manufactured in the USA by the Birdwell Logging Company. The resultant log from this tool was presented as a variable density display from which it was possible to identify the compressional and shear wave arrivals. The Birdwell equipment was first used in the UK by Seismograph Services Ltd (SSL) in 1969.

Christensen (1964) considered in some detail the relationship between the acoustic parameters and the physical properties of the material surrounding the borehole. Following the work of Biot (1952) he noted, in addition to the P and S wave arrivals, later arrivals associated with the generation of a Stonely wave. This occurs at approximately the same time as the compressional wave arrival through the fluid depending on the geometry of the borehole and the properties of the rock mass surrounding the borehole so that Stonely wave is often difficult to identify. Christensen derived two empirical equations. In Christensen (1964) he posed the possibility of predicting formation density and porosity via the acoustic parameters and arrived at an equation for formation density as follows:

$$\rho = f(w) + 1.75\left(\frac{V_P^2 + V_S^2}{V_P^2 - V_S^2}\right) \exp\left[-1.20\left(\frac{V_P - V_S}{V_P + V_S}\right)\right] \quad (5)$$

where $f(w)$ ~0.4 is a function of the degree of anisotropy.

In a later paper (Christensen 1966), the application of the 3D log to the determination of the elastic moduli of rock salt is discussed, and use is made of an empirical equation to derive both V_p and V_s from the measured arrival times on the logs. This equation is quoted as

$$V_\beta = \{Z - 2F(d)\tan[\sin^{-1}(\tau_\beta/\tau_f)]\}\, T_\beta \quad (6)$$

where Z is the separation between source and receiver in the sonde, τ_f is the fluid travel time over (the distance Z), τ_β is the recorded travel time; β refers to the P wave or S wave, τ_β the difference between recorded time and the fluid time and $F(d)$ is a constant which is a function of the borehole diameter and sonde length.

This empirical equation was shown by Christensen (1966) to give a very close approximation to the formation velocities provided that the lithology does not change very rapidly and the borehole diameter is less than $L/10$ where L is the path length of the signal along the borehole wall.

A third empirical equation, attributed to Christensen, is a relationship between V_s and V_p and formation density, which appears to be related to equation (5) above:

$$V_s = V_p\{1 - [1.15(1/\rho + 1/\rho^3)/\exp(1/\rho)]\}^{3/2}. \quad (7)$$

This equation was not published explicitly by Christensen, although it was used in the Birdwell logs of the late 1960s and is now widely used for situations where the shear wave arrival cannot be identified, either because of its low signal level or because it is obscured by the water wave, the Stonely wave, or interference from acoustic noise. Again, the only references to the equation appear in unpublished internal reports from the Birdwell company and as far as the authors can ascertain Christensen did not publish this equation in the scientific literature, although it is possible that our literature searches failed to identify the one vital reference which gives a derivation of both equations (6) and (7). However, the equation for V_s has been widely used and has a direct equivalence to the Wyllie equation relating V_p to the formation porosity (Wyllie *et al.* 1956).

Assessment of Christensen's equation

Christensen's equation for V_s was evaluated by expressing equation (7) in terms of a ratio R such that

$$R = V_s/V_p = \{1 - [1.15(1/\rho + 1/\rho^3)/\exp(1/\rho)]\}^{3/2} \quad (8)$$

which relates R directly to the formation density; the theoretical curve is shown in Fig. 3.

Using data from full wave train sonic and formation density logs from various British Geological Survey (BGS) boreholes, laboratory measurements and results from the literature, a range of lithologies was assessed. The sources of data are given in Table 1.

Processing of data

The data obtained from the above sources were used to determine:

(1) the relationship between R (V_s/V_p) and formation density (equation (8)):
(2) the measured and calculated shear wave velocities.

Using equations (1)–(4) a comparison of the following parameters was made, substituting in both measured and calculated values of shear wave velocity:

(1) Bulk modulus,
(2) Poisson's ratio
(3) Young's modulus,
(4) Shear modulus.

Discussion of results

V_s/V_p against formation density

The data shown plotted in Fig. 3 exhibit considerable variation about the predicted line from Christensen's equation. The majority of the lithological units plot below the line while only chalk and quartz and a few siltstone results plot above. The mudstones are intermediate between the uncemented sediments and the theoretical line. The other lithologies all plot close to

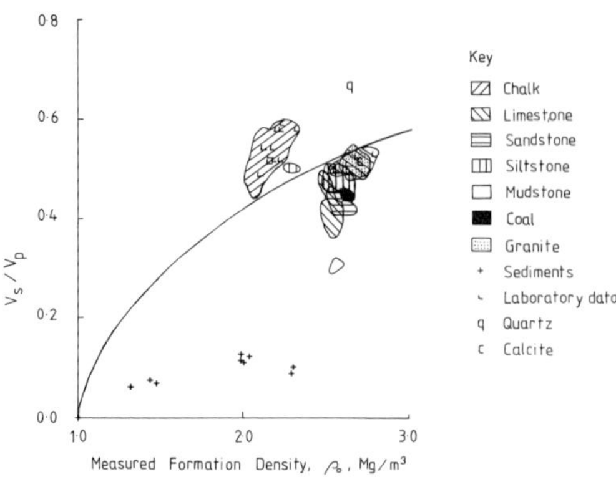

Fig. 3. Measured R (V_s/V_p) and formation density of a number of lithologies and the curve of Christensen's equation.

the theoretical line. The laboratory data are in good agreement with the field data.

The two common rock minerals, calcite and quartz, fall on either side of the predicted line; the calcite plots below the line within the grouping of the laboratory tested Dorset coast limestones.

Measured and predicted shear wave velocity

The measured and predicted (from equation (7)) values of V_s are plotted against each other (Fig. 4). In the majority of cases the predicted value is greater than the measured one, the only exception noted being chalk. The slope of V_s (predicted) against V_s (measured) for each lithology is listed in Table 2.

The values in Table 2 show that the predicted value of V_s is up to 5.5 times too high in the uncemented sediments and about 1.7 times too high in the mudstones. By contrast the granite, some of the coal samples, siltstones and limestones, are near to the 1:1 ratio.

The siltstones separate into three groups, one lower velocity which straddles the 1:1 line, the second which is just above the line and the third which plots well above the predicted line.

The sandstones from the Triassic and from the Coal Measures are within the same cluster whereas the limestones vary from the laboratory measurements of the Upper Jurassic limestones, about the 1:1 line, to the Great Oolite, well above the line.

The predicted value for a majority of the rocks vary between 0.74–1.54 times the

Table 1. *Sources of data*

Source of data	Site	Lithostratigraphic Unit
(a) BGS Field Measurements		
Borehole data using full wave train sonic and formation density log.	Harwell (Oxfordshire)	Middle Jurassic Great Oolite Limestone
		Lower Lias Limestone
		Triassic Sandstone
		Coal Measures: Siltstone
		Sandstone
		Coal
	Altnabreac (Caithness)	Strath Haladale Granite
	West Lulworth (Dorset)	Cretaceous Chalk
BGS cross hole (water saturated) (McCann *et al* 1986)	Newborough Warren (Anglesey)	Recent Sand (water saturated)
(b) Laboratory Measurements	Dorset Coast	Upper Jurassic Limestone
	Cocking (Sussex)	Cretaceous Chalk
	West Lulworth (Dorset)	Cretaceous Chalk
(c) Literature		
In situ measurements	San Diego, California (Hamilton *et al.* 1970)	Recent Sand and Clayey Silts
	Eastern England (Davies 1977)	Tertiary London Clay and Till
Reference Book	(Carmichael 1984)	Minerals

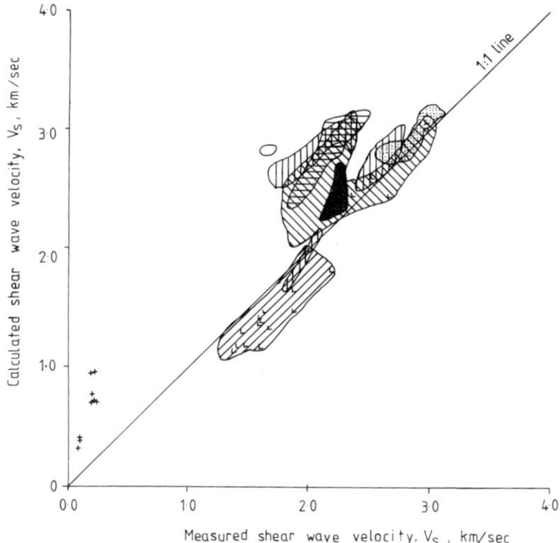

Fig. 4. Measured shear wave velocity against calculated shear wave velocity for a number of lithologies (see Fig. 3 for key).

Table 2. *Slope of predicted V_s measured V_s for each lithology*

Lithology		V_s predicted / V_s measured
Sediments:	Till (mean)	5.5
	Others	2.92 – 4.00
Mudstones		1.68 – 1.74
Siltstone		0.94–1.10; 1.23–1.54
Sandstone		1.24 – 1.42
Coal		1.09 – 1.21
Chalk: field		0.76 – 0.94
lab		0.74 – 1.00
Limestone:	Great Oolite	1.16 – 1.41
	Lower Lias	1.04 – 1.22
	Dorset coast lab.	0.93 – 1.12
Granite		0.99 – 1.12

measured value. It is assumed that Christensen must have worked on hydrocarbon source rocks for his experimental work and it is noticeable that these lithological types lie close to the theoretical line in Fig. 4.

Dynamic elastic moduli

The predicted and measured values of V_s were substituted into equations (1)–(4) to assess their effect on the dynamic elastic moduli values; the results are summarized in Table 3. Graphic representation of the predicted and measured values of bulk modulus, Poisson's ratio, Young's modulus, and shear modulus are also shown in Figs 5–8 respectively.

Good correlation is observed between the predicted and measured values of bulk modulus (Fig. 5). Young's modulus and shear modulus are similar in calculation and the distribution of the lithological clusters in Figs 7 & 8 reflects this similarity; in both cases variations in V_s have far more influence than in the case of the bulk modulus. Little correlation is observed between the predicted and measured values of Poisson's ratio where variations in V_s have a considerable effect (Fig. 6).

It is clear from Figs 5–8 that the dynamic elastic modulus for uncemented sediments cannot be predicted from a V_s calculated from Christensen's equation. Although the bulk modulus values are similar (less than a factor of two between the predicted and measured), the Young's modulus and shear wave modulus values are over-estimated by factors from 8–31. For engineering classification the V_s should not be calculated from Christensen's equation for uncemented sediments, even those which are dense and stiff such as the till.

General discussion

The spread of data presented in this paper may mean that differences between the predicted values of V_s derived from Christensen's equation and the actual measured values are lithologically based. In a majority of cases the predicted values of V_s are higher than the

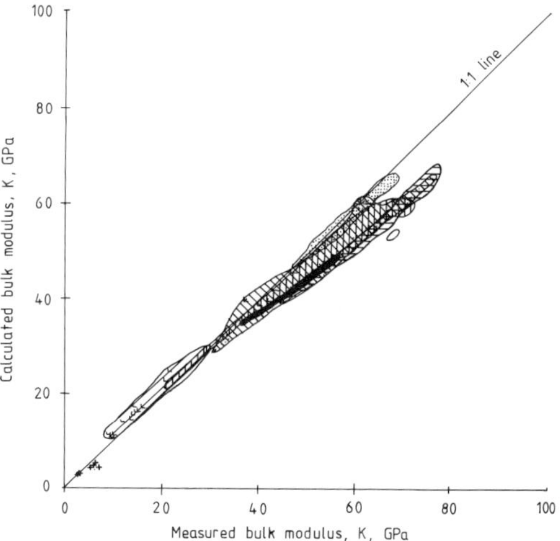

Fig. 5. A plot of bulk density, K, derived from measured and calculated shear wave velocity of a number of lithologies (see Fig. 3 for key).

Table 3. *The predicted dynamic elastic moduli/measured dynamic elastic modulae*

Lithology	Bulk modulus	Predicted/measured Poisson's ratio	Young's modulus	Shear Modulus
Sediments:				
Till	0.59	0.59	26	31
Others	0.81–0.95	0.79–0.96	7.9–17.5	8.5–24
Mudstone	0.80	0.69	2.55–2.74	2.82–3.03
Siltstone:				
i	0.82–0.88	0.67–0.80	0.91–1.20	0.90–1.24
ii	0.93–1.04	0.89–1.06	1.41–2.16	1.50–2.37
Sandstone	0.84–0.88	0.73–0.81	1.40–1.84	1.55–2.01
Coal	0.85–0.93	0.79–0.89	1.06–1.39	1.07–1.48
Chalk:				
Field	1.01–1.21	1.06–1.52	0.65–0.96	0.59–0.96
Laboratory	1.00–1.28	1.00–1.15	0.63–1.00	0.55–0.99
Limestones:				
Great Oolite	0.84–0.92	0.74–0.84	1.28–1.84	1.33–1.99
Lower Lias	0.84–0.97	0.79–0.94	1.06–1.40	1.08–1.48
Dorset Coast L	0.92–1.07	0.85–1.20	0.90–1.19	0.90–1.24
Granite	0.92–1.01	0.87–1.03	0.98–1.20	0.97–1.24

Key L = Laboratory.

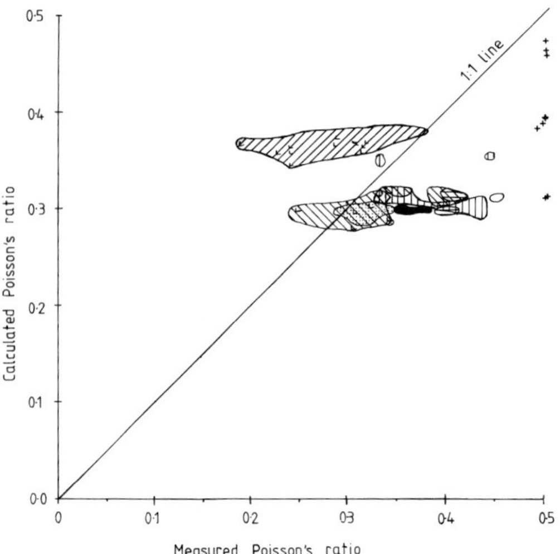

Fig. 6. A plot of Poisson's ratio derived from calculated shear wave velocity of a number of lithologies (see Fig. 3 for key).

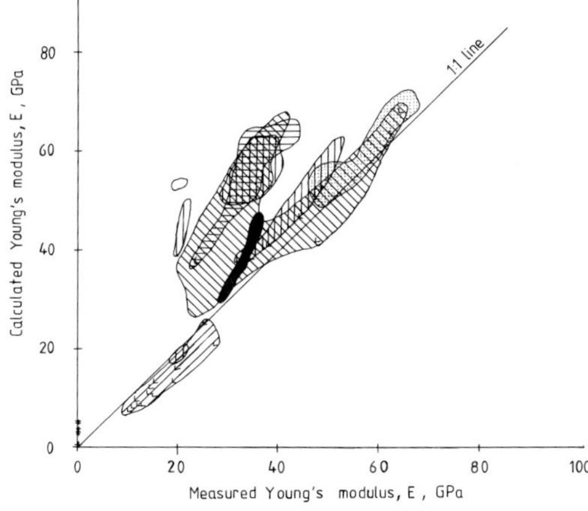

Fig. 7. A plot of Young's modulus, E, derived from measured and calculated shear velocity of a numbe of lithologies (see Fig. 3 for key).

measured ones with the exception of chalk. There is good agreement between predicted and measured V_s for rocks such as granite, and to a lesser degree, in sandstones, chalk, coal, limestones and some of the siltstones.

The lithological differences between the predicted and measured values of V_s are likely to be due to mineral variations in the rocks. This is complicated by the structure of the rocks themselves which are elastic and anisotropic to differing degrees. The materials which plot nearest the Christensen's predicted line are those which are more isotropic. The lithological basis for using Christensen's equation and calibrating from measured values is also supported by the clustering of the Triassic and Coal Measures, sandstone and limestone data. It has been shown that Christensen's equation will not accurately predict V_s for (uncemented) sediments since in most cases the predicted V_s was about four

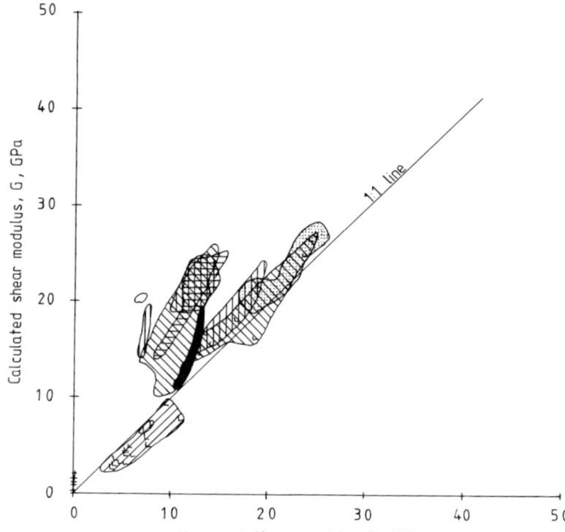

Fig. 8. A plot of shear modulus, G, derived from measured and calculated shear velocity of a number of lithologies (see Fig. 3 for key).

times higher than the measured one; for glacial tills and stiff dense clays, the predicted V_s was up to 5.5 times higher.

The laboratory results plot near the in situ test results on similar lithologies. This is particularly noticeable with the Dorset laboratory samples where the V_s (predicted)/V_s (measured) ratio is close to unity. The process of sample selection from extraction and preparation, in general, produces material which has fewer discontinuities than the rock mass. The laboratory results may be slightly biased, therefore, as they result in lower measured Poisson's ratio and higher measured Young's and shear moduli than the borehole data.

The data presented in this paper have come, in part, from wireline logs. These logs contained zones in which there was no measured V_s. In some cases this was due to the borehole fluid compression wave being greater than V_s as would be the case in all the uncemented sediments. In the Harwell borehole the V_s cut-off point was about 1.5 km s^{-1} (Fig. 3). Therefore, the measurement of the shear wave, especially in uncemented sediments, required a different technique to the full wave train sonic log. Improved techniques of V_s measurement and greater understanding of the movement of shear waves through rocks and sediments will improve the interpretation of dynamic elastic moduli and hence increase the potential for their use by engineers.

The use of Christensen's equation in relation to fissured or highly discontinuous rocks is not covered in this paper. For engineering purposes these types of rock cause greatest concern. However, both V_s and formation density measurement are more difficult to determine in these materials; therefore, V_s derived from Christensen's equation is likely to compare badly with the true value as the equation requires accurate measurement of formation density.

Conclusions

Christensen's equation is often used to derive V_s when this parameter cannot be calculated directly from a full wave train sonic log. The relationship between V_p and measured formation density to produce V_s is probably empirical and the original reference is obscure.

The difference between predicted V_s and measured V_s variation appears to be lithologically based, since most lithologies have higher predicted V_s than measured except for chalk and a few siltstone results. The use of Christensen's equation for uncemented sediments is not valid as the predicted V_s is much higher than that measured.

The best agreement observed between predicted and measured values of dynamic elastic moduli is for the bulk modulus which is least affected by large differences in V_s. Shear modulus and Young's modulus show greater variation between predicted and measured values. There is no relationship between predicted and measured values of Poisson's ratio.

The use of Christensen's equation will be improved if site and lithology corrections are made. The equation cannot be used to predict a value of V_s in uncemented sediments. However, if more measurements of V_s are made in these materials, it would be possible to develop a new equation relating V_s V_p and ρ.

The authors would like to thank Michael Rainsbury for his help in writing the software for the calculations and graphs produced in this paper. The paper is published with the permission of the Director, British Geological Survey, Natural Environment Research Council.

References

Biot, M. A. 1952. Propagation of Elastic Waves in a Cylindrical Bore containing a Fluid. *Journal of Applied Physics*, **23**, 997–1006.

Carmichael, R. S. (ed) *Handbook of physical properties of rocks*, Vol. 2, RC Press, Boca Raton, Fl.

Cheng, C. H. & Toksoz, M. N. 1983. Determination of Shear Wave Velocities in Slow Formations. *Transactions of the 24th Annual Logging Symposium of the Society of Professional Well Log Analysts, Calgary, Alberta*, V1–V18.

Christensen, D. M. 1964. A Theoretical Analysis of Wave Propagation in Fluid Filled Drill Holes for the Interpretation of the 3-Dimensional Velocity Log. *Transactions of the 5th Annual Logging Symposium of the Society of Professional Well Log Analysts, Midland, Texas*.

—— 1966. The determination of the In-Situ Elastic Properties of Rock Salt with a 3-Dimensional Velocity Log. *Proceedings of the Northern Ohio Geological Society, Cleveland, Ohio*, 104–115.

Davies, A. M. 1977. A Technique for the in-situ measurement of Shear Wave Velocity. ABEM Case History. Atlas Copco ABGM AB, Bromma, Sweden.

Geyer, R. L. & Myung, J. I. 1970. The 3-D Velocity Log — A Tool for In-Situ Determination of the Elastic Moduli of Rocks. *In*: *Dynamic Rock Mechanics Proceedings of the 12th Symposium on Rock Mechanics, Rolla, Missouri, 1970*, 71–107.

Hamilton, E. L., Bucker, H. P., Keir, D. L. & Whitney, J. A. 1970. Velocities of compressional and shear waves in marine sediments determined in situ from a research submersible. *Journal of Geophysical Research*, **75**, 4039–4049.

McCann, D. M., Baria, R., Jackson, P. D. & Green, A. S. P. 1986. Application of Cross-hole Seismic Measurements in Site Investigation Surveys. *Geophysics*, **51**, 914–925.

White, J. E. 1965. *Seismic Waves: Radiation, Transmission and Attenuation*. McGraw-Hill, New York.

Wyllie, M. R. J., Gregory, A. R. & Gardner, L. W. 1956. Elastic wave velocities in heterogeneous and porous media. *Geophysics*, **21**, 41–70.

Index

The Quarterly Journal of Engineering Geology

An International Journal Encompassing Engineering Geology and Hydrogeology

Scope

The Quarterly Journal of Engineering Geology is an established international journal published by the Geological Society from its publishing house.

Papers are invited from both Fellows and non-Fellows from all areas of the world on topics concerning geology as applied to civil engineering, mining practice and water resources.

The journal also includes review articles, technical notes, meeting reports, book reviews, notices of meetings and photographic features. Occasionally, the journal is largely given over to a report, e.g. the Geological Society Engineering Group Working Party Report on Engineering Geophysics (1988,vol 21,no 3).

The journal has a wide international circulation, being essential reading for geologists and engineers in both industry and academia, especially those working in the following fields:

- rock mechanics
- applied sedimentology
- geohydrology
- engineering seismology
- engineering application of geophysics
- geotechnics
- soil mechanics
- hydrogeology
- engineering geomorphology

Indexed Abstracted in : Current Contents, Geological Abstracts, GEOREF, Petroleum Abstracts.

A selection of papers

Probabilistic approach for design optimization of rockfall protective barriers, **P. PARONUZZI** (Italy)

Seismic delineation of fissures associated with mining subsidence at Houghton-le-Spring, Co. Durham, **N.R. GOULTY** (UK) & **J.E. KRAGH** (UK)

Stress-strain behaviour of sulphur concrete made with different aggregates and admixtures, **B. CZARNECKI** (Canada) & **J. E. GILLOTT** (Canada)

The 1985 Bairaman landslide dam and resulting debris flow, Papua New Guinea, **J. KING** (Papua New Guinea), **I. LOVEDAY** (USA) & **R.L. SCHUSTER** (USA)

Airborne multispectral scanning of subsidence caused by Permian gypsum dissolution at Ripon, North Yorkshire, **A.H. COOPER** (UK)

Applications of geomorphology to small hydroelectric schemes, **G.P. BIRCH** (UK)

The oxidation of pyrite in cement stabilized colliery shale, **M.D.A. THOMAS** (UK), **R.J. KETTLE** (UK) & **J.A. MORTON** (UK).

The development of layered sediment beds in the laboratory as an illustration of possible field processes, **M.J. EDGE** (UK) & **G.C. SILLS** (UK).

Estimation of groundwater resources of the Berkshire Downs supported by mathematical modelling, **K.R. RUSHTON** (UK), **B.J. CONNORTON** (UK) & **L.M. TOMLINSON** (UK).

Submission procedure

Manuscripts should be prepared as outlined on the inside back cover of the journal, and sent to the Staff Editor: D C Ogden, Geological Society Publishing House, Unit 7, Brassmill Enterprise Centre, Brassmill Lane, Bath BA1 3JN, UK (Tel: 0225 445046 Fax: 0225 442836) There are no page charges. Advertisement enquiries should also be sent to this address.

Subscription rates 1990 (Volume 23)

(The journal is published in January, April, July and October.) ISSN 0481 - 2085

£96.00 (UK)	£115.00 (overseas)	US $198.00 (North America and Japan only)
Single issue:	£28.00 (UK and overseas)	US $ 59.40 (North America and Japan only)

For your **Free sample copy** and all orders and business correspondence relating to subscriptions should be addressed to : Journals Subscriptions Dept., Room Q1, The Geological Society Publishing House, Unit 7, Brassmill Enterprise Centre, Brassmill Lane, Bath BA1 3JN, UK (Tel: 0225 445046 - Fax: 0225 442836)

Orders can be placed directly or through your usual agent.

Journal of the Geological Society

Chief Editor: M. J. Le Bas

The Journal of the Geological Society is among the world's leading references for significant research in geology. It has been continuously published since 1845 and it enjoys a wide circulation throughout the world.

Only material of the highest quality is accepted for publication. As well as major research papers, the Journal publishes Conference Reports, Discussions and rapid publication Short Papers.

Indexed/Abstracted in: Current Contents, Geological Abstracts, GEOREF, Petroleum Abstracts, Mineralogical Abstracts.

Forthcoming Thematic Sets

Palaeoclimates
Nature and Analysis of the Stratigraphic Record
Tectonics and Sedimentation
The Murchison Silurian Symposium
Geology and Geophysics of the Irish Shelf and Continental Margin
Monitoring Active Volcanoes

Forthcoming Papers

The San Nicolas batholith of coastal Peru: early Palaeozoic continental arc or continental rift magmatism? S.B. MUKASA (USA) & D.J. HENRY (USA)
A buried granite batholith beneath the East Midland Shelf of the Southern North Sea Basin, J.A. DONATO (UK) & J.B. MEGSON (UK)
Rapid thermal recovery of thrust related metamorphism in basement windows of the Scandinavian Caledonides, J.E. LINQVIST (SWEDEN)
Geochemistry and origin of the Archaean Beit Bridge Complex, Limpopo Belt, South Africa, K.C. CONDIE (USA) & M. BORYTA (USA)
Geomorphology and surface tilting in an active extensional basin, SW Montana, USA, J. ALEXANDER (UK) & M.R. LEEDER (UK)

Submission procedure

Manuscripts should be sent to the Publications Secretary: Angharad Hills,
The Geological Society Publishing House, Unit 7, Brassmill Enterprise Centre, Brassmill Lane, Bath, Avon BA1 3JN, UK (Tel: 0225 445046 - Fax: 0225 442836) There are no page charges. Advertisement enquiries should also be sent to this address.

Subscription rates 1990 (Volume 147)

(The journal is published in January, March, May, July, September and November)

£198.00 (UK)	£238.00 (overseas)	US $ 400.00 (North America and Japan only)
Single issues:	£39.60 (UK) or £47.60 (overseas)	US $ 80.00 (North America and Japan only)

All orders and business correspondence relating to subscriptions should be addressed to: Journals Subscriptions Dept.,Room J1, The Geological Society Publishing House, Unit 7, Brassmill Enterprise Centre, Brassmill Lane, Bath BA1 3JN, UK (Tel: 0225 445046 - Fax 0225 442836)

Orders can be placed directly or through your usual agent.